Advanced Ceramics and Composites

Volume 1

Series Editor

Longbiao Li, College of Civil Aviation, Nanjing University of Aeronautics and Astronautics, Nanjing, Jiangsu, China

The book series "Advanced Ceramics and Composites" publishes insights and latest research results on advanced ceramics and composites, as well as the applications of these materials. The intent is to cover all the technical contents, applications, and multidisciplinary aspects of Advanced Ceramics and Composites.

The objective of the book series is to publish monographs, reference works, selected contributions from specialized conferences, and textbooks with high quality in the field advanced ceramics and composite materials. The series provides valuable references to a wide audience in research community in materials science, research and development personnel of ceramic and composite materials, industry practitioners and anyone else who are looking to expand their knowledge of ceramics and composites.

More information about this series at http://www.springer.com/series/16543

Longbiao Li

Time-Dependent Mechanical Behavior of Ceramic-Matrix Composites at Elevated Temperatures

Springer

Longbiao Li
College of Civil Aviation
Nanjing University of Aeronautics
and Astronautics
Nanjing, Jiangsu, China

ISSN 2662-9305 ISSN 2662-9313 (electronic)
Advanced Ceramics and Composites
ISBN 978-981-15-3273-3 ISBN 978-981-15-3274-0 (eBook)
https://doi.org/10.1007/978-981-15-3274-0

This Springer imprint is published by the registered company Springer Nature Singapore Pte Ltd.
The registered company address is: 152 Beach Road, #21-01/04 Gateway East, Singapore 189721, Singapore

To Shengning

Preface

At present, superalloy is still the main material of high-temperature structure of aeroengine (such as combustion chamber and turbine). After more than 40 years of development, the temperature resistance of the metal materials represented by the single-crystal alloy has been greatly improved, but the temperature difference between them and the combustion chamber is still large, and the gap is gradually increasing in the new generation of aeroengine. In order to increase the temperature resistance, most designers adopt the active cooling scheme of "thermal barrier coating + film cooling." However, the introduction of cooling gas directly affects the combustion efficiency. Therefore, the development of new material with high temperature resistance is the key technology for next-generation aeroengine.

Ceramic materials with high temperature resistance, good mechanical properties, and low density have long been considered as the potential materials for aeroengine high-temperature structures. However, due to the poor toughness of ceramics, the aeroengine will have disastrous consequences once ceramics damaged, which limits their application. In order to improve the toughness of ceramic materials, material scientists have made efforts to develop ceramic-matrix composites (CMCs). CMCs possess low material density (i.e., only 1/4 $\sim$ 1/3 of high-temperature alloy) and high temperature resistance, which can reduce cooling air and improve structure efficiency. Compared with the monolithic ceramic, the mechanical behavior of CMCs has many different characteristics. Understanding the failure mechanisms and internal damage evolution represents an important step to ensure reliability and safety of CMCs. This book focuses on the time-dependent mechanical behavior of CMCs at elevated temperatures, as follows:

(1) The time-dependent first matrix cracking stress of CMCs is investigated using the energy balance approach. The relationships between the first matrix cracking stress, interface debonding and slip, fiber fracture, and oxidation time and temperature are established. The first matrix cracking stresses of C/SiC with strong and weak interface bonding after unstressed oxidation at 700°C in air atmosphere are predicted for a different oxidation time.

(2) The time-, stress-, and cycle-dependent matrix multicracking of CMCs with the interface debonding, interface wear, interface oxidation, and fiber fracture is investigated. The experimental matrix multicracking evolution of unidirectional C/SiC, SiC/SiC, mini-SiC/SiC, SiC/CAS, SiC/CAS-II, and SiC/borosilicate composites is predicted.
(3) The strength degradation of non-oxide and oxide/oxide CMCs subjected to multiple fatigue loading at room temperature, oxidation environment at elevated temperature, and cyclic loading at elevated temperatures in oxidative environments is investigated. The relationships between the composite residual strength, fatigue peak stress, interface debonding, fiber failure, oxidation time and temperature, and applied cycle number are established. The evolution of residual strength versus oxidation temperature and time and applied cycle number curves of non-oxide and oxide/oxide CMCs is predicted.
(4) The time-dependent tensile damage and fracture of CMCs subjected to pre-exposure at elevated temperatures and thermal fatigue are investigated. The experimental tensile damage and fracture process of CMCs with different fiber preforms are predicted for a different pre-exposure temperature and time.
(5) The time-dependent static fatigue and cyclic fatigue behavior of CMCs are investigated. The stress-strain relationships considering interface oxidation and interface wear in the interface debonding region under static and cyclic fatigue loading are developed to establish the relationships between the peak strain, the interface debonding length, the interface oxidation length, and the interface slip lengths. The experimental fatigue hysteresis loops, interface slip lengths, peak strain, and interface oxidation length of C/[Si-B-C] and SiC/MAS composite at 566°C, 1093°C, and 1200°C in air atmosphere are predicted.

I hope this book can help the material scientists and engineering designers to understand and master the time-dependent mechanical behavior of ceramic-matrix composites.

Nanjing, China
December 2019

Longbiao Li

Acknowledgements

I am grateful to my mum Zhou Yuping, my wife Peng Li, and my son Li Shengning for their encouragement.

A special thanks to Huang Mengchu and Chandra Sekaran Arjunan for their help with my original manuscript.

I am also grateful to the team at Springer for their professional assistance.

Contents

Chapter 1
Time-Dependent First Matrix Cracking Stress of Ceramic-Matrix Composites at Elevated Temperatures

Abstract In this chapter, the time-dependent first matrix cracking stress of fiber-reinforced ceramic-matrix composites (CMCs) is investigated using the energy balance approach. The shear-lag model combined with the interface oxidation model, fiber oxidation model, and fiber failure model is adopted to analyze the microstress distributions in fiber-reinforced CMCs. The relationships between the first matrix cracking stress, interface debonding and slip, fiber fracture, and oxidation time and temperature are established. The effects of the fiber volume, the interface shear stress, the interface debonding energy, the fiber Weibull modulus, the fiber strength on the first matrix cracking stress, the interface debonding length, and the fiber broken fraction are analyzed. The first matrix cracking stresses of C/SiC with strong and weak interface bonding after unstressed oxidation at 700 °C in air atmosphere are predicted for different oxidation time.

Keywords Ceramic-matrix composites (CMCs) · First matrix cracking stress · Oxidation · Interface debonding · Interface oxidation · Fiber failure

1.1 Introduction

Ceramic materials possess high strength and modulus at elevated temperature. But their use as structural components is severely limited because of their brittleness. Continuous fiber-reinforced ceramic-matrix composites (CMCs), by incorporating fibers in ceramic matrices, however, not only exploit their attractive high-temperature strength but also reduce the propensity for catastrophic failure. These materials have already been implemented on some aeroengines' components (Naslain 2004; Li 2018, 2019). When the composite material is subjected to a stress along the fiber direction, a critical stress at which the composites exhibit first evidence of matrix cracking is defined as first matrix cracking stress, which is considered as the maximum allowable design stress for fiber-reinforced CMCs for components subjected to oxidizing environment (Li 2017a, b, c; d).

Many researchers performed experimental and theoretical investigations on the first matrix cracking stress of fiber-reinforced CMCs. The theoretical models can be divided into two types, i.e., (1) the steady-state cracking models based on energy

L. Li, *Time-Dependent Mechanical Behavior of Ceramic-Matrix Composites at Elevated Temperatures*, Advanced Ceramics and Composites 1,
https://doi.org/10.1007/978-981-15-3274-0_1

balance analysis, including the model by Aveston, Cooper, and Kelly (ACK) (1971); Budiansky, Hutchinson, and Evans (BHE) (1986); and (2) the non-state cracking models based on stress intensity analysis, including the model by Marshall, Cox, and Evans (MCE) (1985), and McCartney (1987). The analytical results show that the first matrix cracking stress was closely related to interface friction stress. Cox (1952) investigated the effect of orientation of the fibers on the stiffness and strength of fibrous composites considering the load transfer between fibers. Brighenti et al. (2014) investigated the fiber/matrix interface debonding through the fracture mechanics approach. The stress intensity factors (SIFs) are used to assess the detachment initiation and determined for different remote loadings and composite characteristics. Chaudhuri (2006) presented the asymptotic solutions for three-dimensional singular stress field near a partially debonded cylindrical rigid fiber, subjected to far-field extension-bending (mode I), in-plane shear twisting (mode II), and torsional (mode III) loadings. Tvergaard and Hutchinson (2008) investigated the effects of combined modes I, II and III at the crack tip along an interface between dissimilar materials for the conditions of small-scale yielding, with the fracture process at the interface represented by a cohesive zone model (CZM). Chiang (2000) investigated the effect of interfacial debonding on the first matrix cracking stress in unidirectional fiber-reinforced ceramics. It was found that the interface properties, i.e., the interface shear stress and the interface debonded energy, have profound influences on the first matrix cracking stress. Rajan and Zok (2014) investigate the mechanics of a fully bridged steady-state matrix cracking in unidirectional CMCs under shear loading. Curtin (1993) developed a theory to describe the evolution of multiple matrix cracking in fiber-reinforced CMCs considering the statistical distribution of initial flaws in the matrix and the interface shear stress. The tensile stress-strain relations and unload/reload hysteresis behavior during the evolution of multiple matrix cracking in unidirectional CMCs have been predicted (Ahn and Curtin 1997). Guillaumat and Lamon (1996) investigated the matrix cracking process and fiber failure based on the fracture statistics of SiC/SiC microcomposite consisted of a single fiber coated with interfacial material and matrix and fabricated by chemical vapor deposition (CVD). The Weibull approach is applied to brittle failure of the uncracked matrix fragments, and the matrix strength increases with the number of cracks resulting from the decrease in the volume of fragments. The influence of interfacial properties, i.e., the interface shear stress and the interface debonded length, on the matrix cracking has been analyzed. It was found that when the interface debonded length is large, i.e., $l_d \geq 1$ mm, the propagation of interface debonding did not affect the matrix cracking. Lissart and Lamon (1997) investigated the evolution of matrix multicracking as a function of interfacial properties, and the mechanical properties of minicomposite fabricated by chemical vapor infiltration (CVI) of bundles of 500 SiC fibers by a SiC matrix. It was found that the strong fiber/matrix interface favors matrix cracking, high stresses and a high stress at saturation, and the fibers strength degradation during minicomposite infiltration. The mechanical behavior of the composite would be affected by matrix cracking and interface debonding. Brighenti and Scorza (2012) developed an energy-based homogenization approach to model the mechanical behavior of fiber-reinforced composites considering interface debonding and fibers breaking. Venkat et al. (2008) investigated effect of interfacial debonding and matrix cracking on mechanical properties of multidirectional composites, i.e., three-directional orthogonal, three-directional eight-harness stain weave, and four-directional in-plane composites, using a cohesive zone model (CZM) for interface debonding and an octahedral shear stress failure criterion for matrix cracking. With increasing of strain, the composite moduli significantly degraded due to the matrix cracking and interface debonding. Romanowicz (2010) investigated the effect of interface debonding on composite damage under transverse tension using homogenization approach. It was found that the failure of the composite under transverse tension is mainly controlled by the interface strength and the interphase stiffness. Li et al. (2014) developed a micromechanical model to predict the tensile stress-strain curve of unidirectional C/SiC composite using a statistical matrix multiple cracking model, a fracture mechanics interface debonding criterion and a statistical fiber fracture model. However, the models mentioned above do not consider fibers fracture on the first matrix cracking stress in fiber-reinforced CMCs.

In this chapter, the time-dependent first matrix cracking stress of fiber-reinforced CMCs is investigated using the energy balance approach. The shear-lag model combined with the interface oxidation model, fiber oxidation model, and fiber failure model is adopted to analyze the stress distributions in fiber-reinforced CMCs. The relationships between the first matrix cracking stress, interface debonding and slip, fiber fracture, oxidation time and temperature are established. The effects of the fiber volume, the interface shear stress, the interface debonding energy, the fiber Weibull modulus, the fiber strength on the first matrix cracking stress, the interface debonding length, and the fiber broken fraction are analyzed. The experimental first matrix cracking stress of three different CMCs, i.e., SiC/borosilicate, SiC/LAS, and C/borosilicate, with different fiber volume fraction is predicted. The first matrix cracking stresses of C/SiC with strong and weak interface bonding after unstressed oxidation at $T = 700$ °C in air atmosphere are predicted for different oxidation time.

1.2 First Matrix Cracking Stress of Fiber-Reinforced Ceramic-Matrix Composites Considering Fiber Fracture

In this section, the first matrix cracking stress of fiber-reinforced CMCs is investigated using the energy balance approach considering fiber fracture. The shear-lag model combined with fiber failure model and interface debonding criteria is adopted to analyze the microstress distributions in CMCs. The relationships between the first matrix cracking stress, interface debonding and slip, and fiber fracture are established. The effects of the fiber volume fraction, interface shear stress, interface debonding energy, fiber Weibull modulus, fiber strength on the first matrix cracking stress, interface debonding length, and fiber broken fraction are analyzed. The experimental first matrix cracking stress of three different CMCs, i.e., SiC/borosilicate, SiC/LAS, and C/borosilicate, with different fiber volume fraction are predicted.

1.2.1 Stress Analysis

Marshall and Cox (1987) extended the Marshall, Cox, and Evans (MCE) (1985) model to analyze tensile fracture of unidirectional fiber-reinforced ceramics. In these analyses, the fiber strength is assumed to be single-valued and may be weak enough to be broken in the propagation of a matrix crack. The possibility of fiber failure within the matrix due to the statistical nature of fiber strength can be accounted for by using the Weibull analysis. The two-parameter Weibull model is adopted to describe the fiber strength distribution, and the global load sharing (GLS) assumption is used to determine the stress carried by the intact and fracture fibers (Curtin 1991a).

$$\frac{\sigma}{V_{\mathrm{f}}} = \Phi[1 - P(\Phi)] + \langle \Phi_{\mathrm{b}} \rangle P(\Phi) \tag{1.1}$$

where V_{f} denotes the fiber volume fraction; σ denotes the applied stress; Φ denotes the stress carried by the intact fibers; $\langle \Phi_{\mathrm{b}} \rangle$ denotes the stress carried by broken fibers; and $P(\Phi)$ denotes the fiber failure probability.

$$P(\Phi) = 1 - \exp\left[-\left(\frac{\Phi}{\sigma_{\mathrm{c}}} \right)^{m+1} \right] \tag{1.2}$$

where m denotes the fiber Weibull modulus, which describes the variation in fiber strength; and σ_{c} denotes the fiber characteristic strength of a length δ_{c} of fiber (Curtin 1991a).

$$\sigma_{\mathrm{c}} = \left(\frac{l_0 \sigma_0^m \tau_{\mathrm{i}}}{r_{\mathrm{f}}} \right)^{\frac{1}{m+1}}, \quad \delta_{\mathrm{c}} = \left(\frac{\sigma_0 r_{\mathrm{f}} l_0^{1/m}}{\tau_{\mathrm{i}}} \right)^{\frac{m}{m+1}} \tag{1.3}$$

where σ_0 denotes the fiber strength of a length of l_0.

When a fiber breaks, the stress carried by the fiber drops to zero at the position of break. Similar to the case of matrix cracking, the fiber/matrix interface debonds and the stress builds up in the fiber through the interface shear stress. During the process of loading, the stress in a broken fiber Φ_{b} as a function of the distance z from the break can be written as

$$\Phi_{\mathrm{b}}(z) = \frac{2\tau_{\mathrm{i}}}{r_{\mathrm{f}}} z \tag{1.4}$$

In order to calculate the average stress carried by the broken fibers $\langle \Phi_{\mathrm{b}} \rangle$, it is necessary to construct the probability distribution $F(z)$ of the distance z of a fiber break from the reference matrix crack plane, provided that a break occurs within a distance $\pm l_{\mathrm{f}}$. For this conditional probability distribution, Phoenix and Raj (1992) deduced the following from based on Weibull statistics, as

$$F(z) = \frac{1}{P(\Phi)l_f}\left(\frac{\Phi}{\sigma_c}\right)^{m+1} \exp\left[-\left(\frac{x}{l_f}\right)\left(\frac{\Phi}{\sigma_c}\right)^{m+1}\right], \quad z \in [0, l_f] \tag{1.5}$$

where

$$l_f = \frac{r_f \Phi}{2\tau_i} \tag{1.6}$$

The averaging stress carried by broken fibers $\langle \Phi_b \rangle$ during the process of loading using Eqs. (1.4) and (1.5) leads to

$$\langle \Phi_b \rangle = \Phi\left[\left(\frac{\sigma_c}{\Phi}\right)^{m+1} - \frac{1 - P(\Phi)}{P(\Phi)}\right] \tag{1.7}$$

Substituting Eqs. (1.2) and (1.7) into Eq. (1.1), it leads into the form of,

$$\frac{\sigma}{V_f} = \Phi\left(\frac{\sigma_c}{\Phi}\right)^{m+1}\left\{1 - \exp\left[-\left(\frac{\Phi}{\sigma_c}\right)^{m+1}\right]\right\} \tag{1.8}$$

Using Eq. (1.8), the stress Φ carried by intact fibers at the matrix cracking plane can be determined. Substituting the intact fiber stress Φ into Eq. (1.2), the relationship between fiber failure probability and applied stress can be determined.

1.2.1.1 Downstream Stresses

The composite with the fiber volume fraction V_f is loaded by a remote uniform stress σ normal to a long crack plane, and the fiber bridging occurs in the crack wake as the crack propagates across the composite, as shown in Fig. 1.1. The unit cell in the downstream region I contained a single fiber surrounded by a hollow cylinder of matrix is extracted from the ceramic composite system, as shown in Fig. 1.2. In the present analysis, the CMCs possess low interface shear stress, and the global load sharing (GLS) criterion was adopted to describe the load distribution between the intact and broken fibers, and the reciprocal fibers' interaction has not been considered. The fiber radius is r_f, and the matrix radius is R ($R = r_f/V_f^{1/2}$). The length of the unit cell is half matrix crack spacing $l_c/2$, and the interface debonded length is l_d. In the debonded region, the interface is resisted by τ_i. For the debonded region in region I, the force equilibrium equation of the fiber is given by Eq. (1.9).

$$\frac{d\sigma_f(z)}{dz} = -\frac{2\tau_i(z)}{r_f} \tag{1.9}$$

The boundary condition at the crack plane $z = 0$ is given by

$$\sigma_f(z = 0) = \Phi \tag{1.10}$$

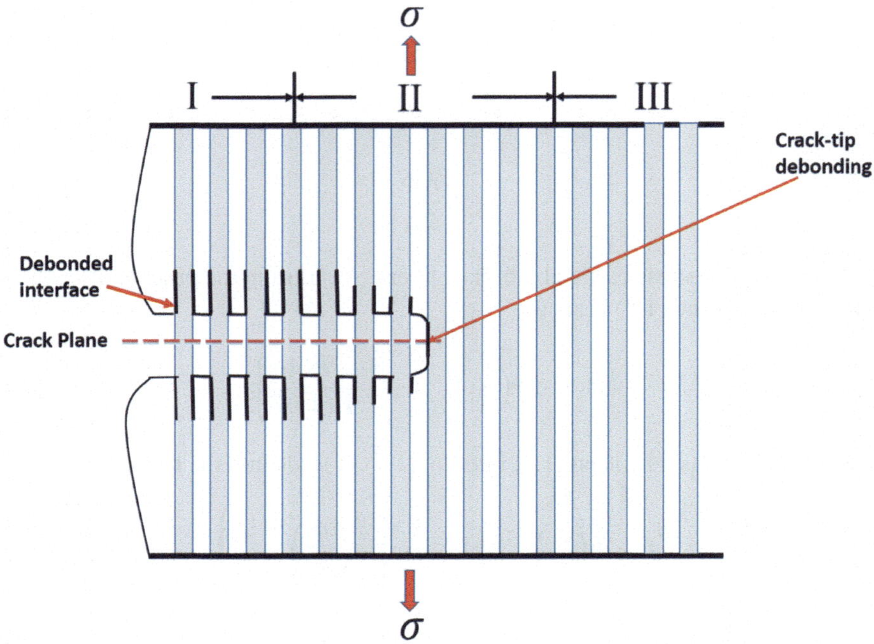

Fig. 1.1 Schematic of crack tip and interface debonding. Reprinted with permission from Li (2017a). Copyright 2017, Elsevier Ltd.

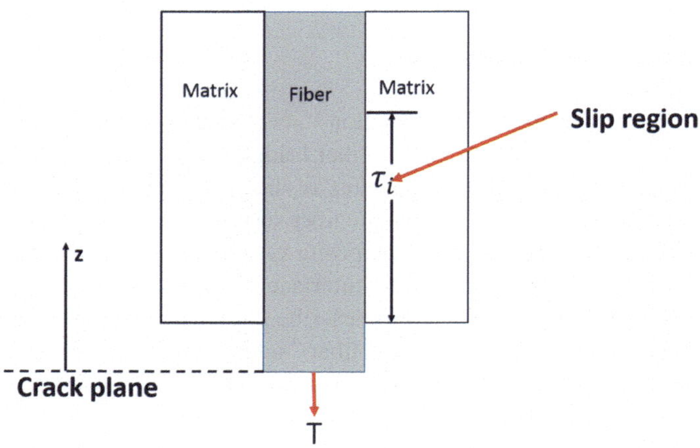

Fig. 1.2 Schematic of shear-lag model considering interface debonding in the region I. Reprinted with permission from Li (2017a). Copyright 2017, Elsevier Ltd.

$$\sigma_{\mathrm{m}}(z=0)=0 \tag{1.11}$$

The total axial stresses in region I satisfy Eq. (1.12).

$$V_{\mathrm{f}}\sigma_{\mathrm{f}}(z)+V_{\mathrm{m}}\sigma_{\mathrm{m}}(z)=\sigma \tag{1.12}$$

Solving Eqs. (1.9) and (1.12) with the boundary conditions given by Eqs. (1.10) and (1.11), and the interface shear stress in the debonded region, the fiber and matrix axial stresses in the interface debonded region, i.e., $0 < z < l_{\mathrm{d}}$, can be determined by Eqs. (1.13) and (1.14).

$$\sigma_{\mathrm{f}}^{\mathrm{D}}(z)=\Phi-\frac{2\tau_{\mathrm{i}}}{r_{\mathrm{f}}}z,\quad z\in(0,l_{\mathrm{d}}) \tag{1.13}$$

$$\sigma_{\mathrm{m}}^{\mathrm{D}}(z)=2\frac{V_{\mathrm{f}}}{V_{\mathrm{m}}}\frac{\tau_i}{r_{\mathrm{f}}}z,\quad z\in(0,l_{\mathrm{d}}) \tag{1.14}$$

For the bonded region ($l_{\mathrm{d}} < z$) in the downstream region I, the fiber, matrix, and composite have the same strains

$$\varepsilon_{\mathrm{f}}=\varepsilon_{\mathrm{m}}=\varepsilon_{\mathrm{c}}=\frac{\sigma}{E_{\mathrm{c}}} \tag{1.15}$$

The fiber and matrix axial stresses in the bonded region ($l_{\mathrm{d}} < z$) become

$$\sigma_{\mathrm{f}}^{\mathrm{D}}=\frac{E_{\mathrm{f}}}{E_{\mathrm{c}}}\sigma \tag{1.16}$$

$$\sigma_{\mathrm{m}}^{\mathrm{D}}=\frac{E_{\mathrm{m}}}{E_{\mathrm{c}}}\sigma \tag{1.17}$$

1.2.1.2 Upstream Stresses

The upstream region III as shown in Fig. 1.1 is so far away from the crack tip that the stress and strain fields are also uniform. The fiber and matrix have the same displacements, and the fiber and matrix stresses are given by

$$\sigma_{\mathrm{f}}^{\mathrm{U}}=\frac{E_{\mathrm{f}}}{E_{\mathrm{c}}}\sigma \tag{1.18}$$

$$\sigma_{\mathrm{m}}^{\mathrm{U}}=\frac{E_{\mathrm{m}}}{E_{\mathrm{c}}}\sigma \tag{1.19}$$

1.2.2 Interface Debonding

When the first matrix cracking propagates to the fiber/matrix interface, it deflects along the interface. The fracture mechanics approach is adopted in the present analysis. The interface debonding criterion is given by Eq. (1.20) (Gao et al. 1988).

$$\xi_{\mathrm{d}} = -\frac{F}{4\pi r_{\mathrm{f}}}\frac{\partial w_{\mathrm{f}}(0)}{\partial l_{\mathrm{d}}} - \frac{1}{2}\int_{0}^{l_{\mathrm{d}}} \tau_{\mathrm{i}}\frac{\partial v(z)}{\partial l_{\mathrm{d}}}\mathrm{d}z \tag{1.20}$$

where ζ_{d} denotes the interface debonded energy; $F(=\pi r_{\mathrm{f}}^2\sigma/V_{\mathrm{f}})$ denotes the fiber load at the matrix cracking plane; $w_{\mathrm{f}}(0)$ denotes the fiber axial displacement on the matrix cracking plane; and $v(z)$ denotes the relative displacement between the fiber and the matrix.

The axial displacements of the fiber and matrix, i.e., $w_{\mathrm{f}}(z)$ and $w_{\mathrm{m}}(z)$, are given by Eqs. (1.21) and (1.22).

$$w_{\mathrm{f}}(z) = \int_{\infty}^{l_{\mathrm{d}}} \frac{\sigma}{E_{\mathrm{c}}}\mathrm{d}z + \frac{\Phi}{E_{\mathrm{f}}}(z - l_{\mathrm{d}}) - \frac{\tau_{\mathrm{i}}}{r_{\mathrm{f}}E_{\mathrm{f}}}\left(z^2 - l_{\mathrm{d}}^2\right) \tag{1.21}$$

$$w_{\mathrm{m}}(z) = \int_{\infty}^{l_{\mathrm{d}}} \frac{\sigma}{E_{\mathrm{c}}}\mathrm{d}z + \frac{V_{\mathrm{f}}\tau_i}{r_{\mathrm{f}}V_{\mathrm{m}}E_{\mathrm{m}}}\left(z^2 - l_{\mathrm{d}}^2\right) \tag{1.22}$$

The relative displacement between the fiber and the matrix, i.e., $v(z)$, is given by Eq. (1.23).

$$v(z) = |w_{\mathrm{f}}(z) - w_{\mathrm{m}}(z)| = \frac{\Phi}{E_{\mathrm{f}}}(l_{\mathrm{d}} - z) - \frac{E_{\mathrm{c}}}{V_{\mathrm{m}}E_{\mathrm{f}}E_{\mathrm{m}}}\frac{\tau_{\mathrm{i}}}{r_{\mathrm{f}}}\left(l_{\mathrm{d}}^2 - z^2\right) \tag{1.23}$$

Substituting $w_{\mathrm{f}}(z = 0)$ and $v(z)$ into Eq. (1.20), it leads to the form of Eq. (1.24).

$$l_{\mathrm{d}}^2 - \frac{r_{\mathrm{f}}V_{\mathrm{m}}E_{\mathrm{m}}\Phi}{E_{\mathrm{c}}\tau_{\mathrm{i}}}l_{\mathrm{d}} + \frac{r_{\mathrm{f}}^2V_{\mathrm{m}}^2E_{\mathrm{m}}^2}{4E_{\mathrm{c}}^2\tau_{\mathrm{i}}^2}\frac{\sigma\Phi}{V_{\mathrm{f}}} - \frac{r_{\mathrm{f}}V_{\mathrm{m}}E_{\mathrm{f}}E_{\mathrm{m}}}{E_{\mathrm{c}}\tau_{\mathrm{i}}^2}\xi_{\mathrm{d}} = 0 \tag{1.24}$$

Solving Eq. (1.24), the interface debonding length is determined by Eq. (1.25).

$$l_{\mathrm{d}} = \frac{r_{\mathrm{f}}V_{\mathrm{m}}E_{\mathrm{m}}\Phi}{2E_{\mathrm{c}}\tau_{\mathrm{i}}} - \sqrt{\frac{r_{\mathrm{f}}V_{\mathrm{m}}E_{\mathrm{f}}E_{\mathrm{m}}}{E_{\mathrm{c}}\tau_{\mathrm{i}}^2}\xi_{\mathrm{d}} - \left(\frac{r_{\mathrm{f}}V_{\mathrm{m}}E_{\mathrm{m}}\Phi}{2E_{\mathrm{c}}\tau_{\mathrm{i}}}\right)^2\left(1 - \frac{\sigma}{V_{\mathrm{f}}\Phi}\right)} \tag{1.25}$$

1.2.3 First Matrix Cracking Stress

The first steady-state matrix cracking means that the stresses at the crack front remain unchanged during the crack growth, and the upstream and downstream stress state, far ahead of and behind the crack front, do not change. During the first matrix cracking propagation, the energy release rate must be balanced by the critical matrix crack extension energy release rate $V_m\xi_m$, where ξ_m is the matrix fracture energy, and the total interface debonding energy release rate of $4V_f l_d \xi_d / r_f$. The energy relationship to evaluate the first matrix cracking stress is expressed by Eq. (1.26) (Budiansky et al. 1986).

$$\frac{1}{2}\int_{-l_d}^{l_d}\left[\frac{V_f}{E_f}\left(\sigma_f^U-\sigma_f^D\right)^2+\frac{V_m}{E_m}\left(\sigma_m^U-\sigma_m^D\right)^2\right]dz + \frac{1}{2\pi R^2 G_m}\int_{-l_d}^{l_d}\int_{r_f}^{R}\left(\frac{r_f\tau_i(z)}{r}\right)2\pi r dr dz = V_m\xi_m+\left(\frac{4V_f l_d}{r_f}\right)\xi_d \quad (1.26)$$

where G_m is the matrix shear modulus. The contribution of the shear energy term in Eq. (1.26) was neglected in the ACK model (Aveston et al. 1971). It was verified that this negligence is well accepted for the interface slip length larger than a few fiber radii. Following the ACK model, the contribution of shear energy is neglected in the present analysis. Substituting the fiber and matrix stresses of Eqs. (1.13), (1.14), (1.16)–(1.19) and the debonding length of Eq. (1.25) into Eq. (1.26), the energy balance equation leads to the form of,

$$\eta_1\sigma^2+\eta_2\sigma+\eta_3=0 \quad (1.27)$$

where

$$\eta_1=\frac{l_d}{E_c} \quad (1.28a)$$

$$\eta_2=-\frac{2V_f l_d\Phi}{E_c} \quad (1.28b)$$

$$\eta_3=\frac{4}{3}\left(\frac{\tau_i}{r_f}\right)^2\left(\frac{V_f E_c}{V_m E_m E_f}\right)l_d^3+\frac{V_f l_d\Phi^2}{E_f}-\frac{2V_f\tau_i l_d^2}{r_f E_f}\Phi-V_m\xi_m-\frac{4V_f l_d}{r_f}\xi_d \quad (1.28c)$$

1.2.4 Results and Discussion

During the process of first matrix cracking propagation, matrix cracks deflect along the fiber/matrix interface, not a crack is preset first in the interface. The interface properties, i.e., the interface debonded energy and the interface shear stress, affect the interface debonding and also the first matrix cracking stress. Guillaumat and Lamon (1996), and Lissart and Lamon (1997) found that the strong fiber/matrix interface favors matrix cracking, high matrix cracking stresses, and a high stress at matrix cracking saturation. The possibility of fiber failure within the matrix due to the statistical nature of fiber strength can affect the first matrix cracking propagation (Marshall and Cox 1987) and be accounted for by using the two-parameter Weibull distribution and the GLS assumption. The ceramic composite system of SiC/borosilicate is used for the case study, and its material properties are given by (Chiang 2000): $V_f = 40\%$, $E_f = 400$ GPa, $E_m = 63$ GPa, $r_f = 70$ μm, $\xi_m = 8.92$ J/m^2, $\xi_d = 0.8$ J/m^2, $\tau_i = 8$ MPa, $\sigma_c = 2$ GPa, and $m = 4$. The effects of the fiber volume, interface properties, and fiber parameters on the first matrix cracking stress of fiber-reinforced CMCs are analyzed.

1.2.4.1 Effect of Fiber Volume on Interface Debonding, Fiber Failure, and First Matrix Cracking Stress

The first matrix cracking stress, the interface debonding length, and the broken fiber fraction versus the fiber volume curves are shown in Fig. 1.3. When the fiber volume increases, the fiber/matrix interface debonding length decreases, the broken fiber fraction increases, leading to the increase of the first matrix cracking stress.

When the fiber volume increases from $V_f = 20$ to 50%, the first matrix cracking stress increases from $\sigma_{mc} = 96$ to 335 MPa; the interface debonding length decreases from $l_d/r_f = 6.4$ to 2.6; and the broken fiber fraction increases from $P = 0.08$ to 0.43%.

1.2.4.2 Effect of Interface Shear Stress on Interface Debonding, Fiber Failure, and First Matrix Cracking Stress

The first matrix cracking stress, the interface debonding length, and the fiber broken fraction versus the interface shear stress curves are shown in Fig. 1.4. When the interface shear stress increases, the fiber/matrix interface debonding length decreases, the fiber broken fraction increases, and the first matrix cracking stress also increases.

When the interface shear stress increases from $\tau_i = 2$ to 11 MPa, the first matrix cracking stress increases from $\sigma_{mc} = 184$ to 255 MPa; the interface debonding length decreases from $l_d/r_f = 5$ to 2.8; and the fiber broken fraction increases from $P = 0.06$ to 0.336%.

1.2.4.3 Effect of Interface Debonding Energy on Interface Debonding, Fiber Failure, and First Matrix Cracking Stress

The first matrix cracking stress, the interface debonding length, and the fiber broken fraction versus the interface debonding energy curves are shown in Fig. 1.5. He and Hutchinson (1989) found that the ratio between the interface debonded energy and matrix fracture energy should be less than 1/4, i.e., $\xi_d/\xi_m < 1/4$; otherwise, the

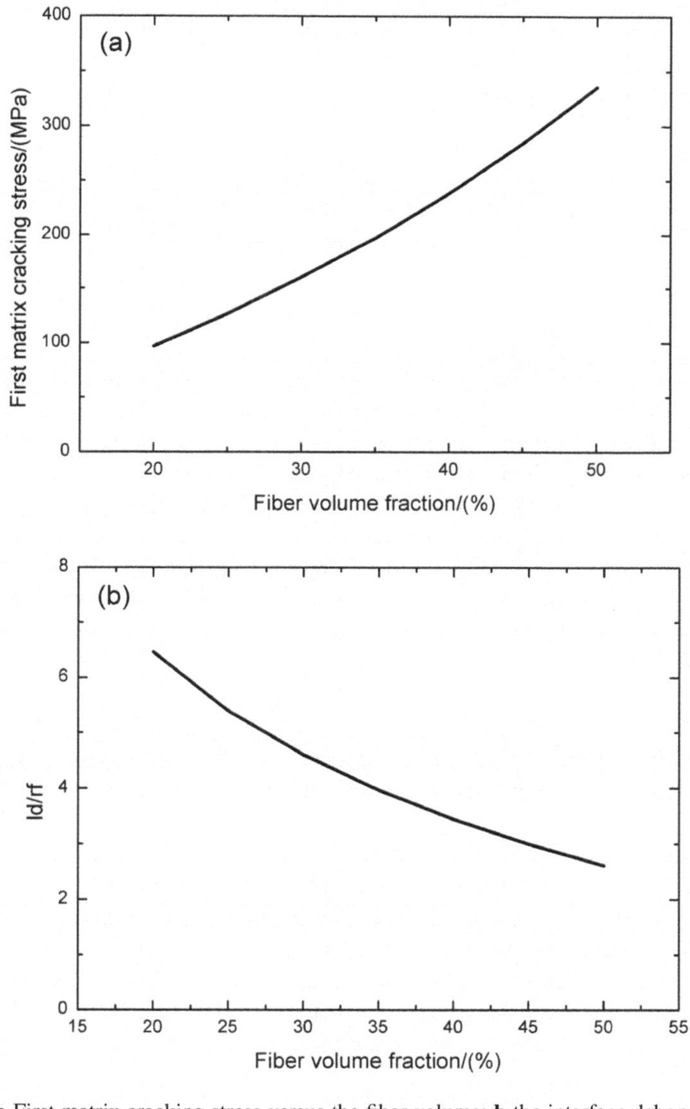

Fig. 1.3 **a** First matrix cracking stress versus the fiber volume; **b** the interface debonding length versus the fiber volume; and **c** the broken fiber fraction versus the fiber volume

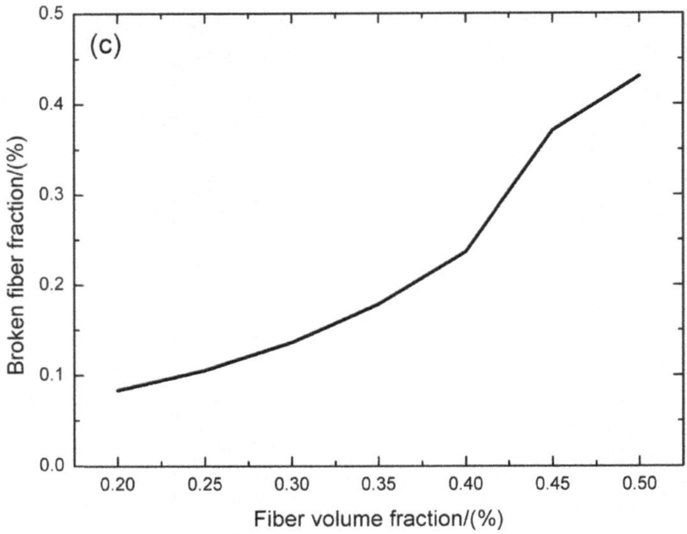

Fig. 1.3 (continued)

matrix cracking tip could propagate through fibers, not deflect along the fiber/matrix interface. Budiansky et al. (1986) found that during the process of matrix crack propagating, the ratio between the interface debonded energy and the matrix fracture energy should be approximately 1/5. When the interface debonding energy increases, the interface debonding length decreases, the fiber broken fraction increases, and the first matrix cracking stress also increases.

When the interface debonding energy increases from $\xi_d/\xi_m = 0.1$ to 0.2, the first matrix cracking stress increases from $\sigma_{mc} = 243$ to 283 MPa; the interface debonding length decreases from $l_d/r_f = 3.3$ to 2.9; and the broken fiber fraction increases from $P = 0.26$ to 0.56%.

1.2.4.4 Effect of Fiber Weibull Modulus on Interface Debonding, Fiber Failure, and First Matrix Cracking Stress

The first matrix cracking stress, the interface debonding length, and the fiber broken fraction versus the fiber Weibull modulus curves are shown in Fig. 1.6. When the fiber Weibull modulus increases, the interface debonding length decreases, the fiber broken fraction decreases, and the first matrix cracking stress increases.

When the fiber Weibull modulus increases from $m = 2$ to 6, the first matrix cracking stress increases from $\sigma_{mc} = 240$ to 244 MPa; the interface debonding length decreases from $l_d/r_f = 3.45$ to 3.37; and the fiber broken fraction decreases from $P = 2.7$ to 0.02%.

1.2.4.5 Effect of Fiber Strength on Interface Debonding, Fiber Failure, and First Matrix Cracking Stress

The first matrix cracking stress, the interface debonding length, and the fiber broken fraction versus the fiber strength curves are shown in Fig. 1.7. When the fiber strength increases, the interface debonding length decreases, the fiber broken fraction decreases, and the first matrix cracking stress increases.

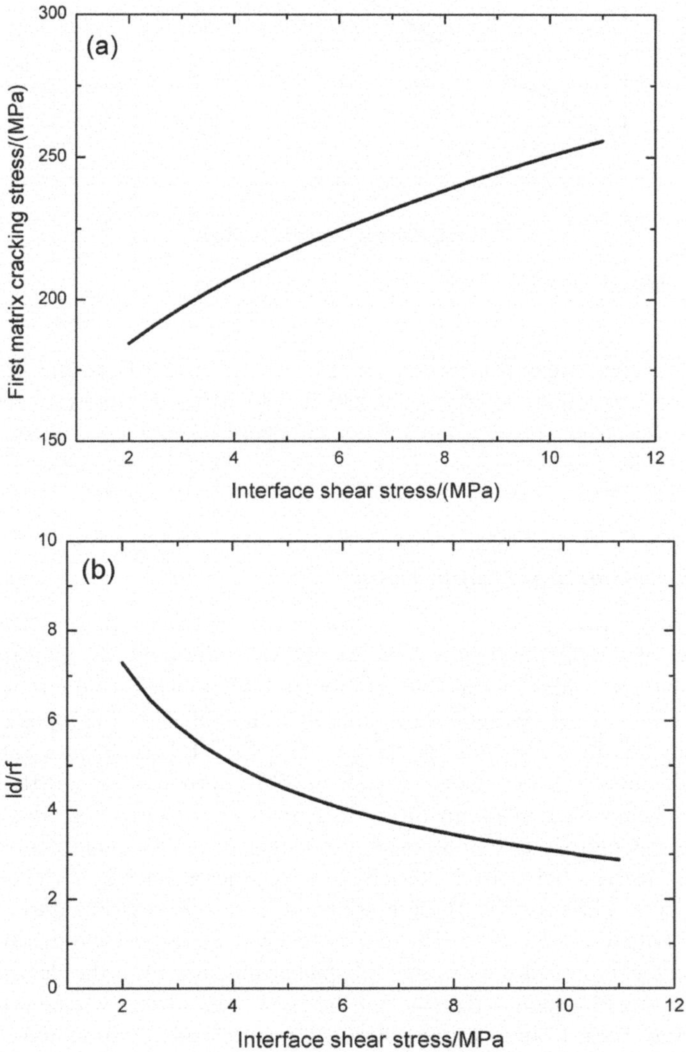

Fig. 1.4 **a** First matrix cracking stress versus the interface shear stress; **b** the interface debonding length versus the interface shear stress; and **c** the broken fiber fraction versus the interface shear stress

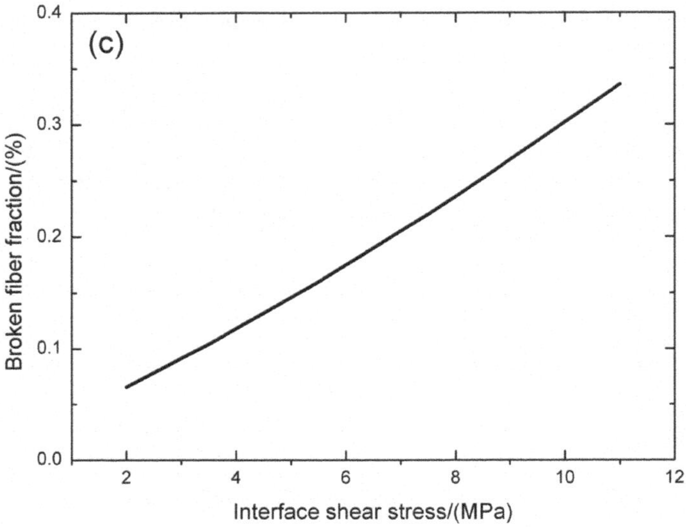

Fig. 1.4 (continued)

When the fiber strength increases from σ_c = 1.5 to 2.5 GPa, the first matrix cracking stress increases from σ_{mc} = 242 to 243 MPa; the interface debonding length decreases from l_d/r_f = 3.4 to 3.3; and the fiber broken fraction decreases from P = 1.1 to 0.08%.

1.2.5 Experimental Comparisons

Barsoum et al. (1992) investigated the matrix crack initiation in fiber-reinforced CMCs. The three unidirectional fiber-reinforced CMCs tested for matrix crack initiation include (1) the SiC/lithium aluminosilicate (SiC/LAS) composite, which comprises a SiC fiber (Nicalon, Nippon Carbon Co., Tokyo, Japan); (2) the carbon/borosilicate composite, which is made of HMU carbon fibers (HMU Hercules carbon fiber) embedded in a borosilicate glass matrix (7740 Corning Glass Works); and (3) the SiC/borosilicate composite, consisting of a SiC monofilament (SCS-6 SiC fiber, Textron Specialty Materials, Lowell, Massachusetts, USA) embedded in a borosilicate glass matrix. Three-point-bend tests were performed to obtain the matrix cracking initiation stress. The maximum beam deflection was measured with an extensometer mounted at the beam mid-span; at the same time, the electrical resistance of the gold film sputtered on the polished specimen surface was measured with a digital Ohm meter. Both measurements were recorded as a continuous function of loading and were then used to determine the matrix crack initiation stress.

The experimental and theoretical first matrix cracking stress versus the fiber volume corresponding to different interface debonding energy of SiC/borosilicate, SiC/LAS, and C/borosilicate composites are shown in Figs. 1.8, 1.9 and 1.10.

For the SiC/borosilicate composite, the predicted first matrix cracking stress with the interface debonding energy range of $\xi_d/\xi_m = 0.05$, 0.1, and 0.2 agrees with experimental data corresponding to the fiber volume changing from $V_f = 10$ to 50%, as shown in Fig. 1.8. When the interface debonding energy is $\xi_d/\xi_m = 0.05$, the

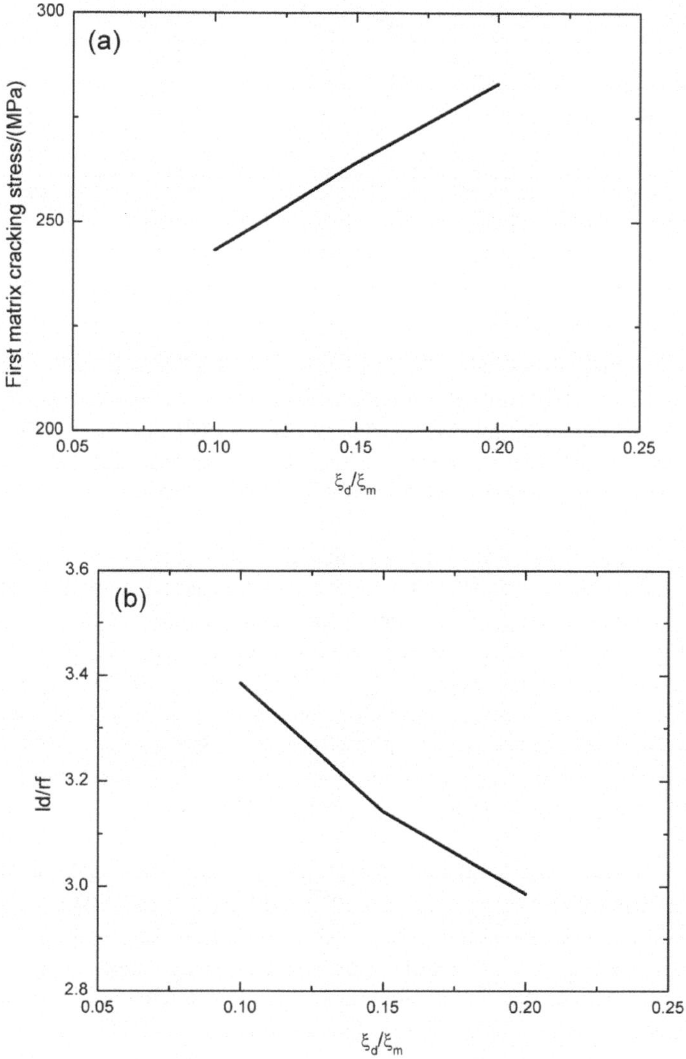

Fig. 1.5 (a) First matrix cracking stress versus the interface debonding energy; (b) the interface debonding length versus the interface debonding energy; and (c) the broken fiber fraction versus the interface debonding energy

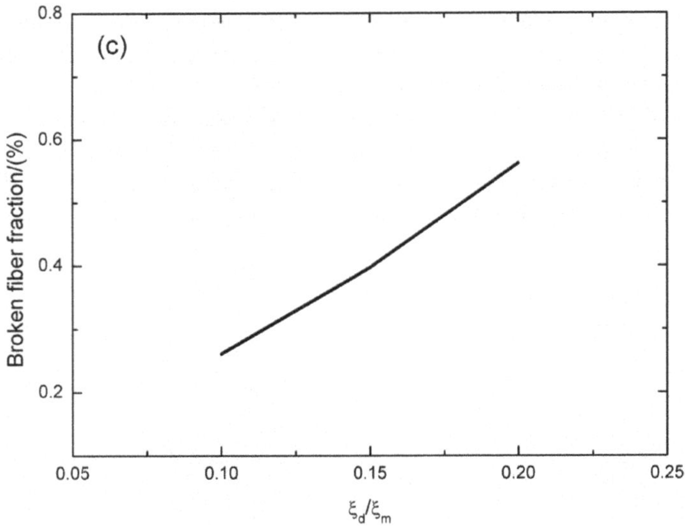

Fig. 1.5 (continued)

first matrix cracking stress increases from $\sigma_{mc} = 49.4$ MPa at $V_f = 10\%$ to $\sigma_{mc} = 334$ MPa at $V_f = 50\%$; when the interface debonding energy is $\xi_d/\xi_m = 0.1$, the first matrix cracking stress increases from $\sigma_{mc} = 52.2$ MPa at $V_f = 10\%$ to $\sigma_{mc} = 371$ MPa at $V_f = 50\%$; and when the interface debonding energy is $\xi_d/\xi_m = 0.2$, the first matrix cracking stress increases from $\sigma_{mc} = 57.1$ MPa at $V_f = 10\%$ to $\sigma_{mc} = 428$ MPa at $V_f = 50\%$.

For the SiC/LAS composite, the first matrix cracking stress with the interface debonding energy range of $\xi_d/\xi_m = 0.01, 0.02$, and 0.03 agrees with the experimental data corresponding to the fiber volume fraction changing from $V_f = 30$ to 50%, as shown in Fig. 1.9. When the interface debonding energy is $\xi_d/\xi_m = 0.01$, the first matrix cracking stress increases from $\sigma_{mc} = 228$ MPa at $V_f = 30\%$ to $\sigma_{mc} = 423$ MPa at $V_f = 50\%$; when the interface debonding energy is $\xi_d/\xi_m = 0.02$, the first matrix cracking stress increases from $\sigma_{mc} = 257$ MPa at $V_f = 30\%$ to $\sigma_{mc} = 489$ MPa at $V_f = 50\%$; and when the interface debonding energy is $\xi_d/\xi_m = 0.03$, the first matrix cracking stress increases from $\sigma_{mc} = 281$ MPa at $V_f = 30\%$ to $\sigma_{mc} = 542$ MPa at $V_f = 50\%$.

For the C/borosilicate composite, the predicted first matrix cracking stress with the interface debonding energy range of $\xi_d/\xi_m = 0.01, 0.02$, and 0.03 agrees with the experimental data corresponding to the fiber volume fraction changing from 30 to 55%, as shown in Fig. 1.10. When the interface debonding energy is $\xi_d/\xi_m = 0.01$, the first matrix cracking stress increases from $\sigma_{mc} = 136$ MPa at $V_f = 30\%$ to $\sigma_{mc} = 438$ MPa at $V_f = 55\%$; when the interface debonding energy is $\xi_d/\xi_m = 0.02$, the first matrix cracking stress increases from $\sigma_{mc} = 156$ MPa at $V_f = 30\%$ to $\sigma_{mc} = 471$ MPa at $V_f = 55\%$; and when the interface debonding energy is $\xi_d/\xi_m = 0.03$, the first matrix cracking stress increases from $\sigma_{mc} = 173$ MPa at $V_f = 30\%$ to $\sigma_{mc} = 504$ MPa at $V_f = 55\%$.

1.3 Time-Dependent First Matrix Cracking Stress of Fiber-Reinforced Ceramic-Matrix Composites Considering Interface Oxidation

In this section, the time-dependent first matrix cracking stress of fiber-reinforced CMCs is investigated using the energy balance approach considering the interface oxidation at elevated temperature. The shear-lag model combined with the

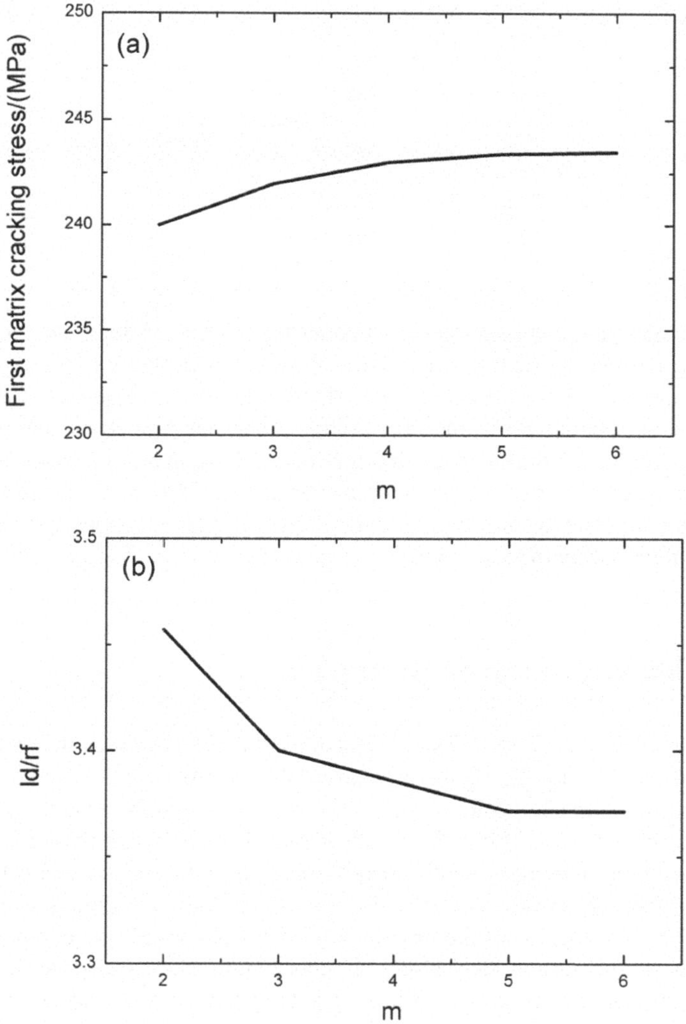

Fig. 1.6 **a** First matrix cracking stress versus the fiber Weibull modulus; **b** the interface debonding length versus the fiber Weibull modulus; and **c** the broken fiber fraction versus the fiber Weibull modulus

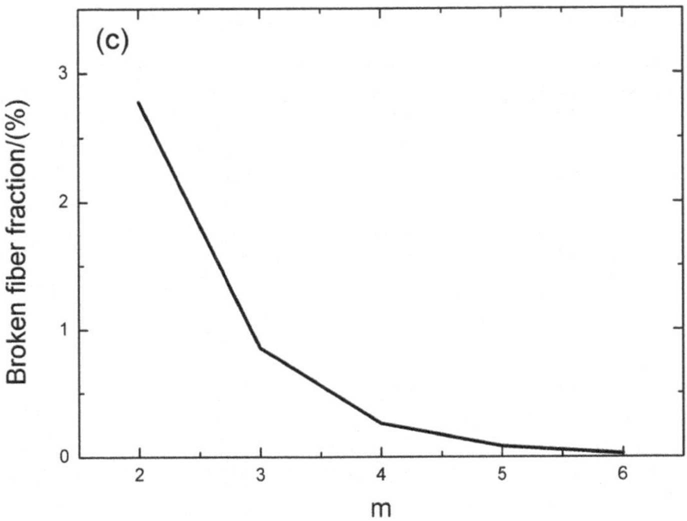

Fig. 1.6 (continued)

interface oxidation model and interface debonding criteria is adopted to analyze the microstress field in the CMCs. The relationships between the first matrix cracking stress, the interface debonding and slip, oxidation temperature, and oxidation time are established. The effects of the fiber volume, interface properties, and oxidation temperature on the evolution of the first matrix cracking stress versus the oxidation time are analyzed. The first matrix cracking stresses of C/SiC with strong and weak interface bonding after unstressed oxidation at 700 °C in air atmosphere are predicted for different oxidation time.

1.3.1 Time-Dependent Stress Analysis

As the mismatch of the axial thermal expansion coefficient between the carbon fiber and silicon carbide matrix, there are unavoidable microcracks existed within the SiC matrix when the composite was cooled down from high fabricated temperature to ambient temperature. These processing-induced microcracks mainly existed at the surface of the material, which do not propagate through the entire thickness of the composite. However, at elevated temperatures the microcracks would serve as avenues for the ingress of the environment atmosphere into the composite. The oxygen reacts with carbon layer along the fiber length at a certain rate of $d\zeta/dt$, in which ζ is the length of carbon lost in each side of the crack (Casas and Martinez-Esnaola 2003).

$$\zeta = \varphi_1 \left[1 - \exp\left(-\frac{\varphi_2 t}{b}\right)\right] \tag{1.29}$$

where b is a delay factor considering the deceleration of reduced oxygen activity; and φ_1 and φ_2 are parameters dependent on temperature and described using the Arrhenius-type laws (Casas and Martinez-Esnaola 2003).

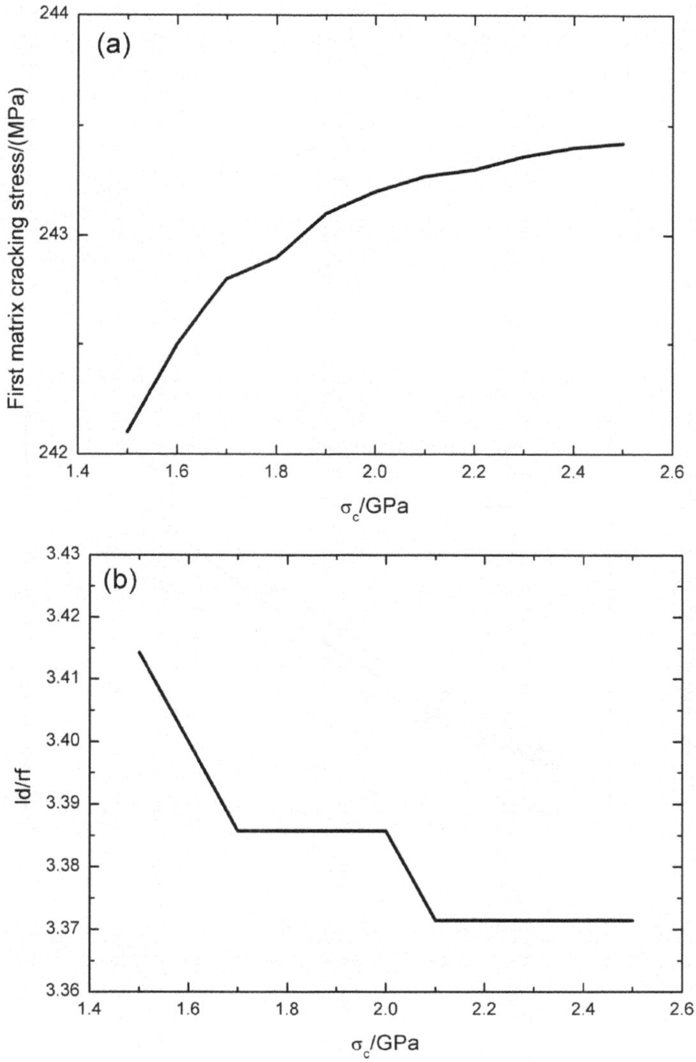

Fig. 1.7 **a** First matrix cracking stress versus the fiber strength; **b** the interface debonding length versus the fiber strength; and **c** the broken fiber fraction versus the fiber strength

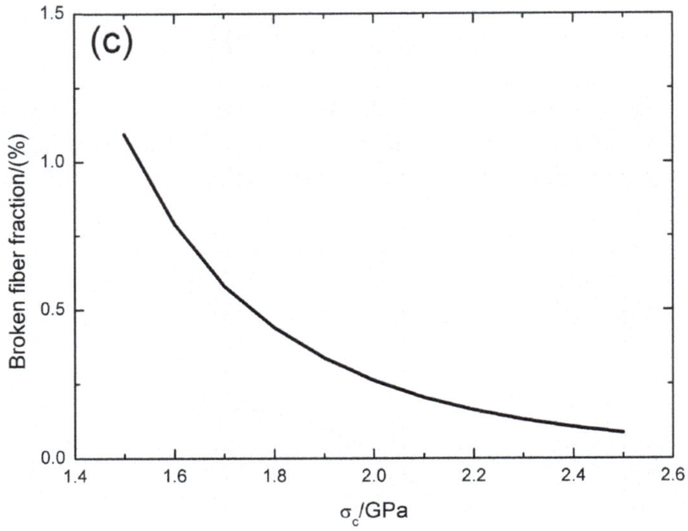

Fig. 1.7 (continued)

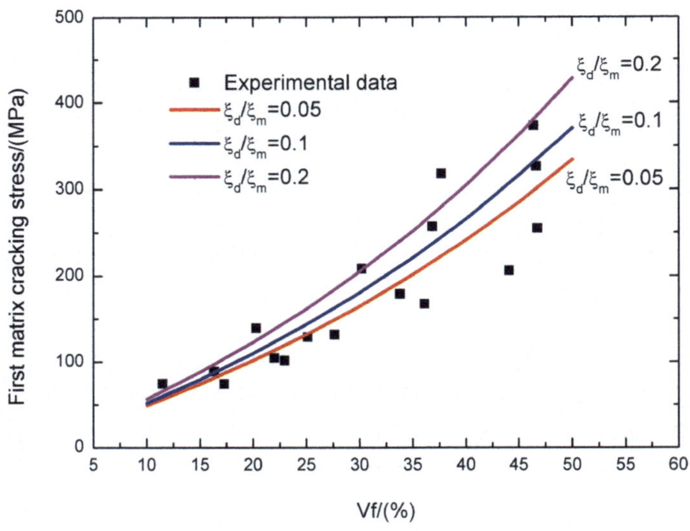

Fig. 1.8 Experimental and theoretical first matrix cracking stress versus the fiber volume corresponding to different interface debonding energy ($\xi_d/\xi_m = 0.05$, 0.1, and 0.2) of SiC/borosilicate composite

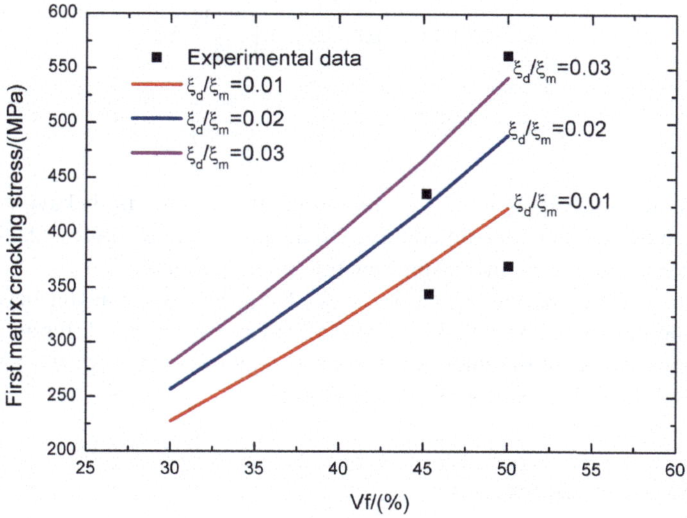

Fig. 1.9 Experimental and theoretical first matrix cracking stress versus fiber volume fraction corresponding to different interface debonding energy ($\xi_d/\xi_m = 0.01$, 0.02, and 0.03) of SiC/LAS composite

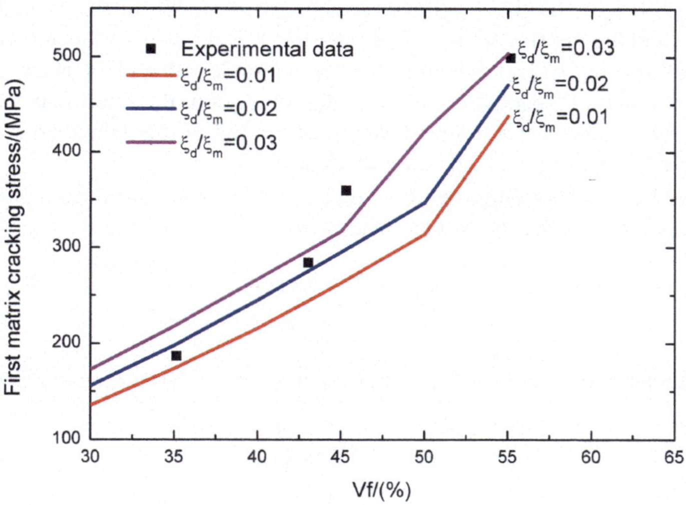

Fig. 1.10 Experimental and theoretical first matrix cracking stress versus fiber volume fraction corresponding to different interface debonding energy ($\xi_d/\xi_m = 0.01$, 0.02, and 0.03) of C/borosilicate composite

$$\varphi_1 = 7.021 \times 10^{-3} \times \exp\left(\frac{8231}{T}\right) \tag{1.30}$$

$$\varphi_2 = 227.1 \times \exp\left(-\frac{17,090}{T}\right) \tag{1.31}$$

where φ_1 is in mm and φ_2 in s^{-1}; φ_1 represents the asymptotic behavior for long times, which decreases with temperature; and the product $\varphi_1\varphi_2$ represents the initial oxidation rate, which is an increasing function of temperature.

In the interface oxidation region, i.e., $z \in [0, \zeta]$, the stress transfer between the fiber and the matrix is controlled by a sliding stress $\tau_i(z) = \tau_f$, different from the interface shear stress in the interface debonded region, i.e., $z \in [\zeta, l_d]$, $\tau_i(z) = \tau_i$. This new interface shear stress τ_f is lower than τ_i.

1.3.1.1 Downstream Stresses

The composite with fiber volume fraction V_f is loaded by a remote uniform stress σ normal to a long crack plane, as shown in Fig. 1.1. The unit cell in the downstream region I is extracted from the ceramic composite system, which contained a single fiber surrounded by a hollow cylinder of matrix, as shown in Fig. 1.2. The fiber radius is r_f, and the matrix radius is R ($R = r_f/V_f^{1/2}$). The length of the unit cell is half matrix crack spacing $l_c/2$, and the interface oxidation length and interface debonded length are ζ and l_d, respectively. In the oxidation region, the fiber/matrix interface is resisted by a constant frictional shear stress τ_f; and in the debonded region, the interface is resisted by τ_i, which is higher than τ_f.

In the interface debonding region of region I, the force equilibrium equation of the fiber is determined by the following equation.

$$\frac{d\sigma_f(z)}{dz} = -\frac{2\tau_i(z)}{r_f} \tag{1.32}$$

The boundary conditions at the matrix crack plane ($z = 0$) are given by:

$$\sigma_f(z = 0) = \frac{\sigma}{V_f} \tag{1.33}$$

$$\sigma_m(z = 0) = 0 \tag{1.34}$$

The total axial stresses in the region I satisfy the following equation.

$$V_f\sigma_f(z) + V_m\sigma_m(z) = \sigma \tag{1.35}$$

Solving Eqs. (1.32) and (1.35) with the boundary conditions given by Eqs. (1.33) and (1.34), and the interface shear stress in the oxidation and debonded region, the fiber and matrix axial stresses in the interface oxidation and debonding region (i.e.,

$0 < z < l_d$) can be determined as:

$$\sigma_f^D(z) = \begin{cases} \frac{\sigma}{V_f} - \frac{2\tau_f}{r_f}z, & z \in [0, \zeta] \\ \frac{\sigma}{V_f} - \frac{2\tau_f}{r_f}\zeta - \frac{2\tau_i}{r_f}(z - \zeta), & z \in [\zeta, l_d] \end{cases} \tag{1.36}$$

$$\sigma_m^D(z) = \begin{cases} 2\frac{V_f}{V_m}\frac{\tau_f}{r_f}z, & z \in (0, \zeta) \\ 2\frac{V_f}{V_m}\frac{\tau_f}{r_f}\zeta + 2\frac{V_f}{V_m}\frac{\tau_i}{r_f}(z - \zeta), & z \in (\zeta, l_d) \end{cases} \tag{1.37}$$

In the interface bonding region ($l_d < z$) of the downstream region I, the fiber, matrix, and composite have the same strain.

$$\varepsilon_f = \varepsilon_m = \varepsilon_c = \frac{\sigma}{E_c} \tag{1.38}$$

The fiber and matrix axial stresses in the interface bonding region ($l_d < z$) are determined using the following equations.

$$\sigma_f^D = \frac{E_f}{E_c}\sigma \tag{1.39}$$

$$\sigma_m^D = \frac{E_m}{E_c}\sigma \tag{1.40}$$

1.3.1.2 Upstream Stresses

The upstream region III, as shown in Fig. 1.1, is so far away from the crack tip that the stress and strain fields are also uniform. The fiber and matrix have the same displacement, and the fiber and matrix stresses are determined as:

$$\sigma_f^U = \frac{E_f}{E_c}\sigma \tag{1.41}$$

$$\sigma_m^U = \frac{E_m}{E_c}\sigma \tag{1.42}$$

1.3.2 Time-Dependent Interface Debonding

The fracture mechanics approach is adopted in the present analysis. The interface debonding criterion is given by the following equation (Gao et al. 1988).

$$\xi_{\mathrm{d}} = -\frac{F}{4\pi r_{\mathrm{f}}}\frac{\partial w_{\mathrm{f}}(0)}{\partial l_{\mathrm{d}}} - \frac{1}{2}\int\limits_{0}^{l_{\mathrm{d}}} \tau_{\mathrm{i}}\frac{\partial v(z)}{\partial l_{\mathrm{d}}}\mathrm{d}z \tag{1.43}$$

in which $F(=\pi r_{\mathrm{f}}^2\sigma/V_{\mathrm{f}})$ denotes the fiber load at the matrix cracking plane; $w_{\mathrm{f}}(0)$ denotes the fiber axial displacement at the matrix cracking plane; and $v(z)$ denotes the relative displacement between the fiber and the matrix.

The axial displacements of the fiber and matrix, i.e., $w_{\mathrm{f}}(z)$ and $w_{\mathrm{m}}(z)$, are determined as:

$$\begin{aligned} w_{\mathrm{f}}(z) &= \int_{\infty}^{z} \frac{\sigma_{\mathrm{f}}(z)}{E_{\mathrm{f}}} dz \\ &= \int_{\infty}^{l_{\mathrm{d}}} \frac{\sigma}{E_{\mathrm{c}}} dz + \frac{\sigma}{V_{\mathrm{f}}E_{\mathrm{f}}}(z - l_{\mathrm{d}}) + \frac{\tau_{\mathrm{i}}}{r_{\mathrm{f}}E_{\mathrm{f}}}(l_{\mathrm{d}} - \zeta)^2 \\ &\quad - \frac{\tau_{\mathrm{f}}}{r_{\mathrm{f}}E_{\mathrm{f}}}\left(\zeta^2 - 2\zeta l_{\mathrm{d}} + z^2\right) \end{aligned} \tag{1.44}$$

$$\begin{aligned} w_{\mathrm{m}}(z) &= \int\limits_{\infty}^{z} \frac{\sigma_{\mathrm{m}}(z)}{E_{\mathrm{m}}}\mathrm{d}z \\ &= \int\limits_{\infty}^{l_{\mathrm{d}}} \frac{\sigma}{E_{\mathrm{c}}}\mathrm{d}z - \frac{V_{\mathrm{f}}\tau_{\mathrm{i}}}{r_{\mathrm{f}}V_{\mathrm{m}}E_{\mathrm{m}}}(l_{\mathrm{d}} - \zeta)^2 + \frac{V_{\mathrm{f}}\tau_{\mathrm{f}}}{r_{\mathrm{f}}V_{\mathrm{m}}E_{\mathrm{m}}}\left(\zeta^2 - 2\zeta l_{\mathrm{d}} + z^2\right) \end{aligned} \tag{1.45}$$

The relative displacement between the fiber and the matrix, i.e., $v(z)$, is determined as:

$$\begin{aligned} v(z) &= |w_{\mathrm{f}}(z) - w_{\mathrm{m}}(z)| \\ &= \frac{\sigma}{V_{\mathrm{f}}E_{\mathrm{f}}}(l_{\mathrm{d}} - z) - \frac{E_{\mathrm{c}}\tau_{\mathrm{i}}}{r_{\mathrm{f}}V_{\mathrm{m}}E_{\mathrm{f}}E_{\mathrm{m}}}(l_{\mathrm{d}} - \zeta)^2 \\ &\quad + \frac{E_{\mathrm{c}}\tau_{\mathrm{f}}}{r_{\mathrm{f}}V_{\mathrm{m}}E_{\mathrm{f}}E_{\mathrm{m}}}\left(\zeta^2 - 2\zeta l_{\mathrm{d}} + z^2\right) \end{aligned} \tag{1.46}$$

Substituting $w_{\mathrm{f}}(z = 0)$ and $v(z)$ into Eq. (1.43), it leads to the form of the following equation.

$$\frac{E_c\tau_i^2}{r_f V_m E_f E_m}(l_d - \zeta)^2 - \frac{\tau_i\sigma}{V_f E_f}(l_d - \zeta) + \frac{2E_c\tau_f\tau_i}{r_f V_m E_f E_m}\zeta(l_d - \zeta)$$
$$- \frac{\tau_f\sigma}{V_f E_f}\zeta + \frac{r_f V_m E_m}{E_f E_c}\frac{\sigma^2}{4V_f^2} + \frac{E_c\tau_f^2}{r_f V_m E_f E_m}\zeta^2 - \xi_d = 0 \quad (1.47)$$

Solving Eq. (1.47), the interface debonding length is determined as:

$$l_d = \left(1 - \frac{\tau_f}{\tau_i}\right)\zeta + \frac{r_f}{2}\frac{V_m E_m\sigma}{V_f E_c\tau_i} - \sqrt{\frac{r_f V_m E_m E_f}{E_c\tau_i^2}\xi_d} \quad (1.48)$$

1.3.3 Time-Dependent First Matrix Cracking Stress

The energy relationship to evaluate the first matrix cracking stress is expressed by the following equation (Budiansky et al. 1986).

$$\frac{1}{2}\int_{-\infty}^{\infty}\left[\frac{V_f}{E_f}\left(\sigma_f^U - \sigma_f^D\right)^2 + \frac{V_m}{E_m}\left(\sigma_m^U - \sigma_m^D\right)^2\right]dz + \frac{1}{2\pi R^2 G_m}$$
$$\int_{-l_d}^{l_d}\int_{r_f}^{R}\left(\frac{r_f\tau_i(z)}{r}\right)2\pi r\,dr\,dz = V_m\xi_m + \left(\frac{4V_f l_d}{r_f}\right)\xi_d \quad (1.49)$$

in which ξ_m is the matrix fracture energy; and G_m is the matrix shear modulus. Substituting the time-dependent fiber and matrix stresses of Eqs. (1.36), (1.37), (1.38)–(1.41) and the interface debonding length of Eq. (1.48) into Eq. (1.49), the energy balance equation leads to the form of the following equation.

$$\eta_1\sigma^2 + \eta_2\sigma + \eta_3 = 0 \quad (1.50)$$

where

$$\eta_1 = \frac{V_m E_m l_d}{V_f E_f E_c} \quad (1.51)$$

$$\eta_2 = -\frac{2\tau_i}{r_f E_f}\left[(l_d - \zeta)^2 + \frac{\tau_f}{\tau_i}\zeta(2l_d - \zeta)\right] \quad (1.52)$$

$$\eta_3 = \frac{4}{3}\left(\frac{V_\mathrm{f}E_\mathrm{c}}{V_\mathrm{m}E_\mathrm{f}E_\mathrm{m}}\right)\left(\frac{\tau_\mathrm{i}}{r_\mathrm{f}}\right)^2\left[(l_\mathrm{d}-\zeta)^3 + \left(\frac{\tau_\mathrm{f}}{\tau_\mathrm{i}}\right)^2\zeta^3\right]$$
$$+\left(\frac{4V_\mathrm{f}E_\mathrm{c}}{V_\mathrm{m}E_\mathrm{f}E_\mathrm{m}}\right)\frac{\tau_\mathrm{f}\tau_\mathrm{i}}{r_\mathrm{f}^2}\zeta(l_\mathrm{d}-\zeta)$$
$$\times\left[l_\mathrm{d} - \left(1-\frac{\tau_\mathrm{f}}{\tau_\mathrm{i}}\right)\zeta\right] - V_\mathrm{m}\zeta_\mathrm{m} - \left(\frac{4V_\mathrm{f}l_\mathrm{d}}{r_\mathrm{f}}\right)\zeta_\mathrm{d} \quad (1.53)$$

1.3.4 Results and Discussion

The effects of the fiber volume, interface debonding energy, interface shear stress, and oxidation temperature on the time-dependent first matrix cracking stress and the fiber/matrix interface debonding are discussed. The C/SiC composite is used for the case study, and its material properties are given by: $V_\mathrm{f} = 35\%$, $r_\mathrm{f} = 3.5$ μm, $E_\mathrm{f} =$ 230 GPa, $E_\mathrm{m} = 350$ GPa, $\tau_\mathrm{i} = 8$ MPa, and $\tau_\mathrm{f} = 2$ MPa, $\alpha_\mathrm{f} = -0.38 \times 10^{-6}$/°C, $\alpha_\mathrm{m} = 4.6 \times 10^{-6}$/°C, and $\Delta T = -1000$ °C.

1.3.4.1 Effect of Fiber Volume on Time-Dependent Interface Debonding and First Matrix Cracking Stress

At elevated temperature of T = 800°C in air atmosphere, the time-dependent first matrix cracking stress, the interface debonding length, and the interface oxidation length versus the oxidation time curves corresponding to the fiber volume of $V_\mathrm{f} =$ 30 and 35% are shown in Fig. 1.11. When the fiber volume increases, the time-dependent fiber/matrix interface debonding length decreases, the interface oxidation length increases, and the first matrix cracking stress also increases.

When the fiber volume is $V_\mathrm{f} = 30\%$, the time-dependent first matrix cracking stress decreases from $\sigma_\mathrm{mc} = 89$ to 58 MPa after 20 h oxidation; the time-dependent interface debonding length increases from $l_\mathrm{d}/r_\mathrm{f} = 10.5$ to 18.2 after 20 h oxidation; and the time-dependent interface oxidation length increases from $\zeta/l_\mathrm{d} = 0$ to 0.92 after 20 h oxidation.

When the fiber volume is $V_\mathrm{f} = 35\%$, the time-dependent first matrix cracking stress decreases from $\sigma_\mathrm{mc} = 103$ to 69 MPa after 20 h oxidation; the time-dependent interface debonding length increases from $l_\mathrm{d}/r_\mathrm{f} = 9.2$ to 17.5 after 20 h oxidation; and the time-dependent interface oxidation length increases from $\zeta/l_\mathrm{d} = 0$ to 0.96 after 20 h oxidation.

1.3.4.2 Effect of Interface Debonding Energy on Time-Dependent Interface Debonding and First Matrix Cracking Stress

At elevated temperature of T = 800 °C in air atmosphere, the time-dependent first matrix cracking stress, the interface debonding length, and the interface oxidation length versus oxidation time curves corresponding to different interface debonding

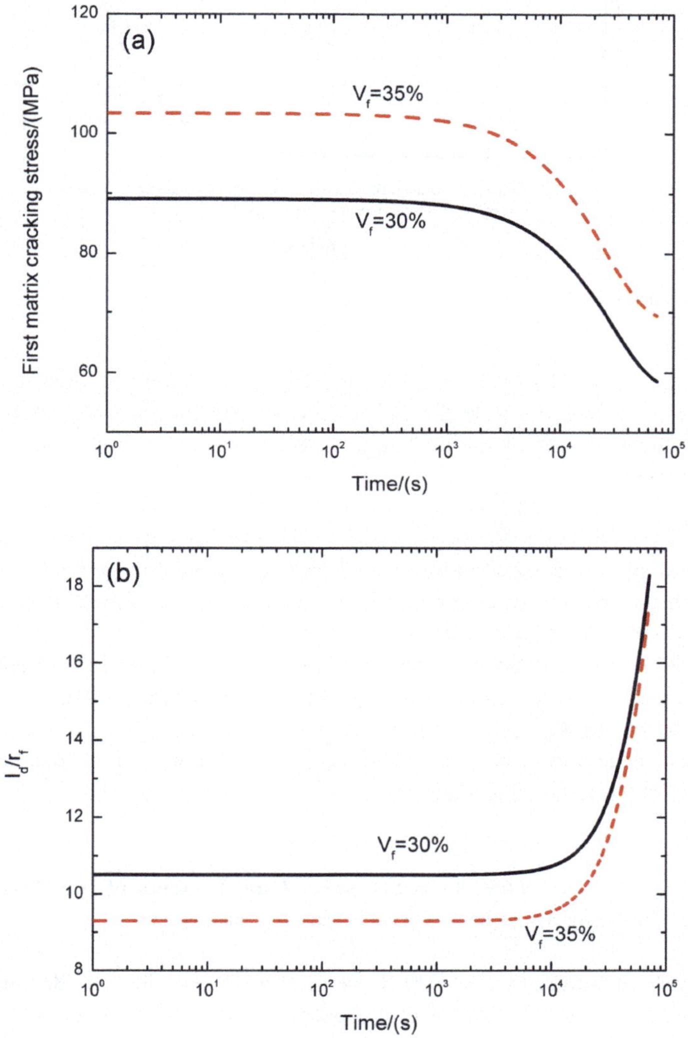

Fig. 1.11 **a** Time-dependent first matrix cracking stress versus the oxidation time; **b** the time-dependent interface debonding length versus the oxidation time; and **c** the time-dependent interface oxidation length versus the oxidation time corresponding to different fiber volume of V_f = 30 and 35% of C/SiC composite at T = 800 °C in air atmosphere

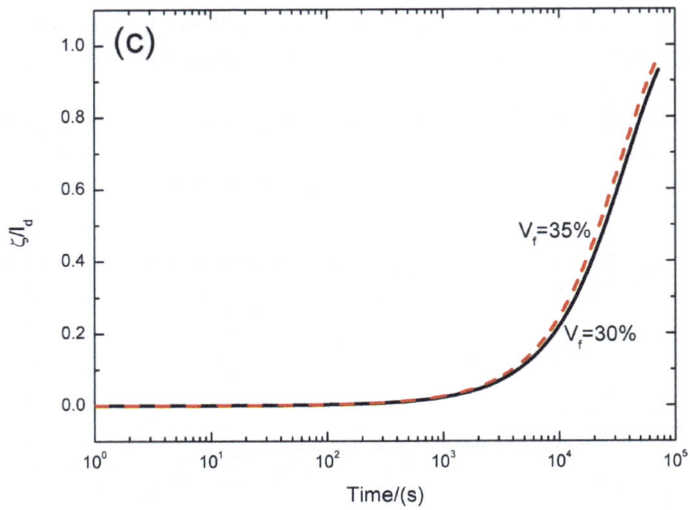

Fig. 1.11 (continued)

energy of $\xi_d = 0.5$ and 1.0 J/m^2 are shown in Fig. 1.12. When the interface debonding energy increases, the time-dependent fiber/matrix interface debonding length decreases, the interface oxidation length increases, and the first matrix cracking stress also increases.

When the interface debonding energy is $\xi_d = 0.5$ J/m^2, the time-dependent first matrix cracking stress decreases from $\sigma_{mc} = 79$ to 48 MPa after 20 h oxidation; the time-dependent interface debonding length increases from $l_d/r_f = 10.8$ to 18.5 after 20 h oxidation; and the time-dependent interface oxidation length increases from $\zeta/l_d = 0$ to 0.91 after 20 h oxidation.

When the interface debonding energy is $\xi_d = 1.0$ J/m^2, the time-dependent first matrix cracking stress decreases from $\sigma_{mc} = 121$ to 93 MPa after 20 h oxidation; the time-dependent interface debonding length increases from $l_d/r_f = 9.4$ to 17.6 after 20 h oxidation; and the time-dependent interface oxidation length increases from $\zeta/l_d = 0$ to 0.96 after 20 h oxidation.

1.3.4.3 Effect of Interface Shear Stress on Time-Dependent Interface Debonding and First Matrix Cracking Stress

At elevated temperature of $T = 800$ °C in air atmosphere, the time-dependent first matrix cracking stress, the interface debonding length, and the interface oxidation length versus the oxidation time curves corresponding to different interface shear stress of $\tau_i = 10$ and 15 MPa are shown in Fig. 1.13. When the interface shear stress in the slip region increases, the time-dependent first matrix cracking stress and the interface oxidation length increase, and the time-dependent interface debonding length decreases.

When the interface shear stress is $\tau_i = 10$ MPa, the time-dependent first matrix cracking stress decreases from $\sigma_{mc} = 128$ to 94 MPa after 20 h oxidation; the time-dependent interface debonding length increases from $l_d/r_f = 8.4$ to 17.5 after 20 h oxidation; and the time-dependent interface oxidation length increases from $\zeta/l_d = 0$ to 0.97 after 20 h oxidation.

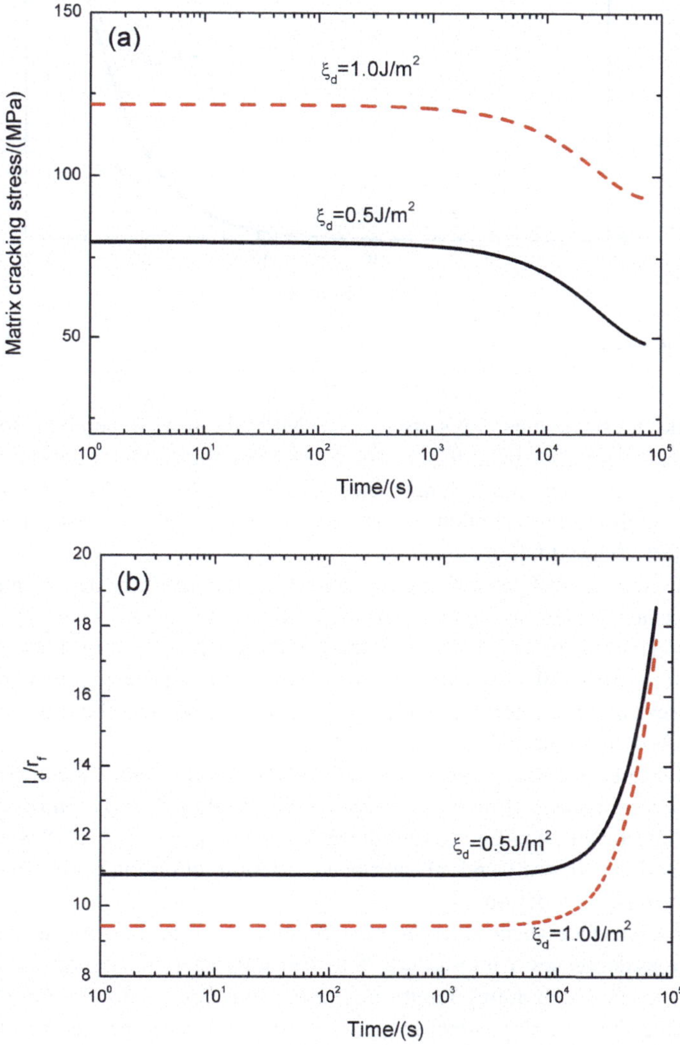

Fig. 1.12 **a** Time-dependent first matrix cracking stress versus the oxidation time; **b** the time-dependent interface debonding length versus the oxidation time; and **c** the time-dependent interface oxidation length versus the oxidation time corresponding to different interface debonding energy of $\xi_d = 0.5$ and 1.0 J/m^2 of C/SiC composite at 800 °C in air atmosphere

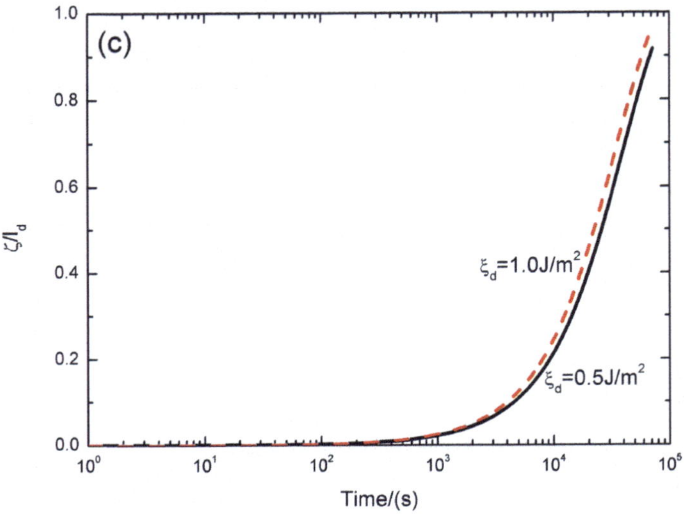

Fig. 1.12 (continued)

When the interface shear stress is $\tau_i = 15$ MPa, the time-dependent first matrix cracking stress decreases from $\sigma_{mc} = 142$ to 94 MPa after 20 h oxidation; the time-dependent interface debonding length increases from $l_d/r_f = 6.7$ to 17.3 after 20 h oxidation; and the time-dependent interface oxidation length increases from $\zeta/l_d = 0$ to 0.98 after 20 h oxidation.

The time-dependent first matrix cracking stress, the interface debonding length, and the interface oxidation length versus the oxidation time curves corresponding to different interface shear stress of $\tau_f = 1$ and 3 MPa are shown in Fig. 1.14. When the interface shear stress in the oxidation region increases, the time-dependent first matrix cracking stress and the interface oxidation length increase, and the time-dependent interface debonding length decreases.

When the interface shear stress is $\tau_f = 1$ MPa, the time-dependent first matrix cracking stress decreases from $\sigma_{mc} = 128$ to 87 MPa after 20 h oxidation; the time-dependent interface debonding length increases from $l_d/r_f = 8.4$ to 18.3 after 20 h oxidation; and the time-dependent interface oxidation length increases from $\zeta/l_d = 0$ to 0.92 after 20 h oxidation.

When the interface shear stress is $\tau_f = 3$ MPa, the time-dependent first matrix cracking stress decreases from $\sigma_{mc} = 128$ to 100 MPa after 20 h oxidation; the time-dependent interface debonding length increases from $l_d/r_f = 8.4$ to 16.6 after 20 h oxidation; and the time-dependent interface oxidation length increases from $\zeta/l_d = 0$ to 1.0 after 20 h oxidation.

1.3.4.4 Effect of Oxidation Temperature on Time-Dependent Interface Debonding and First Matrix Cracking Stress

The time-dependent first matrix cracking stress, the interface debonding length, and the interface oxidation length versus the oxidation time curves corresponding to $T =$ 600 and 800 °C are shown in Fig. 1.15. When the oxidation temperature increases,

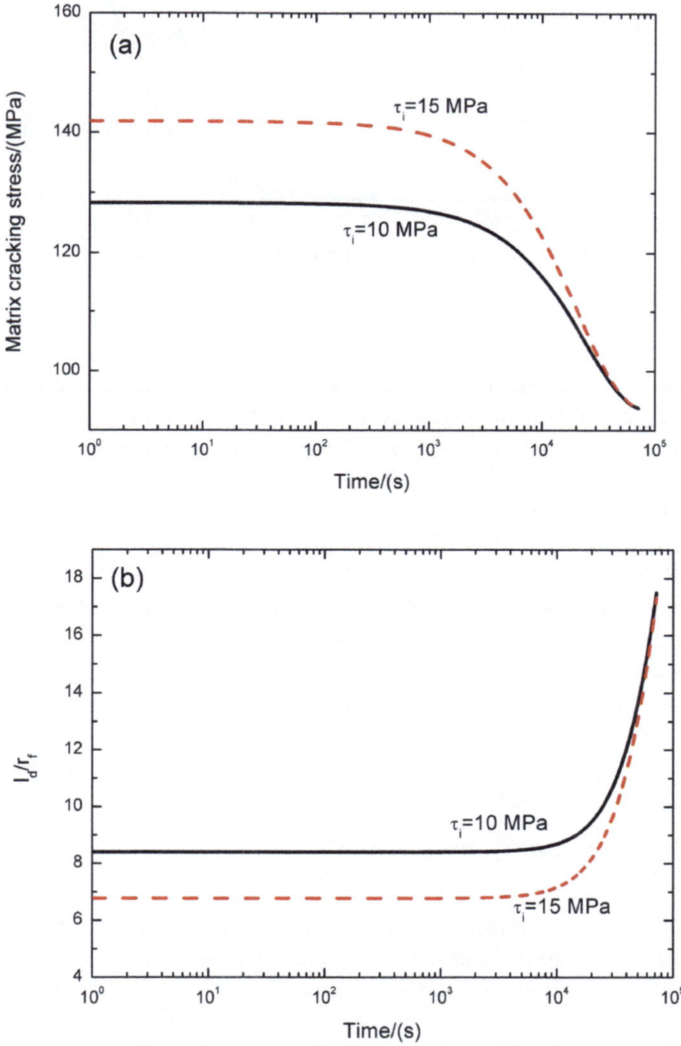

Fig. 1.13 **a** Time-dependent first matrix cracking stress versus the oxidation time; **b** the time-dependent interface debonding length versus the oxidation time; and **c** the time-dependent interface oxidation length versus the oxidation time corresponding to different interface shear stress of $\tau_i =$ 10 and 15 MPa of C/SiC composite at $T =$ 800 °C in air atmosphere

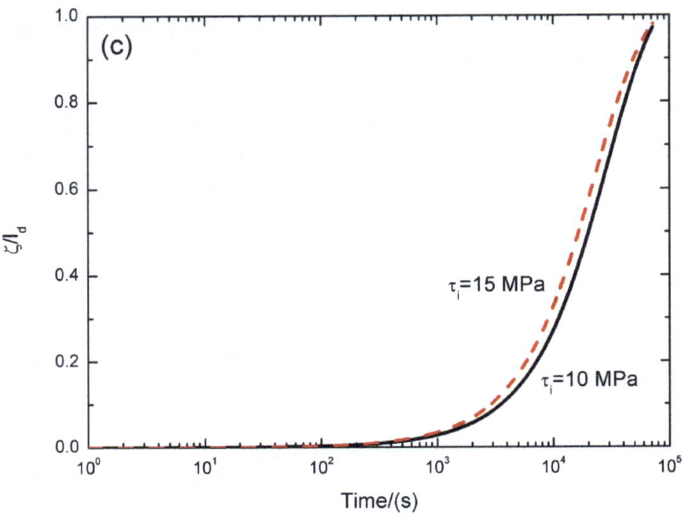

Fig. 1.13 (continued)

the time-dependent first matrix cracking stress decreases, and the time-dependent interface oxidation length and the interface debonding length increase.

At elevated temperature of $T = 600$ °C, the time-dependent first matrix cracking stress decreases from $\sigma_{mc} = 142$ to 119 MPa after 20 h oxidation; the time-dependent interface debonding length increases from $l_d/r_f = 6.7$ to 7.2 after 20 h oxidation; and the time-dependent interface oxidation length increases from $\zeta/l_d = 0$ to 0.35 after 20 h oxidation.

At elevated temperature of $T = 800$ °C, the time-dependent first matrix cracking stress decreases from $\sigma_{mc} = 142$ to 87 MPa after 20 h oxidation; the time-dependent interface debonding length increases from $l_d/r_f = 6.7$ to 17.8 after 20 h oxidation; and the time-dependent interface oxidation length increases from $\zeta/l_d = 0$ to 0.95 after 20 h oxidation.

1.3.5 Experimental Comparisons

At temperature below the processing temperature of C/SiC (i.e., $T < 1100$ °C), the oxidation of the carbon fiber occurs within C/SiC due to microcracks caused by thermal residual tensile stress in the SiC matrix. The microcracks allow a path for the air to seep into C/SiC. At elevated temperature of $T = 1100$ °C, the microcracks within the SiC matrix begin to close due to the match of thermal expansion of the carbon fibers and the SiC matrix. The silica scale (SiO_2) as the SiC matrix oxidizes helps to fill in the matrix cracks, which seal off the fibers from the outside air. The presence of oxidation within C/SiC is drastically reduced between 1000 and 1200 °C at low stresses. At elevated temperature above 1200 °C, these anti-oxidizing occurrences begin to diminish. The over expansion of SiC matrix can cause cracks to widen, and the formation of SiO_2 begins to increase in a parabolic fashion. When designing a system with C/SiC as a main component, oxidation damage of C/SiC must be taken into consideration at low and high temperatures, i.e., less than 1000 °C or greater than 1200 °C.

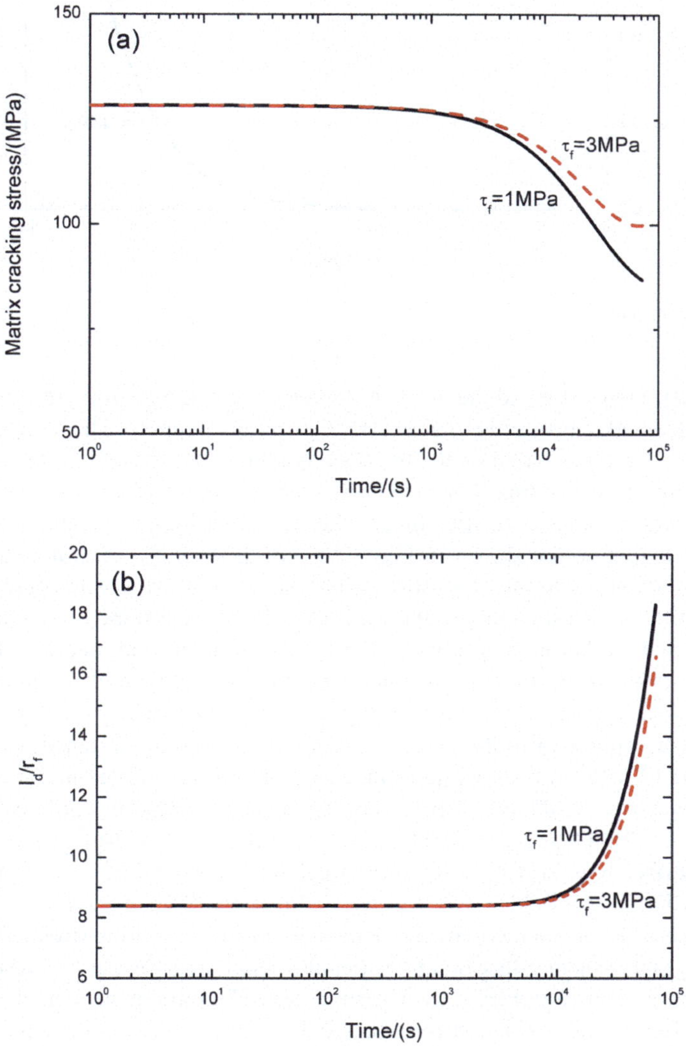

Fig. 1.14 **a** Time-dependent first matrix cracking stress versus the oxidation time; **b** the time-dependent interface debonding length versus the oxidation time; and **c** the time-dependent interface oxidation length versus the oxidation time corresponding to different interface shear stress of $\tau_f =$ 1 and 3 MPa of C/SiC composite at $T = 800$ °C in air atmosphere

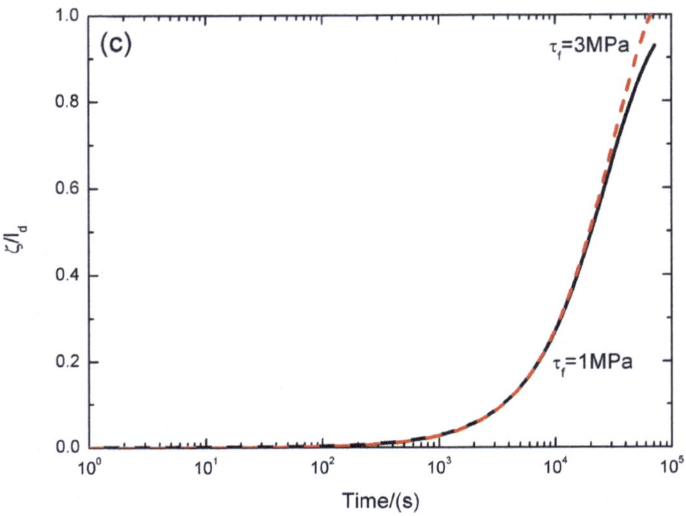

Fig. 1.14 (continued)

Yang (2011) investigated the mechanical behavior of C/SiC after unstressed oxidation at elevated temperature of $T = 700$ °C in air atmosphere. The composite was divided into two types based on the interface bonding, i.e., strong interface bonding and weak interface bonding. For the C/SiC with strong interface bonding, the tensile stress-strain curves corresponding to the cases of without oxidation, 4 and 6 h unstressed oxidation are shown in Fig. 1.16. The first matrix cracking stresses of C/SiC corresponding to the proportional limit stresses in the tensile curves are 37, 30, and 20 MPa corresponding to the cases of without oxidation, 4 h oxidation and 6 h unstressed oxidation, respectively. For the C/SiC with weak interface bonding, the tensile stress-strain curves corresponding to the cases of without oxidation, 2 and 6 h unstressed oxidation are shown in Fig. 1.17. The first matrix cracking stresses of C/SiC corresponding to the proportional limit stresses in the tensile curves are 27, 20, and 13 MPa corresponding to the cases of without oxidation, 2 h oxidation and 6 h unstressed oxidation, respectively. The material properties are given by: $V_f = 20\%$, $E_f = 200$ GPa, $E_m = 350$ GPa, $r_f = 3.5$ μm, $\xi_m = 6$ J/m^2, $\xi_d = 1.2$ J/m^2 (strong interface bonding), $\xi_d = 0.6$ J/m^2 (weak interface bonding), $\tau_i = 6$ MPa, and $\tau_f = 1.0$ MPa, $\alpha_f = -0.38 \times 10^{-6}$/°C, $\alpha_m = 4.6 \times 10^{-6}$/°C, and $\Delta T = -1000$ °C.

The experimental and theoretical predicted first matrix cracking stresses of C/SiC with strong and weak interface bonding after unstressed oxidation at $T = 700$ °C in air atmosphere are shown in Fig. 1.18. The time-dependent first matrix cracking stress decreases 18.9% after oxidation for 4 h, and 46% after oxidation for 6 h for C/SiC with strong interface bonding, and 25.9% after oxidation for 1 h, and 51.8% after oxidation for 6 h for C/SiC with weak interface bonding. The theoretical predicted results agreed with experimental data. The strong interface bonding can be used for oxidation resistant of C/SiC composite at elevated temperatures. Although the strong interfacial bonding can resist the oxidation of CMCs, it can decrease the toughness of CMCs, so the weak interfacial bonding is essential for CMCs.

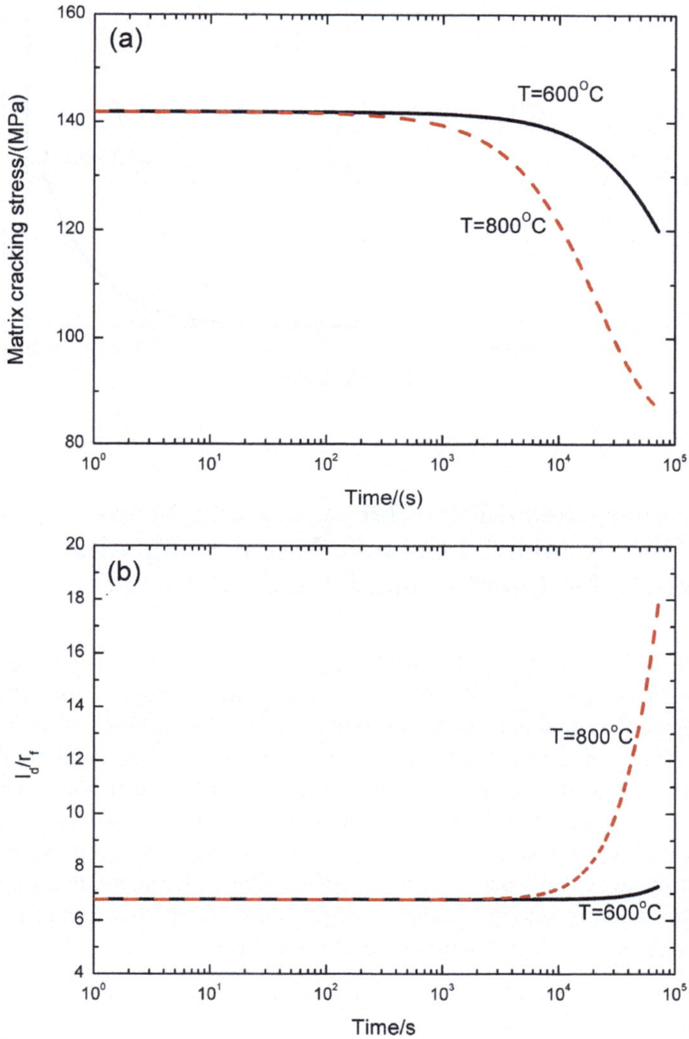

Fig. 1.15 **a** Time-dependent first matrix cracking stress versus the oxidation time; **b** the time-dependent interface debonding length versus the oxidation time; and **c** the time-dependent interface oxidation length versus the oxidation time corresponding to different oxidation temperature of $T =$ 600 and 800 °C of C/SiC composite

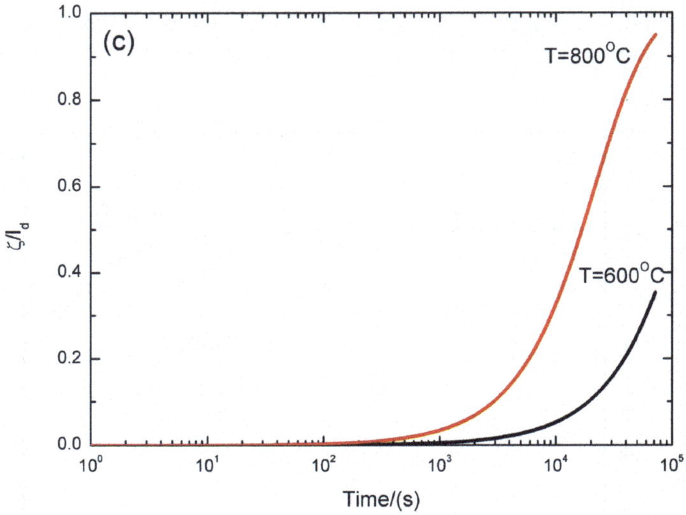

Fig. 1.15 (continued)

1.4 Time-Dependent First Matrix Cracking Stress of Fiber-Reinforced Ceramic-Matrix Composites Considering Interface and Fiber Oxidation

In this section, the synergistic effects of temperature and oxidation on time-dependent first matrix cracking stress in fiber-reinforced CMCs are investigated using the energy balance approach. The shear-lag model combined with the interface oxidation model, the interface debonding model, fiber strength degradation model, and the fiber failure model is adopted to analyze the microstress distribution in the fiber and the matrix. The relationships between the time-dependent first matrix cracking stress, the interface debonding and slip, the fiber failure, the oxidation temperature, and the oxidation time are established. The effects of the fiber volume, the interface properties, the fiber strength, and the oxidation temperature on the evolution of first matrix cracking stress versus the oxidation time are analyzed.

1.4.1 Time-Dependent Stress Analysis

The oxygen reacts with carbon layer along the fiber length at a certain rate of $\mathrm{d}\zeta/\mathrm{d}t$, in which ζ is the length of carbon lost in each side of the crack (Casas and Martinez-Esnaola 2003).

$$\zeta = \varphi_1\left[1 - \exp\left(-\frac{\varphi_2 t}{b}\right)\right] \tag{1.54}$$

where φ_1 and φ_2 are parameters dependent on temperature and described using the Arrhenius-type laws; and b is a delay factor considering the deceleration of reduced oxygen activity. In the interface oxidation region, i.e., $z \in [0, \zeta]$, the stress transfer between the fiber and the matrix is controlled by a sliding stress $\tau_i(x) = \tau_f$, different from the interface shear stress in the interface debonded region, i.e., $z \in [\zeta, l_d]$, $\tau_i(x) = \tau_i$. This new interface shear stress τ_f is lower than τ_i.

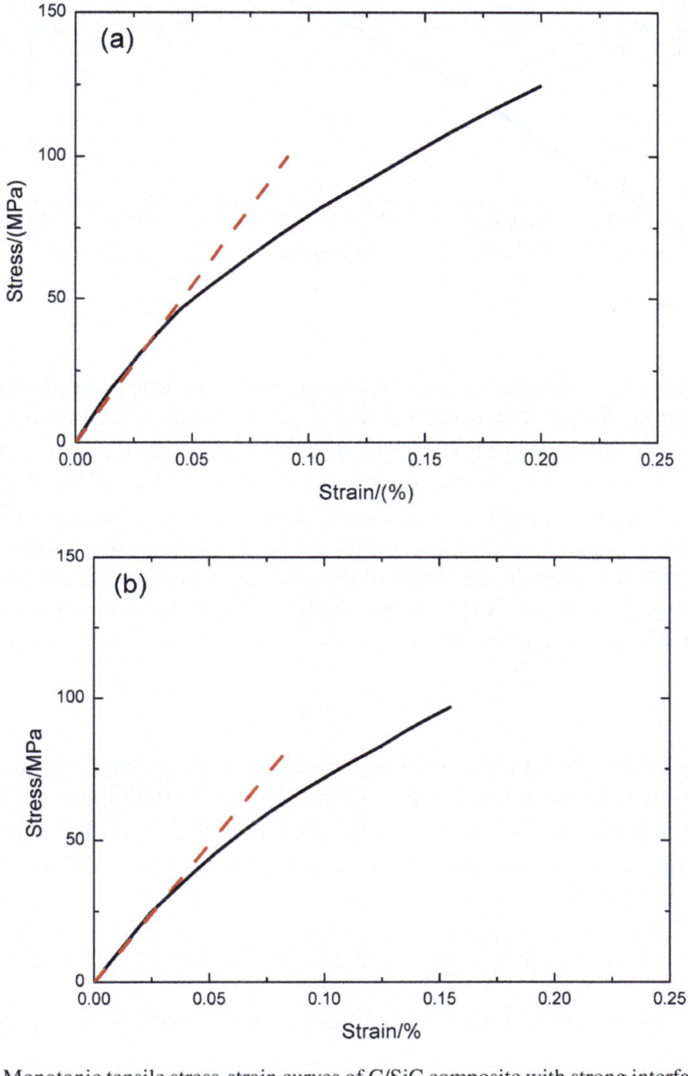

Fig. 1.16 Monotonic tensile stress-strain curves of C/SiC composite with strong interface bonding corresponding to **a** without oxidation; **b** unstressed oxidation of 4 h; and **c** unstressed oxidation of 6 h

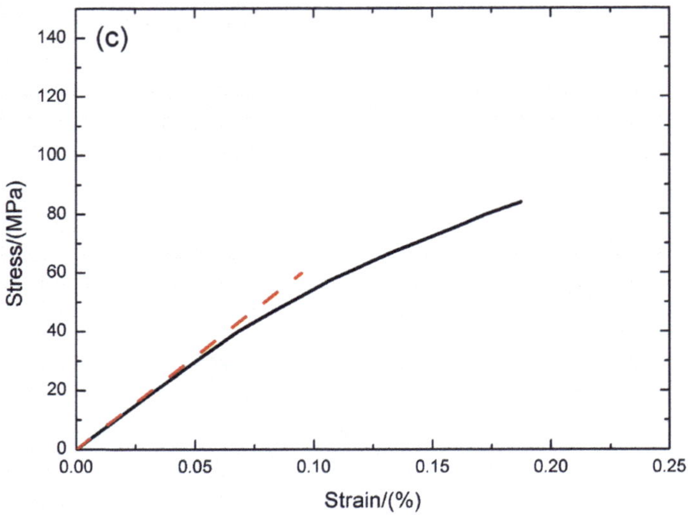

Fig. 1.16 (continued)

The oxidation of fiber is assumed to be controlled by diffusion of oxygen gas through matrix cracks. When the oxidizing gas ingresses into the composite, a sequence of events is triggered starting first with the oxidation of the fiber. For simplicity, it is assumed that both the Weibull and elastic moduli of the fibers remain constant and that the only effect of oxidation is to decrease the strength of fibers. The time-dependent strength of fibers would be controlled by surface defects resulting from the oxidation, with the thickness of the oxidized layer representing the size of the average strength-controlling flaw. According to linear elastic fracture mechanics, the relationship between strength and flaw size is determined as (Lara-Curzio 1999):

$$K_{\mathrm{IC}} = Y\sigma_0\sqrt{a} \tag{1.55}$$

where K_{IC} denotes the critical stress intensity factor; Y is a geometric parameter; σ_0 is the fiber strength; and a is the size of the strength-controlling flaw.

Considering that the oxidation of fibers is controlled by diffusion of oxygen through oxidized layer, the oxidized layer will grow on fiber's surface according to (Lara-Curzio 1999)

$$\alpha = \sqrt{kt} \tag{1.56}$$

where α is the thickness of the oxidized layer at time t; and k is the parabolic rate constant.

By assuming the fracture toughness of the fibers remains constant and that the fiber strength σ_0 is related to the mean oxidized layer thickness according to Eq. (1.55), i.e., $a = \alpha$, then the time dependence of the fiber strength would be determined by the following equations (Lara-Curzio 1999).

$$\sigma_0(t) = \sigma_0, \quad t \leq \frac{1}{k}\left(\frac{K_{IC}}{Y\sigma_0}\right)^4 \tag{1.57}$$

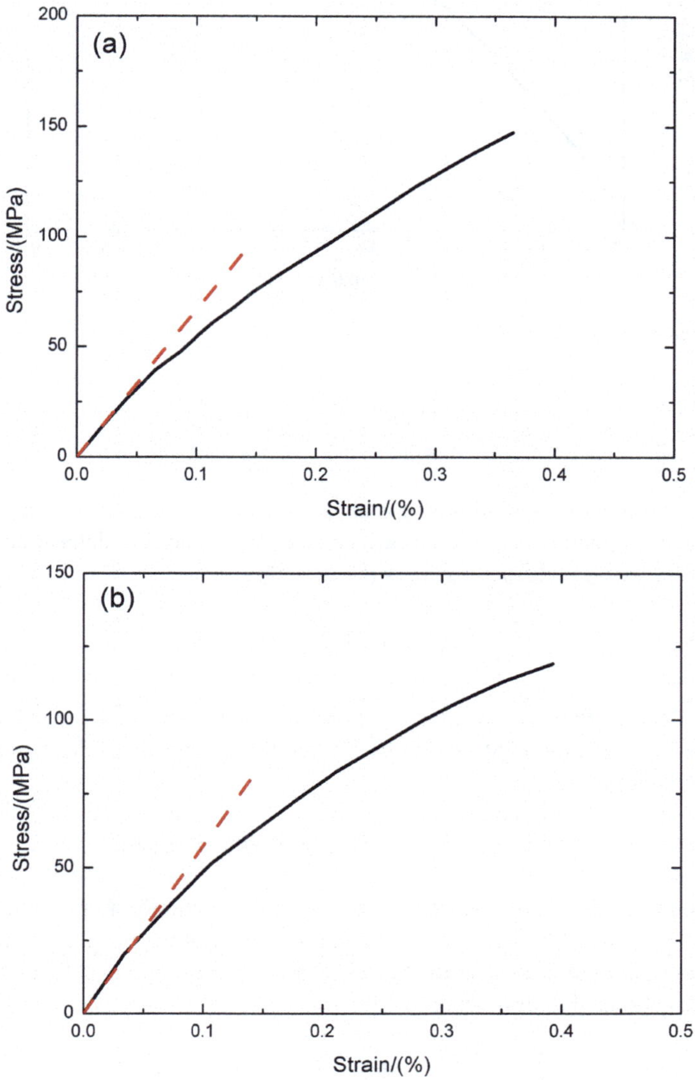

Fig. 1.17 Monotonic tensile stress-strain curves of C/SiC composite with weak interface bonding corresponding to **a** without oxidation; **b** unstressed oxidation of 2 h; and **c** unstressed oxidation of 6 h

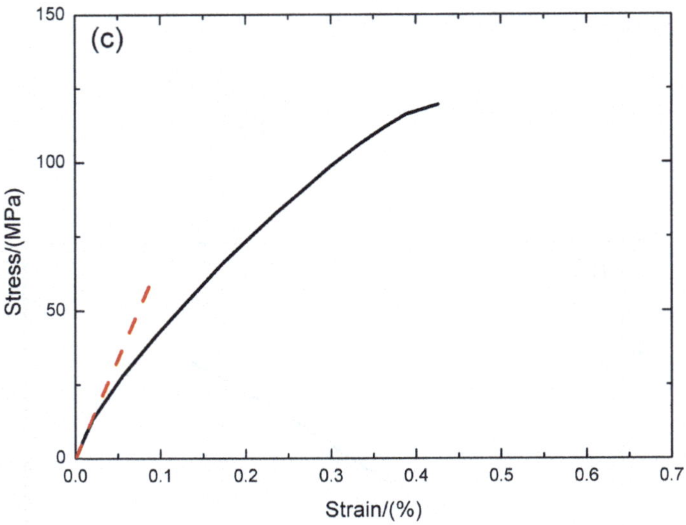

Fig. 1.17 (continued)

$$\sigma_0(t) = \frac{K_{IC}}{Y\sqrt[4]{kt}}, \quad t > \frac{1}{k}\left(\frac{K_{IC}}{Y\sigma_0}\right)^4 \tag{1.58}$$

The two-parameter Weibull model is adopted to describe the fiber strength distribution, and the global load sharing (GLS) assumption is used to determine the load carried by the intact and fracture fibers (Curtin 1991b).

$$\frac{\sigma}{V_\mathrm{f}} = \Phi[1 - P(\Phi)] + \langle\Phi_\mathrm{b}\rangle P(\Phi) \tag{1.59}$$

where V_f denotes the fiber volume fraction; Φ denotes the load carried by intact fibers; $\langle\Phi_\mathrm{b}\rangle$ denotes the load carried by broken fibers; and $P(\Phi)$ denotes the fiber failure probability.

$$P(\Phi) = \chi P_a(\Phi) + (1 - \chi)P_b(\Phi) + P_c(\Phi) \tag{1.60}$$

where χ denotes the oxidation fibers fraction in the oxidation region; and $P_a(\Phi)$, $P_b(\Phi)$, $P_c(\Phi)$, and $P_\mathrm{d}(\Phi)$ denote the fracture probability of oxidized fibers in the oxidation region, unoxidized fibers in the oxidation region, interface debonded region, and interface bonded region, respectively.

$$P_a(\Phi) = 1 - \exp\left\{-\frac{1}{m+1}\frac{r_\mathrm{f}\Phi^{m+1}}{\tau_\mathrm{f} l_0[\sigma_0(t)]^m}\left[1 - \left(1 - \frac{2\tau_\mathrm{f}}{r_\mathrm{f}\Phi}\zeta\right)^{m+1}\right]\right\} \tag{1.61}$$

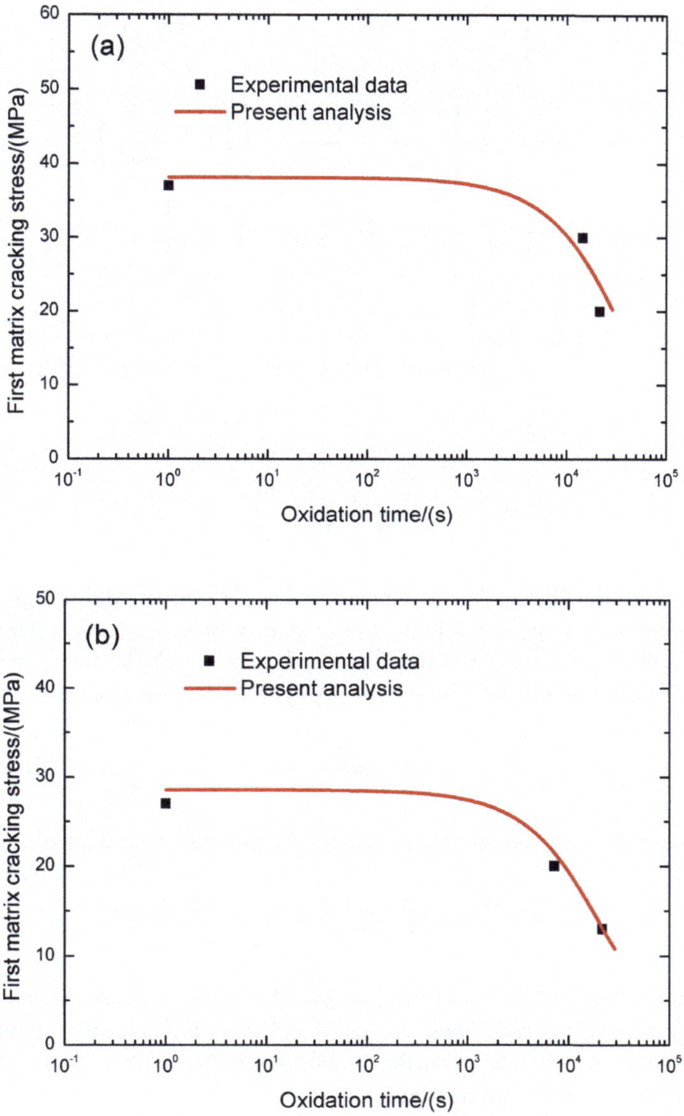

Fig. 1.18 Experimental and predicted first matrix cracking stress versus the oxidation time of C/SiC composite after unstressed oxidation at T = 700 °C in air atmosphere corresponding to **a** strong interface bonding and **b** weak interface bonding

$$P_b(\Phi) = 1 - \exp\left\{-\frac{1}{m+1}\frac{r_f\Phi^{m+1}}{\tau_f l_0(\sigma_0)^m}\left[1 - \left(1 - \frac{2\tau_f}{r_f\Phi}\zeta\right)^{m+1}\right]\right\} \tag{1.62}$$

$$P_c(\Phi) = 1 - \exp\left\{-\frac{1}{m+1}\frac{r_f\Phi^{m+1}}{l_0\tau_i(\sigma_0)^m}\left[\left(1 - \frac{2\tau_f}{r_f\Phi}\zeta\right)^{m+1} - \left(1 - \frac{2\tau_f}{r_f\Phi}\zeta - \frac{2\tau_i}{r_f\Phi}(l_d - \zeta)\right)^{m+1}\right]\right\} \tag{1.63}$$

$$P_d(\Phi) = 1 - \exp\left\{-\frac{2r_f\Phi^m}{\rho l_0\sigma_0^m(m+1)\left(1 - \frac{\tau_f}{\tau_i}\frac{\zeta}{l_s} - \frac{l_d-\zeta}{l_s} - \frac{\sigma_{fo}}{\Phi}\right)} \times\left[\left(1 - \frac{\tau_f}{\tau_i}\frac{\zeta}{l_s} - \frac{l_d - \zeta}{l_s} - \frac{\sigma_{fo}}{\Phi}\right)\left(\rho\frac{l_s - l_d}{r_f}\right) - \left(1 - \frac{\tau_f}{\tau_i}\frac{\zeta}{l_s} - \frac{l_d - \zeta}{l_s}\right)\right]^{m+1}\right\} \tag{1.64}$$

where r_f denotes the fiber radius; m denotes the fiber Weibull modulus; $\sigma_0(t)$ denotes the oxidized fiber strength; t denotes the oxidation time; l_d denotes the interface debonded length; and l_s denotes the slip length over which the fiber stress would decay to zero if not interrupted by the far-field equilibrium stresses.

$$l_s = \frac{r_f\Phi}{2\tau_i} \tag{1.65}$$

The stress carried by broken fibers is determined by the following equation.

$$\langle\Phi_b\rangle = \Phi\left[\left(\frac{\sigma_c}{\Phi}\right)^{m+1} - \frac{1 - P(\Phi)}{P(\Phi)}\right] \tag{1.66}$$

Substituting Eqs. (1.60) and (1.66) into Eq. (1.59), the stress Φ carried by intact fibers at the matrix cracking plane can be determined. Substituting the intact fiber stress Φ into the Eqs. (1.61)–(1.64), the relationship between fiber failure probability and applied stress can be determined.

1.4.2 Downstream Stresses

The composite with fiber volume fraction V_f is loaded by a remote uniform stress σ normal to a long crack plane, as shown in Fig. 1.1. The unit cell in the downstream region I contained a single fiber surrounded by a hollow cylinder of matrix is extracted from the ceramic composite system, as shown in Fig. 1.2. The fiber radius is r_f, and the matrix radius is R ($R = r_f/V_f^{1/2}$). The length of the unit cell is half matrix crack spacing $l_c/2$, and the interface oxidation length and interface debonded length are ζ and l_d, respectively. In the oxidation region, the fiber/matrix interface is resisted by a constant frictional shear stress τ_f; and in the debonded region, the interface is resisted by τ_i, which is higher than τ_f. For the debonded region in region I, the force equilibrium equation of the fiber is given by the following equation.

$$\frac{d\sigma_f(z)}{dz} = -\frac{2\tau_i(z)}{r_f} \tag{1.67}$$

The boundary condition at the crack plane $z = 0$ is given by the following equations.

$$\sigma_f(z = 0) = T \tag{1.68}$$

$$\sigma_m(z = 0) = 0 \tag{1.69}$$

The total axial stresses in region I satisfy the following equation.

$$V_f\sigma_f(z) + V_m\sigma_m(z) = \sigma \tag{1.70}$$

Solving Eqs. (1.67) and (1.70) with the boundary conditions given by Eqs. (1.68) and (1.69), and the interface shear stress in the oxidation and debonded region, the fiber and matrix axial stresses in the interface oxidation and debonded region, i.e., $0 < z < l_d$, can be determined by Eqs. (1.71) and (1.72).

$$\sigma_f^D(z) = \begin{cases} \Phi - \dfrac{2\tau_f}{r_f}z, & z \in [0, \zeta] \\ \Phi - \dfrac{2\tau_f}{r_f}\zeta - \dfrac{2\tau_i}{r_f}(z - \zeta), & z \in [\zeta, l_d] \end{cases} \tag{1.71}$$

$$\sigma_m^D(z) = \begin{cases} 2\dfrac{V_f}{V_m}\dfrac{\tau_f}{r_f}z, & z \in [0, \zeta] \\ 2\dfrac{V_f}{V_m}\dfrac{\tau_f}{r_f}\zeta + 2\dfrac{V_f}{V_m}\dfrac{\tau_i}{r_f}(z - \zeta), & z \in [\zeta, l_d] \end{cases} \tag{1.72}$$

For the bonded region ($l_d < z$) in the downstream region I, the fiber and matrix axial stresses and the interfacial shear stress can be determined using the composite-cylinder model adopted by BHE (Budiansky et al. 1986). The free body diagram of the composite-cylinder model is illustrated in Fig. 1.2, where the fiber closure traction T causes interfacial debonding between the fiber and the matrix over a distance l_d and the crack opening displacement $v(0)$. The radius of the matrix cylinder is determined by the following equation.

$$R = \frac{r_f}{\sqrt{V_f}} \tag{1.73}$$

The model can be further simplified by defining an effective radius $\overline{R}$ ($r_f < \overline{R} < R$) such that the matrix axial load to be concentrated at $\overline{R}$ and the region between r_f and $\overline{R}$ carry only the shear stress.

$$\ln\left(\frac{\overline{R}}{r_f}\right) = -\frac{2\ln V_f + V_m(3 - V_f)}{4V_m^2} \tag{1.74}$$

Considering the equilibrium of the radius force acting on the differential element $dz(dr)(rd\theta)$ in the domain $r_f < r < \overline{R}$ of the bonded matrix region (i.e., $z \geq l_d$) leads to the following differential equation.

$$\frac{\partial \tau_{rz}}{\partial r} + \frac{\tau_{rz}}{r} = 0 \tag{1.75}$$

The shear stress τ_{rz} is given by

$$\tau_{rz}(r, z) = \frac{r_f \tau_i(z)}{r} \tag{1.76}$$

The matrix in the region $r_f < r < \overline{R}$ only carries the shear stress, and the stress-strain relation can be determined as:

$$\tau_{rz} = G_m \frac{\partial w}{\partial r} \tag{1.77}$$

where G_m is the matrix shear modulus; and w is the axial displacement.

Substituting the Eq. (1.76) into Eq. (1.77), the interfacial shear stress $\tau_i(z)$, in the interface bonded region, can be given by the following equation.

$$\tau_i(z) = \frac{G_m(w_m - w_f)}{r_f \ln\left(\overline{R}/r_f\right)} \tag{1.78}$$

where $w_f = w(r_f, z)$ and $w_m = w\left(\overline{R}, z\right)$ denote the fiber and the matrix axial displacement, respectively.

$$\frac{dw_f}{dz} = \frac{\sigma_f}{E_f} \tag{1.79}$$

$$\frac{dw_m}{dz} = \frac{\sigma_m}{E_m} \tag{1.80}$$

where E_f and E_m denote the fiber and matrix elastic modulus.

Substituting Eqs. (1.78)–(1.80) into Eq. (1.67), and applying the boundary condition of Eqs. (1.68) and (1.69), the fiber and matrix axial stresses in the bonded region ($l_d < z$) become,

$$\sigma_{\mathrm{f}}^{\mathrm{D}} = \sigma_{\mathrm{fo}} + \left[\Phi - \frac{2\tau_{\mathrm{f}}}{r_{\mathrm{f}}}\zeta - \frac{2\tau_{\mathrm{i}}}{r_{\mathrm{f}}}(l_{\mathrm{d}} - \zeta) - \sigma_{\mathrm{fo}}\right]\exp\left(-\rho\frac{z - l_{\mathrm{d}}}{r_{\mathrm{f}}}\right) \tag{1.81}$$

$$\sigma_{\mathrm{m}}^{\mathrm{D}} = \sigma_{\mathrm{mo}} + \left[2\frac{V_{\mathrm{f}}}{V_{\mathrm{m}}}\frac{\tau_{\mathrm{f}}}{r_{\mathrm{f}}}\zeta + 2\frac{V_{\mathrm{f}}}{V_{\mathrm{m}}}\frac{\tau_{\mathrm{i}}}{r_{\mathrm{f}}}(l_{\mathrm{d}} - \zeta) - \sigma_{\mathrm{mo}}\right]\exp\left(-\rho\frac{z - l_{\mathrm{d}}}{r_{\mathrm{f}}}\right) \tag{1.82}$$

$$\tau_i^D(z) = \frac{\rho}{2}\left[\Phi - \frac{2\tau_{\mathrm{f}}}{r_{\mathrm{f}}}\zeta - \frac{2\tau_{\mathrm{i}}}{r_{\mathrm{f}}}(l_{\mathrm{d}} - \zeta) - \sigma_{\mathrm{fo}}\right]\exp\left(-\rho\frac{z - l_{\mathrm{d}}}{r_{\mathrm{f}}}\right) \tag{1.83}$$

where ρ denotes the shear-lag model parameter, and

$$\sigma_{\mathrm{fo}} = \frac{E_{\mathrm{f}}}{E_{\mathrm{c}}}\sigma + E_{\mathrm{f}}(\alpha_{\mathrm{c}} - \alpha_{\mathrm{f}})\Delta T \tag{1.84}$$

$$\sigma_{\mathrm{mo}} = \frac{E_{\mathrm{m}}}{E_{\mathrm{c}}}\sigma + E_{\mathrm{m}}(\alpha_{\mathrm{c}} - \alpha_{\mathrm{m}})\Delta T \tag{1.85}$$

where E_{c} denotes the composite elastic modulus; α_{f}, α_{m}, and α_{c} denote the fiber, matrix, and composite thermal expansion coefficient, respectively; and ΔT denotes the temperature difference between the fabricated temperature T_0 and testing temperature T_1 ($\Delta T = T_1 - T_0$).

1.4.3 Upstream Stresses

The upstream region III as shown in Fig. 1.1 is so far away from the crack tip that the stress and strain fields are also uniform. The fiber and matrix have the same displacements, and the fiber and matrix stresses are obtained as:

$$\sigma_{\mathrm{f}}^{\mathrm{U}} = \sigma_{\mathrm{fo}} \tag{1.86}$$

$$\sigma_{\mathrm{m}}^{\mathrm{U}} = \sigma_{\mathrm{mo}} \tag{1.87}$$

1.4.4 Time-Dependent Interface Debonding

The interface debonding criterion is given by the following equation. (Gao et al. 1988)

$$\xi_{\mathrm{d}} = -\frac{F}{4\pi r_{\mathrm{f}}}\frac{\partial w_{\mathrm{f}}(0)}{\partial l_{\mathrm{d}}} - \frac{1}{2}\int_0^{l_{\mathrm{d}}} \tau_{\mathrm{i}}\frac{\partial v(z)}{\partial l_{\mathrm{d}}}\mathrm{d}z \tag{1.88}$$

where $F(=\pi r_f^2\sigma/V_f)$ denotes the fiber load at the matrix cracking plane; $w_f(0)$ denotes the fiber axial displacement on the matrix cracking plane; and $v(z)$ denotes the relative displacement between the fiber and the matrix.

The axial displacements of the fiber and matrix, i.e., $w_f(z)$ and $w_m(z)$, are determined as:

$$
\begin{aligned}
w_f(z) &= \int_{\infty}^{z} \frac{\sigma_f}{E_f}dz \\
&= \int_{\infty}^{l_d} \frac{\sigma_{fo}}{E_f}dz - \frac{\Phi}{E_f}(l_d - z) - \frac{\tau_f}{r_f E_f}\left(\zeta^2 - 2\zeta l_d + z^2\right) + \frac{\tau_i}{r_f E_f}(l_d - \zeta)^2 \\
&\quad - \frac{r_f}{\rho E_f}\left[\Phi - \frac{2\tau_f}{r_f}\zeta - \frac{2\tau_i}{r_f}(l_d - \zeta) - \sigma_{fo}\right]
\end{aligned}
\tag{1.89}
$$

$$
\begin{aligned}
w_m(z) &= \int_{\infty}^{z} \frac{\sigma_m}{E_m}dz \\
&= \int_{\infty}^{l_d} \frac{\sigma_{mo}}{E_m}dz - \frac{V_f\tau_i}{r_f V_m E_m}(l_d - \zeta)^2 + \frac{V_f\tau_f}{r_f V_m E_m}\left(\zeta^2 - 2\zeta l_d + z^2\right) \\
&\quad - \frac{r_f}{\rho E_m}\left[2\frac{\tau_f V_f}{r_f V_m}\zeta + 2\frac{V_f\tau_i}{r_f V_m}(l_d - \zeta) - \sigma_{mo}\right]
\end{aligned}
\tag{1.90}
$$

The relative displacement between the fiber and the matrix, i.e., $v(z)$, is given obtained as:

$$
\begin{aligned}
v(z) &= |w_f(z) - w_m(z)| \\
&= \frac{T}{E_f}(l_d - z) + \frac{E_c\tau_f}{r_f V_m E_m E_f}\left(\zeta^2 - 2\zeta l_d + z^2\right) - \frac{E_c\tau_i}{r_f V_m E_m E_f}(l_d - \zeta)^2 \\
&\quad + \frac{r_f E_c}{\rho V_m E_m E_f}\left[T - \frac{2\tau_f}{r_f}\zeta - \frac{2\tau_i}{r_f}(l_d - \zeta) - \sigma_{fo}\right]
\end{aligned}
\tag{1.91}
$$

Substituting $w_f(z = 0)$ and $v(z)$ into the Eq. (1.88), it leads to the form of the following equation.

$$
\begin{aligned}
&\frac{E_c\tau_i^2}{r_f V_m E_f E_m}(l_d - \zeta)^2 + \frac{2E_c\tau_f\tau_i}{r_f V_m E_f E_m}\zeta(l_d - \zeta) - \frac{\tau_i}{2E_f}\left(\frac{\sigma}{V_f} + \Phi\right)(l_d - \zeta) \\
&+ \frac{E_c\tau_i^2}{\rho V_m E_m E_f}(l_d - \zeta) + \frac{r_f\sigma\Phi}{4V_f E_f} - \frac{r_f\sigma^2}{4V_f E_c} - \frac{\tau_f\sigma\zeta}{2V_f E_f} - \frac{\tau_f\Phi}{2E_f}\zeta + \frac{E_c\tau_f^2}{r_f V_m E_f E_m}\zeta^2 \\
&- \frac{r_f\tau_i\sigma}{2\rho V_f E_f} + \frac{E_c\tau_f\tau_i}{\rho V_m E_m E_f}\zeta - \xi_d = 0
\end{aligned}
\tag{1.92}
$$

Solving Eq. (1.92), the interface debonding length is determined as:

$$
\begin{aligned}
l_{\mathrm{d}} = & \left(1-\frac{\tau_{\mathrm{f}}}{\tau_{\mathrm{i}}}\right)\zeta+\frac{r_{\mathrm{f}}V_{\mathrm{m}}E_{\mathrm{m}}}{4E_{\mathrm{c}}\tau_{\mathrm{i}}}\left(\frac{\sigma}{V_{\mathrm{f}}}+T\right)-\frac{r_f}{2\rho} \\
& -\left\{\frac{r_{\mathrm{f}}^2V_{\mathrm{m}}E_{\mathrm{m}}T^2}{4E_{\mathrm{c}}\tau_{\mathrm{i}}^2}\left[\frac{V_{\mathrm{m}}E_{\mathrm{m}}}{4E_{\mathrm{c}}}\left(\frac{\sigma}{V_{\mathrm{f}}T}+1\right)^2\right.\right. \\
& \left.+\frac{V_{\mathrm{f}}E_{\mathrm{f}}}{E_{\mathrm{c}}}\left(\frac{\sigma}{V_{\mathrm{f}}T}\right)^2-\frac{\sigma}{V_{\mathrm{f}}T}\right]+\frac{r_{\mathrm{f}}^2V_{\mathrm{m}}E_{\mathrm{m}}}{4\rho E_{\mathrm{c}}\tau_{\mathrm{i}}}\left(\frac{\sigma}{V_{\mathrm{f}}}-T\right) \\
& \left.+\left(\frac{r_{\mathrm{f}}}{2\rho}\right)^2+\frac{r_{\mathrm{f}}V_{\mathrm{m}}E_{\mathrm{f}}E_{\mathrm{m}}}{E_{\mathrm{c}}\tau_i^2}\xi_{\mathrm{d}}\right\}^{\frac{1}{2}}
\end{aligned}
\tag{1.93}
$$

1.4.5 Time-Dependent First Matrix Cracking Stress

The energy relationship to evaluate the first matrix cracking stress is determined using the following equation (Budiansky et al. 1986).

$$
\begin{aligned}
& \frac{1}{2}\int_{-\infty}^{\infty}\left[\frac{V_{\mathrm{f}}}{E_{\mathrm{f}}}\left(\sigma_{\mathrm{f}}^{\mathrm{U}}-\sigma_{\mathrm{f}}^{\mathrm{D}}\right)^2+\frac{V_{\mathrm{m}}}{E_{\mathrm{m}}}\left(\sigma_{\mathrm{m}}^{\mathrm{U}}-\sigma_{\mathrm{m}}^{\mathrm{D}}\right)^2\right]\mathrm{d}z \\
& +\frac{1}{2\pi R^2G_{\mathrm{m}}}\int_{-l_{\mathrm{d}}}^{l_{\mathrm{d}}}\int_{r_{\mathrm{f}}}^{R}\left(\frac{r_{\mathrm{f}}\tau_{\mathrm{i}}(z)}{r}\right)2\pi r\mathrm{d}r\mathrm{d}z \\
& = V_{\mathrm{m}}\xi_{\mathrm{m}}+\left(\frac{4V_{\mathrm{f}}l_{\mathrm{d}}}{r_{\mathrm{f}}}\right)\xi_{\mathrm{d}}
\end{aligned}
\tag{1.94}
$$

where ξ_{m} is the matrix fracture energy; and G_{m} is the matrix shear modulus. Substituting the time-dependent fiber and matrix stresses of Eqs. (1.71), (1.72), (1.81)–(1.83) and the interface debonding length of Eq. (1.93) into Eq. (1.94), the energy balance equation leads to the form of

$$
\eta_1\sigma^2+\eta_2\sigma+\eta_3=0 \tag{1.95}
$$

where

$$
\eta_1=\frac{l_{\mathrm{d}}}{E_{\mathrm{c}}}+\frac{r_{\mathrm{f}}V_{\mathrm{f}}E_{\mathrm{f}}}{2\rho V_{\mathrm{m}}E_{\mathrm{m}}E_{\mathrm{c}}} \tag{1.96}
$$

$$
\eta_2=-\frac{2V_{\mathrm{f}}\Phi l_{\mathrm{d}}}{E_{\mathrm{c}}}-\frac{r_{\mathrm{f}}V_{\mathrm{f}}\Phi}{\rho V_{\mathrm{m}}E_{\mathrm{m}}}+\frac{2V_{\mathrm{f}}\tau_{\mathrm{f}}\zeta}{\rho V_{\mathrm{m}}E_{\mathrm{m}}}+\frac{2V_{\mathrm{f}}\tau_{\mathrm{i}}}{\rho V_{\mathrm{m}}E_{\mathrm{m}}}(l_{\mathrm{d}}-\zeta) \tag{1.97}
$$

$$\eta_3 = \frac{V_f l_d \Phi^2}{E_f} - \frac{V_f}{E_f}\left(\frac{2\tau_f}{r_f}\right)\zeta(2l_d - \zeta)\Phi - \frac{V_f}{E_f}\left(\frac{2\tau_i}{r_f}\right)(l_d - \zeta)^2\Phi$$
$$+ \frac{4}{3}\left(\frac{V_f E_c}{V_m E_f E_m}\right)\left(\frac{\tau_i}{r_f}\right)^2\left[(l_d - \zeta)^3 + \left(\frac{\tau_f}{\tau_i}\right)^2\zeta^3\right]$$
$$+ \left(\frac{4V_f E_c}{V_m E_f E_m}\right)\frac{\tau_f\tau_i}{r_f^2}\zeta(l_d - \zeta)\left[l_d - \left(1 - \frac{\tau_f}{\tau_i}\right)\zeta\right]$$
$$+ \frac{r_f V_f E_c \Phi^2}{2\rho V_m E_m E_f} + \frac{2V_f E_c \tau_f^2}{\rho r_f V_m E_m E_f}\zeta^2 + \frac{2V_f E_c \tau_i^2}{\rho r_f V_m E_m E_f}(l_d - \zeta)^2$$
$$- \frac{2V_f E_c \Phi \tau_f}{\rho V_m E_m E_f}\zeta - \frac{2V_f E_c \tau_i \Phi}{\rho V_m E_m E_f}(l_d - \zeta)$$
$$+ \frac{4V_f E_c \tau_f \tau_i}{\rho r_f V_m E_m E_f}\zeta(l_d - \zeta)$$
$$- V_m \xi_m - \left(\frac{4V_f l_d}{r_f}\right)\xi_d \tag{1.98}$$

1.4.6 *Results and Discussion*

The effects of the fiber volume, interface debonding energy, interface shear stress, fiber strength, oxidation temperature on the time-dependent first matrix cracking stress, the interface debonding, and fiber failure are discussed. The ceramic composite system of C/SiC is used for the case study, and its material properties are given by: $V_f = 30\%$, $E_f = 230$ GPa, $E_m = 350$ GPa, $r_f = 3.5$ μm, $\xi_m = 6$ J/m^2, $\xi_d = 0.6$ J/m^2, $\tau_i = 15$ MPa, and $\tau_f = 5$ MPa, $\alpha_f = -0.38 \times 10^{-6}$/°C, $\alpha_m = 4.6 \times 10^{-6}$/°C, $\Delta T = -1000$ °C, $\sigma_0 = 2.6$ GPa, and $m = 5$.

1.4.6.1 Effect of Fiber Volume on Time-Dependent Interface Debonding and First Matrix Cracking Stress

The time-dependent first matrix cracking stress, the interface debonding length, and the interface oxidation length versus the oxidation time curves corresponding to different fiber volume of $V_f = 30$ and 40% are shown in Fig. 1.19. When the fiber volume increases, the time-dependent first matrix cracking stress and the interface oxidation length increase, and the time-dependent interface debonding length decreases.

When the fiber volume is $V_f = 30\%$, the time-dependent first matrix cracking stress decreases from $\sigma_{mc} = 83$ to 43 MPa after 10 h oxidation at 800 °C; the time-dependent interface debonding length first decreases from $l_d/r_f = 8.4$ to 8.2 after 1.9 h oxidation at 800 °C, and then increases to $l_d/r_f = 10.6$ after 10 h oxidation at

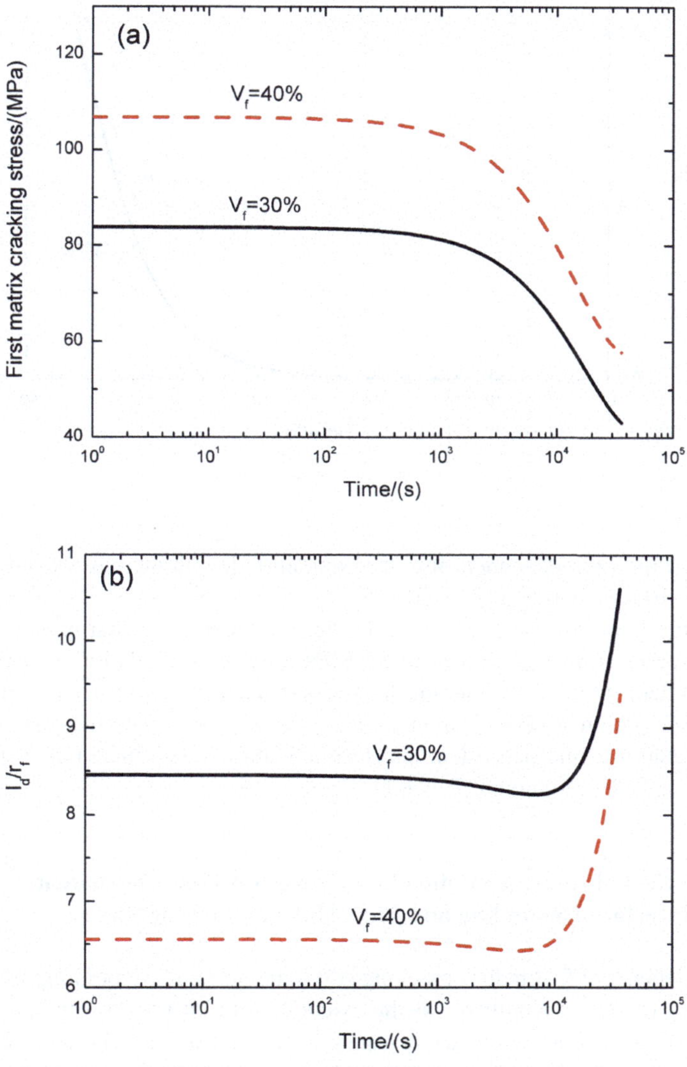

Fig. 1.19 **a** Time-dependent first matrix cracking stress versus the oxidation time; **b** the time-dependent interface debonding length versus the oxidation time; and **c** the time-dependent interface oxidation length versus the oxidation time corresponding to different fiber volume of $V_f = 30$ and 40% of C/SiC composite at $T = 800$ °C in air atmosphere

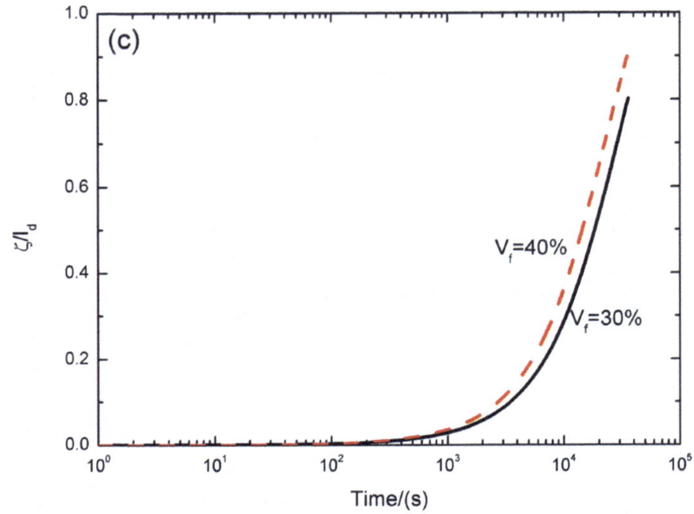

Fig. 1.19 (continued)

800 °C; and the time-dependent interface oxidation length increases from $\zeta/l_d = 0$ to 0.8 after 10 h oxidation at $T = 800$ °C.

When the fiber volume is $V_f = 40\%$, the time-dependent first matrix cracking stress decreases from $\sigma_{mc} = 106$ to 57 MPa after 10 h oxidation at 800 °C; the time-dependent interface debonding length first decreases from $l_d/r_f = 6.5$ to 6.4 after 1.3 h oxidation at 800 °C, and then increases to $l_d/r_f = 9.4$ after 10 h oxidation at 800 °C; and the time-dependent interface oxidation length increases from $\zeta/l_d = 0$ to 0.9 after 10 h oxidation at $T = 800$ °C.

1.4.6.2 Effect of Interface Debonding Energy on Time-Dependent Interface Debonding and First Matrix Cracking Stress

The time-dependent first matrix cracking stress, the interface debonding length, and the interface oxidation length versus the oxidation time curves corresponding to different interface debonding energy of $\xi_d/\xi_m = 0.1$ and 0.2 are shown in Fig. 1.20. When the interface debonding energy increases, the time-dependent first matrix cracking stress and the interface oxidation length increase, and the time-dependent interface debonding length decreases.

When the interface debonding energy is $\xi_d/\xi_m = 0.1$, the time-dependent first matrix cracking stress decreases from $\sigma_{mc} = 94$ to 49 MPa after 10 h oxidation at 800 °C; the time-dependent interface debonding length first decreases from $l_d/r_f = 7.4$ to 7.2 after 1.5 h oxidation at 800 °C, and then increases to $l_d/r_f = 9.9$ after 10 h oxidation at 800 °C; and the time-dependent interface oxidation length increases from $\zeta/l_d = 0$ to 0.85 after 10 h oxidation at 800 °C.

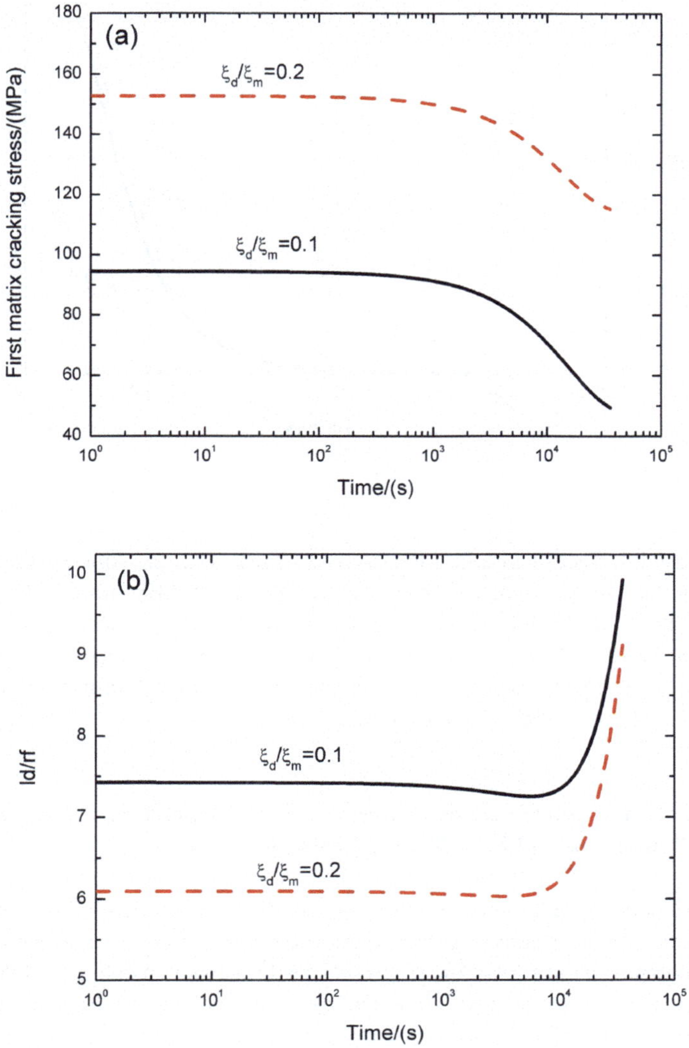

Fig. 1.20 **a** Time-dependent first matrix cracking stress versus the oxidation time; **b** the time-dependent interface debonding length versus the oxidation time; and **c** the time-dependent interface oxidation length versus the oxidation time corresponding to different interface debonding energy of $\xi_d/\xi_m = 0.1$ and 0.2 of C/SiC composite at $T = 800$ °C in air atmosphere

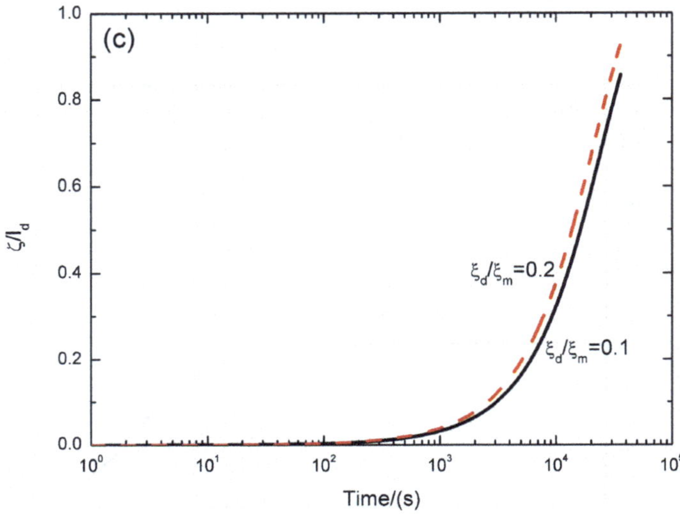

Fig. 1.20 (continued)

When the interface debonding energy is $\xi_d/\xi_m = 0.2$, the time-dependent first matrix cracking stress decreases from $\sigma_{mc} = 152$ to 115 MPa after 10 h oxidation at 800 °C; the time-dependent interface debonding length first decreases from $l_d/r_f = 6.1$ to 6.0 after 1 h oxidation at 800 °C, and then increases to $l_d/r_f = 9.1$ after 10 h oxidation at 800 °C; and the time-dependent interface oxidation length increases from $\zeta/l_d = 0$ to 0.93 after 10 h oxidation at 800 °C.

1.4.6.3 Effect of Interface Shear Stress on Time-Dependent Interface Debonding and First Matrix Cracking Stress

The time-dependent first matrix cracking stress, the interface debonding length, and the interface oxidation length versus the oxidation time curves corresponding to different interface shear stress of $\tau_i = 15$ and 25 MPa are shown in Fig. 1.21. When the interface shear stress in the slip region of τ_i increases, the time-dependent first matrix cracking stress and the interface oxidation length increase, and the time-dependent interface debonding length decreases.

When the interface shear stress is $\tau_i = 15$ MPa, the time-dependent first matrix cracking stress decreases from $\sigma_{mc} = 94$ to 49 MPa after 10 h oxidation at $T =$ 800 °C; the time-dependent interface debonding length first decreases from $l_d/r_f =$ 7.4 to 7.2 after 1.5 h oxidation at $T = 800$ °C and then increases to $l_d/r_f = 9.9$ after 10 h oxidation at $T = 800$ °C; and the time-dependent interface oxidation length increases from $\zeta/l_d = 0$ to 0.85 after 10 h oxidation at $T = 800$ °C.

When the interface shear stress is $\tau_i = 25$ MPa, the time-dependent first matrix cracking stress decreases from $\sigma_{mc} = 125$ MPa to 50 MPa after 10 h oxidation at T

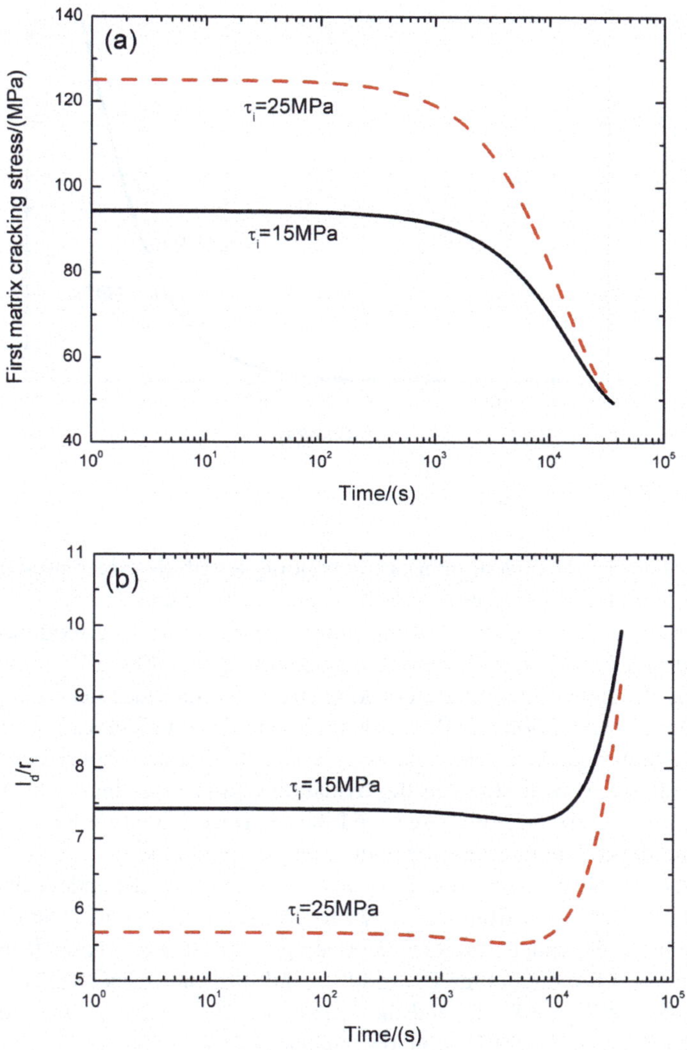

Fig. 1.21 **a** Time-dependent first matrix cracking stress versus the oxidation time; **b** the time-dependent interface debonding length versus the oxidation time; and **c** the time-dependent interface oxidation length versus the oxidation time corresponding to different interface shear stress of $\tau_i =$ 15 and 25 MPa of C/SiC composite at $T = 800$ °C in air atmosphere

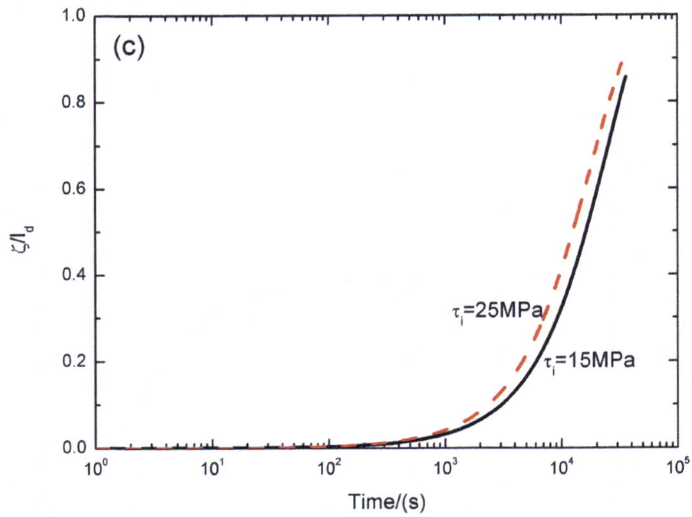

Fig. 1.21 (continued)

= 800 °C; the time-dependent interface debonding length first decreases from l_d/r_f = 5.6 to 5.5 after 1.1 h oxidation at 800 °C and then increases to $l_d/r_f = 9.3$ after 10 h oxidation at $T = 800$ °C; and the time-dependent interface oxidation length increases from $\zeta/l_d = 0$ to 0.91 after 10 h oxidation at $T = 800$ °C.

The time-dependent first matrix cracking stress, the interface debonding length, and the interface oxidation length versus the oxidation time curves corresponding to different interface shear stress of $\tau_f = 1$ and 5 MPa are shown in Fig. 1.22. When the interface shear stress in the oxidation region of τ_f increases, the time-dependent first matrix cracking stress and the interface oxidation length increase, and the time-dependent interface debonding length decreases.

When the interface shear stress is $\tau_f = 1$ MPa, the time-dependent first matrix cracking stress decreases from $\sigma_{mc} = 74$ to 28 MPa after 10 h oxidation at $T =$ 800 °C; the time-dependent interface debonding length first decreases from $l_d/r_f =$ 9.1 to 9 after 1.5 h oxidation at $T = 800$ °C and then increases to l_d/r_f 11.9 after 10 h oxidation at $T = 800$ °C; and the time-dependent interface oxidation length increases from $\zeta/l_d = 0$ to 0.71 after 10 h oxidation at $T = 800$ °C.

When the interface shear stress is $\tau_f = 5$ MPa, the time-dependent first matrix cracking stress decreases from $\sigma_{mc} = 74$ to 48 MPa after 10 h oxidation at $T =$ 800 °C; the time-dependent interface debonding length first decreases from $l_d/r_f =$ 9.1 to 8.9 after 2.2 h oxidation at $T = 800$ °C, and then increases to $l_d/r_f = 10.6$ after 10 h oxidation at $T = 800$ °C; and the time-dependent interface oxidation length increases from $\zeta/l_d = 0$ to 0.79 after 10 h oxidation at $T = 800$ °C.

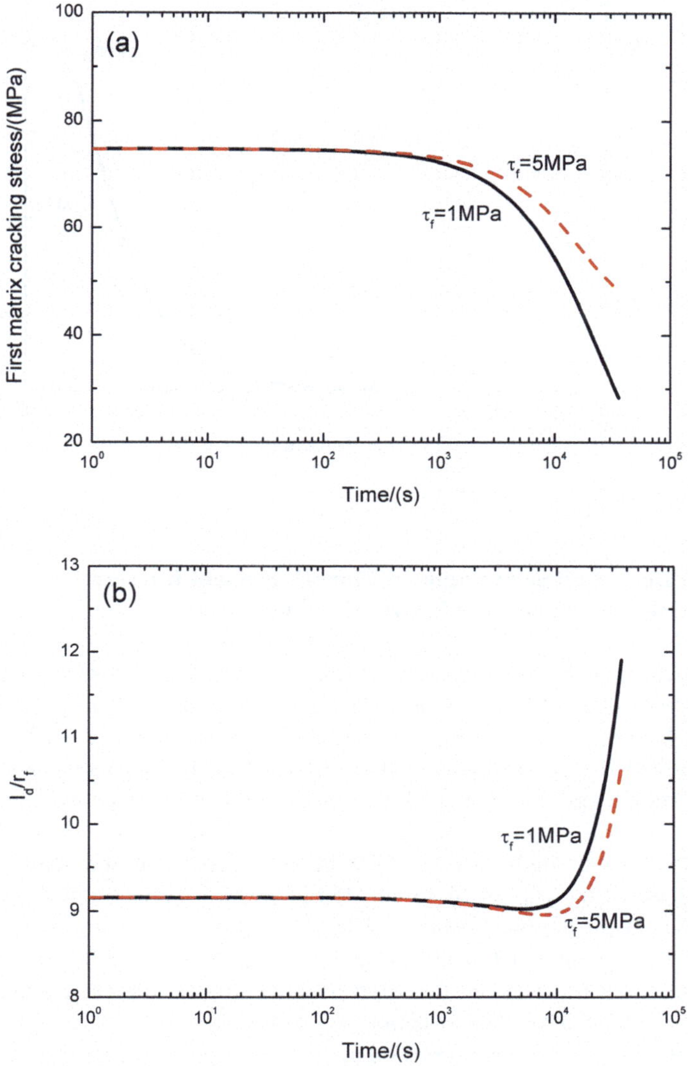

Fig. 1.22 **a** Time-dependent first matrix cracking stress versus the oxidation time; **b** the time-dependent interface debonding length versus the oxidation time; and **c** the time-dependent interface oxidation length versus the oxidation time corresponding to different interface shear stress of $\tau_f =$ 1 and 5 MPa of C/SiC composite at $T = 800$ °C

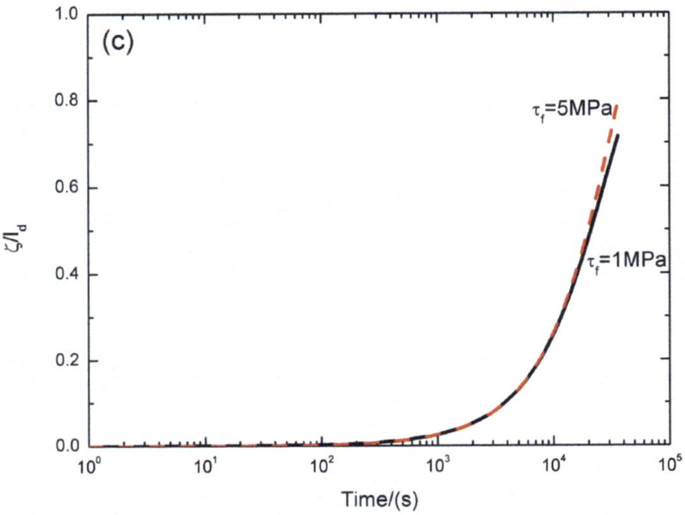

Fig. 1.22 (continued)

1.4.6.4 Effect of Fiber Strength on Time-Dependent Interface Debonding and First Matrix Cracking Stress

The time-dependent first matrix cracking stress, the interface debonding length, and the interface oxidation length versus the oxidation time curves corresponding to different interface shear stress of $\sigma_0 = 1$ and 2 GPa are shown in Fig. 1.23. When the fiber strength increases, the time-dependent first matrix cracking stress and the interface debonding length increase, and the time-dependent interface oxidation length decreases.

When the fiber strength is $\sigma_0 = 1$ GPa, the time-dependent first matrix cracking stress decreases from $\sigma_{mc} = 68$ to 44 MPa after 10 h oxidation at $T = 800$ °C; the time-dependent interface debonding length first decreases from $l_d/r_f = 8.9$ to 8.8 after 1.9 h oxidation at $T = 800$ °C and then increases to $l_d/r_f = 10.5$ after 10 h oxidation at $T = 800$ °C; and the time-dependent interface oxidation length increases from $\zeta/l_d = 0$ to 0.8 after 10 h oxidation at $T = 800$ °C.

When the fiber strength is $\sigma_0 = 2$ GPa, the time-dependent first matrix cracking stress decreases from $\sigma_{mc} = 72$ to 47 MPa after 10 h oxidation at $T = 800$ °C; the time-dependent interface debonding length first decreases from $l_d/r_f = 9.1$ to 8.9 after 2.1 h oxidation at $T = 800$ °C and then increases to $l_d/r_f = 10.6$ after 10 h oxidation at $T = 800$ °C; and the time-dependent interface oxidation length increases from $\zeta/l_d = 0$ to 0.79 after 10 h oxidation at $T = 800$ °C.

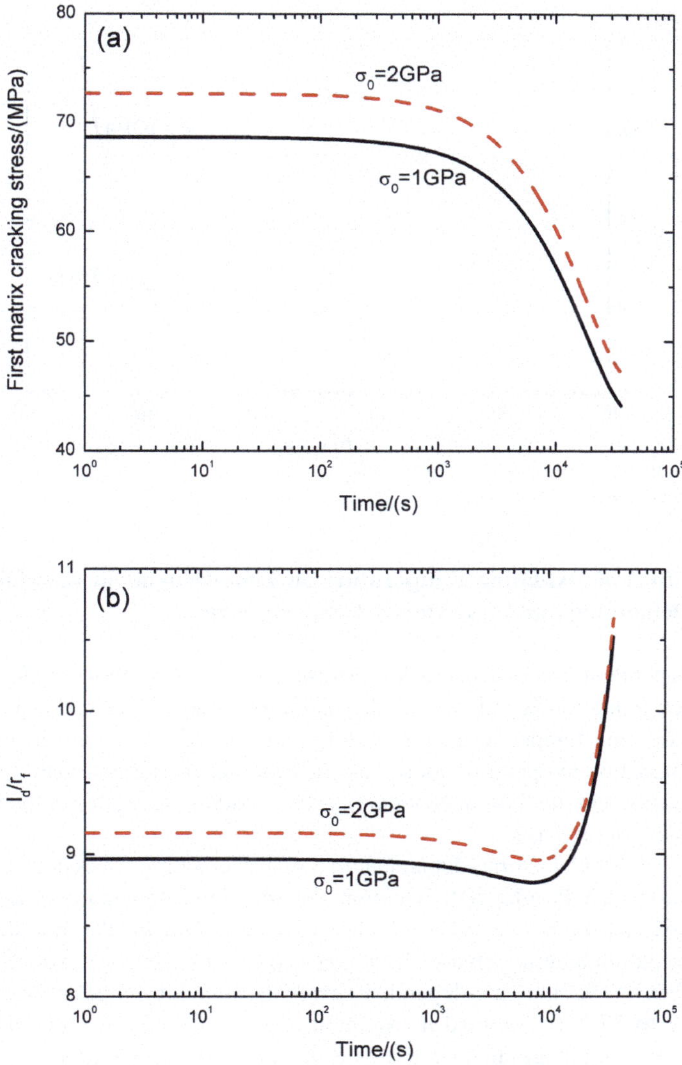

Fig. 1.23 **a** Time-dependent first matrix cracking stress versus the oxidation time; **b** the time-dependent interface debonding length versus the oxidation time; and **c** the time-dependent interface oxidation length versus the oxidation time corresponding to different fiber strength of $\sigma_0 = 1$ and 2 GPa of C/SiC composite at $T = 800$ °C

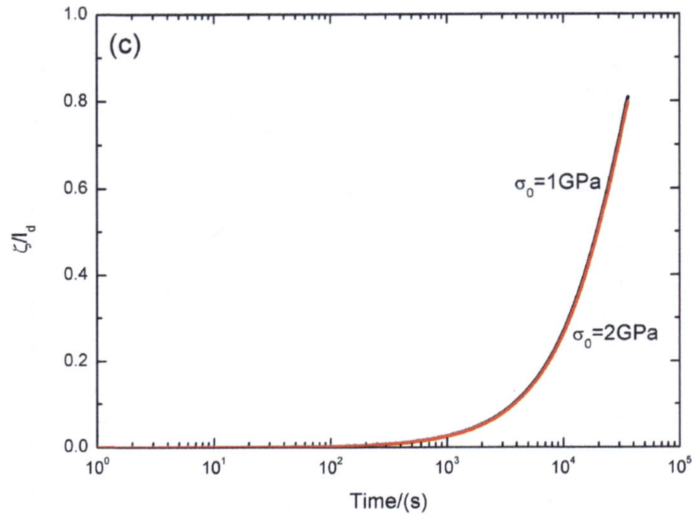

Fig. 1.23 (continued)

1.4.6.5 Effect of Oxidation Temperature on Time-Dependent Interface Debonding and First Matrix Cracking Stress

The time-dependent first matrix cracking stress, the interface debonding length, and the interface oxidation length versus the oxidation time curves corresponding to different oxidation temperature of T = 600 and 700 °C are shown in Fig. 1.24. When the oxidation temperature increases, the time-dependent first matrix cracking stress decreases, and the time-dependent interface oxidation length and the interface debonding length increase.

When $T = 600$ °C, the time-dependent first matrix cracking stress decreases from $\sigma_{mc} = 74.7$ to 66.9 MPa after 10 h oxidation; the time-dependent interface debonding length decreases from $l_d/r_f = 9.1$ to 8.9 after 10 h oxidation; and the time-dependent interface oxidation length increases from $\zeta/l_d = 0$ to 0.14 after 10 h oxidation.

When $T = 700$ °C, the time-dependent first matrix cracking stress decreases from σ_{mc} = 74.7 to 57 MPa after 10 h oxidation at T = 800 °C; the time-dependent interface debonding length first decreases from $l_d/r_f = 9.1$ to 8.9 after 5 h oxidation and then increases to $l_d/r_f = 9.1$ after 10 h oxidation; and the time-dependent interface oxidation length increases from $\zeta/l_d = 0$ to 0.4 after 10 h oxidation.

1.4.7 Experimental Comparisons

Yang (2011) investigated the mechanical behavior of C/SiC composite after unstressed oxidation at 700 °C in air. The composite was divided into two types based

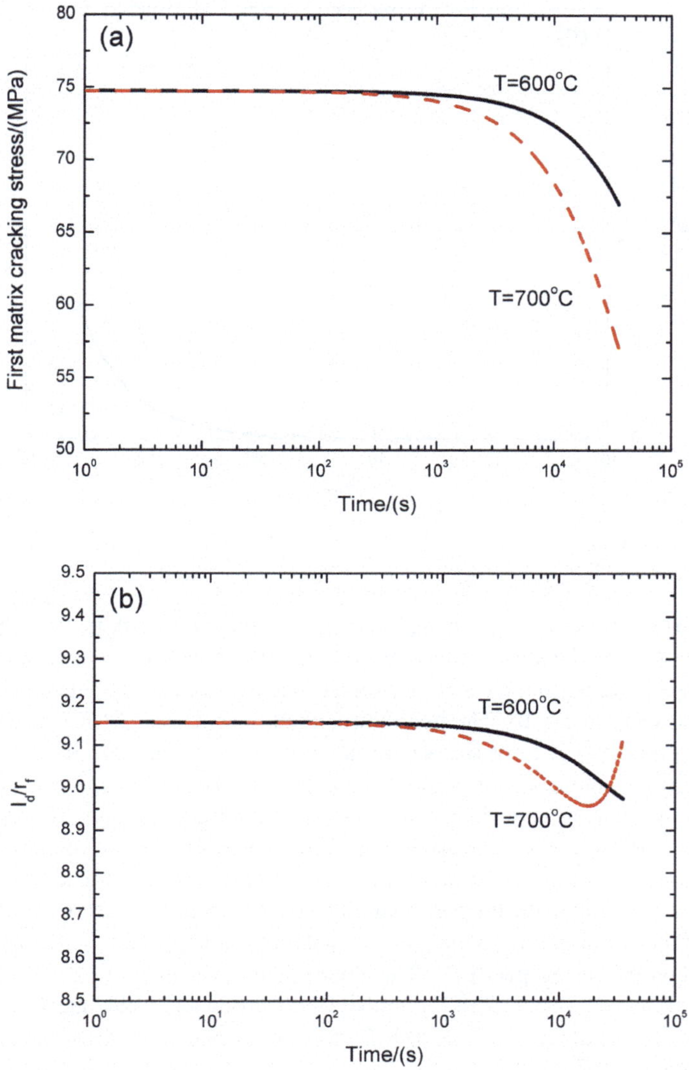

Fig. 1.24 **a** Time-dependent first matrix cracking stress versus the oxidation time; **b** the time-dependent interface debonding length versus the oxidation time; and **c** the time-dependent interface oxidation length versus the oxidation time corresponding to different oxidation temperature of $T =$ 600 and 700 °C

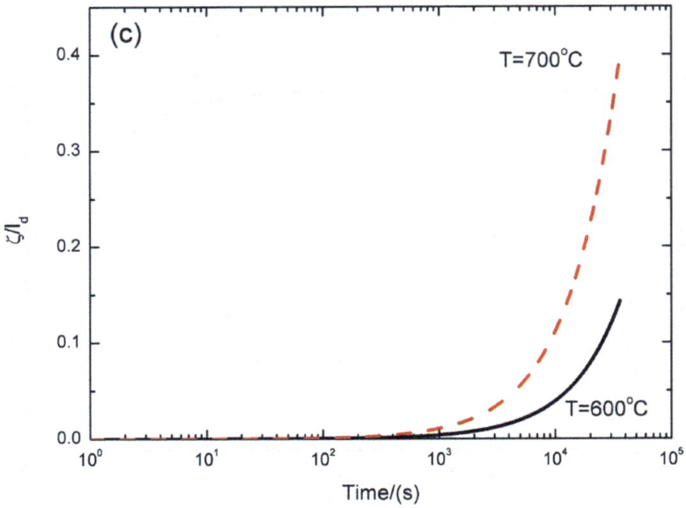

Fig. 1.24 (continued)

on the interface bonding, i.e., strong interface bonding and weak interface bonding. For the C/SiC with the strong interface bonding, the monotonic tensile stress-strain curves corresponding to the cases of without oxidation, 4 and 6 h unstressed oxidation are shown in Fig. 1.16. The first matrix cracking stresses of C/SiC corresponding to the proportional limit stresses in the tensile curves are 37, 30, and 20 MPa corresponding to the cases of without oxidation, 4 h oxidation and 6 h unstressed oxidation, respectively. For the C/SiC with weak interface bonding, the monotonic tensile stress-strain curves corresponding to the cases of without oxidation, 2 and 6 h unstressed oxidation are shown in Fig. 1.17. The first matrix cracking stresses of C/SiC corresponding to the proportional limit stresses in the tensile curves are 27, 20, and 13 MPa corresponding to the cases of without oxidation, 2 h oxidation and 6 h unstressed oxidation, respectively. The experimental and theoretical predicted first matrix cracking stresses of C/SiC composite with strong and weak interface bonding after unstressed oxidation at T = 700 °C in air atmosphere are shown in Figs. 1.25 and 1.26, respectively.

For the C/SiC with strong bonding, the time-dependent first matrix cracking stress decreases 18.9% after oxidation for 4 h, and 46% after oxidation for 6 h, and the predicted results agreed with experimental data; the time-dependent interface debonding length first decreases from $l_d/r_f = 15$ to 14.8 after 0.9 h oxidation and then increases to $l_d/r_f = 21.8$ after 10 h oxidation; and the time-dependent interface oxidation length ζ/l_d increases from zero to 0.83 after 10 h oxidation.

For the C/SiC with weak bonding, the time-dependent first matrix cracking stress decreases 25.9% after oxidation for 1 h, and 51.8% after oxidation for 6 h, and the predicted results agreed with experimental data; the time-dependent interface debonding length first decreases from l_d/r_f = 18 to 17.7 after 1.5 h oxidation and

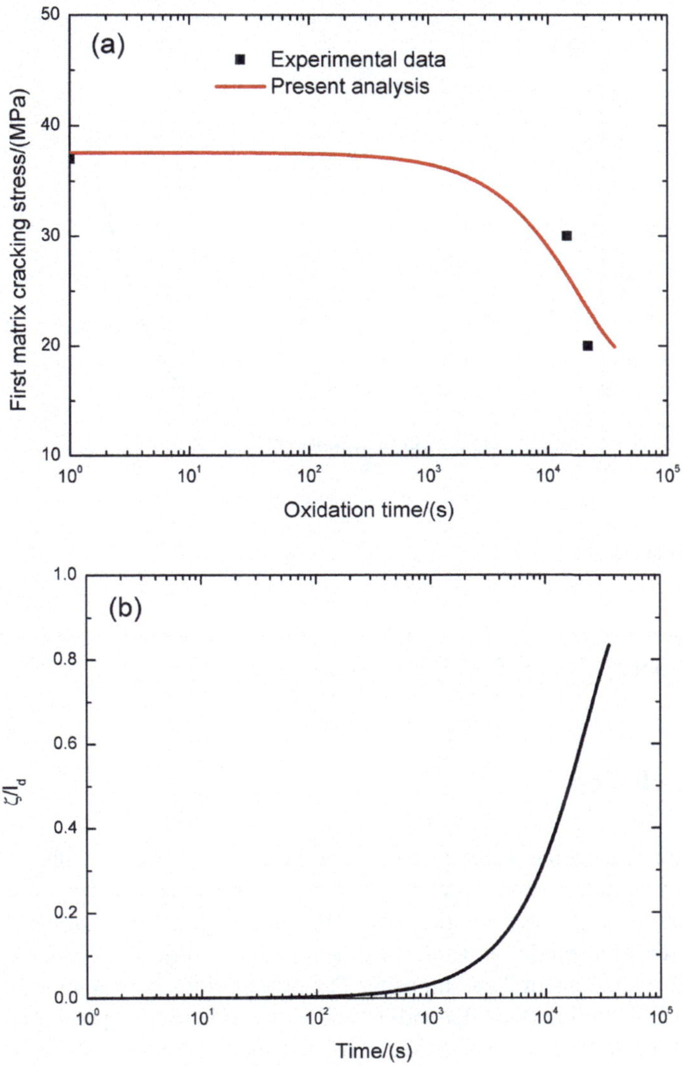

Fig. 1.25 **a** Experimental and predicted time-dependent first matrix cracking stress versus the oxidation time; **b** the time-dependent interface debonding length versus oxidation time; and **c** the time-dependent interface oxidation length versus oxidation time of C/SiC composite after unstressed oxidation at $T = 700$ °C in air atmosphere corresponding to the strong interface bonding

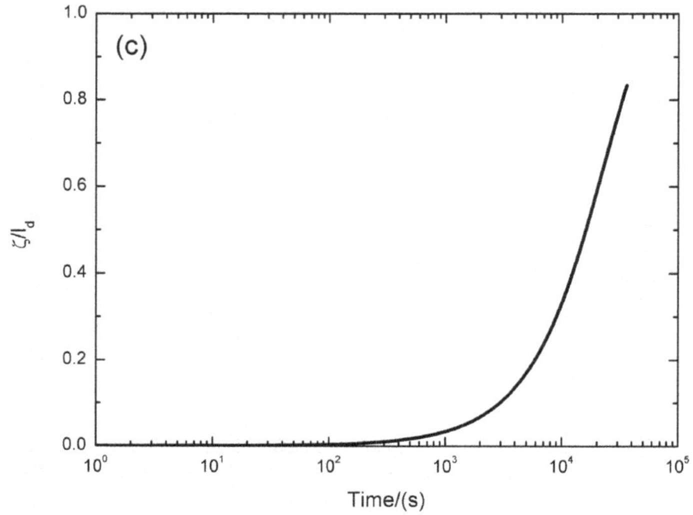

Fig. 1.25 (continued)

then increases to $l_d/r_f = 23.7$ after 10 h oxidation; and the interface oxidation length increases from $\zeta/l_d = 0$ to 0.76 after 10 h oxidation.

1.5 Conclusion

In this chapter, the time-dependent first matrix cracking stress of fiber-reinforced CMCs is investigated using the energy balance approach considering interface and fiber oxidation at elevated temperature. The shear-lag model combined with the interface oxidation model and interface debonding criteria is adopted to analyze the stress distributions in the composite. The relationships between the first matrix cracking stress, the interface debonding and slip, oxidation temperature, oxidation time, and fiber failure are established. The first matrix cracking stresses of C/SiC with strong and weak interface bonding after unstressed oxidation at 700 °C in air atmosphere are predicted for different oxidation time. It was found that the strong interface bonding can be used for oxidation resistant of C/SiC at elevated temperature. The effects of fiber volume fraction, interface properties, and oxidation temperature on the evolution of matrix cracking stress versus oxidation time are analyzed.

(1) With increasing fiber volume, interface debonding energy, and interface shear stress, the time-dependent first matrix cracking stress and the interface oxidation length increase, and the time-dependent interface debonding length decreases.
(2) With increasing oxidation temperature, the time-dependent first matrix cracking stress decreases, and the time-dependent interface oxidation length and the interface debonding length increase.

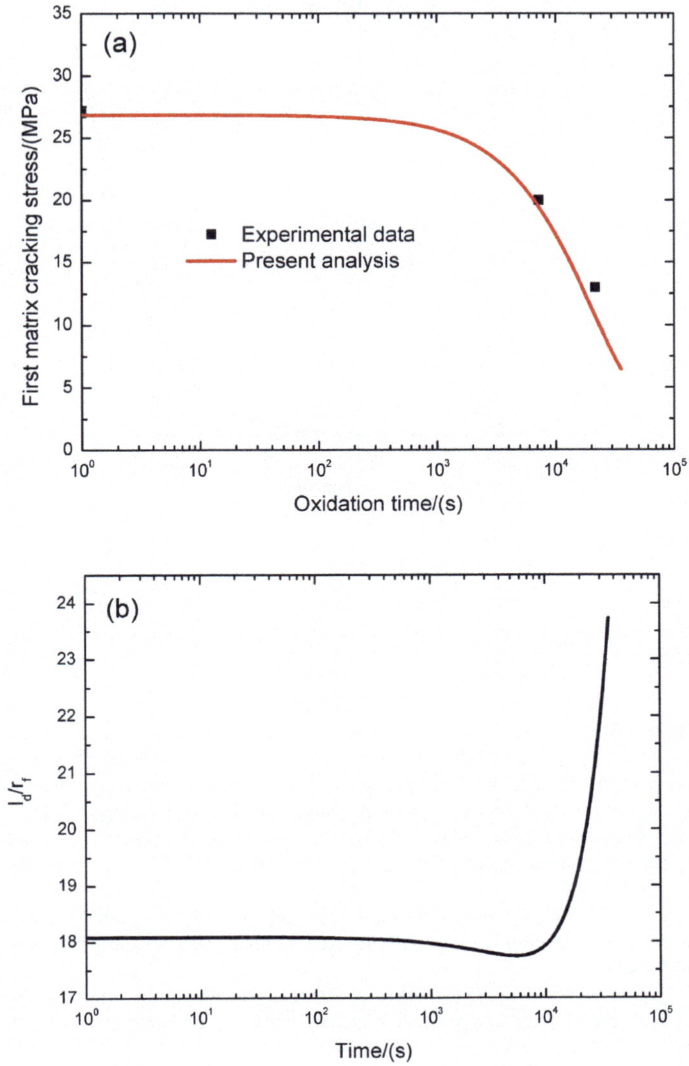

Fig. 1.26 **a** Experimental and predicted time-dependent first matrix cracking stress versus the oxidation time; **b** the time-dependent interface debonding length versus oxidation time; and **c** the time-dependent interface oxidation length versus oxidation time of C/SiC composite after unstressed oxidation at $T = 700$ °C in air atmosphere corresponding to weak interface bonding

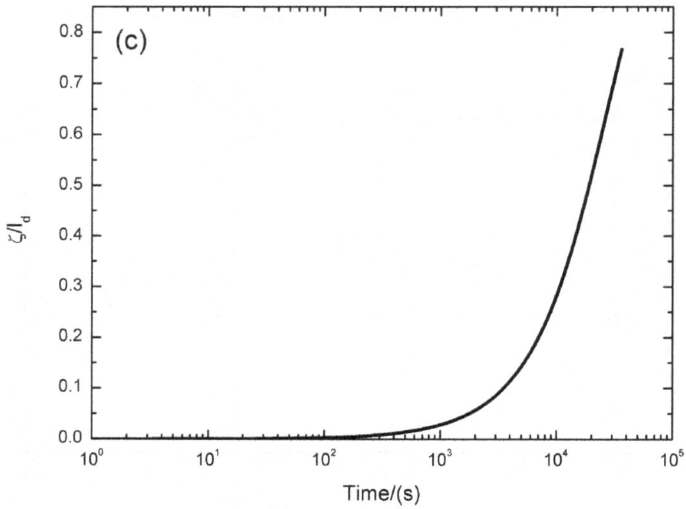

Fig. 1.26 (continued)

References

Ahn BK, Curtin WA (1997) Strain and hysteresis by stochastic matrix cracking in ceramic matrix composites. J Mech Phys Solids 45:177–209. https://doi.org/10.1016/S0022-5096(96)00081-6

Aveston J, Cooper GA, Kelly A (1971) Single and multiple fracture. Properties of fiber composites: conference on proceedings. National Physical Laboratory, IPC, England, pp 15–26

Barsoum MW, Kangutkar P, Wang ASD (1992) Matrix crack initiation in ceramic matrix composites Part I: experiments and test results. Compos Sci Technol 44:257–269. https://doi.org/10.1016/0266-3538(92)90016-V

Brighenti R, Scorza D (2012) A micro-mechanical model for statistically unidirectional and randomly distributed fibre-reinforced solids. Math Mech Solids 17:876–893. https://doi.org/10.1177/1081286512454447

Brighenti R, Carpinteri A, Scorza D (2014) Stress-intensity factors at the interface edge of a partially detached fibre. Theoret Appl Fract Mech 67–68:1–13. https://doi.org/10.1016/j.tafmec.2014.01.005

Budiansky B, Hutchinson JW, Evans AG (1986) Matrix fracture in fiber-reinforced ceramics. J Mech Phys Solids 34(2):167–189. https://doi.org/10.1016/0022-5096(86)90035-9

Casas L, Martinez-Esnaola JM (2003) Modelling the effect of oxidation on the creep behavior of fiber-reinforced ceramic matrix composites. Acta Mater 51:3745–3757. https://doi.org/10.1016/S1359-6454(03)00189-7

Chaudhuri RA (2006) Three-dimensional singular stress field near a partially debonded cylindrical rigid fibre. Compos Struct 72:141–150. https://doi.org/10.1016/j.compstruct.2004.11.017

Chiang YC (2000) On crack-wake debonding in fiber reinforced ceramics. Eng Frac Mech 65:15–28

Cox HL (1952) The elasticity and strength of paper and other fibrous materials. Br J Appl Phys 3(3):72–79

Curtin WA (1991a) Theory of mechanical properties of ceramic matrix composites. J Am Ceram Soc 74(11):2837–2845. https://doi.org/10.1111/j.1151-2916.1991.tb06852.x

Curtin WA (1991b) Theory of mechanical properties of ceramic-matrix composites. J Am Ceram Soc 74:2837–2845. https://doi.org/10.1111/j.1151-2916.1991.tb06852.x
Curtin WA (1993) Multiple matrix cracking in brittle matrix composites. Acta Metal Mater 41:1369–1377. https://doi.org/10.1016/0956-7151(93)90246-O
Gao YC, Mai YW, Cotterell B (1988) Fracture of fiber-reinforced materials. Z Angew Math Phys 39(4):550–572. https://doi.org/10.1007/BF00948962
Guillaumat L, Lamon J (1996) Fracture statistics applied to modelling the non-linear stress-strain behvior in microcomposites: Influence of interfacial parameters. Int J Fract 82:297–316. https://doi.org/10.1007/BF00013235
He M, Hutchinson J (1989) Kinking of a crack out of an interface. J Appl Mech 56:270–278. https://doi.org/10.1115/1.3176078
Lara-Curzio E (1999) Analysis of oxidation-assisted stress-rupture of continuous fiber-reinforced ceramic matrix composites at intermediate temperatures. Compos A 30:549–554. https://doi.org/10.1016/S1359-835X(98)00148-1
Li L (2017a) Modeling first matrix cracking stress of fiber-reinforced ceramic-matrix composites considering fiber fracture. Theoret Appl Fract Mech 92:24–32. https://doi.org/10.1016/j.tafmec.2017.05.004
Li L (2017b) Synergistic effects of temperature and oxidation on matrix cracking in fiber-reinforced ceramic-matrix composites. Appl Compos Mater 24:691–715. https://doi.org/10.1007/s10443-016-9535-y
Li L (2017c) Modeling matrix cracking of fiber-reinforced ceramic-matrix composites under oxidation environment at elevated temperature. Theoret Appl Fract Mech 87:110–119. https://doi.org/10.1016/j.tafmec.2016.11.003
Li L (2017d) Synergistic effects of fiber debonding and fracture on matrix cracking in fiber-reinforced ceramic-matrix composites. Mater Sci Eng A 682:482–490. https://doi.org/10.1016/j.msea.2016.11.077
Li L. (2018). Damage, fracture and fatigue of ceramic-matrix composites. Springer Nature Singapore Pte Ltd., ISBN: 978-981-13-1782-8. https://doi.org/10.1007/978-981-13-1783-5
Li L (2019) Thermomechanical fatigue of ceramic-matrix composites. Wiley-VCH. ISBN: 978-3-527-34637-0. https://onlinelibrary.wiley.com/doi/book/10.1002/9783527822614
Li L, Song Y, Sun Y (2014) Modeling the tensile behaviour of unidirectional C/SiC ceramic-matrix composites. Mech Compos Mater 49:659–672. https://doi.org/10.1007/s11029-013-9382-y
Lissart N, Lamon J (1997) Damage and failure in ceramic matrix minicomposites: experimental study and model. Acta Mater 45:1025–1044. https://doi.org/10.1016/S1359-6454(96)00224-8
Marshall DB, Cox BN (1987) Tensile fracture of brittle matrix composites: influence of fiber strength. Acta Metall 35:2607–2619. https://doi.org/10.1016/0001-6160(87)90260-4
Marshall DB, Cox BN, Evans AG (1985) The mechanics of matrix cracking in brittle-matrix fiber composites. Acta Metall 33(11):2013–2021. https://doi.org/10.1016/0001-6160(85)90124-5
McCartney LN (1987) Mechanics of matrix cracking in brittle-matrix fiber-reinforced composites. Proc R Soc A 409:329–350. https://doi.org/10.1098/rspa.1987.0019
Naslain R (2004) Design, preparation and properties of non-oxide CMCs for application in engines and nuclear reactors: an overview. Compos Sci Technol 64(2):155–170. https://doi.org/10.1016/S0266-3538(03)00230-6
Phoenix SL, Raj R (1992) Scalings in fracture probabilities for a brittle matrix fiber composite. Acta Metal Mater 40:2813–2828. https://doi.org/10.1016/0956-7151(92)90447-M
Rajan VP, Zok FW (2014) Matrix cracking of fiber-reinforced ceramic composites in shear. J Mech Phys Solids 73:3–21. https://doi.org/10.1016/j.jmps.2014.08.007
Romanowicz M (2010) Progressive failure analysis of unidirectional fiber-reinforced polymers with inhomogeneous interphase and randomly distributed fibers under transverse tensile loading. Compos A 41:1829–1838. https://doi.org/10.1016/j.compositesa.2010.09.001
Tvergaard V, Hutchinson JW (2008) Mode III effects on interface delamination. J Mech Phys Solids 56:215–229. https://doi.org/10.1016/j.jmps.2007.04.013

Venkat MR, Mahajan P, Mittal RK (2008) Effect of interfacial debonding and matrix cracking on mechanical properties of multidirectional composites. Compos Interfaces 15(4):379–409. https://doi.org/10.1163/156855408784514739

Yang C (2011) Mechanical characterization and oxidation damage modeling of ceramic matrix composites. PhD thesis. Northwestern Polytechnical University, Xi'an

Chapter 2
Time-, Stress-, and Cycle-Dependent Matrix Multicracking of Fiber-Reinforced Ceramic-Matrix Composites at Elevated Temperatures

Abstract In this chapter, the time-, stress-, and cycle-dependent matrix multicracking of fiber-reinforced ceramic-matrix composites (CMCs) with the interface debonding, interface wear, interface oxidation, and fiber fracture is investigated. The shear-lag model combined with the interface debonding, interface wear, interface oxidation, fiber fracture models, and the fiber/matrix interface debonding criterion is adopted to determine the microstress field of the damaged fiber-reinforced CMCs. The effects of the fiber volume and interface shear stress in the debonding and oxidation region, the interface debonding energy, the oxidation temperature, and time on the matrix multicracking, interface debonding and oxidation, and fiber fracture are discussed. The experimental matrix multicracking evolution of unidirectional C/SiC, SiC/SiC, mini-SiC/SiC, SiC/CAS, SiC/CAS-II, and SiC/borosilicate composites is predicted.

Keywords Ceramic-matrix composites (CMCs) · Matrix multicracking · Oxidation · Interface debonding · Interface oxidation · Fiber failure

2.1 Introduction

Ceramic materials possess high specific strength and specific modulus at elevated temperatures. But their use as structural components is severely limited because of their brittleness. Continuous fiber-reinforced ceramic-matrix composites, by incorporating fibers in ceramic matrices, however, not only exploit their attractive high-temperature strength but also reduce the propensity for catastrophic failure. These materials have already been implemented on some aeroengines' components (Naslain 2004; Schmidt et al. 2004; Li 2018a). The environment inside the hot section components is harsh, and the composite is typically subjected to the complex thermo-mechanical loading, which can lead to matrix multicracking (Sevener et al. 2017). These matrix cracks form paths for the ingress of the environment oxidizing the fibers and leading to the premature failure (Verrilli et al. 2004; Halbig et al. 2008). It is important to develop an understanding of matrix multicracking damage mechanisms to analyze the oxidation behavior inside of CMCs (Parthasarathy et al. 2018).

L. Li, *Time-Dependent Mechanical Behavior of Ceramic-Matrix Composites at Elevated Temperatures*, Advanced Ceramics and Composites 1,
https://doi.org/10.1007/978-981-15-3274-0_2

Many researchers performed experimental and theoretical investigations on matrix multicracking evolution of fiber-reinforced CMCs. Pryce and Smith (1992) investigated the quasi-static tensile behavior of unidirectional and cross-ply SiC/calcium aluminosilicate (CAS) glass-ceramic composites. The first matrix cracking stress is predicted using the Aveston–Cooper–Kelly (ACK) theory (Aveston et al. 1971), and the relationship between the matrix cracking density and stiffness reduction is analyzed with increasing strain. Morscher et al. (2007) investigated the matrix cracking of 2D woven Hi-Nicalon and Sylramic-iBN SiC fiber-reinforced chemical vapor-infiltrated (CVI) SiC matrix composites using modal acoustic emission (AE). The stress-dependent matrix cracking could be related to the stress in the load-bearing CVI SiC matrix for low- and high-density composite. Gowayed et al. (2015) investigated the feasibility of utilizing the shear-lag theory to estimate the matrix crack density in fabric reinforced 2D SiC/SiC composite. The matrix cracking density was highly sensitive to the fiber volume fraction in the load direction and the fiber/matrix interface shear strength between the fibers and the matrix. Ogasawara et al. (2001) investigated experimental multiple matrix cracking of an orthogonal 3D woven Si–Ti–C–O fiber/Si–Ti–C–O matrix composite using microscopic observation. The inelastic tensile stress-strain behavior is governed by matrix cracking in transverse fiber bundles at low stress, matrix cracking in longitudinal fiber bundles at intermediate stress and fiber fragmentation at high stress. Morscher et al. (2005) investigated the occurrence of matrix cracks in melt-infiltrated 3D orthogonal architecture SiC/SiC composite under tension parallel to the Y-direction which is perpendicular to Z-bundle weave direction using the acoustic emission (AE). The matrix cracking stress range depended upon the Z-direction bundle size and the local architecture. Solti et al. (1995) developed an approach of critical matrix strain energy (CMSE) criterion to analyze matrix multicracking evolution, in which the maximum interface shear strength criterion was adopted to determine the fiber/matrix interface debonded length during matrix multicracking. However, following the arguments of Gao et al. (1988) and Stang and Shah (1986), the fracture mechanics approach is preferred to the shear strength approach for the fiber/matrix interface debonding problem. Meng and Wang (2015a, b) proposed a theoretical model to predict the interfacial debonding length and fiber pull-out length in fiber-reinforced polymer-matrix composites and hybrid-fiber-reinforced brittle matrix composites. The interface debonding criterion is given based on the energy release rate relation in an interface debonding process. Meng and Wang (2015c) predicted the interfacial strength and failure mechanisms of particle-reinforced metal-matrix composites using a micromechanical model. The plastic strain of composite increases with the increasing of interfacial strength when the interface debonding begins. Meng et al. (2017, 2018) established a multiscale crack-bridging model to reveal the toughness mechanisms in cellulose nanopaper. A cohesive law is developed to characterize the interfacial properties between cellulose nanofibrils. A unified law is proposed to correlate the fracture toughness of cellulose nanopaper with its microstructure and material parameters. Chiang (2000, 2001) investigated the effect of fiber/matrix interface debonding on steady-state matrix cracking in fiber-reinforced CMCs under tensile loading. The fracture mechanics approach developed by Gao et al. (1988) is used to determine the fiber/matrix interface debonded length. Rajan and Zok (2014) investigate the mechanics of a fully bridged steady-state matrix cracking in unidirectional CMCs under shear loading. Li (2017, 2018b, 2019a, b, c) investigated the synergistic effects of fiber debonding and fracture on the first matrix cracking in fiber-reinforced CMCs under tensile loading. The first matrix cracking stress of three different CMCs has been predicted. However, the models mentioned above do not consider the effect of fiber debonding on matrix multicracking development in fiber-reinforced CMCs.

In this chapter, the time-, stress-, and cycle-dependent matrix multicracking of fiber-reinforced CMCs with the interface debonding, interface wear, interface oxidation, and fiber fracture is investigated. The shear-lag model combined with the interface debonding, interface wear, interface oxidation and fiber fracture models, and the fiber/matrix interface debonding criterion is adopted to determine the microstress field of the damaged fiber-reinforced CMCs. The effects of the fiber volume and interface shear stress in the debonding and oxidation region, the interface debonding energy, the oxidation temperature, and time on the matrix multicracking, interface debonding and oxidation, and fiber fracture are discussed. The experimental matrix multicracking evolution of unidirectional C/SiC, SiC/SiC, mini-SiC/SiC, SiC/CAS, SiC/CAS-II, and SiC/borosilicate composites is predicted.

2.2 Stress-Dependent Matrix Multicracking Evolution of Fiber-Reinforced Ceramic-Matrix Composites Considering Fiber Debonding

In this section, the effect of fiber debonding on matrix multicracking evolution of different fiber-reinforced CMCs is investigated using the micromechanical approach. The shear-lag model is adopted to analyze the fiber and matrix stress distributions of the damaged composite. The fracture mechanics approach is used to determine the fiber/matrix interface debonding length. Combining the critical matrix strain energy criterion and fracture mechanics fiber/matrix interface debonding criterion, the stress-dependent matrix multicracking development is analyzed for different fiber volume fraction, fiber/matrix interface properties, and matrix cracking characteristic stress. The experimental matrix multicracking development of unidirectional C/Si_3N_4, SiC/Si_3N_4, SiC/CAS, SiC/CAS-II, SiC/SiC, SiC/Borosilicate, and mini-SiC/SiC composites is predicted.

2.2.1 Stress Analysis

To analyze the stress distributions in the fiber and the matrix of the damaged composite, a unit cell is extracted from the fiber-reinforced CMCs. The unit cell contains a single fiber surrounded by a hollow cylinder of matrix. The fiber radius is r_f, and the matrix radius is R ($R = r_f/V_f^{1/2}$). The length of the unit cell is $l_c/2$, which is half of matrix crack spacing. The fiber/matrix interface debonding length is l_d. At the matrix cracking plane, fibers carry all the applied stress (σ/V_f, where σ denotes the far-field applied stress and V_f denotes the fiber volume fraction). The shear-lag model developed by Budiansky et al. (1986) is adopted to perform the stress and strain calculations in the fiber/matrix interface debonding region ($z \in [0, l_d]$) and fiber/matrix interface debonding region ($z \in [l_d, l_c/2]$). The fiber axial stress $\sigma_f(z)$, matrix axial stress $\sigma_m(z)$, and the fiber/matrix interface shear stress $\tau_i(z)$ are determined using the following equations.

$$\sigma_f(z) = \begin{cases} \dfrac{\sigma}{V_f} - \dfrac{2\tau_i}{r_f} z, \ z \in [0, l_d] \\ \sigma_{fo} + \left(\dfrac{V_m}{V_f}\sigma_{mo} - 2\dfrac{l_d}{r_f}\tau_i\right)\exp\left(-\rho\dfrac{z - l_d}{r_f}\right), \ z \in \left[l_d, \dfrac{l_c}{2}\right] \end{cases} \tag{2.1}$$

$$\sigma_m(z) = \begin{cases} 2\tau_i \dfrac{V_f}{V_m}\dfrac{z}{r_f}, \ z \in [0, l_d] \\ \sigma_{mo} - \left(\sigma_{mo} - 2\tau_i \dfrac{V_f}{V_m}\dfrac{l_d}{r_f}\right)\exp\left[-\dfrac{\rho(z - l_d)}{r_f}\right], \ z \in \left[l_d, \dfrac{l_c}{2}\right] \end{cases} \tag{2.2}$$

$$\tau_i(z) = \begin{cases} \tau_i, \ z \in [0, l_d] \\ \dfrac{\rho}{2}\left(\dfrac{V_m}{V_f}\sigma_{mo} - 2\tau_i\dfrac{l_d}{r_f}\right)\exp\left[-\dfrac{\rho(z - l_d)}{r_f}\right], \ z \in \left[l_d, \dfrac{l_c}{2}\right] \end{cases} \tag{2.3}$$

where V_f and V_m denote the fiber and matrix volume fraction, respectively; τ_i denotes the fiber/matrix constant interface shear stress; ρ denotes the shear-lag model parameter; and σ_{fo} and σ_{mo} denote the fiber and matrix axial stress in the fiber/matrix interface bonded region, respectively.

$$\sigma_{fo} = \frac{E_f}{E_c}\sigma + E_f(\alpha_c - \alpha_f)\Delta T \tag{2.4}$$

$$\sigma_{mo} = \frac{E_m}{E_c}\sigma + E_m(\alpha_c - \alpha_m)\Delta T \tag{2.5}$$

where E_f, E_m, and E_c denote the fiber, matrix, and composite elastic modulus, respectively; α_f, α_m, and α_c denote the fiber, matrix, and composite thermal expansion coefficient, respectively; ΔT denotes the temperature difference between the fabricated temperature T_0 and testing temperature T_1 ($\Delta T = T_1 - T_0$).

2.2.2 Interface Debonding

The fiber/matrix interface debonding criterion is determined using the following equation (Gao et al. 1988).

$$\xi_{\mathrm{d}} = -\frac{F}{4\pi r_{\mathrm{f}}}\frac{\partial w_{\mathrm{f}}(0)}{\partial l_{\mathrm{d}}} - \frac{1}{2}\int_{0}^{l_{\mathrm{d}}} \tau_{\mathrm{i}}\frac{\partial v(z)}{\partial l_{\mathrm{d}}}\mathrm{d}z \tag{2.6}$$

where $F(= \pi r_{\mathrm{f}}^2\sigma/V_{\mathrm{f}})$ denotes the fiber load at the matrix cracking plane; $w_{\mathrm{f}}(0)$ denotes the fiber axial displacement at the matrix cracking plane; and $v(z)$ denotes the relative displacement between the fiber and the matrix.

The axial displacements of the fiber and matrix, i.e., $w_{\mathrm{f}}(z)$ and $w_{\mathrm{m}}(z)$, are determined using the following equations.

$$\begin{aligned} w_{\mathrm{f}}(z) = \int_{x}^{l_{\mathrm{c}}/2} \frac{\sigma_{\mathrm{f}}(z)}{E_{\mathrm{f}}}\mathrm{d}z = \frac{\sigma}{V_{\mathrm{f}}E_{\mathrm{f}}}(l_{\mathrm{d}} - z) - \frac{\tau_{\mathrm{i}}}{r_{\mathrm{f}}E_{\mathrm{f}}}\left(l_{\mathrm{d}}^2 - z^2\right) \\ - \frac{2\tau_{\mathrm{i}}}{\rho E_{\mathrm{f}}}l_{\mathrm{d}} + \frac{r_{\mathrm{f}}V_{\mathrm{m}}E_{\mathrm{m}}}{\rho V_{\mathrm{f}}E_{\mathrm{f}}E_{\mathrm{c}}}\sigma + \frac{\sigma}{E_{\mathrm{c}}}(l_{\mathrm{c}}/2 - l_{\mathrm{d}}) \end{aligned} \tag{2.7}$$

$$\begin{aligned} w_{\mathrm{m}}(z) = \int_{x}^{l_{\mathrm{c}}/2} \frac{\sigma_{\mathrm{m}}(z)}{E_{\mathrm{m}}}\mathrm{d}z = \frac{V_{\mathrm{f}}\tau_{i}}{V_{\mathrm{m}}E_{\mathrm{m}}r_{\mathrm{f}}}\left(l_{\mathrm{d}}^2 - z^2\right) + \frac{2V_{\mathrm{f}}\tau_{\mathrm{i}}}{\rho V_{\mathrm{m}}E_{\mathrm{m}}}l_{\mathrm{d}} \\ - \frac{r_{\mathrm{f}}}{\rho E_{\mathrm{c}}}\sigma + \frac{\sigma}{E_{\mathrm{c}}}(l_{\mathrm{c}}/2 - l_{\mathrm{d}}) \end{aligned} \tag{2.8}$$

The relative displacement between the fiber and the matrix, i.e., $v(z)$, is determined using the following equation.

$$\begin{aligned} v(z) = |w_{\mathrm{f}}(z) - w_{\mathrm{m}}(z)| = \frac{\sigma}{V_{\mathrm{f}}E_{\mathrm{f}}}(l_{\mathrm{d}} - z) - \frac{\tau_{\mathrm{i}}E_{\mathrm{c}}}{V_{\mathrm{m}}E_{\mathrm{m}}E_{\mathrm{f}}r_{\mathrm{f}}}\left(l_{\mathrm{d}}^2 - z^2\right) \\ - \frac{2\tau_{\mathrm{i}}E_{\mathrm{c}}l_{\mathrm{d}}}{\rho V_{\mathrm{m}}E_{\mathrm{m}}E_{\mathrm{f}}} + \frac{r_{\mathrm{f}}}{\rho V_{\mathrm{f}}E_{\mathrm{f}}}\sigma \end{aligned} \tag{2.9}$$

Substituting $w_{\mathrm{f}}(x = 0)$ and $v(z)$ into Eq. (2.6), it leads to the following equation.

$$\frac{E_{\mathrm{c}}\tau_{\mathrm{i}}^2}{V_{\mathrm{m}}E_{\mathrm{m}}E_{\mathrm{f}}r_{\mathrm{f}}}l_{\mathrm{d}}^2 + \left(\frac{E_{\mathrm{c}}\tau_{\mathrm{i}}^2}{\rho V_{\mathrm{m}}E_{\mathrm{m}}E_{\mathrm{f}}} - \frac{\tau_{\mathrm{i}}\sigma}{V_{\mathrm{f}}E_{\mathrm{f}}}\right)l_{\mathrm{d}} + \frac{r_{\mathrm{f}}V_{\mathrm{m}}E_{\mathrm{m}}\sigma^2}{4V_{\mathrm{f}}^2E_{\mathrm{f}}E_{\mathrm{c}}} - \frac{r_{\mathrm{f}}\tau_{\mathrm{i}}}{2\rho V_{\mathrm{f}}E_{\mathrm{f}}}\sigma - \xi_{\mathrm{d}} = 0 \tag{2.10}$$

Solving Eq. (2.10), the fiber/matrix interface debonded length is determined using the following equation.

$$l_{\mathrm{d}} = \frac{r_{\mathrm{f}}}{2}\left(\frac{V_{\mathrm{m}}E_{\mathrm{m}}\sigma}{V_{\mathrm{f}}E_{\mathrm{c}}\tau_{\mathrm{i}}} - \frac{1}{\rho}\right) - \sqrt{\left(\frac{r_{\mathrm{f}}}{2\rho}\right)^2 + \frac{r_{\mathrm{f}}V_{\mathrm{m}}E_{\mathrm{m}}E_{\mathrm{f}}}{E_{\mathrm{c}}\tau_{\mathrm{i}}^2}\xi_{\mathrm{d}}} \tag{2.11}$$

2.2.3 *Matrix Multicracking*

Solti et al. (1995) developed the critical matrix strain energy (CMSE) criterion to predict the matrix multicracking evolution in fiber-reinforced CMCs. The concept of critical matrix strain energy presupposes the existence of an ultimate or critical strain energy. Beyond the critical value of matrix strain energy, as more energy is entered into the composite with increasing applied stress, the matrix cannot support the extra load and continues to fail. The failure is assumed to consist of the formation of new matrix crackings and fiber/matrix interface debonding, to make the total energy within the matrix remain constant and equal to its critical value.

The matrix strain energy is determined using the following equation.

$$U_{\mathrm{m}} = \frac{1}{2E_{\mathrm{m}}}\int\limits_{A_{\mathrm{m}}}\int\limits_{0}^{l_{\mathrm{c}}} \sigma_{\mathrm{m}}^2(z)\mathrm{d}z\mathrm{d}A_{\mathrm{m}} \tag{2.12}$$

where A_{m} is the cross-sectional area of the matrix in the unit cell. Substituting the matrix axial stresses in Eq. (2.2) into Eq. (2.12), the matrix strain energy considering the matrix multicracking and fiber/matrix interface partially debonding is described using the following equation.

$$\begin{aligned} U_{\mathrm{m}} = \frac{A_{\mathrm{m}}}{E_{\mathrm{m}}}\Bigg\{ & \frac{4}{3}\left(\frac{V_{\mathrm{f}}\tau_{\mathrm{i}}}{V_{\mathrm{m}}r_{\mathrm{f}}}l_{\mathrm{d}}\right)^2 l_{\mathrm{d}} + \sigma_{\mathrm{mo}}^2(l_{\mathrm{c}}/2 - l_{\mathrm{d}}) \\ & + 2\sigma_{\mathrm{mo}}\left(\frac{2V_{\mathrm{f}}\tau_{\mathrm{i}}l_{\mathrm{d}}}{V_{\mathrm{m}}r_{\mathrm{f}}} - \sigma_{\mathrm{mo}}\right)\left(-\frac{r_{\mathrm{f}}}{\rho}\right)\left[\exp\left(-\rho\frac{l_{\mathrm{c}}/2 - l_{\mathrm{d}}}{r_{\mathrm{f}}}\right) - 1\right] \\ & + \left(\frac{2V_{\mathrm{f}}\tau_{\mathrm{i}}l_{\mathrm{d}}}{V_{\mathrm{m}}r_{\mathrm{f}}} - \sigma_{\mathrm{mo}}\right)^2\left(-\frac{r_{\mathrm{f}}}{2\rho}\right)\left[\exp\left(-2\rho\frac{l_{\mathrm{c}}/2 - l_{\mathrm{d}}}{r_{\mathrm{f}}}\right) - 1\right]\Bigg\} \end{aligned} \tag{2.13}$$

When the fiber/matrix interface completely debonds, the matrix strain energy is described using the following equation.

$$U_{\mathrm{m}}(\sigma, l_{\mathrm{c}}, l_{\mathrm{d}} = l_{\mathrm{c}}/2) = \frac{A_{\mathrm{m}}l_{\mathrm{c}}^3}{6E_{\mathrm{m}}}\left(\frac{\tau_{\mathrm{i}}V_{\mathrm{f}}}{r_{\mathrm{f}}V_{\mathrm{m}}}\right)^2 \tag{2.14}$$

By evaluating the matrix strain energy at a critical stress of σ_{cr}, the critical matrix strain energy of U_{crm} can be obtained. The critical matrix strain energy is described using the following equation.

$$U_{\mathrm{crm}} = \frac{1}{2}kA_{\mathrm{m}}l_0\frac{\sigma_{\mathrm{mocr}}^2}{E_{\mathrm{m}}} \tag{2.15}$$

where k ($k \in [0, 1]$) is the critical matrix strain energy parameter; l_0 is the initial matrix crack spacing; and σ_{mocr} is determined using the following equation.

$$\sigma_{\mathrm{mocr}} = \frac{E_{\mathrm{m}}}{E_{\mathrm{c}}}\sigma_{\mathrm{cr}} + E_{\mathrm{m}}(\alpha_{\mathrm{c}} - \alpha_{\mathrm{m}})\Delta T \tag{2.16}$$

where σ_{cr} is the critical stress corresponding to composite's proportional limit stress, i.e., the stress at which the stress-strain curve starts to deviate from linearity due to the damage accumulation of matrix crackings (Li 2017). The critical stress is defined to be Aveston–Cooper–Kelly matrix cracking stress (Aveston et al. 1971), which was determined using the energy balance criterion, involving the calculation of energy balance relationship before and after the formation of a single dominant cracking. The Aveston–Cooper–Kelly model can be used to describe the long-steady-state matrix cracking stress, corresponding to the proportional limit stress of tensile stress-strain curve. The Aveston–Cooper–Kelly matrix cracking stress is determined using the following equation (Aveston et al. 1971).

$$\sigma_{\mathrm{cr}} = \left(\frac{6V_{\mathrm{f}}^2E_{\mathrm{f}}E_{\mathrm{c}}^2\tau_{\mathrm{i}}\xi_{\mathrm{m}}}{r_{\mathrm{f}}V_{\mathrm{m}}E_{\mathrm{m}}^2}\right)^{\frac{1}{3}} - E_{\mathrm{c}}(\alpha_{\mathrm{c}} - \alpha_{\mathrm{m}})\Delta T \tag{2.17}$$

where ξ_{m} denotes the matrix fracture energy. However, as microcracks exist in the matrix when the fiber-reinforced CMCs are cooled down from high fabrication temperature to room temperature, due to the thermal expansion coefficient misfit between the fiber and the matrix, these microcracks are short-matrix-cracking, and the cracking stresses of these microcracks lie in the linear region of the tensile stress-strain curve (Li et al. 2014, 2015). With increasing applied stress, matrix micro-cracks can propagate into the long-matrix-cracking. The matrix cracking stress of Aveston–Cooper–Kelly model is used to determine the critical matrix strain energy.

The matrix multicracking evolution is determined using the following equation.

$$U_{\mathrm{m}}(\sigma > \sigma_{\mathrm{cr}}, l_{\mathrm{c}}, l_{\mathrm{d}}) = U_{\mathrm{crm}}(\sigma_{\mathrm{cr}}, l_0) \tag{2.18}$$

When the fiber/matrix interface partially debonds, the matrix strain energy is determined by Eq. (2.13); and when the fiber/matrix interface completely debonds, the matrix strain energy is determined by Eq. (2.14). The critical matrix strain energy is given by Eq. (2.15). Substituting Eqs. (2.13), (2.14), and (2.15) into Eq. (2.18) when the critical matrix cracking stress of σ_{cr} and the fiber/matrix interface debonded length of l_{d} are determined by Eqs. (2.11) and (2.17), the matrix multicracking evolution versus applied stress can be obtained.

The matrix cracking density is determined using the following equation.

$$\varphi = \frac{1000}{l_c} \tag{2.19}$$

2.2.4 *Results and Discussion*

The effects of the fiber volume, interface shear stress, and interface debonding energy on matrix multicracking and interface debonding are discussed. The ceramic composite system of SiC/CAS is used for the case study, and its material properties are given by Beyerle et al. (1992): $V_f = 30\%$, $E_f = 200$ GPa, $E_m = 97$ GPa, $r_f = 7.5$ μm, $\xi_m = 6$ J/m^2, $\xi_d = 0.4$ J/m^2, $\tau_i = 20$ MPa, $\alpha_f = 4 \times 10^{-5}$ °C, $\alpha_m = 5 \times 10^{-5}$ °C, and $\Delta T = -1000$ °C.

2.2.4.1 Effect of Fiber Volume Fraction on Matrix Multicracking and Interface Debonding

The matrix multicracking density and fiber/matrix interface debonding length for different fiber volume (i.e., $V_f = 30$ and 35%) are shown in Fig. 2.1. With increasing fiber volume, the first matrix cracking stress, matrix cracking saturation stress, and saturation matrix cracking density increase; the critical matrix strain energy decreases; and the matrix cracking evolves with higher applied stress.

When the fiber volume increases from $V_f = 30$ to 35%, the matrix first cracking stress increases from $\sigma_{mc} = 102$ to 119 MPa; the matrix cracking saturation stress increases from $\sigma_{sat} = 180$ to 208 MPa; the critical matrix strain energy decreases from $U_{crm} = 0.113$ to 0.093 μJ; the saturation matrix cracking density increases from $\varphi = 6.9$ to 8.0 mm; and the fiber/matrix interface debonding length increases from $2l_d/l_c = 0.7$ to 64.6% when $V_f = 30\%$, and $2l_d/l_c = 0.6$ to 64.4% when $V_f =$ 35%.

2.2.4.2 Effect of Interface Shear Stress on Matrix Multicracking and Interface Debonding

The matrix multicracking density and fiber/matrix interface debonding length for different fiber/matrix interface shear stress (i.e., $\tau_i = 10$ and 15 MPa) are shown in Fig. 2.2. With increasing interface shear stress, the first matrix cracking stress, matrix cracking saturation stress, and saturation matrix cracking density increase; the critical matrix strain energy increases; and the matrix cracking evolves with higher applied stress.

When the fiber/matrix interface shear stress increases from $\tau_i = 10$ to 15 MPa, the matrix first cracking stress increases from $\sigma_{mc} = 69$ to 87 MPa; the matrix cracking saturation stress increases from $\sigma_{sat} = 106$ to 149 MPa; the critical matrix strain

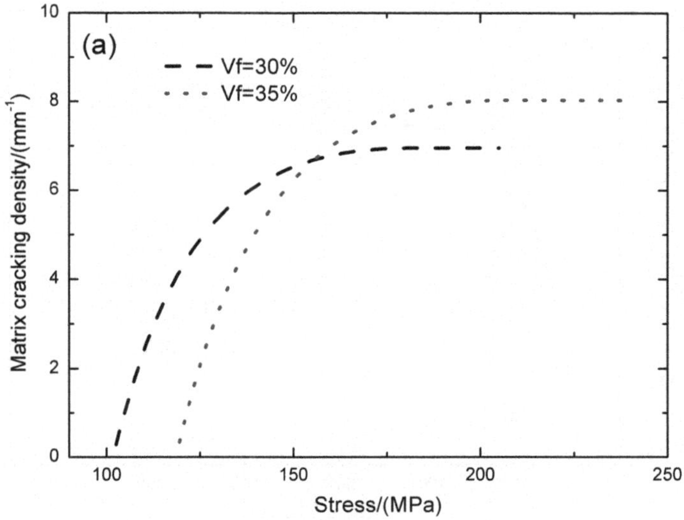

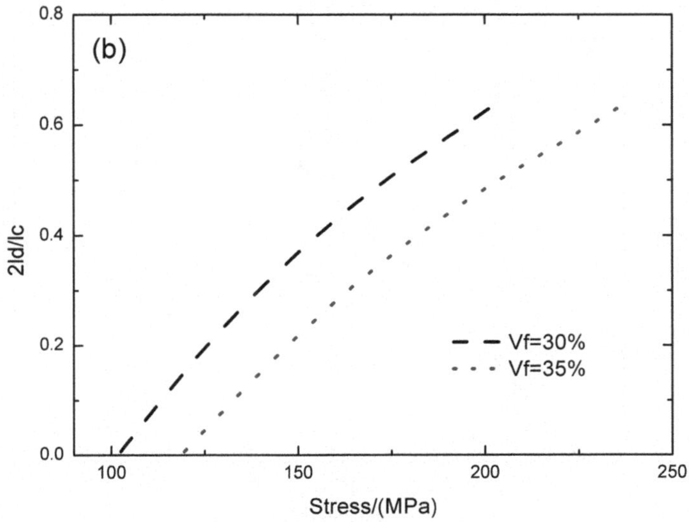

Fig. 2.1 Effect of the fiber volume on **a** the matrix multicracking density versus the applied stress curves and **b** the fiber/matrix interface debonding length versus the applied stress curves

energy increases from $U_{\mathrm{crm}} = 0.013$ to 0.086 μJ; the saturation matrix cracking density increases from $\varphi = 5.9$ to 6.3 mm; and the fiber/matrix interface debonded length increases from $2l_d/l_c = 0.4$ to 60% when $\tau_i = 10$ MPa, and $2l_d/l_c = 0.6$ to 62.2% when $\tau_i = 15$ MPa.

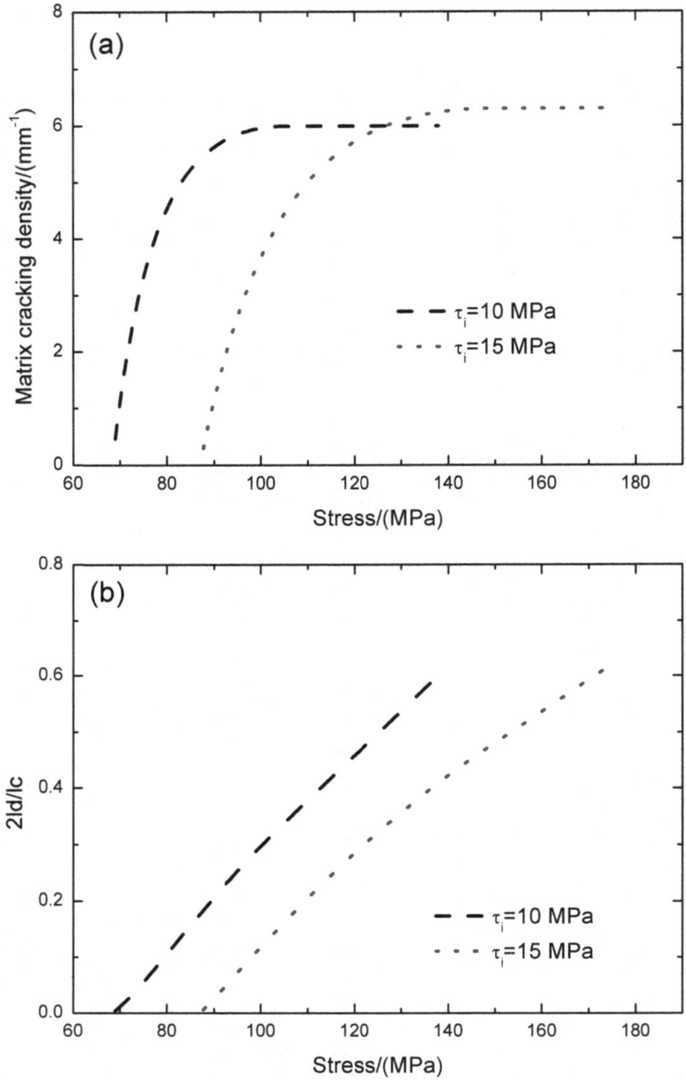

Fig. 2.2 Effect of the interface shear stress on **a** the matrix multicracking density versus applied stress curves; and **b** the interface debonding length versus applied stress curves

2.2.4.3 Effect of Interface Debonding Energy on Matrix Multicracking and Interface Debonding

The matrix multicracking density and fiber/matrix interface debonding length for different fiber/matrix interface debonding energy (i.e., $\xi_d = 0.1$ and $0.6\ \text{J/m}^2$) are shown in Fig. 2.3. With increasing interface debonding energy, the matrix cracking saturation stress decreases; the saturation matrix cracking density increases; the

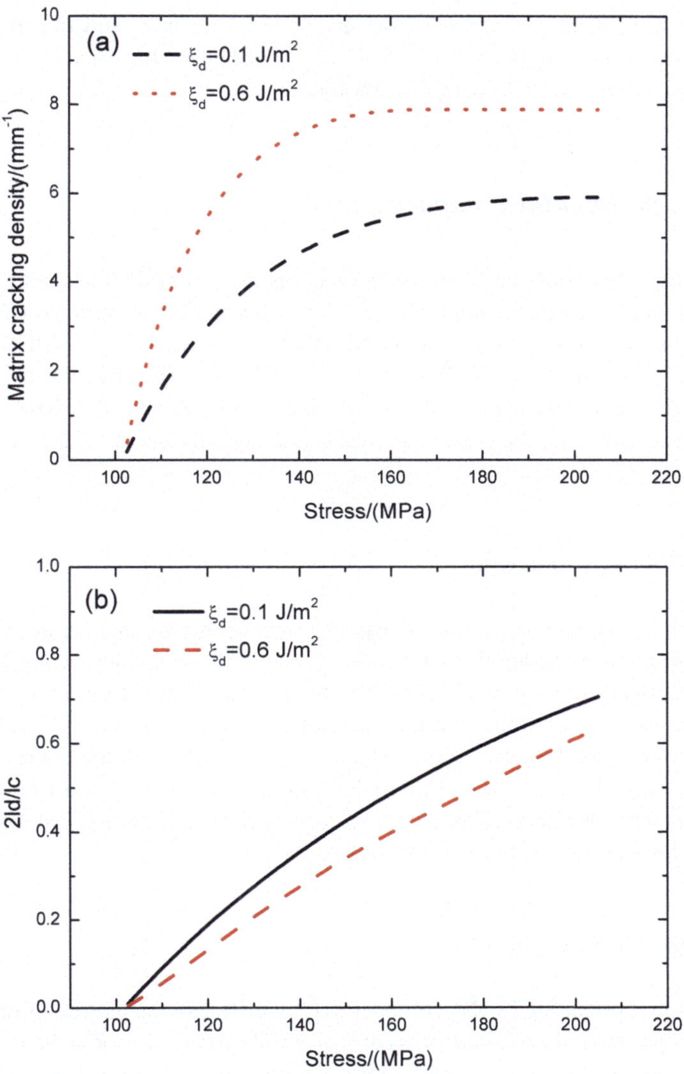

Fig. 2.3 Effect of fiber/matrix interface debonding energy on **a** the matrix multicracking density versus applied stress curves; and **b** the fiber/matrix interface debonding length versus applied stress curves

critical matrix strain energy decreases; and the rate of matrix cracking development increases due to the decrease of fiber/matrix interface debonding ratio.

When the fiber/matrix interface debonding energy increases from $\xi_d = 0.1$ to $0.6\ \text{J/m}^2$, the matrix first cracking stress remains the same value of $\sigma_{mc} = 102$ MPa; the matrix cracking saturation stress decreases from $\sigma_{sat} = 205$ to 168 MPa; the critical matrix strain energy decreases from $U_{crm} = 0.117$ to 0.079 μJ; the saturation matrix cracking density increases from $\varphi = 5.9$ to 7.9 mm; and the fiber/matrix interface debonding length increases from $2l_d/l_c = 0.9$ to 70% when $\xi_d = 0.1$ J/m^2, and $2l_d/l_c = 0.5$ to 63.7% when $\xi_d = 0.6$ J/m^2.

2.2.5 Experimental Comparisons

The experimental and theoretical matrix cracking density and fiber/matrix interface debonding length versus the applied stress for different fiber-reinforced CMCs, i.e., unidirectional C/Si_3N_4 (Olivier 1998), SiC/Si_3N_4 (Olivier 1998), SiC/CAS (Pryce and Smith 1992), SiC/CAS-II (Beyerle et al. 1992), SiC/SiC (Beyerle et al. 1992), SiC/Borosilicate (Okabe et al. 1999), and mini-SiC/SiC (Zhang et al. 2016) composites, are predicted using the present analysis, as shown in Figs. 2.4, 2.5, 2.6, 2.7, 2.8, 2.9 and 2.10.

2.2.5.1 C/Si_3N_4 Composite

For the unidirectional C/Si_3N_4 composite (Olivier 1998), the matrix multicracking evolution starts from the applied stress of $\sigma_{mc} = 100$ MPa and approaches the saturation at the applied stress of $\sigma_{sat} = 200$ MPa; the critical matrix strain energy (CMSE) criterion is used to determine the matrix multicracking evolution in C/Si_3N_4 composite, and the critical matrix strain energy is $U_{crm} = 0.047$ μJ; the matrix cracking density increases from $\varphi = 0.85$ mm to the saturation value of $\varphi = 6.1$ mm; and the fiber/matrix interface debonding length increases from $2l_d/l_c = 0.6\%$ at 100 MPa to $2l_d/l_c = 90.8\%$ at 320 MPa, as shown in Fig. 2.4.

2.2.5.2 SiC/Si_3N_4 Composite

For the unidirectional SiC/Si_3N_4 composite (Olivier 1998), the matrix multicracking evolution starts from the applied stress of $\sigma_{mc} = 350$ MPa and approaches the saturation at the applied stress of $\sigma_{sat} = 650$ MPa; the critical matrix strain energy (CMSE) criterion is used to determine the matrix multicracking evolution in SiC/Si_3N_4 composite, and the critical matrix strain energy is $U_{crm} = 0.667$ μJ; the matrix cracking density increases from $\varphi = 0.25$ mm to the saturation value of $\varphi = 6.8$ mm; and the fiber/matrix interface debonding length increases from $2l_d/l_c = 0.8\%$ at 350 MPa to $2l_d/l_c = 73.1\%$ at 650 MPa, as shown in Fig. 2.5.

2.2.5.3 SiC/CAS Composite

For the unidirectional SiC/CAS composite (Pryce and Smith 1992), the matrix multicracking evolution starts from the applied stress of $\sigma_{mc} = 140$ MPa and approaches

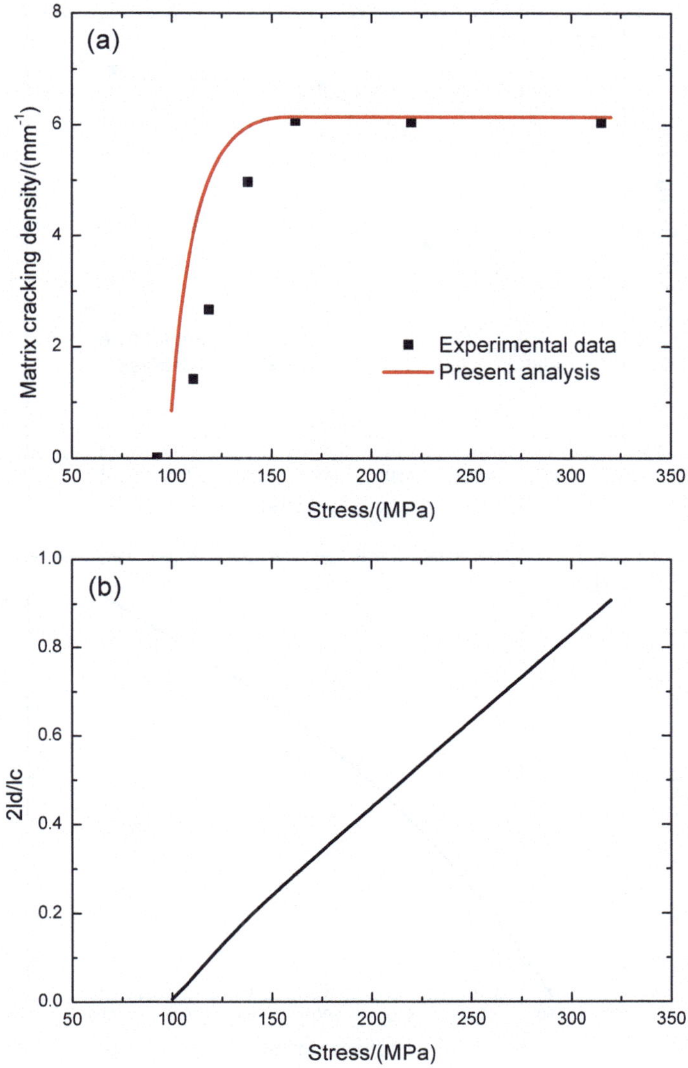

Fig. 2.4 **a** Experimental and predicted matrix multicracking density versus applied stress curves; and **b** the fiber/matrix interface debonding length versus applied stress curve of C/Si_3N_4 composite

saturation at the applied stress of $\sigma_{sat} = 288$ MPa; the critical matrix strain energy (CMSE) criterion is used to determine the matrix multicracking evolution in SiC/CAS composite, and the critical matrix strain energy is $U_{crm} = 0.219$ μJ; the matrix cracking density increases from $\varphi = 0.7$ mm to the saturation value of $\varphi = 7$ mm; and the fiber/matrix interface debonding length increases from $2l_d/l_c = 0.7\%$ at 160 MPa to $2l_d/l_c = 79\%$ at 360 MPa, as shown in Fig. 2.6.

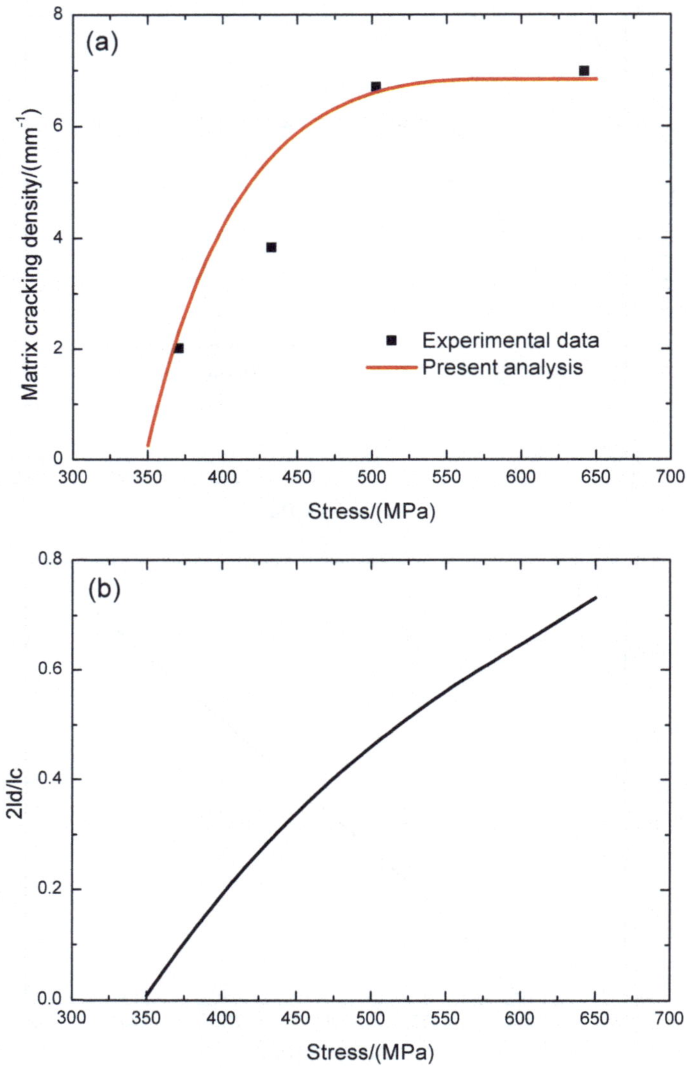

Fig. 2.5 **a** Experimental and predicted matrix multicracking density versus applied stress curves; and **b** the fiber/matrix interface debonded length versus applied stress curve of SiC/Si_3N_4 composite

2.2.5.4 SiC/CAS-II Composite

For the unidirectional SiC/CAS-II composite (Beyerle et al. 1992), the matrix multicracking evolution starts from the applied stress of σ_{mc} = 260 MPa and approaches saturation at the applied stress of σ_{sat} = 360 MPa; the critical matrix strain energy (CMSE) criterion is used to determine the matrix multicracking evolution in SiC/CAS-II composite, and the critical matrix strain energy is U_{crm} =

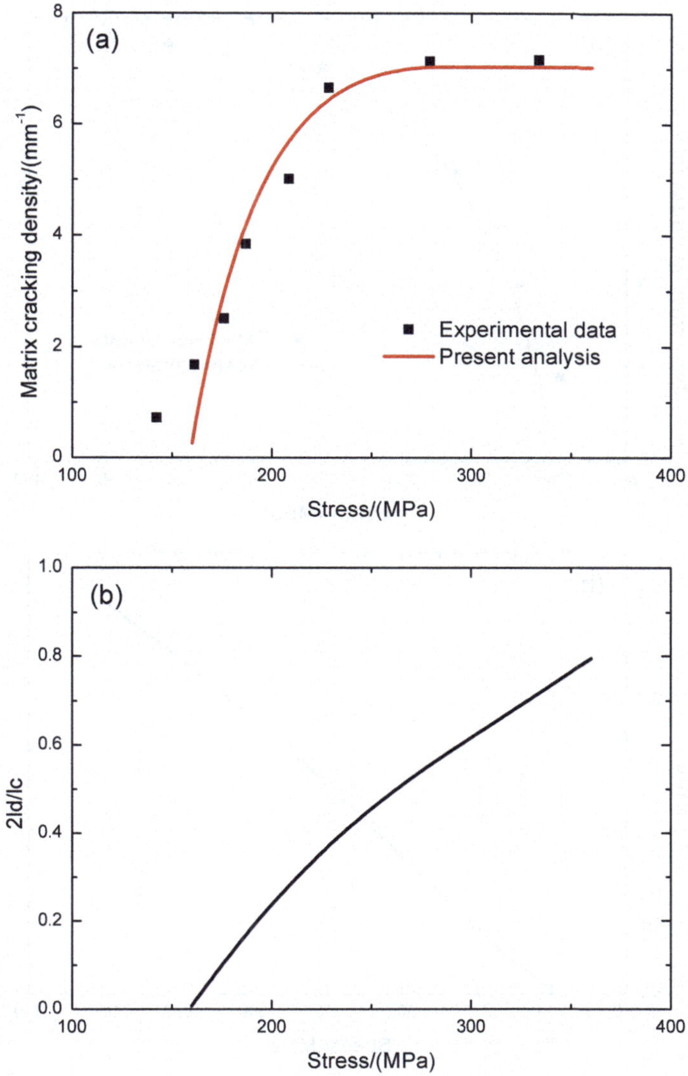

Fig. 2.6 **a** Experimental and predicted matrix multicracking density versus applied stress curves; and **b** the fiber/matrix interface debonding length versus applied stress curve of SiC/CAS composite

0.117 μJ; the matrix cracking density increases from $\varphi = 1.9$ mm to the saturation value of $\varphi = 9.2$ mm; and the fiber/matrix interface debonded length increases from $2l_d/l_c = 0.3\%$ at 260 MPa to $2l_d/l_c = 40\%$ at 380 MPa, as shown in Fig. 2.7.

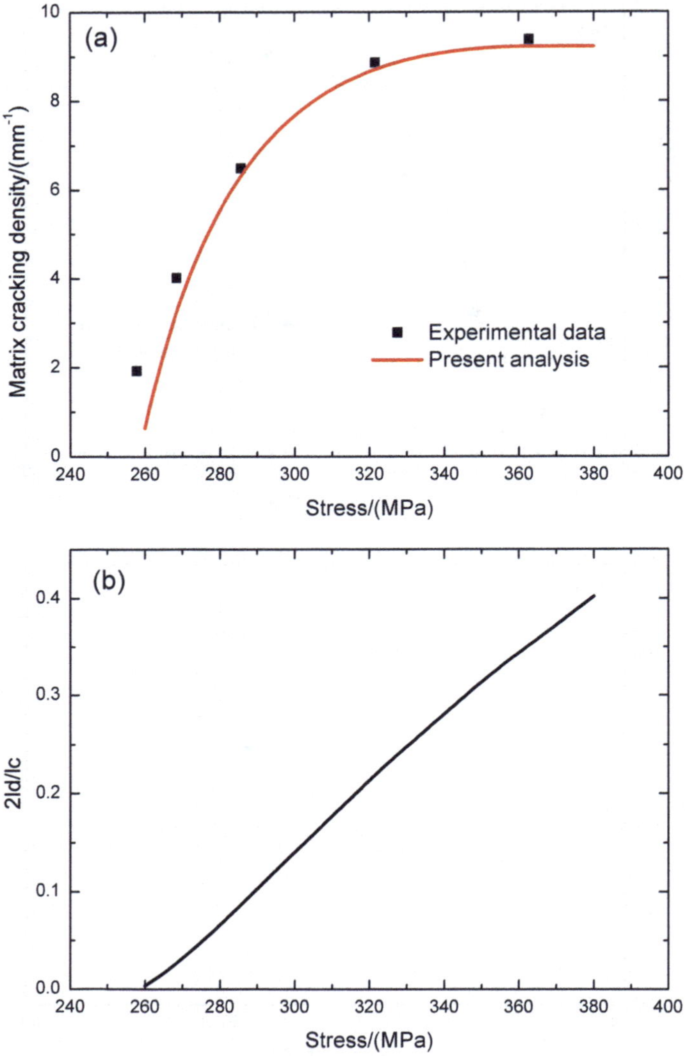

Fig. 2.7 **a** Experimental and predicted matrix multicracking density versus applied stress curves; and **b** the fiber/matrix interface debonding length versus applied stress curve of SiC/CAS-II composite

2.2.5.5 SiC/SiC Composite

For the unidirectional SiC/SiC composite (Beyerle et al. 1992), the matrix multicracking evolution starts from the applied stress of $\sigma_{mc} = 240$ MPa and approaches saturation at the applied stress of $\sigma_{sat} = 320$ MPa; the critical matrix strain energy (CMSE) criterion is used to determine the matrix multicracking evolution in SiC/SiC

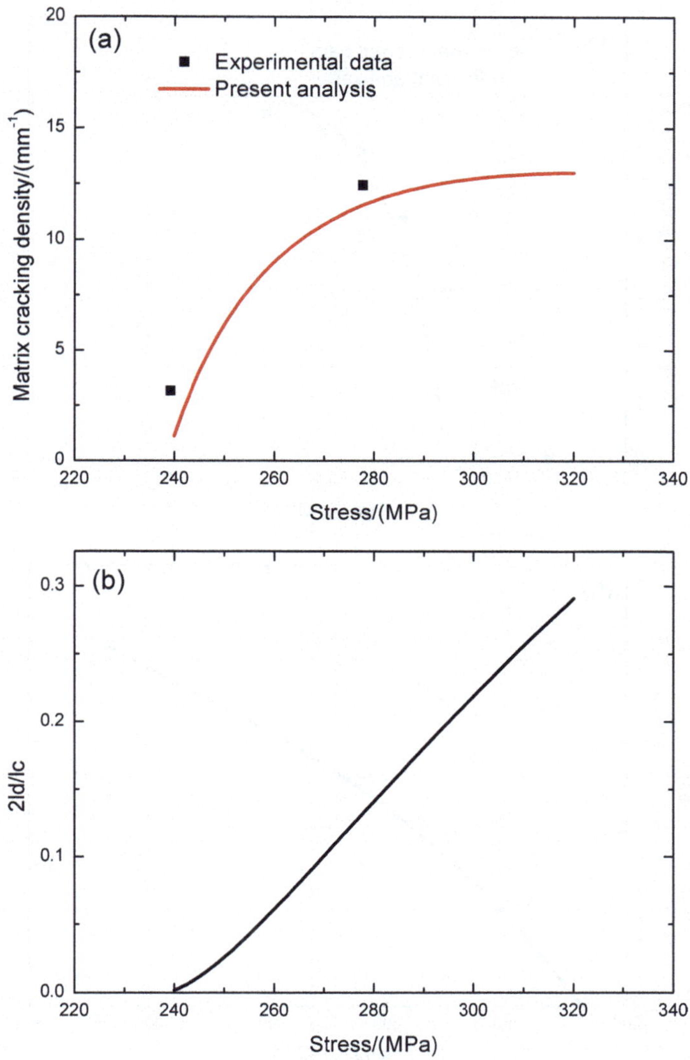

Fig. 2.8 **a** Experimental and predicted matrix multicracking density versus applied stress curves; and **b** the fiber/matrix interface debonded length versus applied stress curve of SiC/SiC composite

composite, and the critical matrix strain energy is $U_{\mathrm{crm}} = 0.052$ μJ; the matrix cracking density increases from $\varphi = 1.1$ mm to the saturation value of $\varphi = 13$ mm; and the fiber/matrix interface debonded length increases from $2l_d/l_c = 0.1\%$ at 240 MPa to $2l_d/l_c = 29\%$ at 320 MPa, as shown in Fig. 2.8.

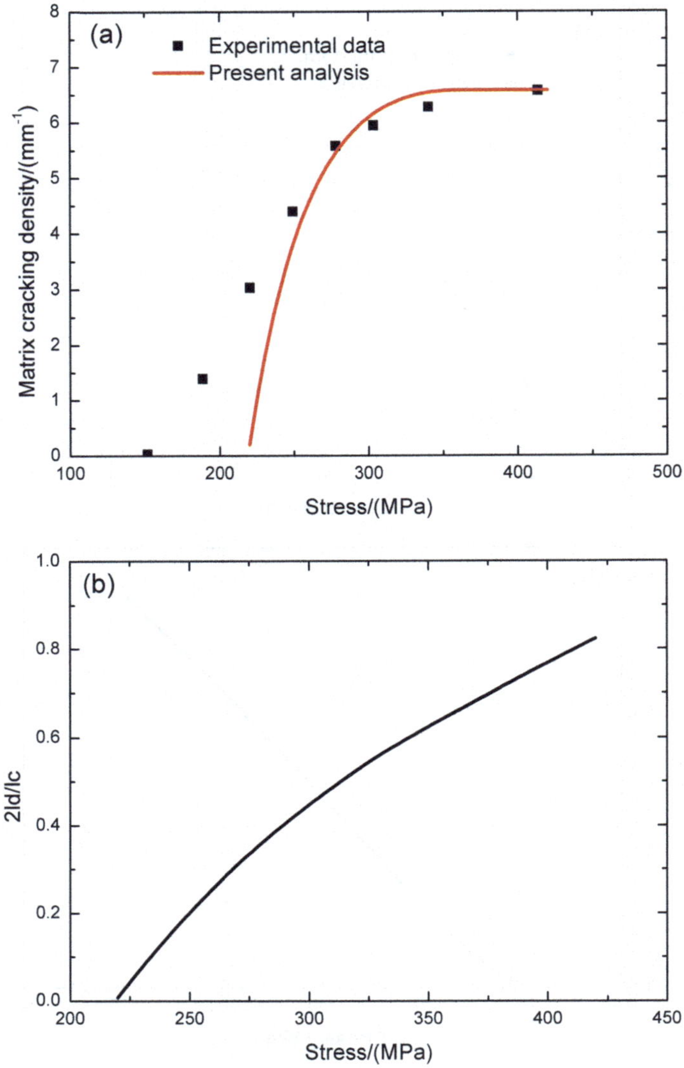

Fig. 2.9 **a** Experimental and predicted matrix multicracking density versus applied stress curves; and **b** the fiber/matrix interface debonded length versus applied stress curve of SiC/borosilicate composite

2.2.5.6 SiC/Borosilicate Composite

For the unidirectional SiC/Borosilicate composite (Okabe et al. 1999), the matrix multicracking evolution starts from the applied stress of $\sigma_{mc} = 220$ MPa and approaches saturation at the applied stress of $\sigma_{sat} = 360$ MPa; the critical matrix

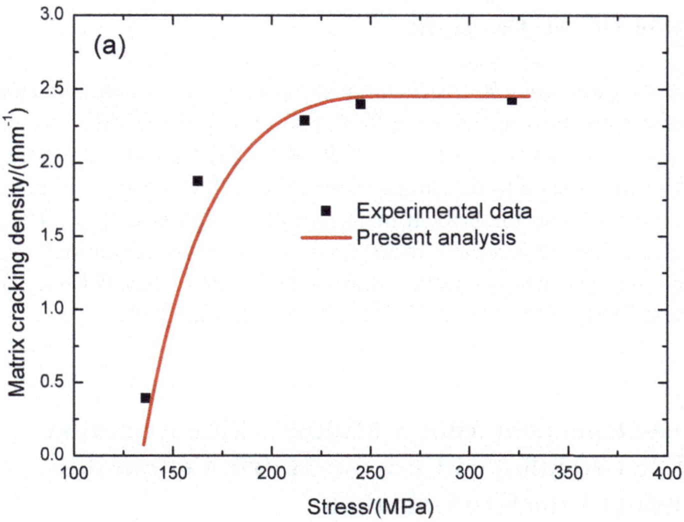

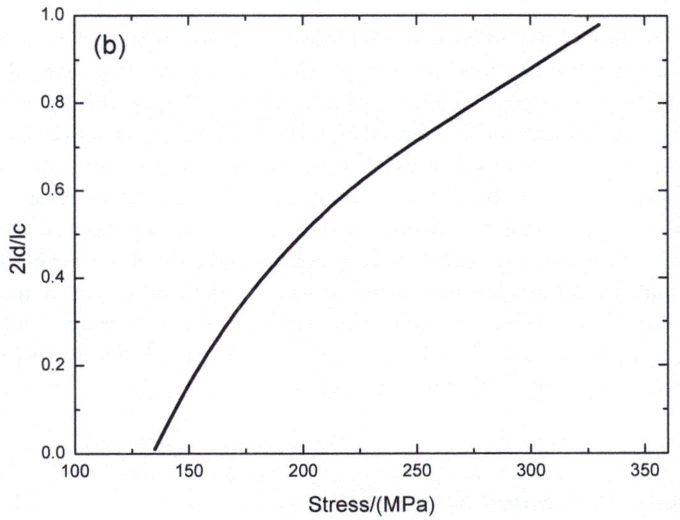

Fig. 2.10 **a** Experimental and predicted matrix multicracking density versus applied stress curves; and **b** the fiber/matrix interface debonding length versus applied stress curve of mini-SiC/SiC composite

strain energy (CMSE) criterion is used to determine the matrix multicracking evolution in SiC/Borosilicate composite, and the critical matrix strain energy is $U_{crm} = 0.266\ \mu J$; the matrix cracking density increases from $\varphi = 0.2$ mm to the saturation value of $\varphi = 6.5$ mm; and the fiber/matrix interface debonding length increases from $2l_d/l_c = 0.8\%$ at 220 MPa to $2l_d/l_c = 82\%$ at 420 MPa, as shown in Fig. 2.9.

2.2.5.7 Mini-SiC/SiC Composite

For the mini-unidirectional SiC/SiC composite (Zhang et al. 2016), the matrix multicracking evolution starts from the applied stress of $\sigma_{mc} = 135$ MPa and approaches saturation at the applied stress of $\sigma_{sat} = 250$ MPa; the critical matrix strain energy (CMSE) criterion is used to determine the matrix multicracking evolution in mini-SiC/SiC composite, and the critical matrix strain energy is $U_{crm} = 0.552$ μJ; the matrix cracking density increases from $\varphi = 0.4$ mm to the saturation value of $\varphi = 2.4$ mm; and the fiber/matrix interface debonding length increases from $2l_d/l_c = 1\%$ at 135 MPa to $2l_d/l_c = 98\%$ at 330 MPa, as shown in Fig. 2.10.

2.3 Time-Dependent Matrix Multicracking Evolution of Fiber-Reinforced Ceramic-Matrix Composites Consider Interface Oxidation

In this section, the effect of fiber/matrix interface oxidation on time-dependent matrix multicracking evolution of fiber-reinforced CMCs is investigated using the critical matrix strain energy criterion. The shear-lag model combined with the fiber/matrix interface oxidation model and fiber/matrix interface debonding criterion is adopted to analyze the fiber and matrix axial stress distribution inside of the damaged composite. The relationships between the matrix multicracking, interface debonding, and interface oxidation are established. The effects of the fiber volume, interface shear stress, interface debonding energy, oxidation temperature, and oxidation time on the stress-dependent matrix multicracking evolution are discussed. Comparisons of matrix multicracking evolution with/without oxidation are analyzed. The experimental matrix multicracking development of unidirectional C/SiC, SiC/CAS, SiC/borosilicate, and mini-SiC/SiC composites with/without oxidation is predicted.

2.3.1 Time-Dependent Stress Analysis

As the mismatch of the axial thermal expansion coefficient between fiber and matrix, there are unavoidable microcracks existed within the matrix when the composite was cooled down from high fabricated temperature to ambient temperature. These processing-induced microcracks mainly existed in the surface of the material, which do not propagate through the entire thickness of the composite. However, at elevated temperature, the microcracks would serve as avenues for the ingress of the environment atmosphere into the composite (Casas and Martinez-Esnaola 2003). The oxygen reacts with carbon layer along the fiber length at a certain rate of $\mathrm{d}\zeta/\mathrm{d}t$, in which ζ is the length of carbon lost in each side of the crack (Li et al. 2015).

$$\zeta = \varphi_1\left[1 - \exp\left(-\frac{\varphi_2 t}{b}\right)\right] \tag{2.20}$$

where φ_1 and φ_2 are parameters dependent on temperature and described using the Arrhenius-type laws; and b is a delay factor considering the deceleration of reduced oxygen activity.

The composite with fiber volume fraction V_f is loaded by a remote uniform stress σ normal to the crack plane. The fiber radius is r_f, and the matrix radius is R ($R = r_f/V_f^{1/2}$). The length of the unit cell is half matrix crack spacing $l_c/2$, and the interface oxidation length and interface debonding length are ζ and l_d, respectively. In the oxidation region, the fiber/matrix interface is resisted by a constant frictional shear stress τ_f; and in the debonded region, the interface is resisted by τ_i, which is higher than τ_f. For the interface debonding region, the force equilibrium equation of the fiber is given by the following equation.

$$\sigma_f(z) = \begin{cases} \dfrac{\sigma}{V_f} - \dfrac{2\tau_f}{r_f}z, \; z \in [0, \zeta] \\ \dfrac{\sigma}{V_f} - \dfrac{2\tau_f}{r_f}\zeta - \dfrac{2\tau_i}{r_f}(z - \zeta), \; z \in [\zeta, l_d] \\ \sigma_{fo} + \left[\dfrac{V_m}{V_f}\sigma_{mo} - \dfrac{2\tau_f}{r_f}\zeta - \dfrac{2\tau_i}{r_f}(l_d - \zeta)\right]\exp\left(-\rho\dfrac{z - l_d}{r_f}\right), \; z \in [l_d, l_c/2] \end{cases} \tag{2.21}$$

$$\sigma_m(z) = \begin{cases} 2\dfrac{V_f}{V_m}\dfrac{\tau_f}{r_f}z, \; z \in [0, \zeta] \\ 2\dfrac{V_f}{V_m}\dfrac{\tau_f}{r_f}\zeta + 2\dfrac{V_f}{V_m}\dfrac{\tau_i}{r_f}(z - \zeta), \; z \in [\zeta, l_d] \\ \sigma_{mo} + \left[2\dfrac{V_f}{V_m}\dfrac{\tau_f}{r_f}\zeta + 2\dfrac{V_f}{V_m}\dfrac{\tau_i}{r_f}(l_d - \zeta) - \sigma_{mo}\right]\exp\left(-\rho\dfrac{z - l_d}{r_f}\right), \; z \in \left[l_d, \dfrac{l_c}{2}\right] \end{cases} \tag{2.22}$$

$$\tau_i(z) = \begin{cases} \tau_f, \; z \in [0, \zeta] \\ \tau_i, \; z \in [\zeta, l_d] \\ \dfrac{\rho}{2}\left[\dfrac{V_m}{V_f}\sigma_{mo} - \dfrac{2\tau_f}{r_f}\zeta - \dfrac{2\tau_i}{r_f}(l_d - \zeta)\right]\exp\left(-\rho\dfrac{z - l_d}{r_f}\right), \; z \in \left[l_d, \dfrac{l_c}{2}\right] \end{cases} \tag{2.23}$$

where ρ denotes the shear-lag model parameter, and

$$\sigma_{fo} = \frac{E_f}{E_c}\sigma + E_f(\alpha_c - \alpha_f)\Delta T \tag{2.24}$$

$$\sigma_{mo} = \frac{E_m}{E_c}\sigma + E_m(\alpha_c - \alpha_m)\Delta T \tag{2.25}$$

where E_f, E_m, and E_c denote the fiber, matrix, and composite elastic modulus, respectively; α_f, α_m, and α_c denote the fiber, matrix, and composite thermal expansion coefficient, respectively; and ΔT denotes the temperature difference between the fabricated temperature T_0 and testing temperature T_1 ($\Delta T = T_1 - T_0$).

2.3.2 Time-Dependent Interface Debonding

The fracture mechanics approach is adopted to determine the fiber/matrix interface debonded length (Gao et al. 1988).

$$\xi_d = -\frac{F}{4\pi r_f}\frac{\partial w_f(0)}{\partial l_d} - \frac{1}{2}\int_0^{l_d}\tau_i\frac{\partial v(z)}{\partial l_d}\mathrm{d}z \tag{2.26}$$

where $F = \pi r_f^2\sigma/V_f$ denotes the fiber load at the matrix cracking plane; $w_f(0)$ denotes the fiber axial displacement at the matrix cracking plane, and $v(z)$ denotes the relative displacement between the fiber and the matrix.

The fiber and matrix axial displacements of $w_f(z)$ and $w_m(z)$ are described using the following equations.

$$\begin{aligned} w_f(z) &= \int_z^{l_c/2}\frac{\sigma_f(z)}{E_f}\mathrm{d}z \\ &= \frac{\sigma}{V_f E_f}(l_d - z) - \frac{\tau_f}{r_f E_f}\left(2\zeta l_d - \zeta^2 - z^2\right) - \frac{\tau_i}{r_f E_f}(l_d - \zeta)^2 + \frac{\sigma_{fo}}{E_f}\left(\frac{l_c}{2} - l_d\right) \\ &\quad + \frac{r_f}{\rho E_f}\left[\frac{V_m}{V_f}\sigma_{mo} - \frac{2\tau_f}{r_f}\zeta - \frac{2\tau_i}{r_f}(l_d - \zeta)\right]\left[1 - \exp\left(-\rho\frac{l_c/2 - l_d}{r_f}\right)\right] \end{aligned} \tag{2.27}$$

$$\begin{aligned} w_m(z) &= \int_z^{l_c/2}\frac{\sigma_m(z)}{E_m}\mathrm{d}z \\ &= \frac{V_f\tau_f}{r_f V_m E_m}\left(2\zeta l_d - \zeta^2 - z^2\right) + \frac{V_f\tau_i}{r_f V_m E_m}(l_d - \zeta)^2 + \frac{\sigma_{mo}}{E_m}\left(\frac{l_c}{2} - l_d\right) \\ &\quad - \frac{r_f}{\rho E_m}\left[\sigma_{mo} - 2\frac{V_f\tau_f}{r_f V_m}\zeta - 2\frac{V_f\tau_i}{r_f V_m}(l_d - \zeta)\right]\left[1 - \exp\left(-\rho\frac{l_c/2 - l_d}{r_f}\right)\right] \end{aligned} \tag{2.28}$$

The relative displacement $v(z)$ between the fiber and the matrix is described using the following equation.

$$
\begin{aligned}
v(x) &= |w_{\mathrm{f}}(z) - w_{\mathrm{m}}(z)| \\
&= \frac{\sigma}{V_{\mathrm{f}}E_{\mathrm{f}}}(l_{\mathrm{d}} - z) - \frac{E_{\mathrm{c}}\tau_{\mathrm{f}}}{r_{\mathrm{f}}V_{\mathrm{m}}E_{\mathrm{m}}E_{\mathrm{f}}}\left(2\zeta l_{\mathrm{d}} - \zeta^2 - z^2\right) - \frac{E_{\mathrm{c}}\tau_{\mathrm{i}}}{r_{\mathrm{f}}V_{\mathrm{m}}E_{\mathrm{m}}E_{\mathrm{f}}}(l_{\mathrm{d}} - \zeta)^2 \\
&\quad + \frac{r_{\mathrm{f}}E_{\mathrm{c}}}{\rho V_{\mathrm{m}}E_{\mathrm{m}}E_{\mathrm{f}}}\left[\sigma_{\mathrm{mo}} - 2\frac{\tau_{\mathrm{f}}}{r_{\mathrm{f}}}\zeta - 2\frac{\tau_{\mathrm{i}}}{r_{\mathrm{f}}}(l_{\mathrm{d}} - \zeta)\right]\left[1 - \exp\left(-\rho\frac{l_{\mathrm{c}}/2 - l_{\mathrm{d}}}{r_{\mathrm{f}}}\right)\right]
\end{aligned}
\tag{2.29}
$$

Substituting $w_{\mathrm{f}}(z = 0)$ and $v(z)$ into Eq. (2.26), it leads to the following equation.

$$
\begin{aligned}
&\frac{E_{\mathrm{c}}\tau_{\mathrm{i}}^2}{r_{\mathrm{f}}V_{\mathrm{m}}E_{\mathrm{m}}E_{\mathrm{f}}}(l_{\mathrm{d}} - \zeta)^2 + \frac{E_{\mathrm{c}}\tau_{\mathrm{i}}^2}{\rho V_{\mathrm{m}}E_{\mathrm{m}}E_{\mathrm{f}}}(l_{\mathrm{d}} - \zeta) - \frac{\tau_{\mathrm{i}}\sigma}{V_{\mathrm{f}}E_{\mathrm{f}}}(l_{\mathrm{d}} - \zeta) + \frac{2E_{\mathrm{c}}\tau_{\mathrm{f}}\tau_{\mathrm{i}}}{r_{\mathrm{f}}V_{\mathrm{m}}E_{\mathrm{m}}E_{\mathrm{f}}}\zeta(l_{\mathrm{d}} - \zeta) \\
&- \frac{r_{\mathrm{f}}\tau_{\mathrm{i}}\sigma}{2\rho V_{\mathrm{f}}E_{\mathrm{f}}} + \frac{E_{\mathrm{c}}\tau_{\mathrm{f}}^2}{r_{\mathrm{f}}V_{\mathrm{m}}E_{\mathrm{m}}E_{\mathrm{f}}}\zeta^2 + \frac{E_{\mathrm{c}}\tau_{\mathrm{f}}\tau_{\mathrm{i}}}{\rho V_{\mathrm{m}}E_{\mathrm{m}}E_{\mathrm{f}}}\zeta - \frac{\tau_{\mathrm{f}}\sigma}{V_{\mathrm{f}}E_{\mathrm{f}}}\zeta + \frac{r_{\mathrm{f}}V_{\mathrm{m}}E_{\mathrm{m}}\sigma^2}{4V_{\mathrm{f}}^2E_{\mathrm{f}}E_{\mathrm{c}}} - \xi_d = 0
\end{aligned}
\tag{2.30}
$$

Solving Eq. (2.30), the fiber/matrix interface debonding length is determined by the following equation.

$$
l_{\mathrm{d}} = \left(1 - \frac{\tau_{\mathrm{f}}}{\tau_{\mathrm{i}}}\right)\zeta + \frac{r_{\mathrm{f}}}{2}\left(\frac{V_{\mathrm{m}}E_{\mathrm{m}}\sigma}{V_{\mathrm{f}}E_{\mathrm{c}}\tau_{\mathrm{i}}} - \frac{1}{\rho}\right) - \sqrt{\left(\frac{r_{\mathrm{f}}}{2\rho}\right)^2 + \frac{r_{\mathrm{f}}V_{\mathrm{m}}E_{\mathrm{m}}E_{\mathrm{f}}}{E_{\mathrm{c}}\tau_{\mathrm{i}}^2}\xi_{\mathrm{d}}} \tag{2.31}
$$

2.3.3 Time-Dependent Matrix Multicracking

Solti et al. (1995) developed the critical matrix strain energy (CMSE) criterion to predict matrix multicracking evolution in fiber-reinforced CMCs. The concept of critical matrix strain energy presupposes the existence of an ultimate or critical strain energy. Beyond the critical value of matrix strain energy, as more energy is entered into the composite with increasing applied stress, the matrix cannot support the extra load and continues to fail. The failure is assumed to consist of the formation of new cracks and fiber/matrix interface debonding, to make the total energy within the matrix remain constant and equal to its critical value.

The matrix strain energy is determined using the following equation.

$$
U_{\mathrm{m}} = \frac{1}{2E_{\mathrm{m}}}\int_{A_{\mathrm{m}}}\int_{0}^{l_{\mathrm{c}}} \sigma_{\mathrm{m}}^2(z)\mathrm{d}z\mathrm{d}A_{\mathrm{m}} \tag{2.32}
$$

where A_{m} is the cross-sectional area of matrix in the unit cell. Substituting the matrix axial stresses in Eq. (2.22) into Eq. (2.32), the matrix strain energy considering matrix multicracking and fiber/matrix interface partial debonding is described using the following equation.

$$
\begin{aligned}
U_{\mathrm{m}} = \frac{A_{\mathrm{m}}}{E_{\mathrm{m}}}\Bigg\{ & \frac{4}{3}\left(\frac{V_{\mathrm{f}}}{V_{\mathrm{m}}}\frac{\tau_{\mathrm{f}}}{r_{\mathrm{f}}}\right)^2 \zeta^3 + 4\left(\frac{V_{\mathrm{f}}}{V_{\mathrm{m}}}\frac{\tau_{\mathrm{f}}}{r_{\mathrm{f}}}\right)^2 \zeta^2 (l_{\mathrm{d}} - \zeta) + 4\left(\frac{V_{\mathrm{f}}}{r_{\mathrm{f}} V_{\mathrm{m}}}\right)^2 \tau_{\mathrm{f}}\tau_{\mathrm{i}}\zeta (l_{\mathrm{d}} - \zeta)^2 \\
& + \frac{4}{3}\left(\frac{V_{\mathrm{f}}}{V_{\mathrm{m}}}\frac{\tau_{\mathrm{i}}}{r_{\mathrm{f}}}\right)^2 (l_{\mathrm{d}} - \zeta)^3 + \sigma_{\mathrm{mo}}^2\left(\frac{l_{\mathrm{c}}}{2} - l_{\mathrm{d}}\right) + \frac{2 r_{\mathrm{f}}\sigma_{\mathrm{mo}}}{\rho}\left[2\frac{V_{\mathrm{f}}}{V_{\mathrm{m}}}\frac{\tau_{\mathrm{f}}}{r_{\mathrm{f}}}\zeta\right. \\
& \left. + 2\frac{V_{\mathrm{f}}}{V_{\mathrm{m}}}\frac{\tau_{\mathrm{i}}}{r_{\mathrm{f}}}(l_{\mathrm{d}} - \zeta) - \sigma_{\mathrm{mo}}\right]\left[1 - \exp\left(-\rho\frac{l_{\mathrm{c}}/2 - l_{\mathrm{d}}}{r_{\mathrm{f}}}\right)\right] \\
& + \frac{r_{\mathrm{f}}}{2\rho}\left[2\frac{V_{\mathrm{f}}}{V_{\mathrm{m}}}\frac{\tau_{\mathrm{f}}}{r_{\mathrm{f}}}\zeta + 2\frac{V_{\mathrm{f}}}{V_{\mathrm{m}}}\frac{\tau_{\mathrm{i}}}{r_{\mathrm{f}}}(l_{\mathrm{d}} - \zeta) - \sigma_{\mathrm{mo}}\right]^2\left[1 - \exp\left(-2\rho\frac{l_{\mathrm{c}}/2 - l_{\mathrm{d}}}{r_{\mathrm{f}}}\right)\right]\Bigg\}
\end{aligned} \tag{2.33}
$$

When the interface completely debonds, the matrix strain energy is described using the following equation.

$$
U_{\mathrm{m}}(\sigma, l_{\mathrm{c}}, l_{\mathrm{d}} = l_{\mathrm{c}}/2) = \frac{A_{\mathrm{m}}}{E_{\mathrm{m}}}\left[\begin{array}{l} \frac{4}{3}\left(\frac{V_{\mathrm{f}}}{V_{\mathrm{m}}}\frac{\tau_{\mathrm{f}}}{r_{\mathrm{f}}}\right)^2 \zeta^3 + 4\left(\frac{V_{\mathrm{f}}}{V_{\mathrm{m}}}\frac{\tau_{\mathrm{f}}}{r_{\mathrm{f}}}\right)^2 \zeta^2 (l_{\mathrm{d}} - \zeta) \\ + 4\left(\frac{V_{\mathrm{f}}}{r_{\mathrm{f}} V_{\mathrm{m}}}\right)^2 \tau_{\mathrm{f}}\tau_{\mathrm{i}}\zeta (l_{\mathrm{d}} - \zeta)^2 + \frac{4}{3}\left(\frac{V_{\mathrm{f}}}{V_{\mathrm{m}}}\frac{\tau_{\mathrm{i}}}{r_{\mathrm{f}}}\right)^2 (l_{\mathrm{d}} - \zeta)^3 \end{array}\right] \tag{2.34}
$$

By evaluating the matrix strain energy at a critical stress of σ_{cr}, the critical matrix strain energy of U_{crm} can be obtained. The critical matrix strain energy is described using the following equation.

$$
U_{\mathrm{crm}} = \frac{1}{2} k A_{\mathrm{m}} l_0 \frac{\sigma_{\mathrm{mocr}}^2}{E_{\mathrm{m}}} \tag{2.35}
$$

where k ($k \in [0, 1]$) is the critical matrix strain energy parameter; and L_0 is the initial matrix crack spacing and σ_{mocr} is determined using the following equation.

$$
\sigma_{\mathrm{mocr}} = \frac{E_{\mathrm{m}}}{E_{\mathrm{c}}}\sigma_{\mathrm{cr}} + E_{\mathrm{m}}(\alpha_{\mathrm{c}} - \alpha_{\mathrm{m}})\Delta T \tag{2.36}
$$

where σ_{cr} is the critical stress corresponding to composite's proportional limit stress, i.e., the stress at which the stress-strain curve starts to deviate from linearity due to damage accumulation of matrix cracks (Li 2017). The critical stress is defined to be Aveston–Cooper–Kelly matrix cracking stress (Aveston et al. 1971), which was determined using the energy balance criterion, involving calculation of energy balance relationship before and after the formation of a single dominant cracking. The Aveston–Cooper–Kelly model can be used to describe long-steady-state matrix cracking stress, corresponding to the proportional limit stress of tensile stress-strain curve. The Aveston–Cooper–Kelly matrix cracking stress is determined using the following equation (Aveston et al. 1971).

$$\sigma_{cr} = \left(\frac{6V_f^2 E_f E_c^2 \tau_i \xi_m}{r_f V_m E_m^2}\right)^{\frac{1}{3}} - E_c(\alpha_c - \alpha_m)\Delta T \tag{2.37}$$

where ξ_m denotes the matrix fracture energy. However, as microcracks exist in the matrix when CMCs were cooled down from the high fabrication temperature to room temperature, due to thermal expansion coefficient misfit between the fiber and the matrix, these microcracks are short-matrix-cracking, and the cracking stresses of these microcracks lie in the linear region of tensile stress-strain curve (Li et al. 2014, 2015). With increasing applied stress, matrix microcracks can propagate into long-matrix-cracking. The matrix cracking stress of Aveston–Cooper–Kelly model was used to determine the critical matrix strain energy.

The energy balance relationship to evaluate matrix multicracking evolution is determined using the following equation.

$$U_m(\sigma > \sigma_{cr}, l_c, l_d) = U_{crm}(\sigma_{cr}, l_0) \tag{2.38}$$

The matrix multicracking evolution versus applied stress can be solved by Eq. (2.38) when the critical matrix cracking stress of σ_{cr} and the fiber/matrix interface debonded length of l_d are determined by Eqs. (2.31) and (2.37).

2.3.4 Results and Discussion

The effects of the fiber volume, interface shear stress, interface debonding energy, oxidation temperature, oxidation time on matrix multicracking, and interface debonding are discussed. The ceramic composite system of SiC/CAS is used for the case study, and its material properties are given by: V_f = 30%, E_f = 200 GPa, E_m = 97 GPa, r_f = 7.5 μm, ξ_m = 6 J/m^2, τ_i = 20 MPa, $\alpha_f = 4 \times 10^{-5}$ °C, $\alpha_m = 5 \times 10^{-5}$ °C, $\Delta T = -1000$ °C.

2.3.4.1 Effect of Fiber Volume on Time-Dependent Matrix Multicracking and Interface Debonding

The time-dependent matrix multicracking density, fiber/matrix interface debonding length, interface debonding ratio, and interface oxidation ratio corresponding to different fiber volume (i.e., V_f = 30 and 35%) are shown in Fig. 2.11. With increasing fiber volume, the first matrix cracking stress, matrix saturation cracking stress and matrix cracking density increase; the interface debonding length and interface debonding ratio decrease; and the interface oxidation ratio increases.

When the fiber volume is $V_f = 30\%$, the time-dependent matrix multicracking density increases from $\varphi = 0.09$ mm at the first matrix cracking stress of $\sigma_{mc} = 201$ MPa to $\varphi = 3.1$ mm at the saturation matrix cracking stress of $\sigma_{sat} = 360$ MPa; the time-dependent interface debonding length increases from $l_d/r_f = 6.5$ to 13.5;

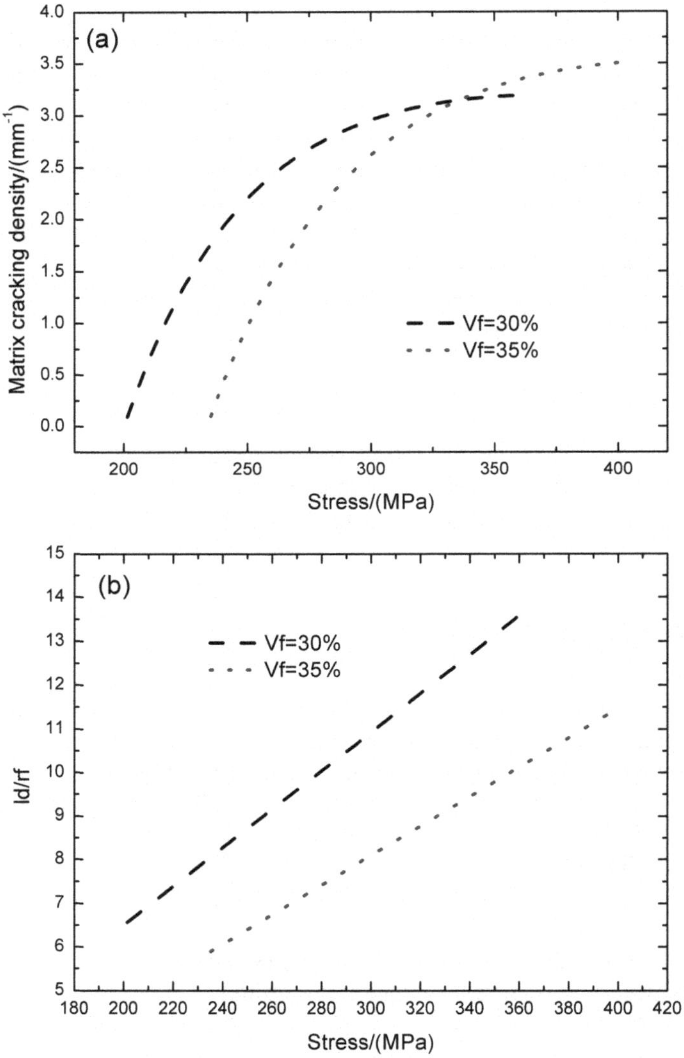

Fig. 2.11 Effect of the fiber volume (i.e., $V_f = 30$ and 35%) on **a** the time-dependent matrix multicracking density versus applied stress curves; **b** the time-dependent interface debonding length versus applied stress curves; **c** the time-dependent interface debonding ratio versus applied stress curves; and **d** the time-dependent interface oxidation ratio versus applied stress curves

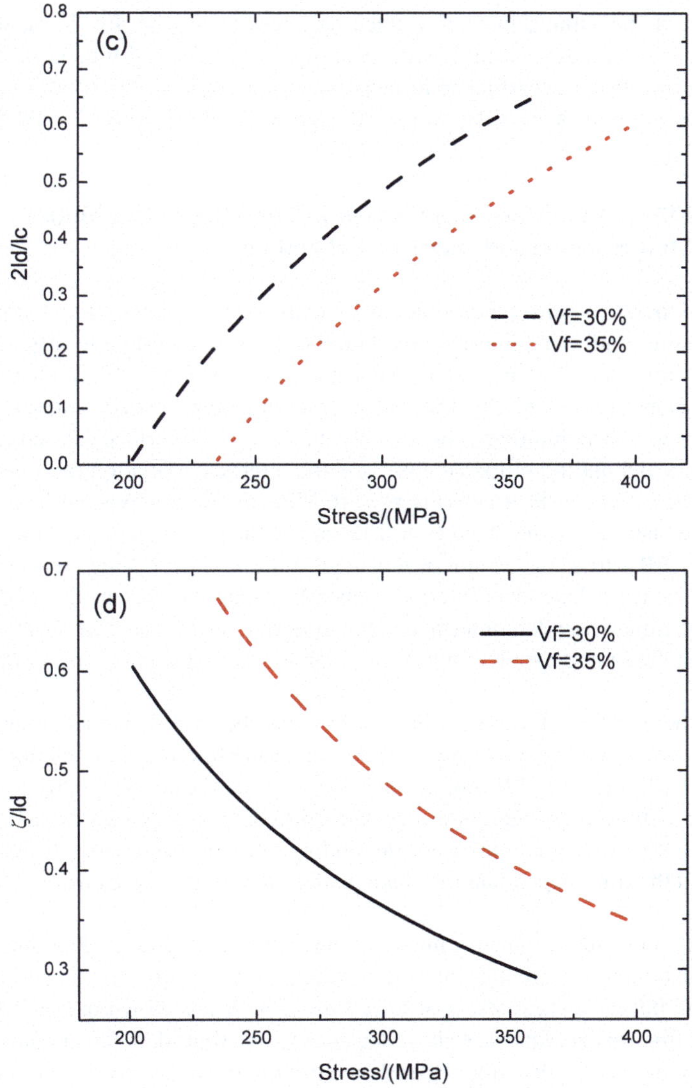

Fig. 2.11 (continued)

the time-dependent interface debonding ratio increases from $2l_d/l_c = 0.9$ to 64.9%; and the time-dependent interface oxidation ratio decreases from $\zeta/l_d = 60$ to 29%.

When the fiber volume is $V_f = 35\%$, the time-dependent matrix cracking density increases from $\varphi = 0.1$ mm at the first matrix cracking stress of $\sigma_{mc} = 235$ MPa to 3.5 mm at the saturation matrix cracking stress of $\sigma_{sat} = 400$ MPa; the time-dependent interface debonding length increases from $l_d/r_f = 5.9$ to 11.4; the time-dependent interface debonding ratio increases from $2l_d/l_c = 0.9$ to 60.3%; and the time-dependent interface oxidation ratio decreases from $\zeta/l_d = 67.2$ to 34.5%.

2.3.4.2 Effect of Interface Shear Stress on Time-Dependent Matrix Multicracking and Interface Debonding

The time-dependent matrix multicracking density, interface debonding length, interface debonding ratio, and interface oxidation ratio corresponding to different interface shear stress of $\tau_i = 10$ and 15 MPa are shown in Fig. 2.12. With increasing interface shear stress of τ_i, the first matrix cracking stress, matrix saturation cracking stress, and matrix multicracking density increase; the interface debonding length and interface debonding ratio decrease; and the interface oxidation ratio increases.

When the interface shear stress is $\tau_i = 10$ MPa, the time-dependent matrix multicracking density increases from $\varphi = 0.12$ mm at the first matrix cracking stress of $\sigma_{mc} = 147$ MPa to $\varphi = 2.7$ mm at the saturation matrix cracking stress of $\sigma_{sat} =$ 238 MPa; the time-dependent interface debonding length increases from $l_d/r_f = 4.4$ to 19.6; the time-dependent interface debonding ratio increases from $2l_d/l_c = 0.8$ to 81.9%; and the time-dependent interface oxidation ratio decreases from $\zeta/l_d = 90.2$ to 20.1%.

When the interface shear stress is $\tau_i = 15$ MPa, the time-dependent matrix multicracking density increases from $\varphi = 0.09$ mm at the first matrix cracking stress of $\sigma_{mc} = 177$ MPa to $\varphi = 2.9$ mm at the saturation matrix cracking stress of $\sigma_{sat} =$ 315 MPa; the time-dependent interface debonding length increases from $l_d/r_f = 6.0$ to 14.4; the time-dependent interface debonding ratio increases from $2l_d/l_c = 0.9$ to 63.9%; and the time-dependent interface oxidation ratio decreases from $\zeta/l_d = 65.8$ to 27.5%.

The time-dependent matrix multicracking density, interface debonding length, interface debonding ratio, and interface oxidation ratio corresponding to different fiber/matrix interface shear stress of $\tau_f = 1$ and 5 MPa are shown in Fig. 2.13. With increasing fiber/matrix interface shear stress of τ_f, the time-dependent matrix cracking density increases; the time-dependent fiber/matrix interface debonding length decreases; and the time-dependent fiber/matrix interface oxidation ratio increases.

When the interface shear stress is $\tau_f = 1$ MPa, the time-dependent matrix multicracking density increases from $\varphi = 0.08$ mm at the first matrix cracking stress of $\sigma_{mc} = 201$ MPa to $\varphi = 2.8$ mm at the saturation matrix cracking stress of σ_{sat} = 320 MPa; the time-dependent interface debonding length increases from $l_d/r_f =$ 7.3 to 12.6; the time-dependent fiber/matrix interface debonding ratio increases from $2l_d/l_c = 0.93$ to 54.5%; and the time-dependent fiber/matrix interface oxidation ratio decreases from $\zeta/l_d = 53.9$ to 31.5%.

When the fiber/matrix interface shear stress is $\tau_f = 5$ MPa, the time-dependent matrix cracking density increases from $\varphi = 0.09$ mm at the first matrix cracking stress of $\sigma_{mc} = 201$ MPa to $\varphi = 3.0$ mm at the saturation matrix cracking stress of σ_{sat}

= 320 MPa; the time-dependent fiber/matrix interface debonding length increases from $l_d/r_f = 6.5$ to 11.8; the time-dependent fiber/matrix interface debonding ratio increases from $2l_d/l_c = 0.92$ to 54.6%; and the time-dependent fiber/matrix interface oxidation ratio decreases from $\zeta/l_d = 60.4$ to 33.6%.

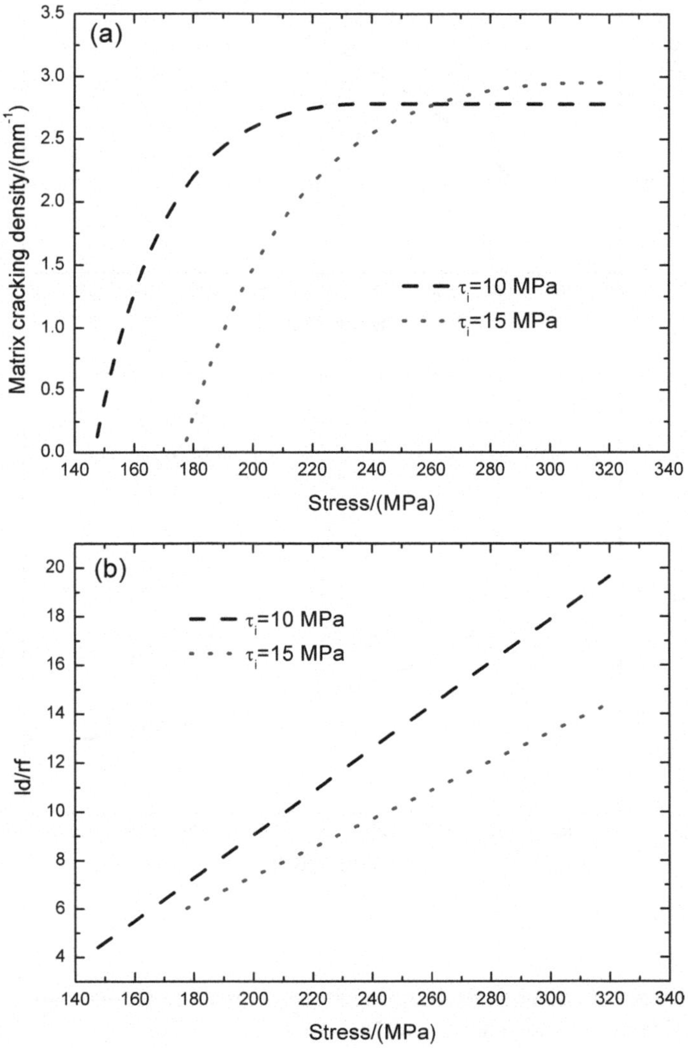

Fig. 2.12 Effect of the interface shear stress (i.e., $\tau_i = 10$ and 15 MPa) on **a** the time-dependent matrix multicracking density versus applied stress curves; **b** the time-dependent interface debonding length versus applied stress curves; **c** the time-dependent interface debonding ratio versus applied stress curves; and **d** the time-dependent fiber/matrix interface oxidation ratio versus applied stress curves

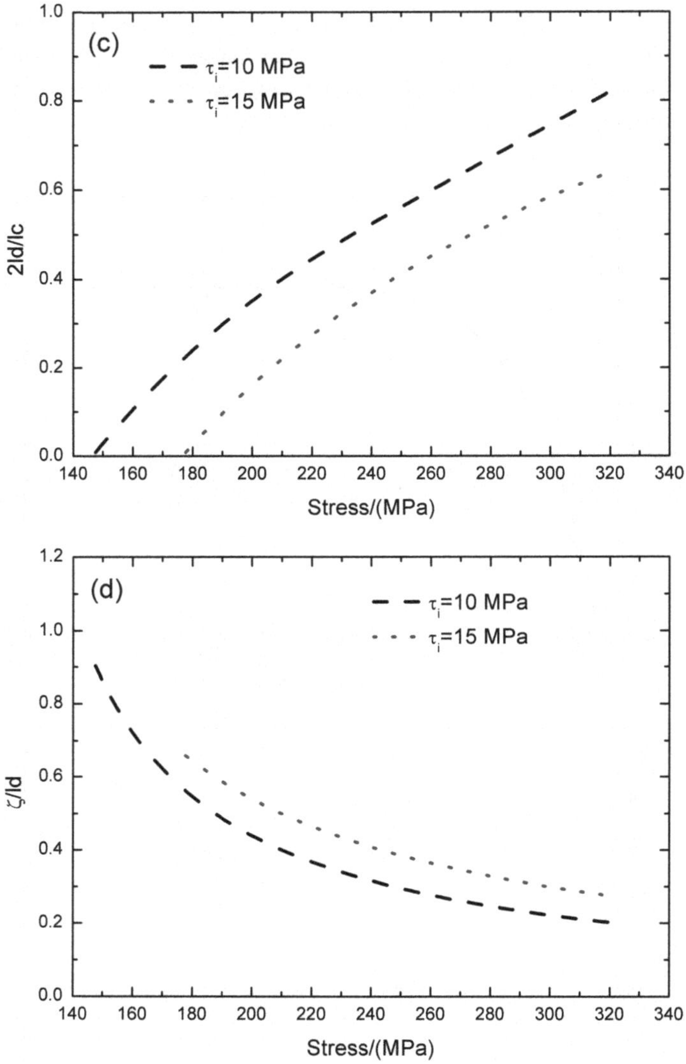

Fig. 2.12 (continued)

2.3.4.3 Effect of Interface Debonding Energy on Time-Dependent Matrix Multicracking and Interface Debonding

The time-dependent matrix multicracking density, fiber/matrix interface debonding length, interface debonding ratio, and interface oxidation ratio corresponding to different fiber/matrix interface debonding energy of $\xi_d = 0.5$ and 1.0 J/m^2 are shown in Fig. 2.14. With increasing fiber/matrix interface debonding energy, the time-dependent matrix multicracking density increases; the time-dependent fiber/matrix interface debonding length and interface debonding ratio decrease; and the time-dependent fiber/matrix interface oxidation ratio increases.

When the fiber/matrix interface debonding energy is $\xi_d = 0.5$ J/m^2, the time-dependent matrix multicracking density increases from $\varphi = 0.08$ mm at the first matrix cracking stress of $\sigma_{mc} = 201$ MPa to $\varphi = 2.9$ mm at the saturation matrix

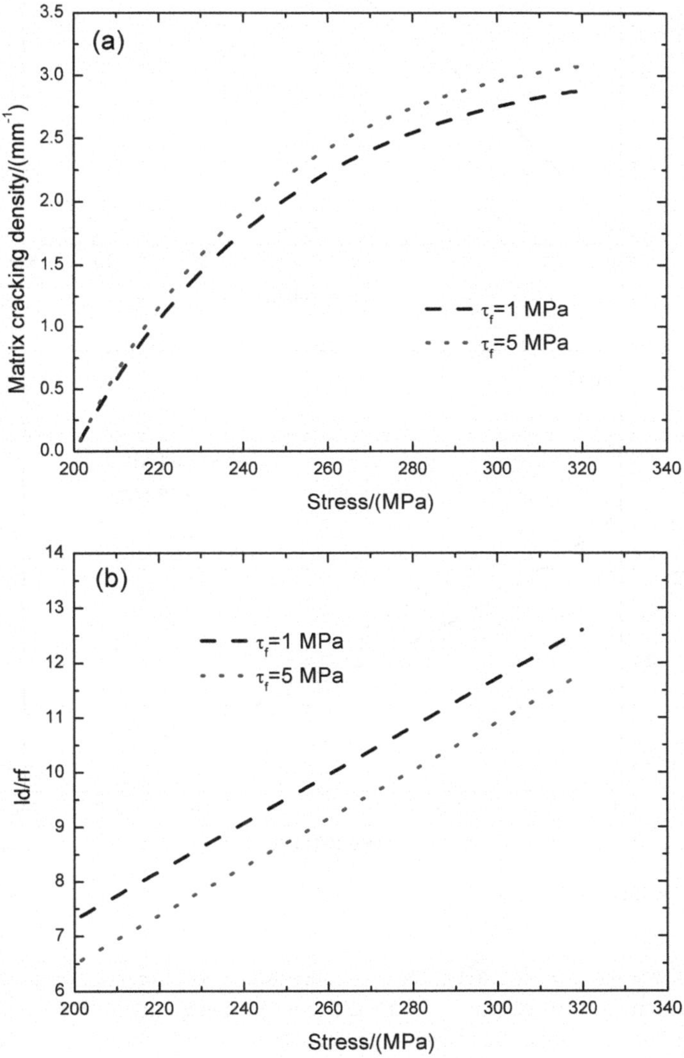

Fig. 2.13 Effect of the interface shear stress (i.e., $\tau_f = 1$ and 5 MPa) on **a** the time-dependent matrix multicracking density versus applied stress curves; **b** the time-dependent interface debonding length versus applied stress curves; **c** the time-dependent interface debonding ratio versus applied stress curves; and **d** the time-dependent interface oxidation ratio versus applied stress curves

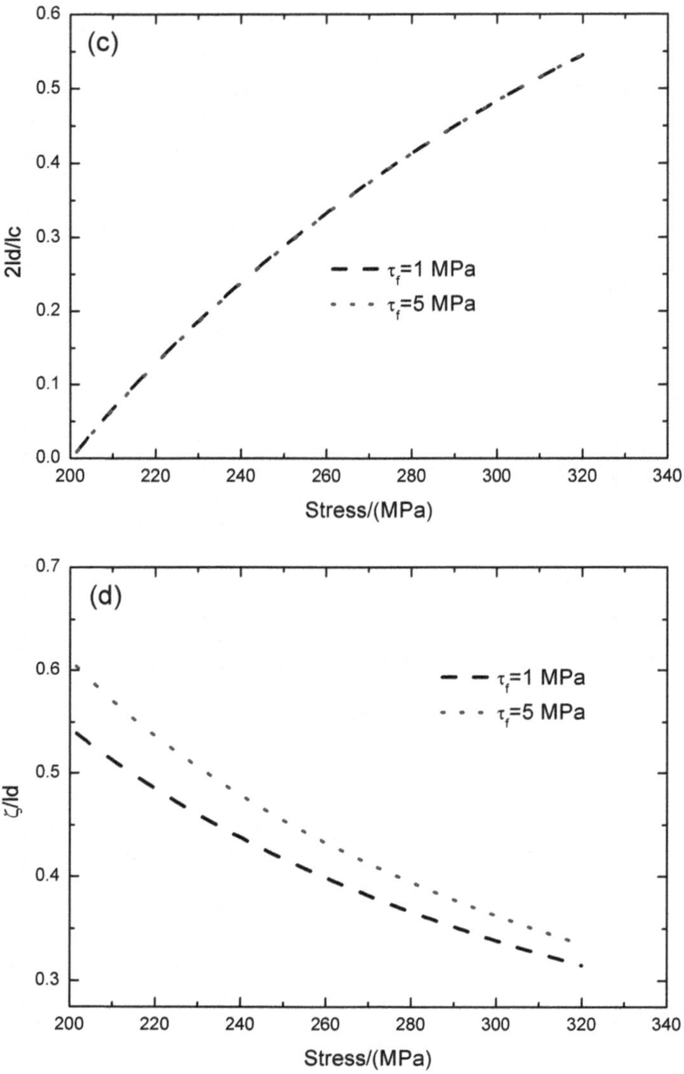

Fig. 2.13 (continued)

cracking stress of $\sigma_{\text{sat}} = 320$ MPa; the time-dependent fiber/matrix interface debonding length increases from $l_{\text{d}}/r_{\text{f}} = 7.6$ to 12.9; the time-dependent fiber/matrix interface debonding ratio increases from $2l_{\text{d}}/l_{\text{c}} = 0.95$ to 56.3%; and the time-dependent fiber/matrix interface oxidation ratio decreases from $\zeta/l_{\text{d}} = 51.6$ to 30.7%.

When the fiber/matrix interface debonded energy is $\xi_{\text{d}} = 1.0$ J/m^2, the time-dependent matrix cracking density increases from $\varphi = 0.1$ mm at the first matrix cracking stress of $\sigma_{\text{mc}} = 201$ MPa to $\varphi = 3.2$ mm at the saturation matrix cracking stress of $\sigma_{\text{sat}} = 320$ MPa; the time-dependent fiber/matrix interface debonding length

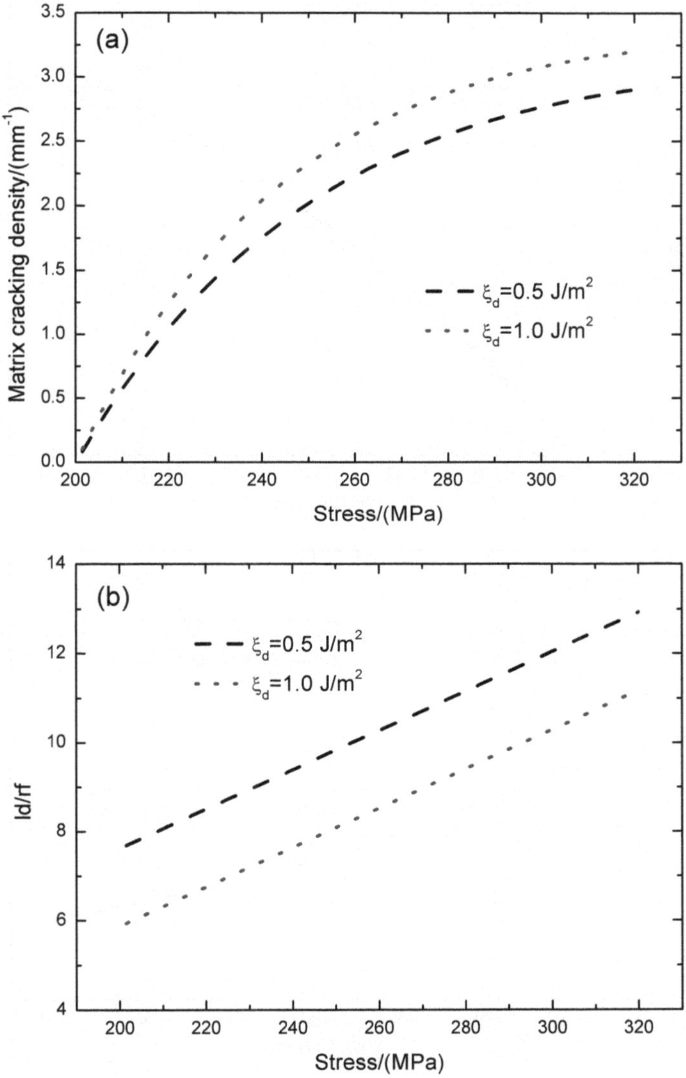

Fig. 2.14 Effect of the fiber/matrix interface debonding energy (i.e., $\xi_d = 0.5$ and 1.0 J/m^2) on **a** the time-dependent matrix multicracking density versus applied stress curves; **b** the time-dependent fiber/matrix interface debonding length versus applied stress curves; **c** the time-dependent fiber/matrix interface debonding ratio versus applied stress curves; and **d** the time-dependent fiber/matrix interface oxidation ratio versus applied stress curves

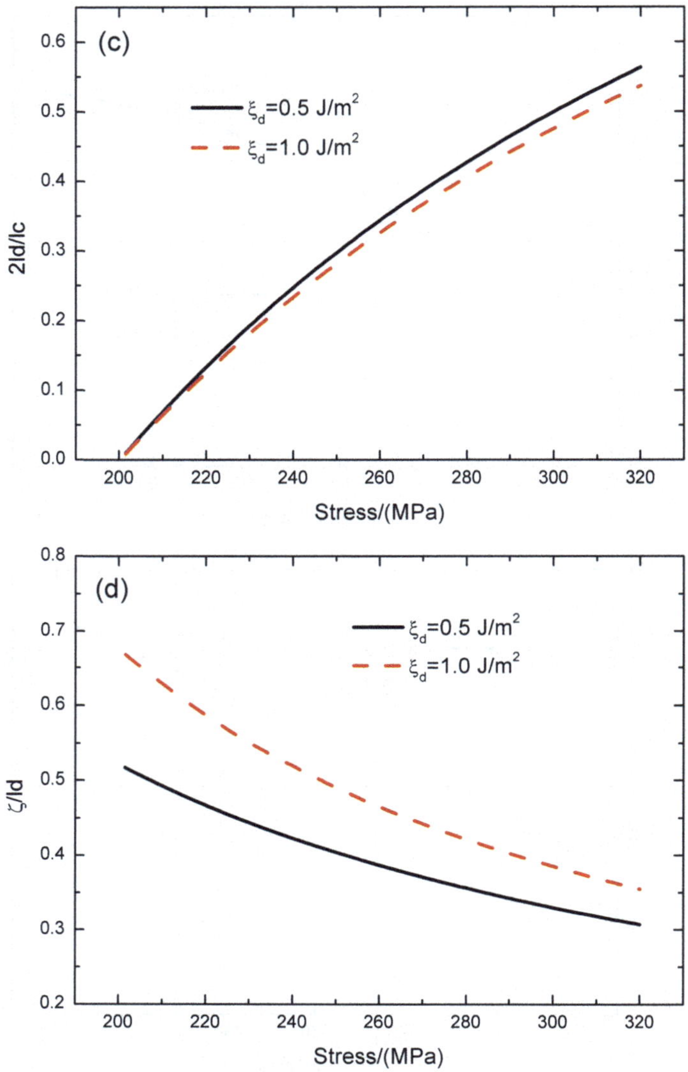

Fig. 2.14 (continued)

increases from $l_d/r_f = 5.9$ to 11.2; the time-dependent fiber/matrix interface debonding ratio increases from $2l_d/l_c = 0.9$ to 53.7%; and the time-dependent fiber/matrix interface oxidation ratio decreases from $\zeta/l_d = 66.8$ to 35.4%.

2.3.4.4 Effect of Oxidation Temperature on Time-Dependent Matrix Multicracking and Interface Debonding

The time-dependent matrix multicracking density, fiber/matrix interface debonding length, interface debonding ratio, and interface oxidation ratio corresponding to different oxidation temperature of T = 600 and 900 °C are shown in Fig. 2.15. With increasing oxidation temperature, the time-dependent matrix multicracking density decreases; the time-dependent fiber/matrix interface debonding length increases; and the time-dependent fiber/matrix interface oxidation ratio increases.

When the oxidation temperature is T = 600 °C, the time-dependent matrix multicracking density increases from $\varphi = 0.14$ mm at the first matrix cracking stress of $\sigma_{\rm mc} = 201$ MPa to $\varphi = 3.9$ mm at the saturation matrix cracking stress of $\sigma_{\rm sat} = 320$ MPa; the time-dependent fiber/matrix interface debonding length increases from $l_{\rm d}/r_{\rm f} = 4.0$ to 9.2; the time-dependent fiber/matrix interface debonding ratio increases from $2l_{\rm d}/l_{\rm c} = 0.8$ to 55%; and the time-dependent fiber/matrix interface oxidation ratio decreases from $\zeta/l_{\rm d} = 14.8$ to 6.4%.

When the oxidation temperature is T = 900 °C, the time-dependent matrix multicracking density increases from $\varphi = 0.06$ mm at the first matrix cracking stress of $\sigma_{\rm mc} = 201$ MPa to $\varphi = 2.4$ mm at the saturation matrix cracking stress of $\sigma_{\rm sat} = 320$ MPa; the time-dependent fiber/matrix interface debonding length increases from $l_{\rm d}/r_{\rm f} = 9.6$ to 14.8; the time-dependent fiber/matrix interface debonding ratio increases from $2l_{\rm d}/l_{\rm c} = 0.95$ to 54.3%; and the time-dependent fiber/matrix interface oxidation ratio decreases from $\zeta/l_{\rm d} = 83.4$ to 53.9%.

2.3.4.5 Effect of Oxidation Time on Time-Dependent Matrix Multicracking and Interface Debonding

The time-dependent matrix multicracking density, fiber/matrix interface debonding length, interface debonding ratio, and interface oxidation ratio corresponding to different oxidation time of t = 1 and 2 h are shown in Fig. 2.16. With increasing oxidation time, the time-dependent matrix multicracking density decreases; the time-dependent fiber/matrix interface debonding length increases; and the time-dependent fiber/matrix interface oxidation ratio increases.

When the oxidation time is t = 1 h, the time-dependent matrix multicracking density increases from $\varphi = 0.09$ mm at the first matrix cracking stress of $\sigma_{\rm mc} = 201$ MPa to $\varphi = 3.0$ mm at the saturation matrix cracking stress of $\sigma_{\rm sat} = 320$ MPa; the time-dependent fiber/matrix interface debonding length increases from $l_{\rm d}/r_{\rm f} = 6.5$ to 11.8; the time-dependent fiber/matrix interface debonding ratio increases from $2l_{\rm d}/l_{\rm c} = 0.9$ to 54.6%; and the time-dependent fiber/matrix interface oxidation ratio decreases from $\zeta/l_{\rm d} = 60.4$ to 33.6%.

When the oxidation time is t = 2 h, the time-dependent matrix multicracking density increases from $\varphi = 0.06$ mm at the first matrix cracking stress of $\sigma_{\rm mc} = 201$ MPa to $\varphi = 2.4$ mm at the saturation matrix cracking stress of $\sigma_{\rm sat} = 320$ MPa; the time-dependent fiber/matrix interface debonding length increases from $l_{\rm d}/r_{\rm f} =$

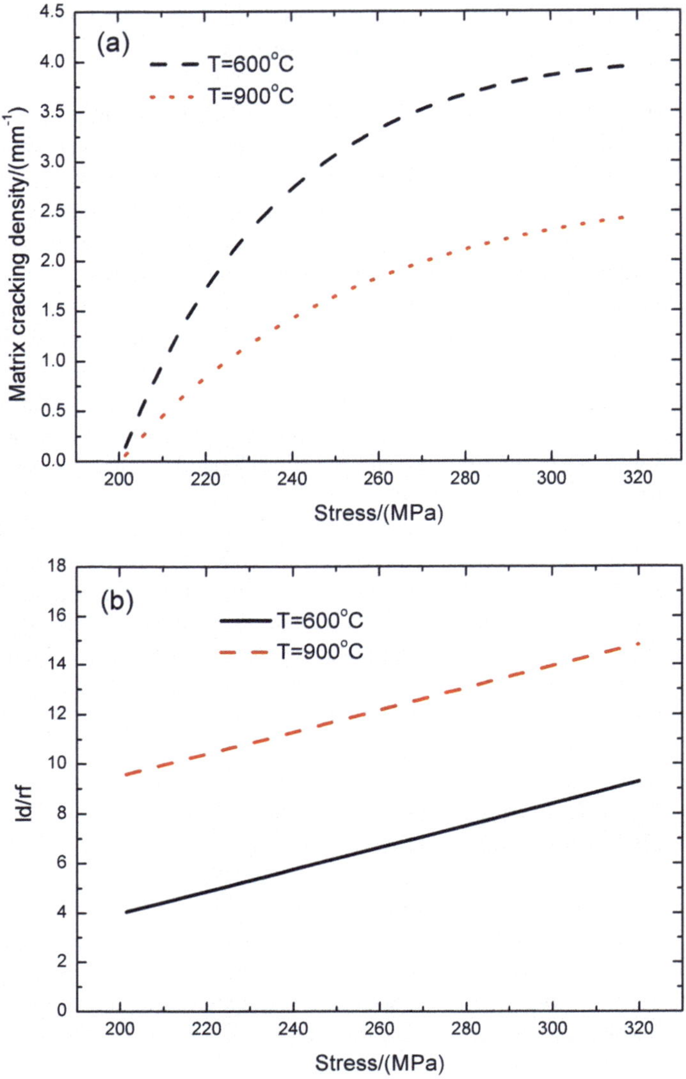

Fig. 2.15 Effect of the oxidation temperature (i.e., $T = 600$ and 900 °C) on **a** the time-dependent matrix multicracking density versus applied stress curves; **b** the time-dependent fiber/matrix interface debonding length versus applied stress curves; **c** the time-dependent fiber/matrix interface debonding ratio versus applied stress curves; and **d** the time-dependent fiber/matrix interface oxidation ratio versus applied stress curves

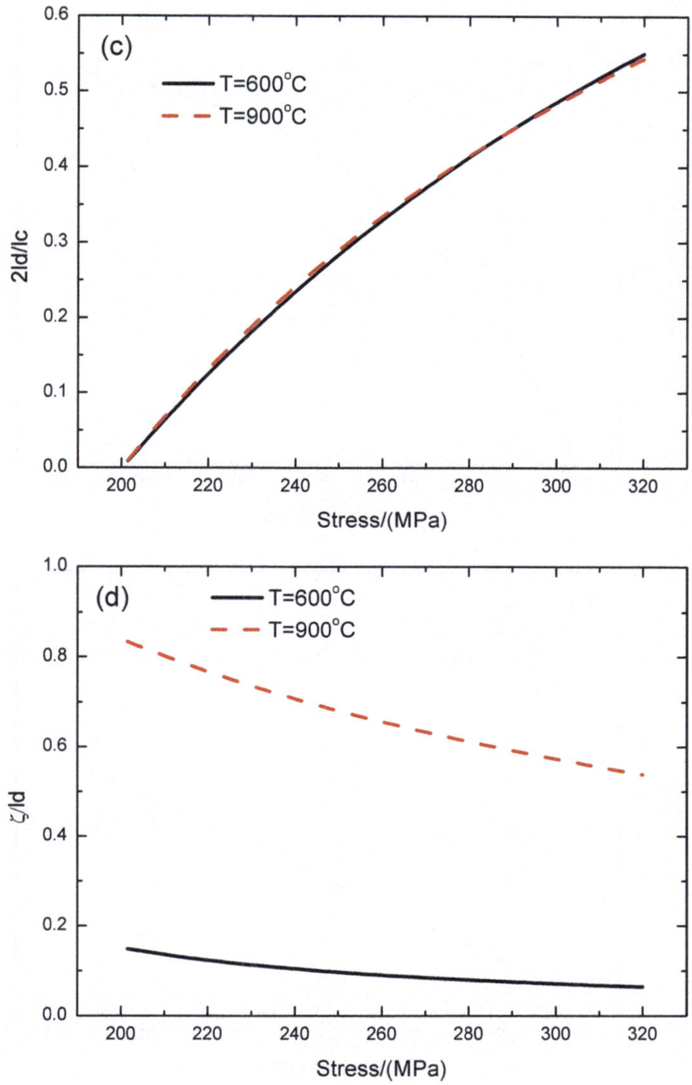

Fig. 2.15 (continued)

9.5 to 14.7; the time-dependent fiber/matrix interface debonding ratio increases from $2l_d/l_c = 0.95$ to 54.3%; and the time-dependent fiber/matrix interface oxidation ratio decreases from $\zeta/l_d = 83.1$ to 53.6%.

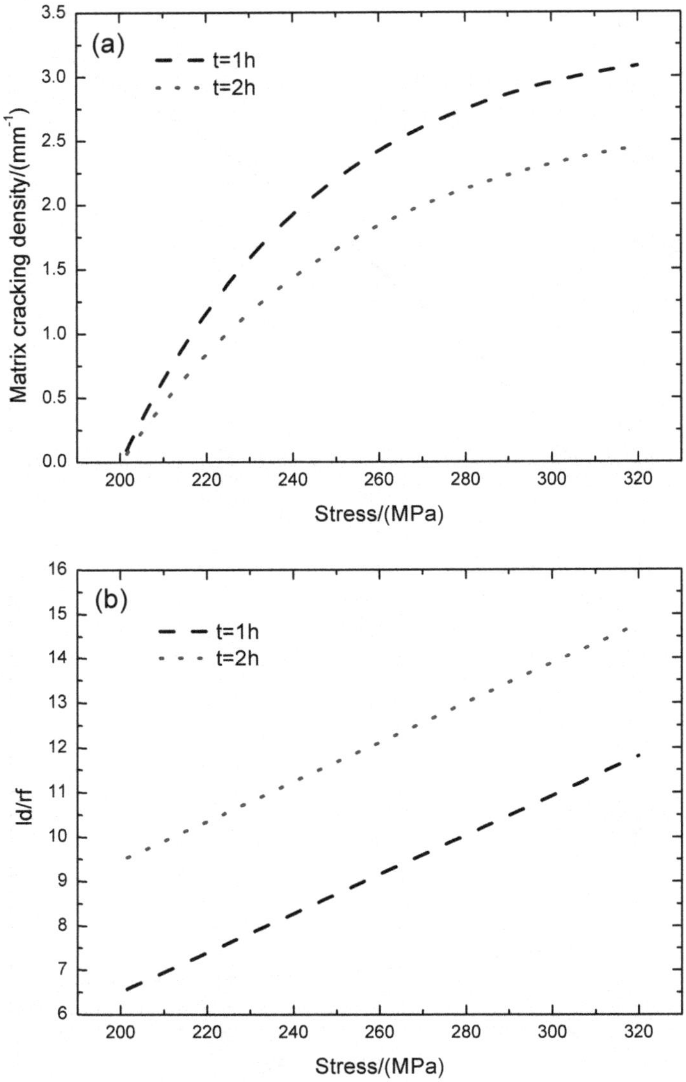

Fig. 2.16 Effect of the oxidation time (i.e., t = 1 and 2 h) on **a** the time-dependent matrix multicracking density versus applied stress curves; **b** the time-dependent fiber/matrix interface debonding length versus applied stress curves; **c** the time-dependent fiber/matrix interface debonding ratio versus applied stress curves; and **d** the time-dependent fiber/matrix interface oxidation ratio versus applied stress curves

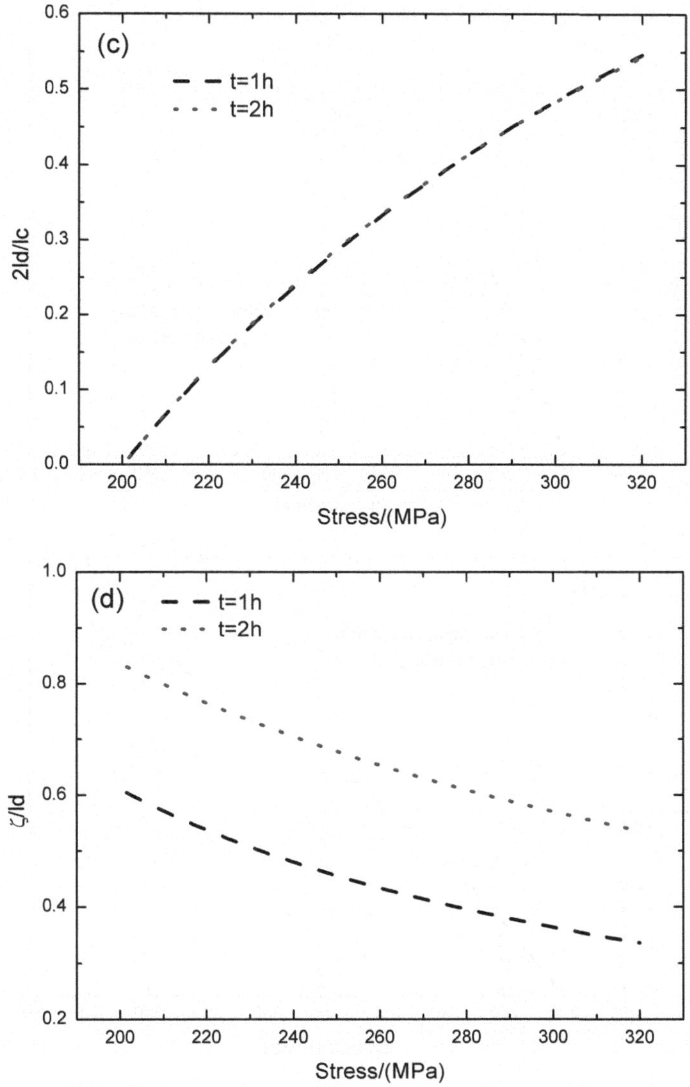

Fig. 2.16 (continued)

2.3.4.6 Comparisons of Matrix Multicracking Evolution With/Without Oxidation

Comparisons of matrix multicracking density and fiber/matrix interface debonding ratio with and without oxidation are shown in Fig. 2.17.

Without considering fiber/matrix interface oxidation, the matrix multicracking density increases from $\varphi = 0.13$ mm at the first matrix cracking stress of $\sigma_{mc} =$

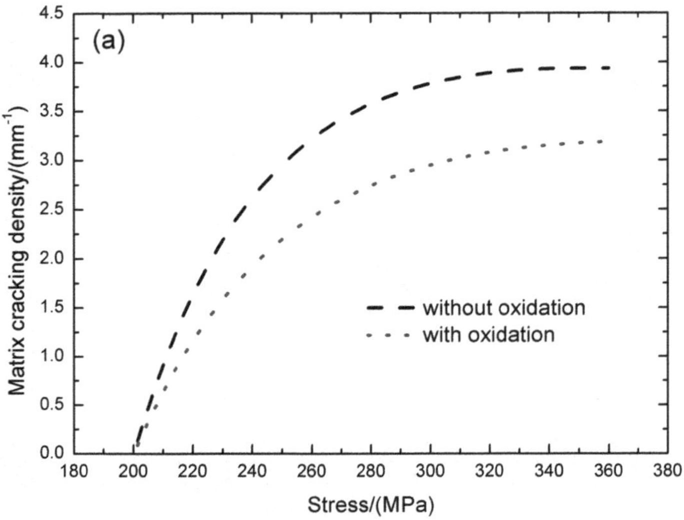

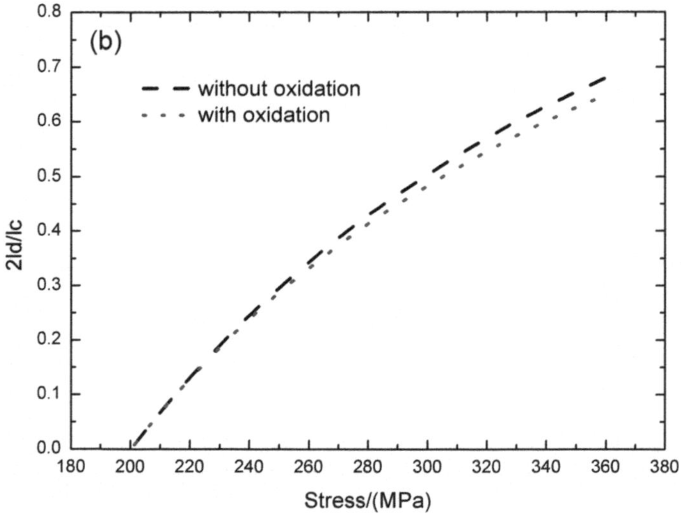

Fig. 2.17 Comparisons of matrix multicracking evolution with/without oxidation for **a** the matrix multicracking density versus applied stress curves; and **b** the fiber/matrix interface debonding ratio versus applied stress curves

201 MPa to $\varphi = 3.9$ mm at the saturation matrix cracking stress of $\sigma_{\text{sat}} = 360$ MPa; and the fiber/matrix interface debonding ratio increases from $2l_d/l_c = 0.92$ to 68%.

With considering fiber/matrix interface oxidation, the time-dependent matrix multicracking density increases from $\varphi = 0.09$ mm at the first matrix cracking stress of $\sigma_{\text{mc}} = 201$ MPa to $\varphi = 3.1$ mm at the saturation matrix cracking stress of $\sigma_{\text{sat}} =$

360 MPa; and the time-dependent fiber/matrix interface debonding ratio increases from $2l_d/l_c = 0.92$ to 64.9%.

With fiber/matrix interface oxidation, the matrix multicracking density and the fiber/matrix interface debonding ratio decrease.

2.3.5 *Experimental Comparisons*

The experimental and predicted time-dependent matrix multicracking density, fiber/matrix interface debonding length, and broken fiber fraction versus applied stress for different CMCs, i.e., unidirectional C/SiC, SiC/CAS (Pryce and Smith 1992), SiC/borosilicate (Okabe et al. 1999), and mini-SiC/SiC (Zhang et al. 2016) composites, are predicted using the present analysis, as shown in Figs. 2.18, 2.19, 2.20, 2.21 and 2.22.

2.3.5.1 C/SiC Composite

For the C/SiC composite without oxidation, the matrix multicracking evolution starts from the applied stress of $\sigma_{mc} = 100$ MPa and approaches to saturation at the applied stress of $\sigma_{sat} = 220$ MPa; the matrix multicracking density increases from $\varphi = 2.0$ mm to the saturation value of $\varphi = 8.5$ mm; and the fiber/matrix interface debonding length increases from $2l_d/l_c = 0.05\%$ at $\sigma_{mc} = 100$ MPa to $2l_d/l_c = 42.7\%$ at $\sigma_{sat} = 220$ MPa. With oxidation, the time-dependent matrix multicracking density increases from $\varphi = 0.6$ mm at $\sigma_{mc} = 100$ MPa to $\varphi = 6.4$ mm at $\sigma_{sat} = 190$ MPa; and the time-dependent fiber/matrix interface debonding ratio increases from $2l_d/l_c = 0.6\%$ at $\sigma_{mc} = 100$ MPa to $2l_d/l_c = 38\%$ at $\sigma_{sat} = 220$ MPa, as shown in Fig. 2.18.

2.3.5.2 SiC/CAS Composite

For the SiC/CAS composite without oxidation, the matrix multicracking evolution starts from the applied stress of $\sigma_{mc} = 140$ MPa and approaches to saturation at the applied stress of $\sigma_{sat} = 288$ MPa; the matrix multicracking density increases from $\varphi = 0.7$ mm to the saturation value of $\varphi = 7$ mm; and the fiber/matrix interface debonding length increases from $2l_d/l_c = 0.7\%$ at 160 MPa to $2l_d/l_c = 79\%$ at 360 MPa. With oxidation, the time-dependent matrix multicracking density increases from $\varphi = 0.13$ mm at 160 MPa to $\varphi = 4.8$ mm at 332 MPa; and the time-dependent fiber/matrix interface debonding ratio increases from $2l_d/l_c = 0.86\%$ at 160 MPa to $2l_d/l_c = 72\%$ at 360 MPa, as shown in Fig. 2.19.

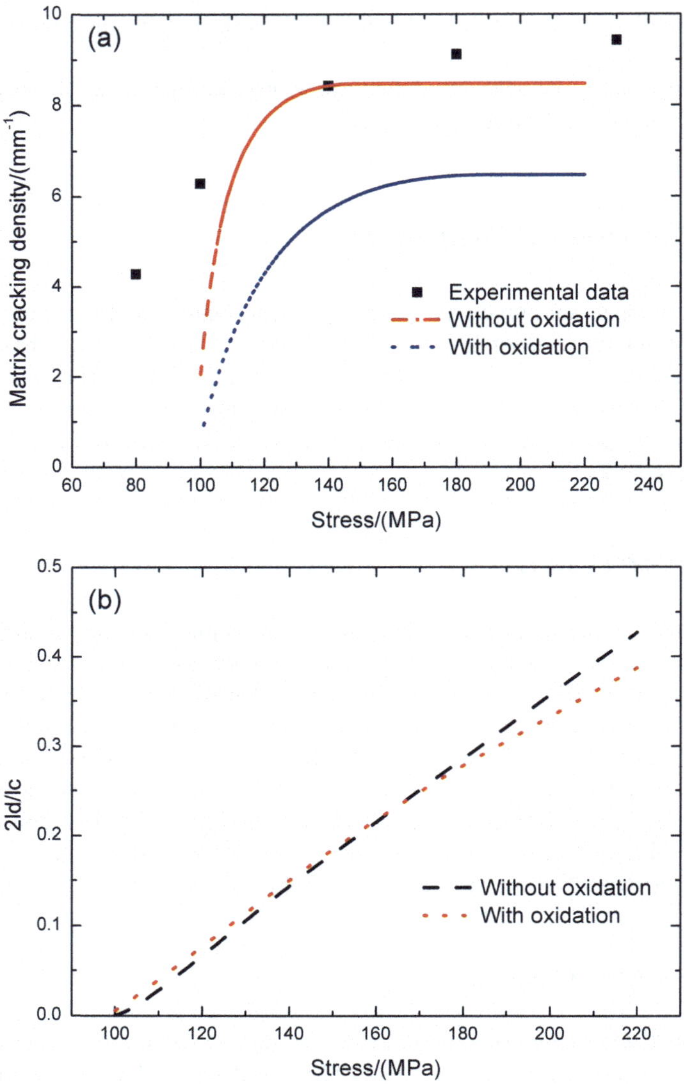

Fig. 2.18 **a** Experimental and predicted matrix multicracking density versus applied stress curves; **b** the fiber/matrix interface debonding ratio versus applied stress curves of unidirectional C/SiC composite with/without oxidation

2.3.5.3 SiC/Borosilicate Composite

For the SiC/borosilicate composite without oxidation, the matrix multicracking evolution starts from the applied stress of $\sigma_{\mathrm{mc}} = 220$ MPa and approaches saturation at the applied stress of $\sigma_{\mathrm{sat}} = 360$ MPa; the matrix cracking density increases from

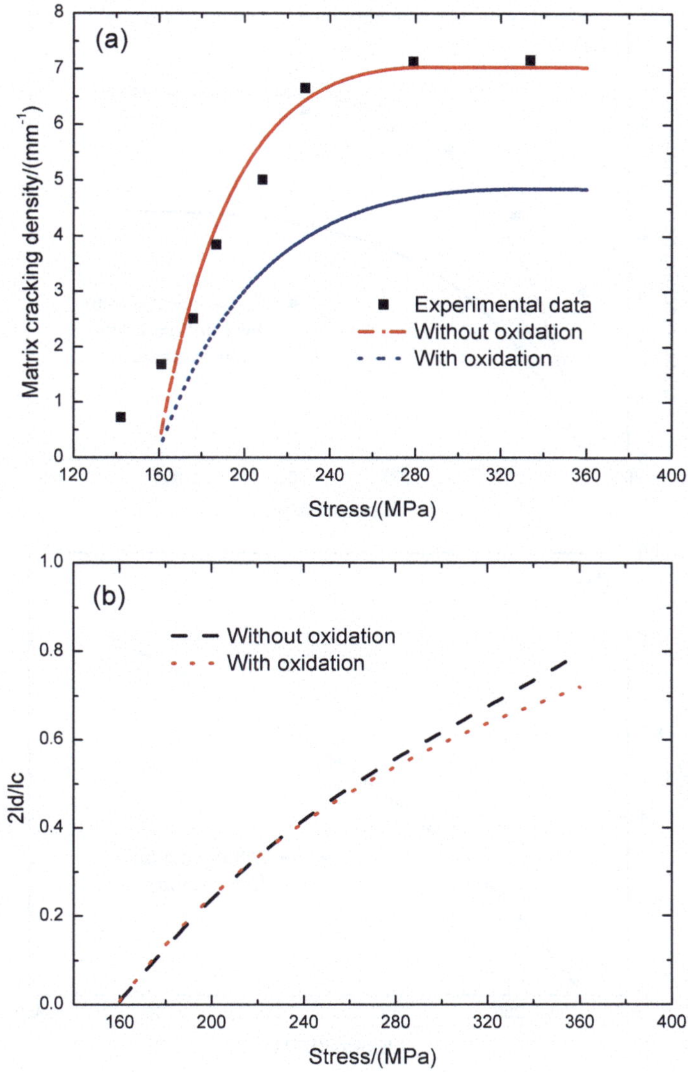

Fig. 2.19 **a** Experimental and predicted matrix multicracking density versus applied stress curves; **b** the fiber/matrix interface debonding ratio versus applied stress curves of unidirectional SiC/CAS composite with/without oxidation

$\varphi = 0.2$ mm to the saturation value of $\varphi = 6.5$ mm; and the fiber/matrix interface debonding length increases from $2l_d/l_c = 0.8\%$ at $\sigma_{mc} = 220$ MPa to 82% at 420 MPa. With oxidation, the time-dependent matrix multicracking density increases from $\varphi = 0.1$ mm at $\sigma_{mc} = 220$ MPa to $\varphi = 4.4$ mm at 414 MPa; and the time-dependent fiber/matrix interface debonding ratio increases from $2l_d/l_c = 0.92\%$ at $\sigma_{mc} = 220$ MPa to $2l_d/l_c = 78\%$ at 420 MPa, as shown in Fig. 2.20.

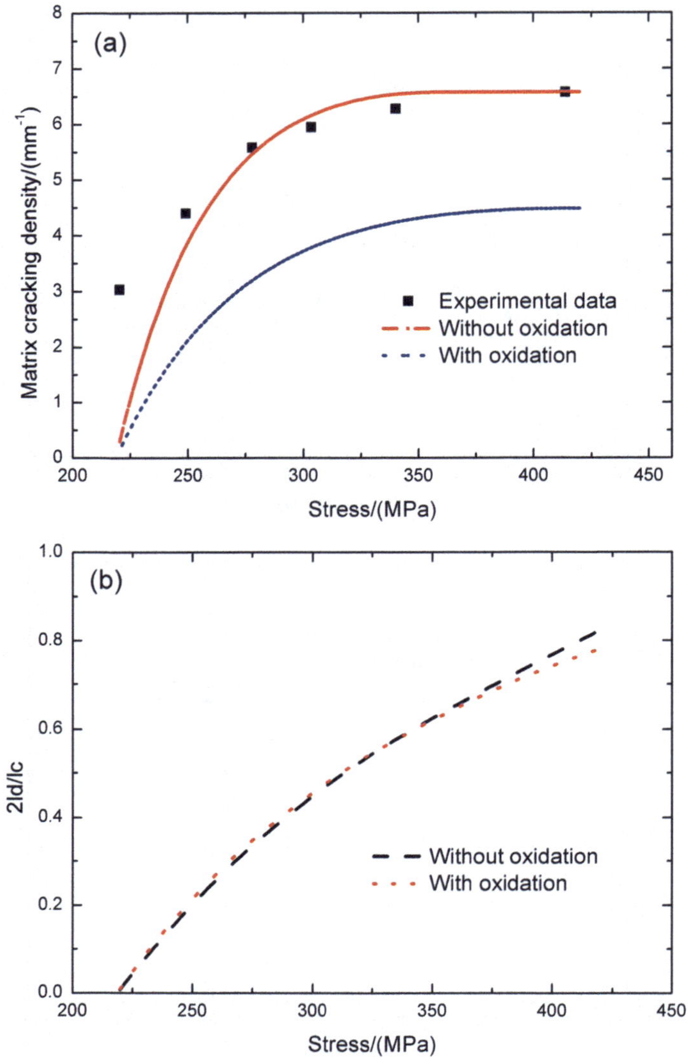

Fig. 2.20 **a** Experimental and predicted matrix multicracking density versus applied stress curves; **b** the fiber/matrix interface debonding ratio versus applied stress curves of unidirectional SiC/borosilicate composite with/without oxidation

2.3.5.4 Mini-SiC/SiC Composite

For the mini-SiC/SiC composite without oxidation, the matrix multicracking evolution starts from the applied stress of $\sigma_{\text{mc}} = 135$ MPa and approaches saturation at the applied stress of $\sigma_{\text{sat}} = 250$ MPa; the matrix cracking density increases from $\varphi = 0.4$ mm to the saturation value of $\varphi = 2.4$ mm; and the fiber/matrix interface

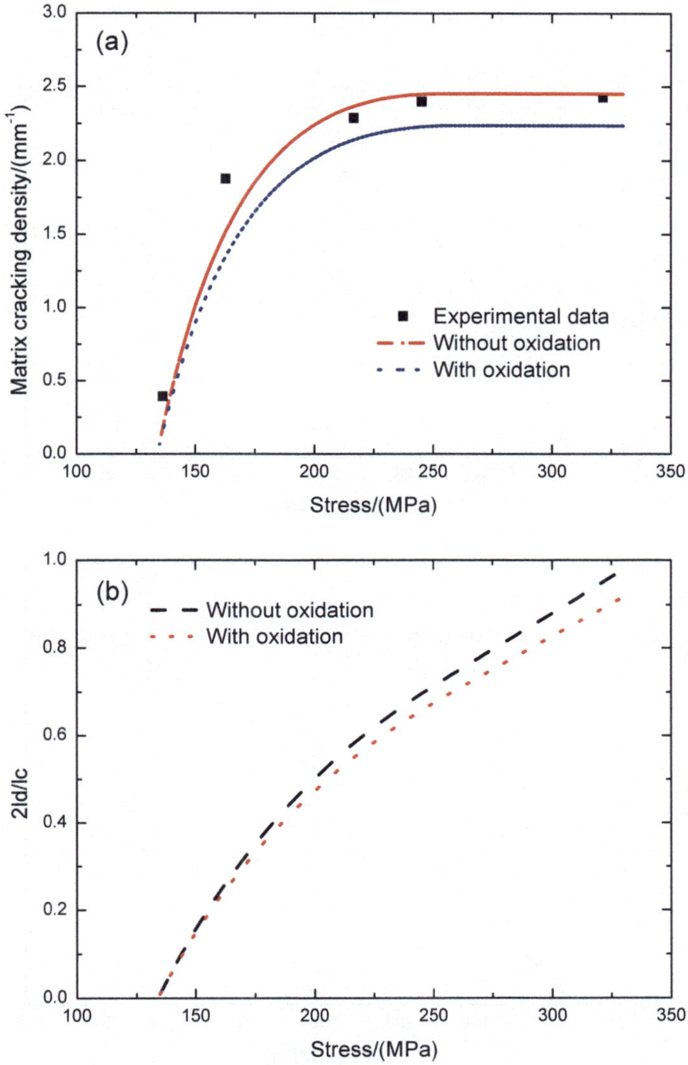

Fig. 2.21 **a** Experimental and predicted matrix multicracking density versus applied stress curves; **b** the fiber/matrix interface debonding ratio versus applied stress curves of mini-SiC/SiC composite with/without oxidation

debonding length increases from $2l_d/l_c = 1\%$ at 135 MPa to $2l_d/l_c = 98\%$ at 330 MPa. With oxidation, the time-dependent matrix multicracking density increases from $\varphi =$ 0.06 mm at 135 MPa to $\varphi = 2.2$ mm at 260 MPa; and the time-dependent fiber/matrix interface debonding ratio increases from $2l_d/l_c = 0.99\%$ at 135 MPa to $2l_d/l_c = 91.8\%$ at 330 MPa, as shown in Fig. 2.21.

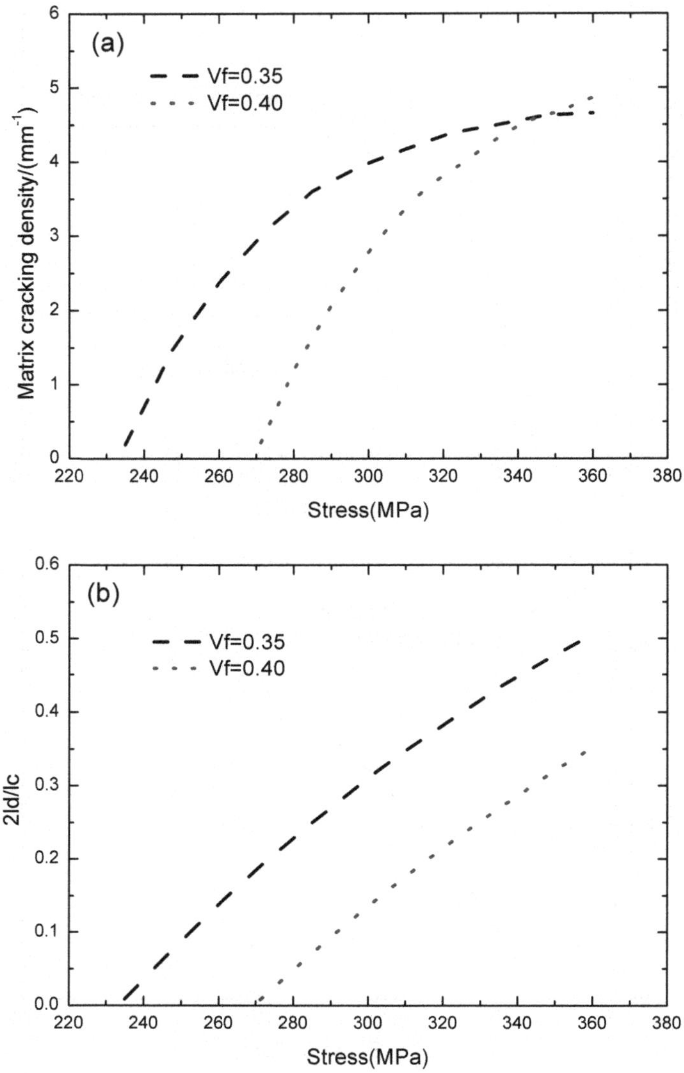

Fig. 2.22 Effect of the fiber volume ($V_f = 0.35$ and 0.4) on **a** the time-dependent matrix multicracking density versus applied stress curves; **b** the time-dependent fiber/matrix interface debonding ratio versus applied stress curves; **c** the time-dependent fiber/matrix interface oxidation ratio versus applied stress curves; and **d** the time-dependent broken fiber fraction versus applied cycles curves

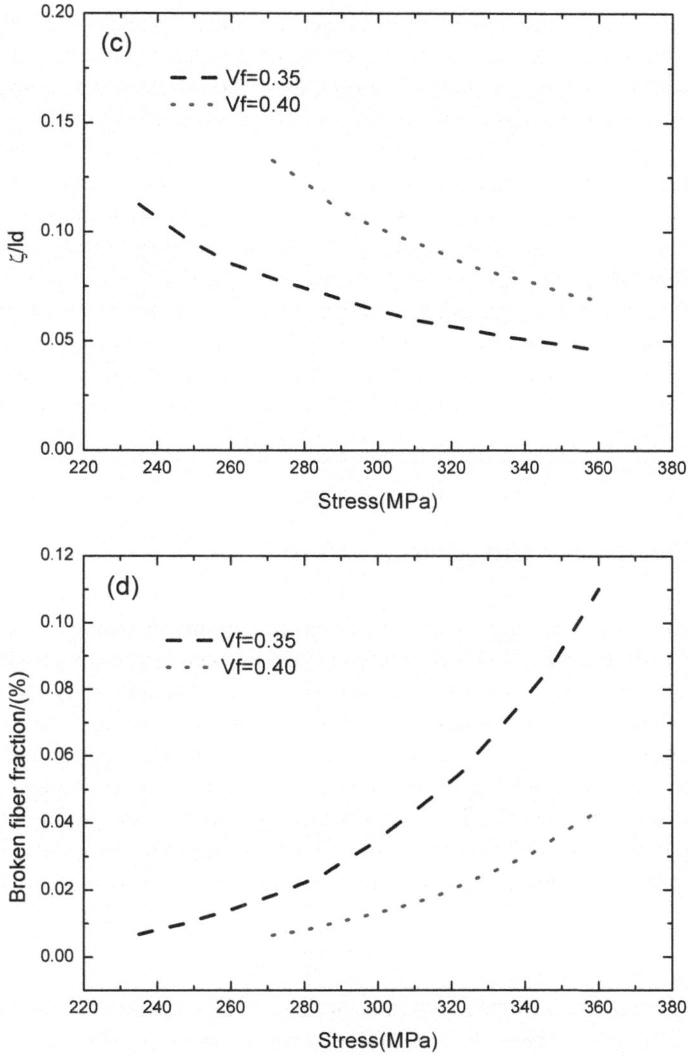

Fig. 2.22 (continued)

2.4 Time-Dependent Matrix Multicracking Evolution of Fiber-Reinforced Ceramic-Matrix Composites Considering Interface and Fiber Oxidation

In this section, the time- and stress-dependent matrix multiple fracture of fiber-reinforced CMCs with fiber oxidation and fracture is investigated. The shear-lag model is combined with the fiber oxidation and fracture models, and the fiber/matrix interface debonding criterion is adopted to determine the microstress field of the damaged fiber-reinforced CMCs. The effects of the fiber volume fraction and interface shear stress in the debonding and oxidation region, the interface debonding energy, the oxidation temperature, and time on the stress-dependent matrix multiple fracture, interface debonding and oxidation, and fiber fracture are discussed. When the fiber volume fraction and the interface shear stress in the debonding region increased, the matrix first cracking stress and saturation cracking stress increased, the saturation matrix cracking space decreased, and the matrix multiple fracture evolution rate increased. When the interface debonding energy and the interface shear stress in the oxidation region increased, the matrix cracking density increased. When the oxidation temperature and oxidation duration increased, the matrix cracking density decreased. The experimental matrix multiple fracture of unidirectional C/SiC, SiC/SiC, mini-SiC/SiC, SiC/CAS, SiC/CAS-II, and SiC/borosilicate composites is predicted.

2.4.1 Time-Dependent Stress Analysis

The oxidation of fibers is assumed to be controlled by the diffusion of oxygen gas through the matrix cracks. When the oxidizing gas enters the composite, a sequence of events is triggered, starting with the oxidation of the fiber. For simplicity, it is assumed that the Weibull and elastic moduli of the fiber remain constant, and the only effect of oxidation is to decrease the strengths of the fibers. The time-dependent strength of the fiber would be controlled by surface defects resulting from the oxidation, with the thickness of the oxidized layer representing the size of the average strength-controlling flaw (Naslain 2004). According to linear elastic fracture mechanics, the relationship between the strength and flaw size is determined as (Lara-Curzio 1999):

$$K_{\mathrm{IC}} = Y\sigma_0\sqrt{a} \tag{2.39}$$

where K_{IC} denotes the critical stress intensity factor, Y is a geometric parameter, σ_0 is the fiber strength, and a is the size of the strength-controlling flaw.

Assuming that the oxidation of the fibers is controlled by the diffusion of oxygen through the oxidized layer, the oxidized layer will grow on the fiber surface according to the following equation (Lara-Curzio 1999):

$$\alpha = \sqrt{kt} \tag{2.40}$$

where α is the thickness of the oxidized layer at time t and k is the parabolic rate constant.

Assuming the fracture toughness of the fiber remains constant and that the fiber strength σ_0 is related to the mean oxidized layer thickness according to Eq. (2.40), i.e., $a = \alpha$, the time dependence of the fiber strength is expressed as (Lara-Curzio 1999):

$$\sigma_0(t) = \sigma_0, t \leq \frac{1}{k}\left(\frac{K_{\mathrm{IC}}}{Y\sigma_0}\right)^4, \tag{2.41}$$

$$\sigma_0(t) = \frac{K_{\mathrm{IC}}}{Y\sqrt[4]{kt}}, t > \frac{1}{k}\left(\frac{K_{\mathrm{IC}}}{Y\sigma_0}\right)^4. \tag{2.42}$$

The two-parameter Weibull model is adopted to describe the fiber strength distribution, and the global load sharing (GLS) assumption is used to determine the load distribution between the intact and fractured fibers (Curtin 1991):

$$\frac{\sigma}{V_{\mathrm{f}}} = \Phi[1 - P(\Phi)] + \langle\Phi_{\mathrm{b}}\rangle P(\Phi) \tag{2.43}$$

where V_{f} denotes the fiber volume fraction, Φ denotes the load carried by the intact fibers, $\langle\Phi_{\mathrm{b}}\rangle$ denotes the load carried by the broken fiber, and $P(\Phi)$ denotes the fiber failure probability,

$$P(\Phi) = \eta P_a(\Phi) + (1 - \eta)P_b(\Phi) + P_c(\Phi), \tag{2.44}$$

where η denotes the fraction of oxidized fibers in the oxidation region, and $P_a(\Phi)$, $P_b(\Phi)$, and $P_c(\Phi)$ denote the fracture probabilities of the oxidized fibers in the oxidation region, unoxidized fibers in the oxidation region, and interface debonding region, respectively (Li 2018c), defined as follows:

$$P_a(\Phi) = 1 - \exp\left\{-\frac{1}{m+1}\frac{r_{\mathrm{f}}\Phi^{m+1}}{\tau_{\mathrm{f}}l_0[\sigma_0(t)]^m}\left[1 - \left(1 - \frac{2\tau_{\mathrm{f}}}{r_{\mathrm{f}}\Phi}\zeta\right)^{m+1}\right]\right\} \tag{2.45}$$

$$P_b(\Phi) = 1 - \exp\left\{-\frac{1}{m+1}\frac{r_{\mathrm{f}}\Phi^{m+1}}{\tau_{\mathrm{f}}l_0(\sigma_0)^m}\left[1 - \left(1 - \frac{2\tau_{\mathrm{f}}}{r_{\mathrm{f}}\Phi}\zeta\right)^{m+1}\right]\right\} \tag{2.46}$$

$$P_c(\Phi) = 1 - \exp\left\{-\frac{1}{m+1}\frac{r_{\mathrm{f}}\Phi^{m+1}}{l_0\tau_{\mathrm{i}}(\sigma_0)^m}\left[\left(1 - \frac{2\tau_{\mathrm{f}}}{r_{\mathrm{f}}\Phi}\zeta\right)^{m+1} - \left(1 - \frac{2\tau_{\mathrm{f}}}{r_{\mathrm{f}}\Phi}\zeta - \frac{2\tau_{\mathrm{i}}}{r_{\mathrm{f}}\Phi}(l_{\mathrm{d}} - \zeta)\right)^{m+1}\right]\right\} \tag{2.47}$$

where r_{f} denotes the fiber radius, m denotes the fiber Weibull modulus, $\sigma_0(t)$ denotes the oxidized fiber strength, t denotes the oxidation duration, l_{d} denotes the interface debonding length, and l_{s} denotes the slip length over which the fiber stress would decay to zero if not interrupted by the far-field equilibrium stresses. l_{s} is defined as:

$$l_{\mathrm{s}} = \frac{r_{\mathrm{f}}\Phi}{2\tau_{\mathrm{i}}} \tag{2.48}$$

The stress carried by the broken fiber can be given by:

$$\langle \Phi_b \rangle = \Phi\left[\left(\frac{\sigma_c}{\Phi}\right)^{m+1} - \frac{1 - P(\Phi)}{P(\Phi)}\right] \tag{2.49}$$

Substituting Eqs. (2.44) and (2.49) into Eq. (2.43), the stress Φ carried by the intact fibers at the matrix cracking plane can be determined. Substituting the intact fiber stress Φ into Eqs. (2.45)–(2.47), the relationship between the fiber failure probability and applied stress can be determined.

The composite with fiber volume fraction V_f is loaded by a remote uniform stress σ normal to the crack plane. The fiber radius is r_f, and the matrix radius is R (where $R = r_f/V_f^{1/2}$). The length of the unit cell is the half matrix crack spacing $l_c/2$. In the interface oxidation and debonding region, the fiber and matrix axial stress distributions can be defined as:

$$\sigma_f(z) = \begin{cases} \Phi - \dfrac{2\tau_f}{r_f}z, \ z \in [0, \zeta] \\ \Phi - \dfrac{2\tau_f}{r_f}\zeta - \dfrac{2\tau_i}{r_f}(z - \zeta), \ z \in [\zeta, l_d] \end{cases}, \tag{2.50}$$

$$\sigma_m(z) = \begin{cases} 2\dfrac{V_f}{V_m}\dfrac{\tau_f}{r_f}z, \ z \in (0, \zeta) \\ 2\dfrac{V_f}{V_m}\dfrac{\tau_f}{r_f}\zeta + 2\dfrac{V_f}{V_m}\dfrac{\tau_i}{r_f}(z - \zeta), \ z \in (\zeta, l_d) \end{cases}. \tag{2.51}$$

For the fiber/matrix bonded region, the fiber and matrix axial stresses and the fiber/matrix interfacial shear stress can be determined using the composite cylinder model adopted by Budiansky et al. (1986). The fiber and matrix axial stresses and the fiber/matrix interface shear stress in the interface bonded region then become:

$$\sigma_f(z) = \sigma_{fo} + \left[T - \frac{2\tau_f}{r_f}\zeta - \frac{2\tau_i}{r_f}(l_d - \zeta) - \sigma_{fo}\right]\exp\left(-\rho\frac{z - l_d}{r_f}\right) \tag{2.52}$$

$$\sigma_m(z) = \sigma_{mo} + \left[2\frac{V_f}{V_m}\frac{\tau_f}{r_f}\zeta + 2\frac{V_f}{V_m}\frac{\tau_i}{r_f}(l_d - \zeta) - \sigma_{mo}\right]\exp\left(-\rho\frac{z - l_d}{r_f}\right) \tag{2.53}$$

$$\tau_i(z) = \frac{\rho}{2}\left[T - \frac{2\tau_f}{r_f}\zeta - \frac{2\tau_i}{r_f}(l_d - \zeta) - \sigma_{fo}\right]\exp\left(-\rho\frac{z - l_d}{r_f}\right), \tag{2.54}$$

where ρ denotes the shear-lag model parameter, and

$$\sigma_{fo} = \frac{E_f}{E_c}\sigma + E_f(\alpha_c - \alpha_f)\Delta T \tag{2.55}$$

$$\sigma_{mo} = \frac{E_m}{E_c}\sigma + E_m(\alpha_c - \alpha_m)\Delta T \tag{2.56}$$

where E_f, E_m, and E_c denote the fiber, matrix, and composite elastic moduli, respectively; α_f, α_m, and α_c denote the fiber, matrix, and composite thermal expansion coefficients, respectively; and ΔT denotes the temperature difference between the fabricated temperature T_0 and testing temperature T_1 ($\Delta T = T_1 - T_0$).

2.4.2 *Interface Debonding*

When matrix cracking propagates to the fiber/matrix interface, it deflects along the interface. There are two approaches to the problem of fiber/matrix interface debonding: the shear stress and the fracture mechanics approaches. The shear stress approach is based upon a maximum shear stress criterion in which interface debonding occurs as the shear stress reaches the shear strength of the interface (Hsueh 1996). In contrast, the fracture mechanics approach treats interface debonding as a particular crack propagation problem in which interface debonding occurs as the strain energy release rate of the fiber/matrix interface reaches the debonding toughness (Gao et al. 1988). It has been proven that the fracture mechanics approach is preferred to the shear stress approach for interface debonding (Sun and Singh 1998). The fracture mechanics approach is adopted to determine the fiber/matrix interface debonding length (Gao et al. 1988), expressed as:

$$\xi_{\mathrm{d}} = -\frac{F}{4\pi r_{\mathrm{f}}}\frac{\partial w_{\mathrm{f}}(z=0)}{\partial l_{\mathrm{d}}} - \frac{1}{2}\int_{0}^{l_{\mathrm{c}}/2} \tau_{\mathrm{i}}(z)\frac{\partial v(z)}{\partial l_{\mathrm{d}}}\mathrm{d}z \tag{2.57}$$

where F (where $F = \pi r_{\mathrm{f}}^2\sigma/V_{\mathrm{f}}$) denotes the fiber load at the matrix cracking plane, $w_{\mathrm{f}}(z=0)$ denotes the fiber axial displacement at the matrix cracking plane, and $v(z)$ denotes the relative displacement between the fibers and the matrix.

The fiber and matrix axial displacements of $w_{\mathrm{f}}(z)$ and $w_{\mathrm{m}}(z)$ can be expressed as:

$$\begin{aligned} w_{\mathrm{f}}(z) &= \int_{z}^{\frac{l_{\mathrm{c}}}{2}} \frac{\sigma_{\mathrm{f}}(z)}{E_{\mathrm{f}}}\mathrm{d}x \\ &= \frac{T}{E_{\mathrm{f}}}(l_{\mathrm{d}} - z) - \frac{\tau_{\mathrm{f}}}{r_{\mathrm{f}}E_{\mathrm{f}}}\left(2\zeta l_{\mathrm{d}} - z^2 - \zeta^2\right) - \frac{\tau_{\mathrm{i}}}{r_{\mathrm{f}}E_{\mathrm{f}}}(l_{\mathrm{d}} - \zeta)^2 + \frac{\sigma_{\mathrm{fo}}}{E_{\mathrm{f}}}\left(\frac{l_{\mathrm{c}}}{2} - l_{\mathrm{d}}\right), \\ &+ \frac{r_{\mathrm{f}}}{\rho E_{\mathrm{f}}}\left[T - \sigma_{\mathrm{fo}} - \frac{2\tau_{\mathrm{f}}}{r_{\mathrm{f}}}\zeta - \frac{2\tau_{\mathrm{i}}}{r_{\mathrm{f}}}(l_{\mathrm{d}} - \zeta)\right]\left[1 - \exp\left(-\rho\frac{l_{\mathrm{c}}/2 - l_{\mathrm{d}}}{r_{\mathrm{f}}}\right)\right] \end{aligned} \tag{2.58}$$

$$\begin{aligned} w_{\mathrm{m}}(z) &= \int\limits_{z}^{\frac{l_c}{2}} \frac{\sigma_{\mathrm{m}}(z)}{E_{\mathrm{m}}} \mathrm{d}z \\ &= \frac{V_{\mathrm{f}}\tau_{\mathrm{f}}}{r_{\mathrm{f}}V_{\mathrm{m}}E_{\mathrm{m}}}\left(2\zeta l_{\mathrm{d}} - \zeta^2 - z^2\right) + \frac{V_{\mathrm{f}}\tau_{\mathrm{i}}}{r_{\mathrm{f}}V_{\mathrm{m}}E_{\mathrm{m}}}(l_{\mathrm{d}} - \zeta)^2 + \frac{\sigma_{\mathrm{mo}}}{E_{\mathrm{m}}}\left(\frac{l_{\mathrm{c}}}{2} - l_{\mathrm{d}}\right) \\ &\quad - \frac{r_{\mathrm{f}}V_{\mathrm{f}}}{\rho V_{\mathrm{m}}E_{\mathrm{m}}}\left[T - \sigma_{\mathrm{fo}} - \frac{2\tau_{\mathrm{f}}}{r_{\mathrm{f}}}\zeta - \frac{2\tau_{\mathrm{i}}}{r_{\mathrm{f}}}(l_{\mathrm{d}} - \zeta)\right]\left[1 - \exp\left(-\rho\frac{l_{\mathrm{c}}/2 - l_{\mathrm{d}}}{r_{\mathrm{f}}}\right)\right] \end{aligned} \tag{2.59}$$

The relative displacement of $v(z)$ between the fiber and the matrix is given by:

$$\begin{aligned} v(z) &= |w_{\mathrm{f}}(z) - w_{\mathrm{m}}(z)| \\ &= \frac{T}{E_{\mathrm{f}}}(l_{\mathrm{d}} - z) - \frac{E_{\mathrm{c}}\tau_{\mathrm{f}}}{r_{\mathrm{f}}V_{\mathrm{m}}E_{\mathrm{f}}E_{\mathrm{m}}}\left(2\zeta l_{\mathrm{d}} - \zeta^2 - z^2\right) - \frac{E_{\mathrm{c}}\tau_{\mathrm{i}}}{r_{\mathrm{f}}V_{\mathrm{m}}E_{\mathrm{f}}E_{\mathrm{m}}}(l_{\mathrm{d}} - \zeta)^2 \\ &\quad + \frac{r_{\mathrm{f}}}{\rho E_{\mathrm{f}}}\left[T - \sigma_{\mathrm{fo}} - \frac{2\tau_{\mathrm{f}}}{r_{\mathrm{f}}}\zeta - \frac{2\tau_{\mathrm{i}}}{r_{\mathrm{f}}}(l_{\mathrm{d}} - \zeta)\right]\left[1 - \exp\left(-\rho\frac{l_{\mathrm{c}}/2 - l_{\mathrm{d}}}{r_{\mathrm{f}}}\right)\right] \\ &\quad + \frac{r_{\mathrm{f}}V_{\mathrm{f}}}{\rho V_{\mathrm{m}}E_{\mathrm{m}}}\left[T - \sigma_{\mathrm{fo}} - \frac{2\tau_{\mathrm{f}}}{r_{\mathrm{f}}}\zeta - \frac{2\tau_{\mathrm{i}}}{r_{\mathrm{f}}}(l_{\mathrm{d}} - \xi)\right]\left[1 - \exp\left(-\rho\frac{l_{\mathrm{c}}/2 - l_{\mathrm{d}}}{r_{\mathrm{f}}}\right)\right] \end{aligned} \tag{2.60}$$

Substituting $w_{\mathrm{f}}(z = 0)$ and $v(z)$ into Eq. (2.57) leads to the equation

$$\begin{aligned} &\frac{r_{\mathrm{f}}T^2}{4E_{\mathrm{f}}} - \frac{T\tau_{\mathrm{f}}}{E_{\mathrm{f}}}\zeta - \frac{T\tau_{\mathrm{i}}}{E_{\mathrm{f}}}(l_{\mathrm{d}} - \zeta) - \frac{r_{\mathrm{f}}\sigma T}{4E_{\mathrm{c}}} - \frac{r_{\mathrm{f}}T\tau_{\mathrm{i}}}{2\rho E_{\mathrm{f}}} + \frac{E_{\mathrm{c}}\tau_{\mathrm{f}}^2}{r_{\mathrm{f}}V_{\mathrm{m}}E_{\mathrm{f}}E_{\mathrm{m}}}\zeta^2 \\ &+ \frac{2E_{\mathrm{c}}\tau_{\mathrm{f}}\tau_{\mathrm{i}}}{r_{\mathrm{f}}V_{\mathrm{m}}E_{\mathrm{f}}E_{\mathrm{m}}}\zeta(l_{\mathrm{d}} - \zeta) + \frac{E_{\mathrm{c}}\tau_{\mathrm{f}}\tau_{\mathrm{i}}}{\rho V_{\mathrm{m}}E_{\mathrm{f}}E_{\mathrm{m}}}\zeta \\ &+ \frac{E_{\mathrm{c}}\tau_{\mathrm{i}}^2}{r_{\mathrm{f}}V_{\mathrm{m}}E_{\mathrm{f}}E_{\mathrm{m}}}(l_{\mathrm{d}} - \zeta)^2 + \frac{E_{\mathrm{c}}\tau_{\mathrm{i}}^2}{\rho V_{\mathrm{m}}E_{\mathrm{f}}E_{\mathrm{m}}}(l_{\mathrm{d}} - \zeta) - \xi_{\mathrm{d}} = 0 \end{aligned} \tag{2.61}$$

Solving Eq. (2.61) yields an expression for the fiber/matrix interface debonding length with fiber oxidation:

$$\begin{aligned} l_{\mathrm{d}} = &\left(1 - \frac{\tau_{\mathrm{f}}}{\tau_{\mathrm{i}}}\right)\zeta + \frac{r_{\mathrm{f}}}{2}\left(\frac{V_{\mathrm{m}}E_{\mathrm{m}}T}{E_{\mathrm{c}}\tau_{\mathrm{i}}} - \frac{1}{\rho}\right) - \left[\left(\frac{r_{\mathrm{f}}}{2\rho}\right)^2 + \frac{r_{\mathrm{f}}^2V_{\mathrm{f}}V_{\mathrm{m}}E_{\mathrm{f}}E_{\mathrm{m}}T^2}{4E_{\mathrm{c}}^2\tau_{\mathrm{i}}^2}\right. \\ &\left.\left(\frac{\sigma}{V_{\mathrm{f}}T} - 1\right) + \frac{r_{\mathrm{f}}V_{\mathrm{m}}E_{\mathrm{f}}E_{\mathrm{m}}}{E_{\mathrm{c}}\tau_{\mathrm{i}}^2}\xi_{\mathrm{d}}\right]^{\frac{1}{2}} \end{aligned} \tag{2.62}$$

2.4.3 Matrix Multicracking

Solti et al. (1995) developed the critical matrix strain energy (CMSE) criterion to predict matrix multiple fracture in fiber-reinforced CMCs. The matrix strain energy is given by:

$$U_{\mathrm{m}} = \frac{1}{2E_{\mathrm{m}}}\int\limits_{A_{\mathrm{m}}}\int\limits_{0}^{l_{\mathrm{c}}} \sigma_{\mathrm{m}}^{2}(z)\mathrm{d}z\mathrm{d}A_{\mathrm{m}} \tag{2.63}$$

where A_{m} denotes the cross-sectional area of the matrix in the unit cell. Substituting the matrix axial stresses given by Eqs. (2.14) and (2.16) into Eq. (2.27), the matrix strain energy including matrix multiple fracture and interface partial debonding becomes

$$\begin{aligned}U_{\mathrm{m}} = \frac{A_{\mathrm{m}}}{E_{\mathrm{m}}}\Bigg\{&\frac{4}{3}\left(\frac{V_{\mathrm{f}}}{V_{\mathrm{m}}}\frac{\tau_{\mathrm{f}}}{r_{\mathrm{f}}}\right)^{2}\zeta^{3} + 4\left(\frac{V_{\mathrm{f}}}{V_{\mathrm{m}}}\frac{\tau_{\mathrm{f}}}{r_{\mathrm{f}}}\right)^{2}\zeta^{2}(l_{\mathrm{d}}-\zeta) + 4\left(\frac{V_{\mathrm{f}}}{r_{\mathrm{f}}V_{\mathrm{m}}}\right)^{2}\tau_{\mathrm{f}}\tau_{\mathrm{i}}\zeta(l_{\mathrm{d}}-\zeta)^{2}\\
&+\frac{4}{3}\left(\frac{V_{\mathrm{f}}}{V_{\mathrm{m}}}\frac{\tau_{\mathrm{i}}}{r_{\mathrm{f}}}\right)^{2}(l_{\mathrm{d}}-\zeta)^{3} + \sigma_{\mathrm{mo}}^{2}\left(\frac{l_{\mathrm{c}}}{2}-l_{\mathrm{d}}\right) - 2\sigma_{\mathrm{mo}}\frac{V_{\mathrm{f}}}{V_{\mathrm{m}}}\frac{r_{\mathrm{f}}}{\rho}\left[T-\sigma_{\mathrm{fo}}-\frac{2\tau_{\mathrm{f}}}{r_{\mathrm{f}}}\zeta-\frac{2\tau_{\mathrm{i}}}{r_{\mathrm{f}}}(l_{\mathrm{d}}-\zeta)\right]\\
&\left[1-\exp\left(-\rho\frac{l_{\mathrm{c}}/2-l_{\mathrm{d}}}{r_{\mathrm{f}}}\right)\right] + \frac{r_{\mathrm{f}}}{2\rho}\left(\frac{V_{\mathrm{f}}}{V_{\mathrm{m}}}\right)^{2}\left[T-\sigma_{\mathrm{fo}}-\frac{2\tau_{\mathrm{f}}}{r_{\mathrm{f}}}\zeta-\frac{2\tau_{\mathrm{i}}}{r_{\mathrm{f}}}(l_{\mathrm{d}}-\zeta)\right]^{2}\\
&\left[1-\exp\left(-2\rho\frac{l_{\mathrm{c}}/2-l_{\mathrm{d}}}{r_{\mathrm{f}}}\right)\right]\Bigg\}\end{aligned} \tag{2.64}$$

When the interface completely debonds, the matrix strain energy can be expressed as:

$$\begin{aligned}U_{\mathrm{m}}(\sigma, l_{\mathrm{c}}, l_{\mathrm{d}} = l_{\mathrm{c}}/2) = \frac{A_{\mathrm{m}}}{E_{\mathrm{m}}}\Bigg[&\frac{4}{3}\left(\frac{V_{\mathrm{f}}}{V_{\mathrm{m}}}\frac{\tau_{\mathrm{f}}}{r_{\mathrm{f}}}\right)^{2}\zeta^{3} + 4\left(\frac{V_{\mathrm{f}}}{V_{\mathrm{m}}}\frac{\tau_{\mathrm{f}}}{r_{\mathrm{f}}}\right)^{2}\zeta^{2}(l_{\mathrm{d}}-\zeta)\\
&+4\left(\frac{V_{\mathrm{f}}}{r_{\mathrm{f}}V_{\mathrm{m}}}\right)^{2}\tau_{\mathrm{f}}\tau_{\mathrm{i}}\zeta(l_{\mathrm{d}}-\zeta)^{2} + \frac{4}{3}\left(\frac{V_{\mathrm{f}}}{V_{\mathrm{m}}}\frac{\tau_{\mathrm{i}}}{r_{\mathrm{f}}}\right)^{2}(l_{\mathrm{d}}-\zeta)^{3}\Bigg]\end{aligned} \tag{2.65}$$

Evaluating the matrix strain energy at a critical stress σ_{cr}, the critical matrix strain energy of U_{mcr} can be obtained. The critical matrix strain energy turns to:

$$U_{\mathrm{mcr}} = \frac{1}{2}kA_{\mathrm{m}}l_{0}\frac{\sigma_{\mathrm{mocr}}^{2}}{E_{\mathrm{m}}} \tag{2.66}$$

where k ($k \in [0, 1]$) is the critical matrix strain energy parameter, l_0 is the initial matrix crack spacing, and σ_{mocr} is expressed as:

$$\sigma_{\mathrm{mocr}} = \frac{E_{\mathrm{m}}}{E_{\mathrm{c}}}\sigma_{\mathrm{cr}} + E_{\mathrm{m}}(\alpha_{\mathrm{c}}-\alpha_{\mathrm{m}})\Delta T \tag{2.67}$$

where σ_{cr} denotes the critical stress corresponding to composite proportional limit stress, i.e., the stress at which the stress-strain curve begins to become nonlinear due to damage accumulation of matrix cracks and interface debonding. The critical stress of σ_{cr} is defined to be the ACK matrix cracking stress (Aveston et al. 1971), which is determined using an energy balance criterion. This involves the calculation of the energy balance relationship before and after the formation of a single dominant crack. The ACK matrix cracking stress can be described as (Aveston et al. 1971):

$$\sigma_{cr}^{ACK} = \left(\frac{6V_f^2 E_f E_c^2 \tau_i \xi_m}{r_f V_m E_m^2}\right)^{\frac{1}{3}} - E_c(\alpha_c - \alpha_m)\Delta T \tag{2.68}$$

where ξ_m denotes the matrix fracture energy.

The energy balance relationship to evaluate the matrix multiple fracture takes the form:

$$U_m(\sigma > \sigma_{\triangleright}, l_c, l_d) = U_{mcr}(\sigma_{cr}, l_0) \tag{2.69}$$

The matrix multiple fracture versus the applied stress relationship can be obtained by solving Eq. (2.69) when the critical matrix cracking stress and interface debonding length are determined by Eqs. (2.62) and (2.68).

2.4.4 *Results and Discussion*

The effects of the fiber volume fraction, interface shear stress in the debonding and oxidation regions, interface debonding energy, oxidation temperature, and oxidation duration on the matrix multiple fracture, interface debonding and oxidation, and fiber failure are discussed. The ceramic composite system of SiC/CAS is used for the case study, and its material properties are as follows: $V_f = 30\%$, $E_f = 200$ GPa, $E_m = 97$ GPa, $r_f = 7.5$ μm, $\xi_m = 25$ J/m^2, $\xi_d = 0.8$ J/m^2, $\tau_i = 20$ MPa, $\tau_f = 5$ MPa, $\alpha_f = 4 \times 10^{-5}$ °C, $\alpha_m = 5 \times 10^{-5}$ °C, $\Delta T = -1000$ °C, and $m = 5$.

2.4.4.1 Effect of Fiber Volume on Time-Dependent Matrix Multiple Fracture, Interface Debonding, and Fiber Failure

The time-dependent matrix cracking density, interface debonding ratio, interface oxidation ratio, and broken fiber fraction corresponding to different fiber volumes are shown in Fig. 2.22. When the fiber volume increased, the matrix first cracking stress and saturation cracking stress increased, the saturation matrix cracking space decreased, and the matrix multiple fracture evolution rate increased. The interface debonding ratio and the broken fiber fraction decreased, and the interface oxidation ratio increased.

When the fiber volume is $V_f = 35\%$, the time-dependent matrix multicracking density increased from $\varphi = 0.19$ mm at the first matrix cracking stress of $\sigma_{mc} = 235$ MPa to $\varphi = 4.6$ mm at the saturation matrix cracking stress of $\sigma_{sat} = 360$ MPa. The time-dependent fiber/matrix interface debonding ratio increased from $2l_d/l_c = 0.8$ to $2l_d/l_c = 50.6\%$, the time-dependent fiber/matrix interface oxidation ratio decreased from $\zeta/l_d = 11.2$ to $\zeta/l_d = 4.5\%$, and the time-dependent broken fiber fraction increased from $P = 0.07$ to $P = 0.1\%$.

When the fiber volume is $V_f = 40\%$, the time-dependent matrix multicracking density increased from $\varphi = 0.22$ mm at the first matrix cracking stress of $\sigma_{mc} = 271$ MPa to $\varphi = 4.8$ mm at the saturation matrix cracking stress of $\sigma_{sat} = 360$ MPa, the time-dependent interface debonding ratio increased from $2l_d/l_c = 0.8$ to $2l_d/l_c = 35.2\%$, the time-dependent interface oxidation ratio decreased from $\zeta/l_d = 13.3$ to $\zeta/l_d = 6.8\%$, and the time-dependent broken fiber fraction increased from $P = 0.006$ to $P = 0.04\%$.

2.4.4.2 Effect of Fiber/Matrix Interface Shear Stress on Time-Dependent Matrix Multiple Fracture, Interface Debonding, and Fiber Failure

The time-dependent matrix multicracking density, interface debonding ratio, interface oxidation ratio, and broken fiber fraction for different interface shear stresses in the debonding region are shown in Fig. 2.23. When the interface shear stress in the debonding region increased, the matrix first cracking stress, and saturation cracking stress increased, the saturation matrix cracking space decreased, and the matrix multiple fracture evolution rate increased. The interface debonding ratio and the broken fiber fraction decreased, and the interface oxidation ratio increased.

When the interface shear stress in the debonding region is $\tau_i = 25$ MPa, the time-dependent matrix multicracking density increased from $\varphi = 0.21$ mm at the first matrix cracking stress of $\sigma_{mc} = 298$ MPa to $\varphi = 4.3$ mm at the saturation matrix cracking stress of $\sigma_{sat} = 360$ MPa, the time-dependent interface debonding ratio increased from $2l_d/l_c = 0.85$ to $2l_d/l_c = 25\%$, the time-dependent interface oxidation ratio decreased from $\zeta/l_d = 12.7$ to $\zeta/l_d = 8.5\%$, and the time-dependent broken fiber fraction increased from $P = 0.01$ to $P = 0.035\%$.

When the interface shear stress in the debonding region is $\tau_i = 30$ MPa, the time-dependent matrix multicracking density increased from $\varphi = 0.22$ mm at the first matrix cracking stress of $\sigma_{mc} = 322$ MPa to $\varphi = 3.2$ mm at the saturation matrix cracking stress of $\sigma_{sat} = 360$ MPa, the time-dependent interface debonding ratio increased from $2l_d/l_c = 0.86$ to $2l_d/l_c = 15\%$, the time-dependent interface oxidation ratio decreased from $\zeta/l_d = 12.7$ to $\zeta/l_d = 10.1\%$, and the time-dependent broken fiber fraction increased from $P = 0.01$ to $P = 0.03\%$.

The time-dependent matrix multicracking density, interface debonding ratio, interface oxidation ratio, and broken fiber fraction for different fiber/matrix interface shear stress in the oxidation region are shown in Fig. 2.24. When the interface shear stress in the oxidation region increased, the matrix cracking density increased, the interface

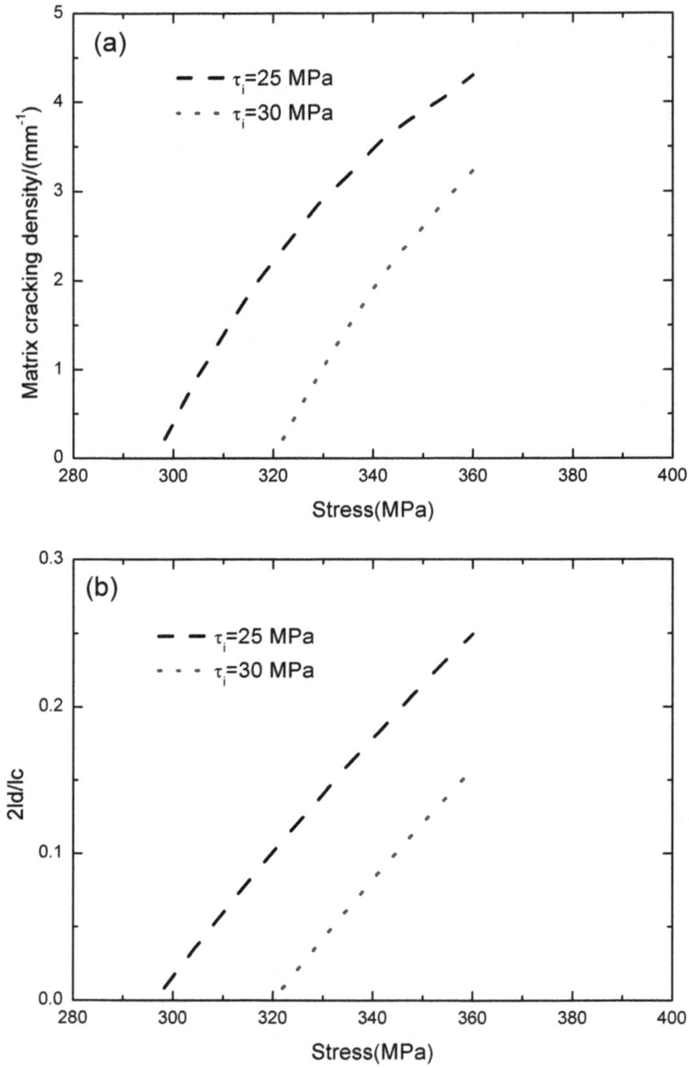

Fig. 2.23 Effect of the fiber/matrix interface shear stress in debonding region ($\tau_i = 25$ and 30 MPa) on **a** the time-dependent matrix multicracking density versus applied stress curves; **b** the time-dependent fiber/matrix interface debonding ratio versus applied stress curves; **c** the time-dependent fiber/matrix interface oxidation ratio versus applied stress curves; and **d** the time-dependent broken fiber fraction versus applied cycles curves

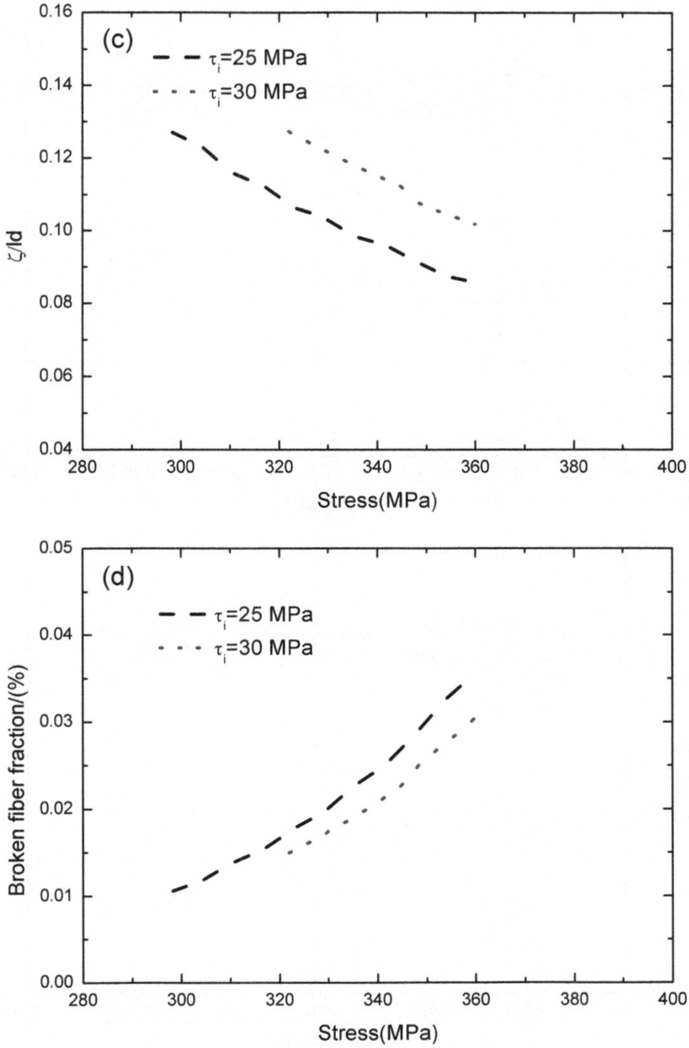

Fig. 2.23 (continued)

debonding ratio and the broken fiber fraction decreased, and the interface oxidation ratio increased.

When the interface shear stress in the oxidation region is $\tau_f = 1$ MPa, the time-dependent matrix multicracking density increased from $\varphi = 0.22$ mm at the first matrix cracking stress of $\sigma_{mc} = 271$ MPa to $\varphi = 4.8$ mm at the saturation matrix cracking stress of $\sigma_{sat} = 360$ MPa, the time-dependent interface debonding ratio increased from $2l_d/l_c = 0.83$ to $2l_d/l_c = 35\%$, the time-dependent interface oxidation

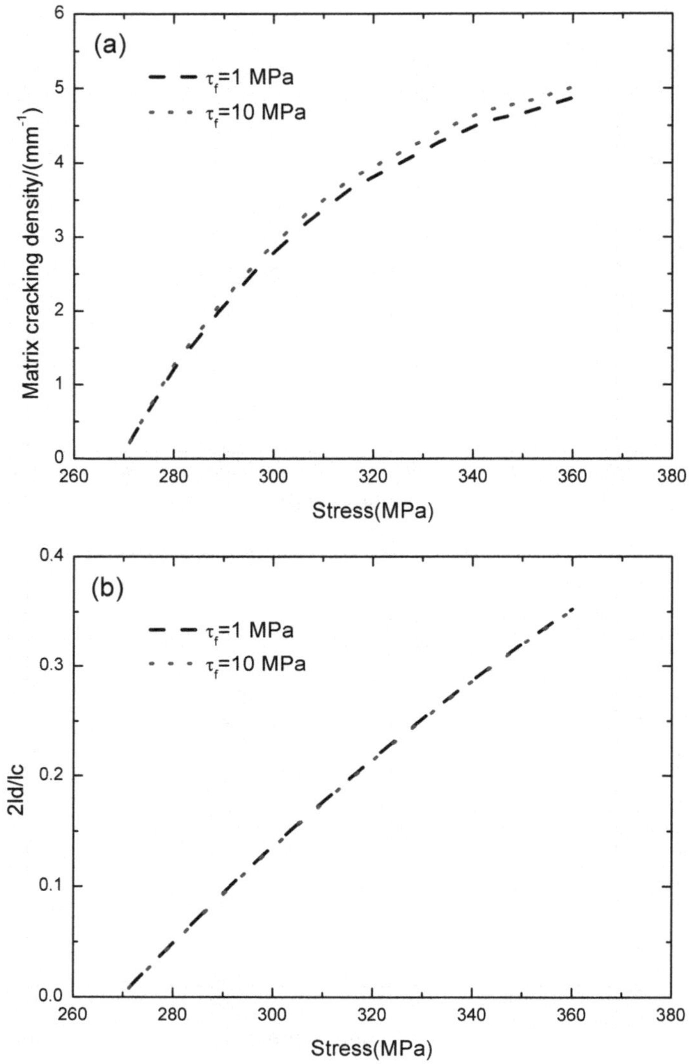

Fig. 2.24 Effect of fiber/matrix interface shear stress in oxidation region ($\tau_f = 1$ and 10 MPa) on **a** the time-dependent matrix cracking density versus applied stress curves; **b** the time-dependent fiber/matrix interface debonding ratio versus applied stress curves; **c** the time-dependent fiber/matrix interface oxidation ratio versus applied stress curves; and **d** the time-dependent broken fiber fraction versus applied cycles curves

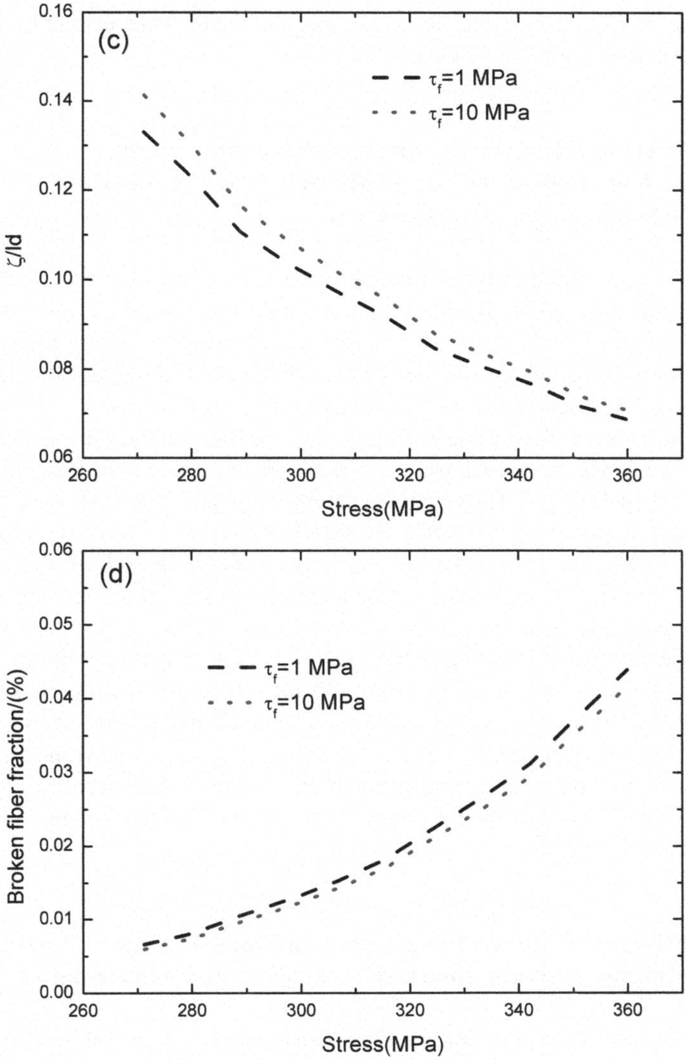

Fig. 2.24 (continued)

ratio decreased from $\zeta/l_d = 13.3$ to $\zeta/l_d = 6.8\%$, and the time-dependent broken fiber fraction increased from $P = 0.006$ to $P = 0.04\%$.

When the interface shear stress in the oxidation region was $\tau_f = 10$ MPa, the time-dependent matrix multicracking density increased from $\varphi = 0.23$ mm at the first matrix cracking stress of $\sigma_{mc} = 271$ MPa to $\varphi = 5.0$ mm at the saturation matrix cracking stress of $\sigma_{sat} = 360$ MPa, the time-dependent interface debonding ratio increased from $2l_d/l_c = 0.82$ to $2l_d/l_c = 35\%$, the time-dependent interface oxidation ratio decreased from $\zeta/l_d = 14.1$ to $\zeta/l_d = 7\%$, and the time-dependent broken fiber fraction increased from $P = 0.006$ to $P = 0.04\%$.

2.4.4.3 Effect of Fiber/Matrix Interface Debonding Energy on Time-Dependent Matrix Multiple Fracture, Interface Debonding, and Fiber Failure

The time-dependent matrix multicracking density, interface debonding ratio, interface oxidation ratio, and broken fiber fraction for different interface debonding energies are shown in Fig. 2.25. When the interface debonding energy increased, the matrix multicracking density increased, the interface debonding ratio and the broken fiber fraction decreased, and the interface oxidation ratio increased.

When the interface debonding energy is $\xi_d = 1.0$ J/m^2, the time-dependent matrix multicracking density increased from $\varphi = 0.27$ mm at the first matrix cracking stress of $\sigma_{mc} = 271$ MPa to $\varphi = 5.4$ mm at the saturation matrix cracking stress of $\sigma_{sat} = 360$ MPa, the time-dependent interface debonding ratio increased from $2l_d/l_c = 0.7$ to $2l_d/l_c = 34.2\%$, the time-dependent interface oxidation ratio decreased from $\zeta/l_d = 17.7$ to $\zeta/l_d = 7.8\%$, and the time-dependent broken fiber fraction increased from $P = 0.005$ to $P = 0.04\%$.

When the interface debonding energy is $\xi_d = 1.5$ J/m^2, the time-dependent matrix multicracking density increased from $\varphi = 0.5$ mm at the first matrix cracking stress of $\sigma_{mc} = 271$ MPa to $\varphi = 7.0$ mm at the saturation matrix cracking stress of $\sigma_{sat} = 360$ MPa, the time-dependent interface debonding ratio increased from $2l_d/l_c = 0.5$ to $2l_d/l_c = 31.7\%$, the time-dependent interface oxidation ratio decreased from $\zeta/l_d = 49.1$ to $\zeta/l_d = 11\%$, and the time-dependent broken fiber fraction increased from $P = 0.002$ to $P = 0.03\%$.

2.4.4.4 Effect of Oxidation Temperature on Time-Dependent Matrix Multiple Fracture, Interface Debonding, and Fiber Failure

The time-dependent matrix multicracking density, interface debonding ratio, interface oxidation ratio, and broken fiber fraction corresponding to different oxidation temperatures of $T = 700$ and 1000 °C are shown in Fig. 2.26. When the oxidation temperature increased, the matrix cracking density decreased, and the interface debonding ratio, oxidation ratio, and broken fiber fraction increased.

When the oxidation temperature was $T = 700$ °C, the time-dependent matrix multicracking density increased from $\varphi = 0.23$ mm at the first matrix cracking stress of $\sigma_{mc} = 271$ MPa to $\varphi = 5.0$ mm at the saturation matrix cracking stress of $\sigma_{sat} = 360$ MPa, the time-dependent interface debonding ratio increased from $2l_d/l_c = 0.8$ to $2l_d/l_c = 35.1\%$, the time-dependent interface oxidation ratio decreased from $\zeta/l_d = 6.2$ to $\zeta/l_d = 3\%$, and the time-dependent broken fiber fraction increased from $P = 0.005$ to $P = 0.04\%$.

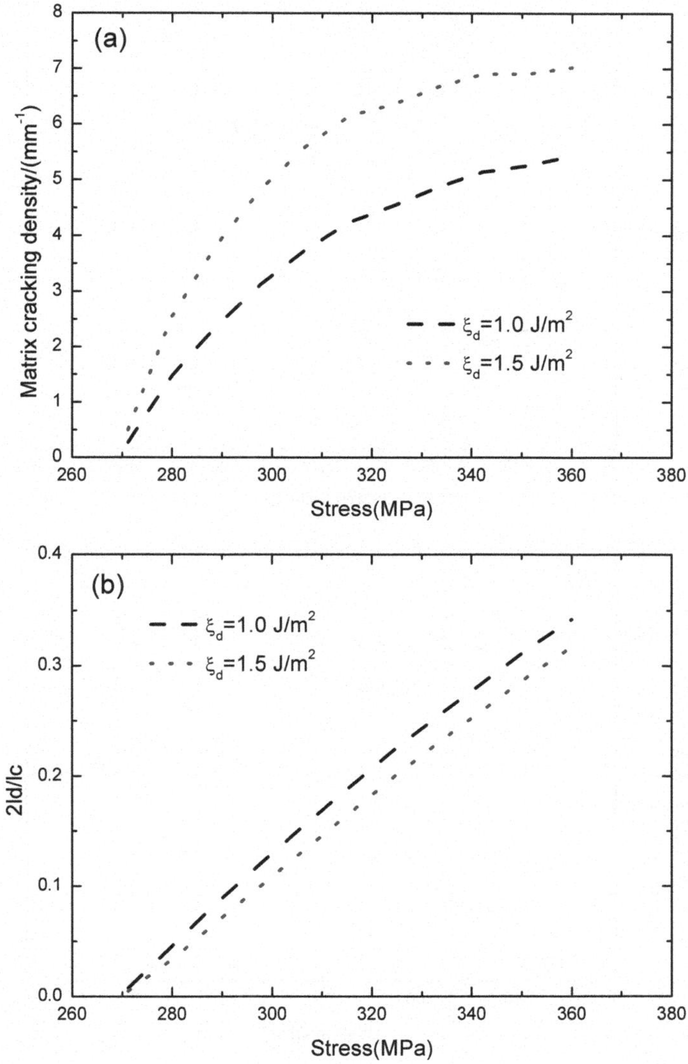

Fig. 2.25 Effect of fiber/matrix interface debonding energy (ξ_d = 1.0 and 1.5 J/m^2) on **a** the time-dependent matrix multicracking density versus applied stress curves; **b** the time-dependent fiber/matrix interface debonding ratio versus applied stress curves; **c** the time-dependent fiber/matrix interface oxidation ratio versus applied stress curves; and **d** the time-dependent broken fiber fraction versus applied cycles curves

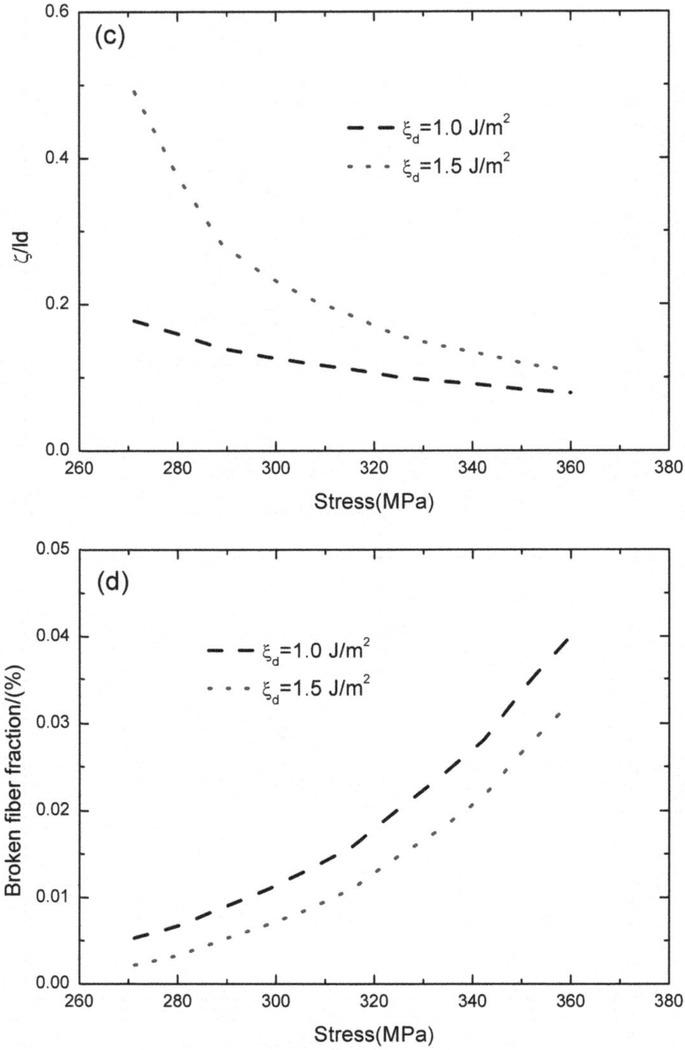

Fig. 2.25 (continued)

When the oxidation temperature is $T = 1000$ °C, the time-dependent matrix multicracking density increased from $\varphi = 0.18$ mm at the first matrix cracking stress of $\sigma_{\text{mc}} = 271$ MPa to $\varphi = 4.3$ mm at the saturation matrix cracking stress of $\sigma_{\text{sat}} = 360$ MPa, the time-dependent interface debonding ratio increased from $2l_d/l_c = 0.8$ to $2l_d/l_c = 35.4\%$, the time-dependent interface oxidation ratio decreased from $\zeta/l_d = 39.2$ to $\zeta/l_d = 22.3\%$, and the time-dependent broken fiber fraction increases from $P = 0.008$ to $P = 0.05\%$.

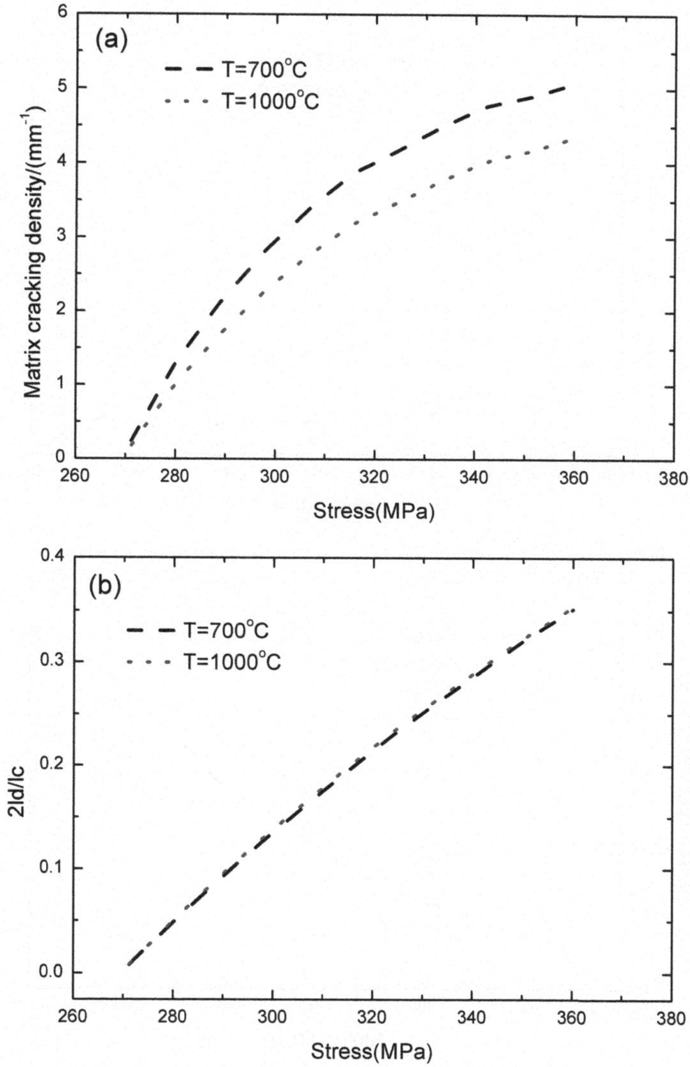

Fig. 2.26 Effect of the oxidation temperature ($T = 700$ and $1000\,°C$) on **a** the time-dependent matrix multicracking density versus applied stress curves; **b** the time-dependent fiber/matrix interface debonding ratio versus applied stress curves; **c** the time-dependent fiber/matrix interface oxidation ratio versus applied stress curves; and **d** the time-dependent broken fiber fraction versus applied cycles curves

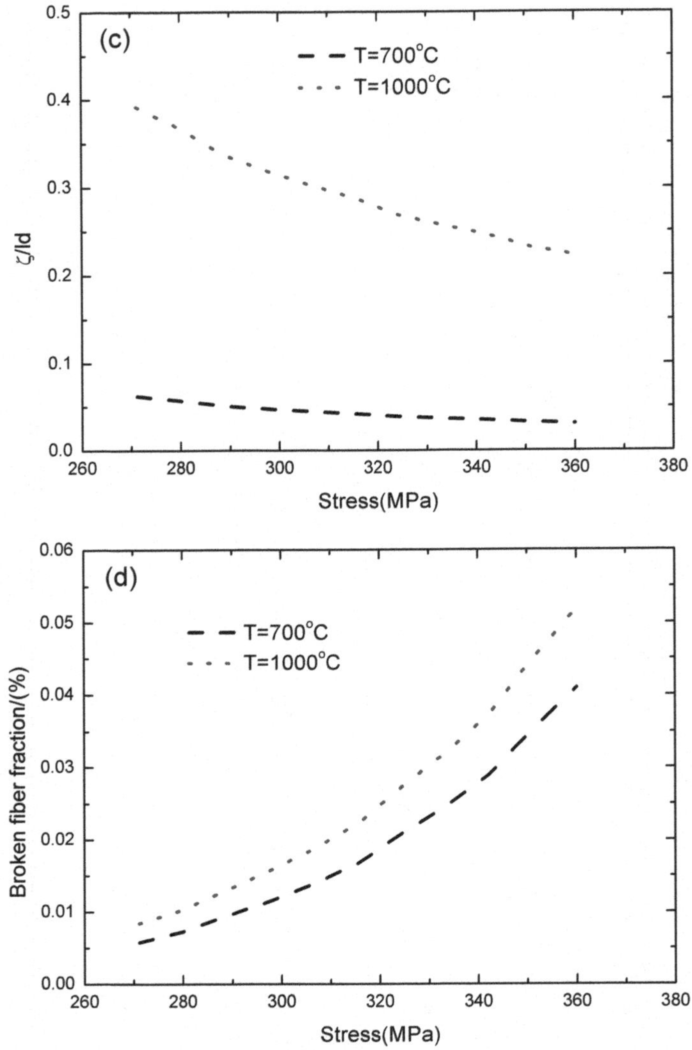

Fig. 2.26 (continued)

2.4.4.5 Effect of Oxidation Duration on Time-Dependent Matrix Multiple Fracture, Interface Debonding, and Fiber Failure

The time-dependent matrix cracking density, interface debonding ratio, interface oxidation ratio, and broken fiber fraction corresponding to different oxidation durations are shown in Fig. 2.27. When the oxidation duration increased, the time-dependent matrix multicracking density decreased, the interface debonding ratio, oxidation ratio, and broken fiber fraction increased.

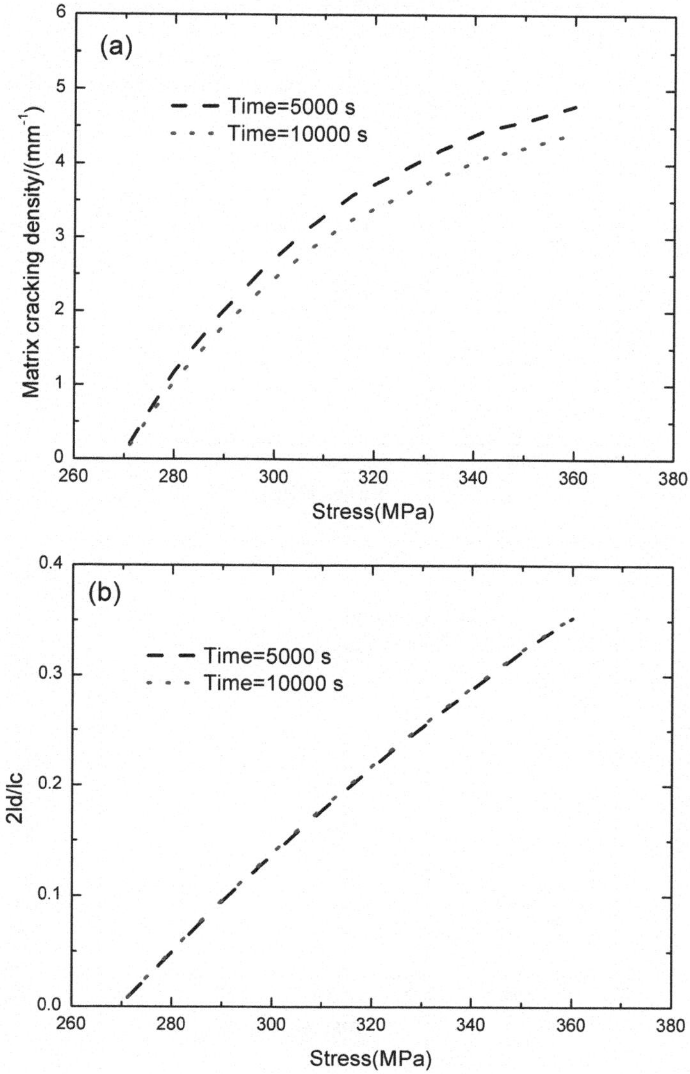

Fig. 2.27 Effect of the oxidation duration ($t = 5000$ and 10,000 s) on **a** the time-dependent matrix multicracking density versus applied stress curves; **b** the time-dependent fiber/matrix interface debonding ratio versus applied stress curves; **c** the time-dependent fiber/matrix interface oxidation ratio versus applied stress curves; and **d** the time-dependent broken fiber fraction versus applied cycles curves

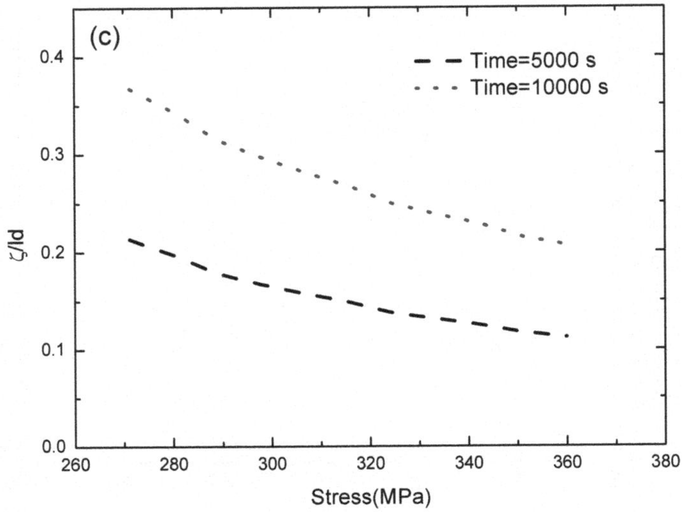

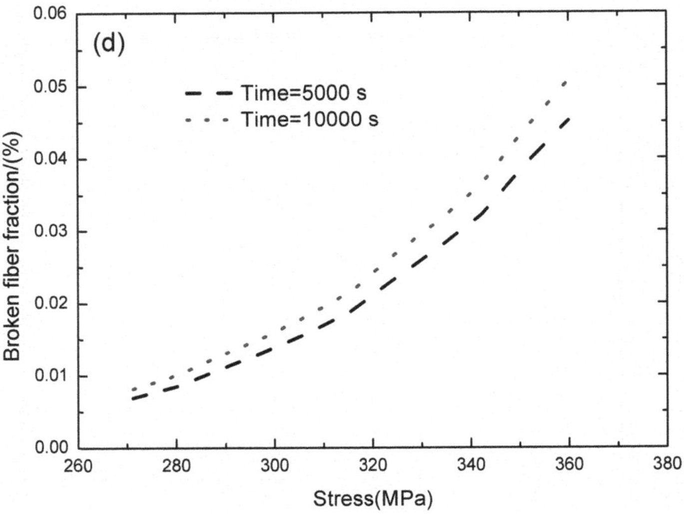

Fig. 2.27 (continued)

When the oxidation duration is $t = 5000$ s, the time-dependent matrix multicracking density increased from $\varphi = 0.21$ mm at the first matrix cracking stress of $\sigma_{\mathrm{mc}} = 271$ MPa to $\varphi = 4.7$ mm at the saturation matrix cracking stress of $\sigma_{\mathrm{sat}} = 360$ MPa, the time-dependent interface debonding ratio increased from $2l_{\mathrm{d}}/l_{\mathrm{c}} = 0.83$ to $2l_{\mathrm{d}}/l_{\mathrm{c}} = 35.2\%$, the time-dependent interface oxidation ratio decreased from $\zeta/l_{\mathrm{d}} = 21.3$ to $\zeta/l_{\mathrm{d}} = 11.2\%$, and the time-dependent broken fiber fraction increased from $P = 0.007$ to $P = 0.045\%$.

When the oxidation duration is $t = 10{,}000$ s, the time-dependent matrix cracking density increased from $\varphi = 0.18$ mm at the first matrix cracking stress of $\sigma_{\mathrm{mc}} =$

271 MPa to $\varphi = 4.4$ mm at the saturation matrix cracking stress of $\sigma_{\text{sat}} = 360$ MPa, the time-dependent interface debonding ratio increased from $2l_d/l_c = 0.85$ to $2l_d/l_c = 35.4\%$, the time-dependent interface oxidation ratio decreased from $\zeta/l_d = 36.7$ to $\zeta/l_d = 20.6\%$, and the time-dependent broken fiber fraction increased from $P = 0.008$ to $P = 0.05\%$.

2.4.5 Experimental Comparisons

The experimental and theoretical time-dependent matrix multicracking density, interface debonding ratio, interface oxidation ratio, and broken fiber fraction versus the applied stress for different fiber-reinforced CMCs, i.e., unidirectional C/SiC (Li et al. 2014), SiC/SiC (Beyerle et al. 1993), mini-SiC/SiC (Zhang et al. 2016), SiC/CAS (Pryce and Smith 1992), SiC/CAS-II (Beyerle et al. 1993), and SiC/borosilicate (Okabe et al. 1999) composites, were predicted, as shown in Figs. 2.28, 2.29, 2.30, 2.31, 2.32 and 2.33.

2.4.5.1 C/SiC Composite

For the unidirectional C/SiC composite, the matrix multiple fracture began at an applied stress of $\sigma_{\text{mc}} = 100$ MPa and approached saturation at the applied stress of $\sigma_{\text{sat}} = 220$ MPa, and the matrix cracking density increased from $\varphi = 4.2$ mm to the saturation value of $\varphi = 9.4$ mm. At the oxidation temperature of $T = 800$ °C for an oxidation duration of $t = 3000$ s, the matrix cracking density increased from $\varphi = 0.5$ mm at the first matrix cracking stress of $\sigma_{\text{mc}} = 100$ MPa to $\varphi = 6.2$ mm at the saturation matrix cracking stress of $\sigma_{\text{sat}} = 220$ MPa, the interface debonding ratio increased from $2l_d/l_c = 0.74$ to $2l_d/l_c = 39.7\%$, the interface oxidation ratio decreased from $\zeta/l_d = 40.3$ to $\zeta/l_d = 22.2\%$, and the broken fiber fraction increased from $P = 0.0004$ to $P = 0.01\%$. At the oxidation temperature of 800 °C for an oxidation duration of $t = 5000$ s, the matrix cracking density increased from $\varphi = 0.3$ mm at the first matrix cracking stress of $\sigma_{\text{cr}} = 100$ MPa to $\varphi = 5.0$ mm at the saturation matrix cracking stress of $\sigma_{\text{sat}} = 220$ MPa, the interface debonding ratio increased from $2l_d/l_c = 0.85$ to $2l_d/l_c = 39.2\%$, the interface oxidation ratio decreased from $\zeta/l_d = 54.6$ to $\zeta/l_d = 32.8\%$, and the broken fiber fraction increased from $P = 0.0008$ to $P = 0.02\%$, as shown in Fig. 2.28.

2.4.5.2 SiC/SiC Composite

For the unidirectional SiC/SiC composite, the matrix multiple fracture began from the applied stress of $\sigma_{\text{mc}} = 240$ MPa and approached saturation at an applied stress of $\sigma_{\text{sat}} = 320$ MPa, and the matrix cracking density increased from $\varphi = 1.1$ mm to the saturation value of $\varphi = 13$ mm. At the oxidation temperature of $T = 800$ °C for

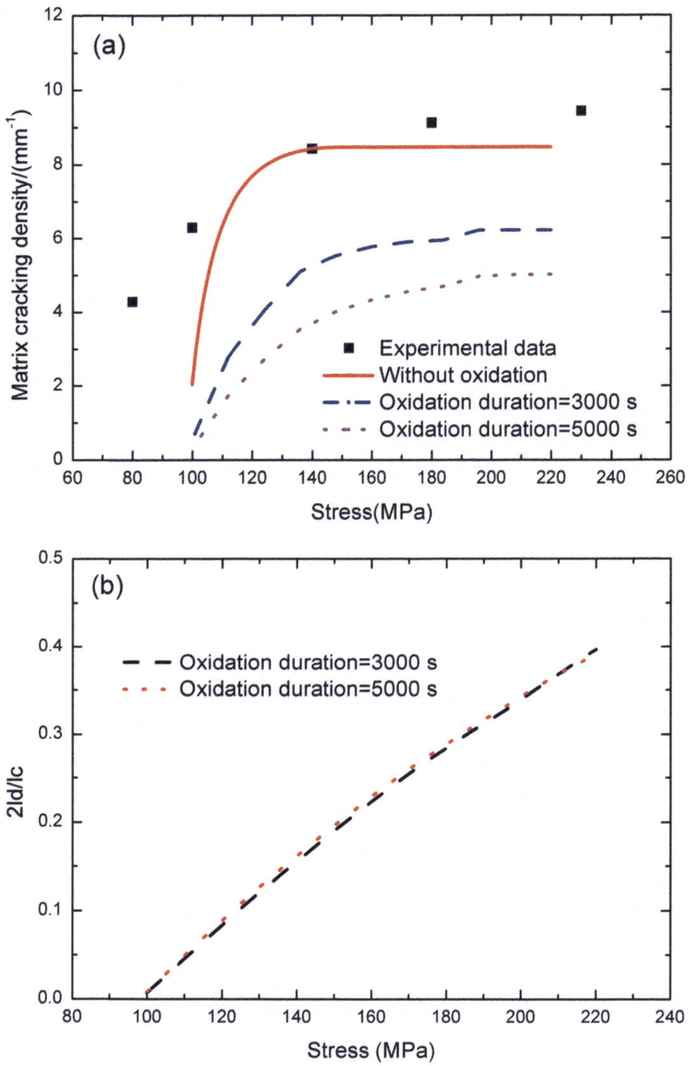

Fig. 2.28 **a** Experimental and predicted time-dependent matrix multicracking density versus applied stress curves; **b** the time-dependent fiber/matrix interface debonding ratio versus applied stress curves; **c** the time-dependent fiber/matrix interface oxidation ratio versus applied stress curves; and **d** the time-dependent broken fiber fraction versus applied stress curve of unidirectional C/SiC composite

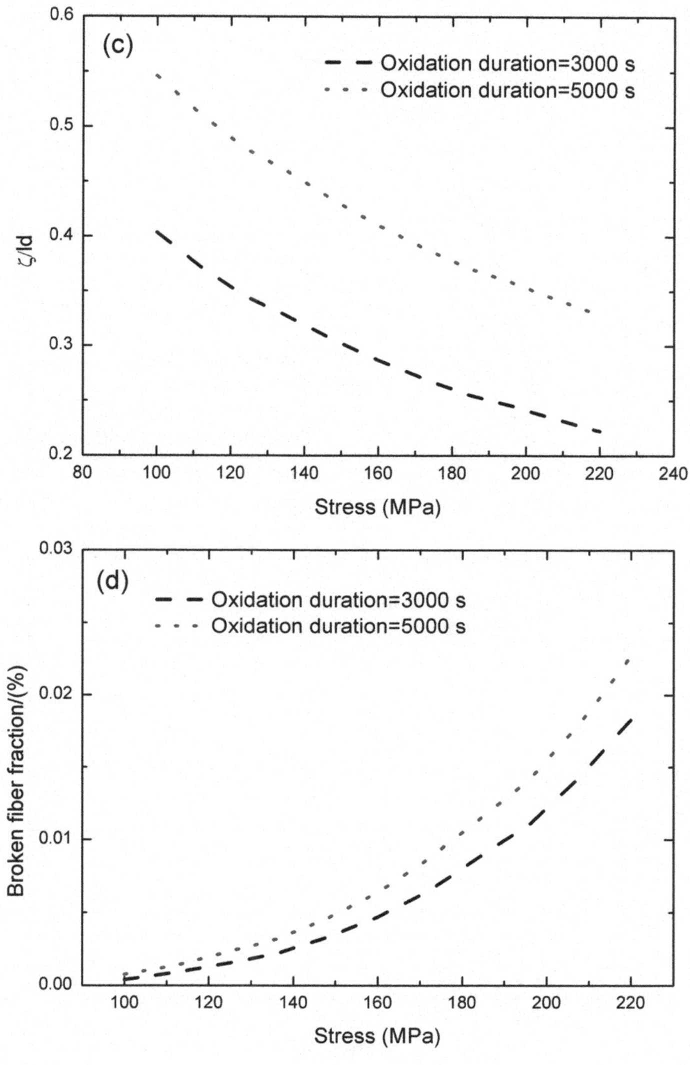

Fig. 2.28 (continued)

the oxidation duration of $t = 3000$ s, the matrix cracking density increased from $\varphi = 0.28$ mm at the first matrix cracking stress of $\sigma_{mc} = 240$ MPa to $\varphi = 6.8$ mm at the saturation matrix cracking stress of $\sigma_{sat} = 320$ MPa, the interface debonding ratio increased from $2l_d/l_c = 0.8$ to $2l_d/l_c = 32.9\%$, the interface oxidation ratio decreased from $\zeta/l_d = 88.8$ to $\zeta/l_d = 51.9\%$, and the broken fiber fraction increased from $P = 0.01$ to $P = 0.04\%$. At the oxidation temperature of $T = 800$ °C for the oxidation duration of $t = 5000$ s, the matrix cracking density increased from $\varphi = 0.2$ mm at the first matrix cracking stress of $\sigma_{mc} = 240$ MPa to $\varphi = 5.3$ mm at the saturation matrix cracking stress of $\sigma_{sat} = 320$ MPa, the interface debonding ratio increased

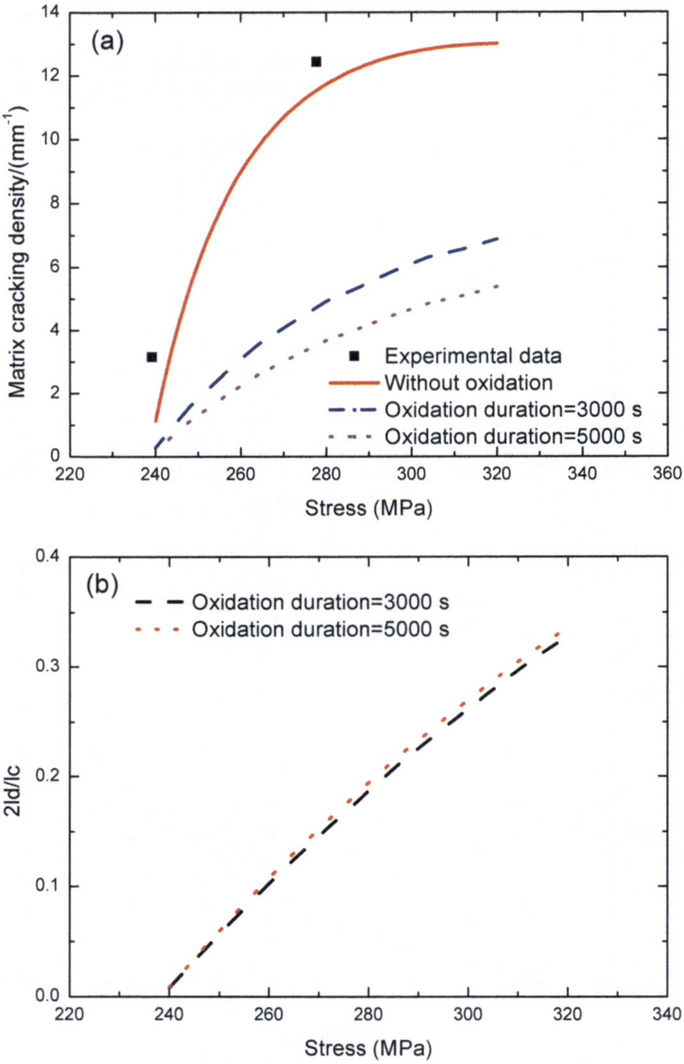

Fig. 2.29 **a** Experimental and predicted time-dependent matrix multicracking density versus applied stress curves; **b** the time-dependent fiber/matrix interface debonding ratio versus applied stress curves; **c** the time-dependent fiber/matrix interface oxidation ratio versus applied stress curves; and **d** the time-dependent broken fiber fraction versus applied stress curve of unidirectional SiC/SiC composite

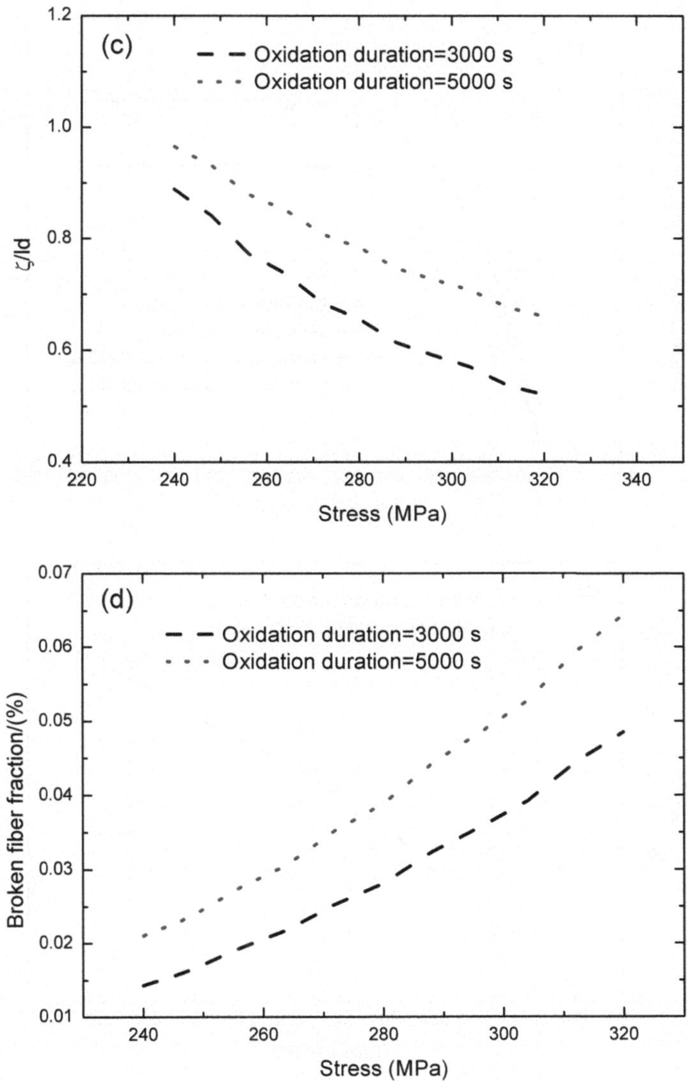

Fig. 2.29 (continued)

from $2l_d/l_c = 0.86$ to $2l_d/l_c = 33.7\%$, the interface oxidation ratio decreased from $\zeta/l_d = 96.6$ to $\zeta/l_d = 66\%$, and the broken fiber fraction increased from $P = 0.02$ to $P = 0.06\%$, as shown in Fig. 2.29.

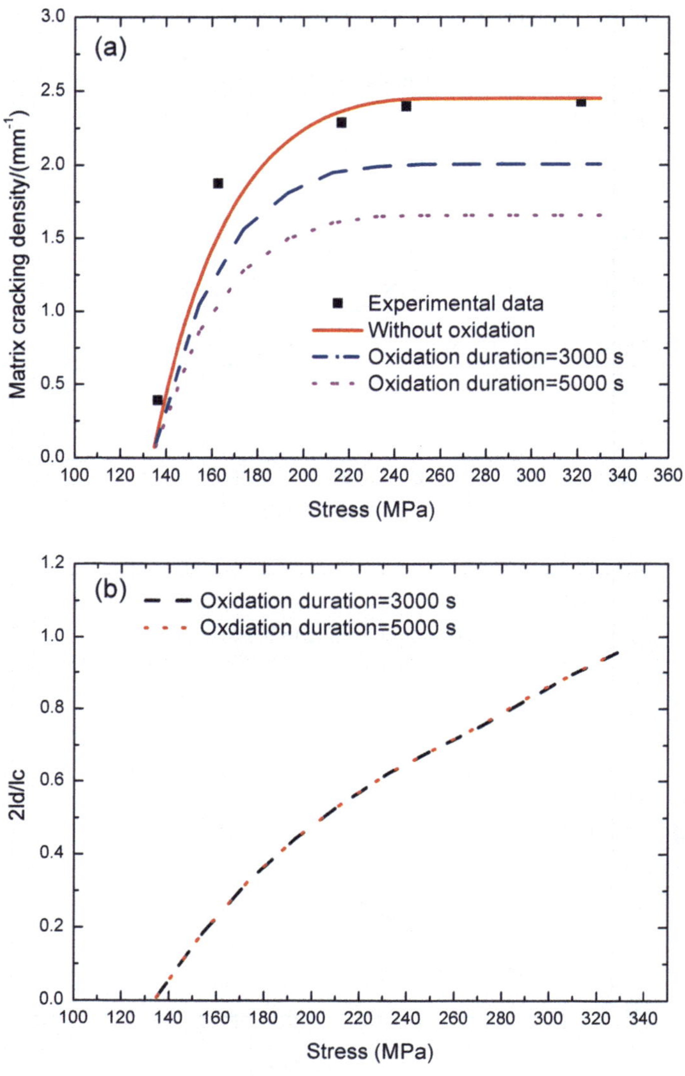

Fig. 2.30 **a** Experimental and predicted time-dependent matrix multicracking density versus applied stress curves; **b** the time-dependent fiber/matrix interface debonding ratio versus applied stress curves; **c** the time-dependent fiber/matrix interface oxidation ratio versus applied stress curves; and **d** the time-dependent broken fiber fraction versus applied stress curve of mini-SiC/SiC composite

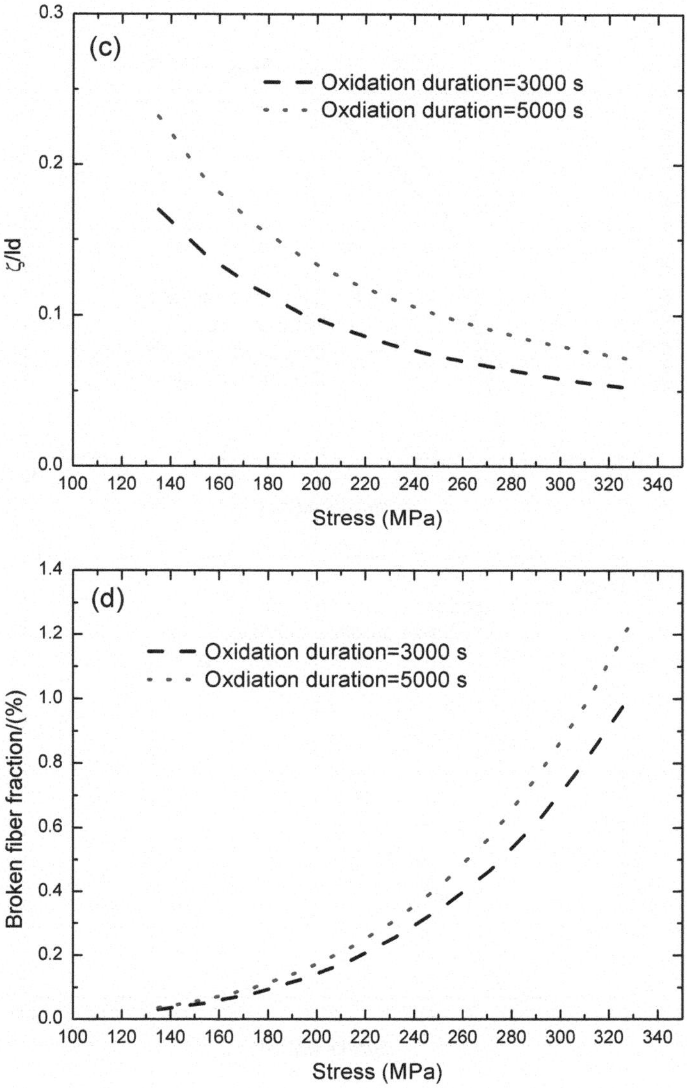

Fig. 2.30 (continued)

2.4.5.3 Mini-SiC/SiC Composite

For the mini-SiC/SiC composite, the matrix multiple fracture began at an applied stress of σ_{mc} = 135 MPa and approached saturation at the applied stress of σ_{sat} = 330 MPa, and the matrix cracking density increased from φ = 0.07 mm to the saturation value of φ = 2.4 mm. At an oxidation temperature of T = 800 °C for the oxidation duration of t = 3000 s, the matrix cracking density increased from φ =

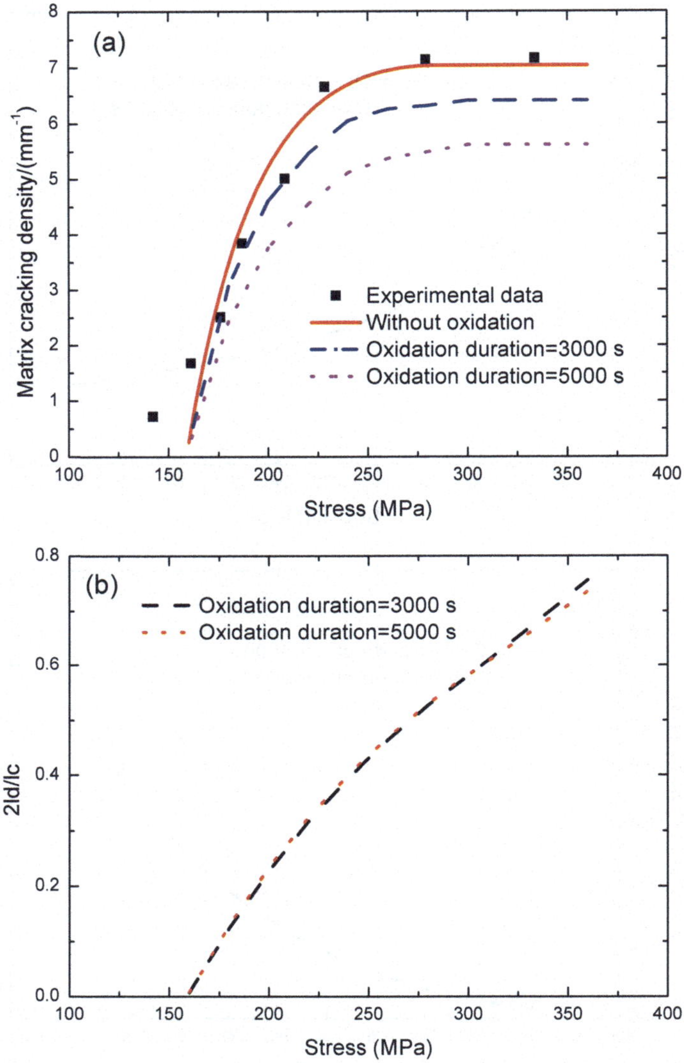

Fig. 2.31 **a** Experimental and predicted time-dependent matrix multicracking density versus applied stress curves; **b** the time-dependent fiber/matrix interface debonding ratio versus applied stress curves; **c** the time-dependent fiber/matrix interface oxidation ratio versus applied stress curves; and **d** the time-dependent broken fiber fraction versus applied stress curve of unidirectional SiC/CAS composite

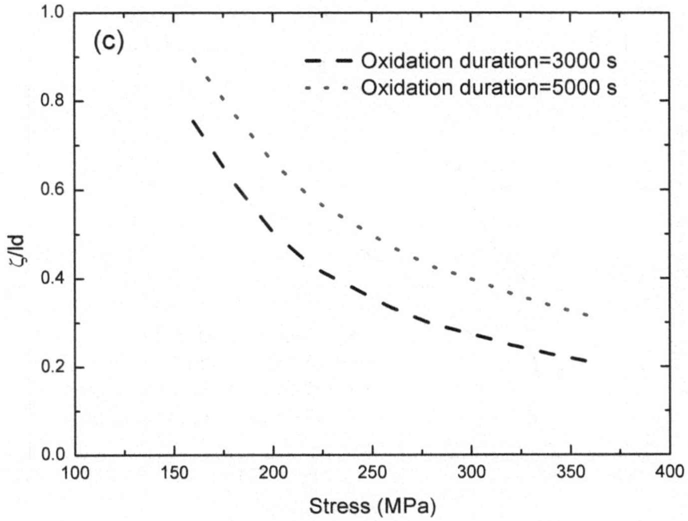

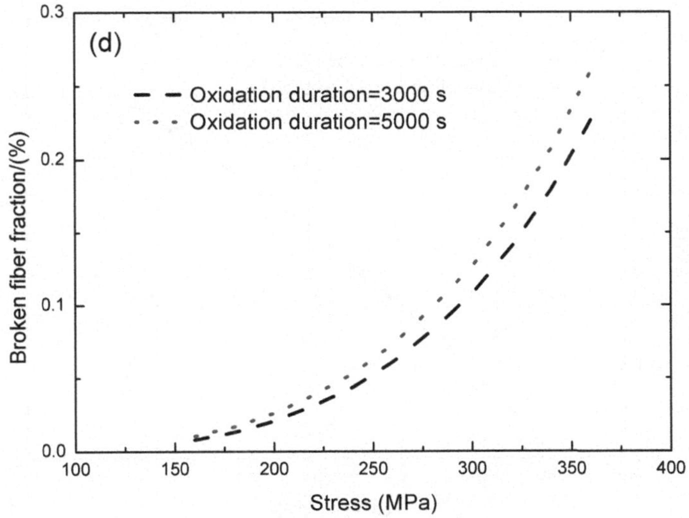

Fig. 2.31 (continued)

0.06 mm at the first matrix cracking stress of $\sigma_{\text{mc}} = 135$ MPa to $\varphi = 0.96$ mm at the saturation matrix cracking stress of $\sigma_{\text{sat}} = 330$ MPa, the interface debonding ratio increased from $2l_d/l_c = 1$ to $2l_d/l_c = 96\%$, the interface oxidation ratio decreased from $\zeta/l_d = 16.9$ to $\zeta/l_d = 5.2\%$, and the broken fiber fraction increased from $P = 0.03$ to $P = 1\%$. At the oxidation temperature of $T = 800$ °C for the oxidation duration of $t = 5000$ s, the matrix cracking density increased from $\varphi = 0.05$ mm at the first matrix cracking stress of $\sigma_{\text{mc}} = 135$ MPa to $\varphi = 1.6$ mm at the saturation matrix cracking stress of $\sigma_{\text{sat}} = 330$ MPa, the interface debonding ratio increased

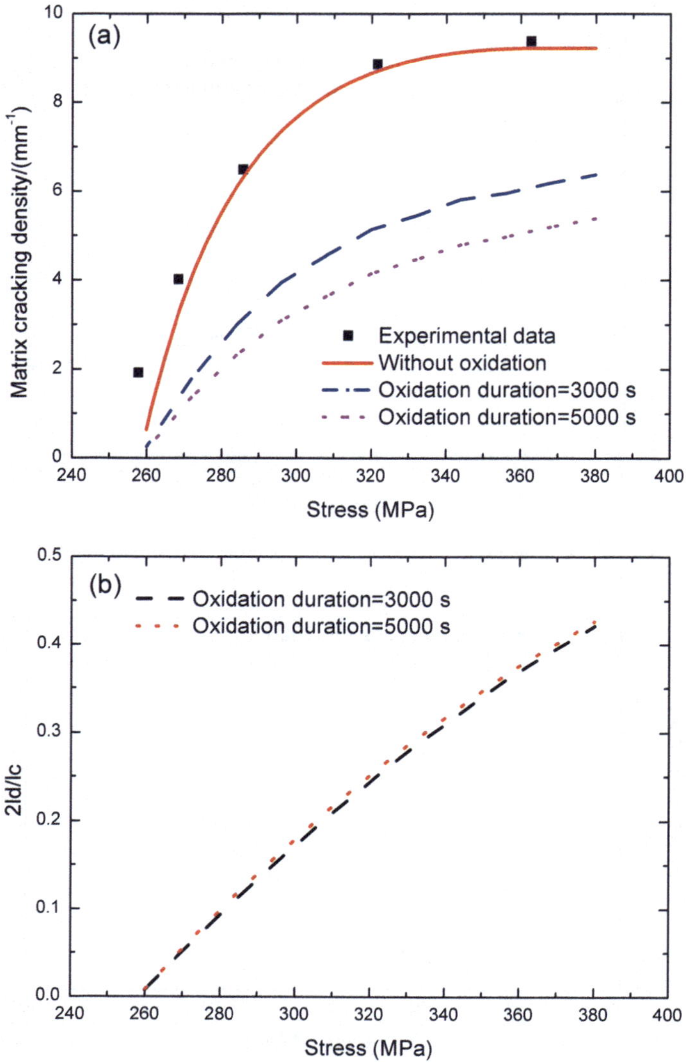

Fig. 2.32 **a** Experimental and predicted time-dependent matrix multicracking density versus applied stress curves; **b** the time-dependent fiber/matrix interface debonding ratio versus applied stress curves; **c** the time-dependent fiber/matrix interface oxidation ratio versus applied stress curves; and **d** the time-dependent broken fiber fraction versus applied stress curve of unidirectional SiC/CAS-II composite

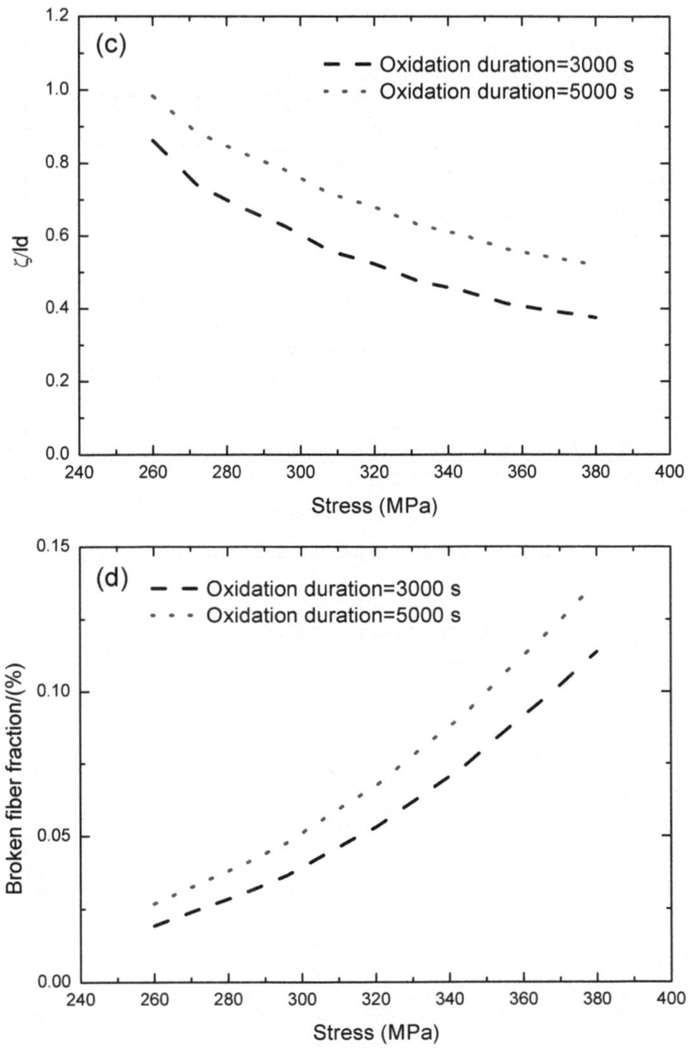

Fig. 2.32 (continued)

from $2l_d/l_c = 1$ to $2l_d/l_c = 96.3\%$, the interface oxidation ratio decreased from $\zeta/l_d = 23.2$ to $\zeta/l_d = 7.1\%$, and the broken fiber fraction increased from $P = 0.04$ to $P = 1.2\%$, as shown in Fig. 2.30.

2.4.5.4 SiC/CAS Composite

For the unidirectional SiC/CAS composite, the matrix multiple fracture evolution began at an applied stress of $\sigma_{mc} = 160$ MPa and approached saturation at an applied

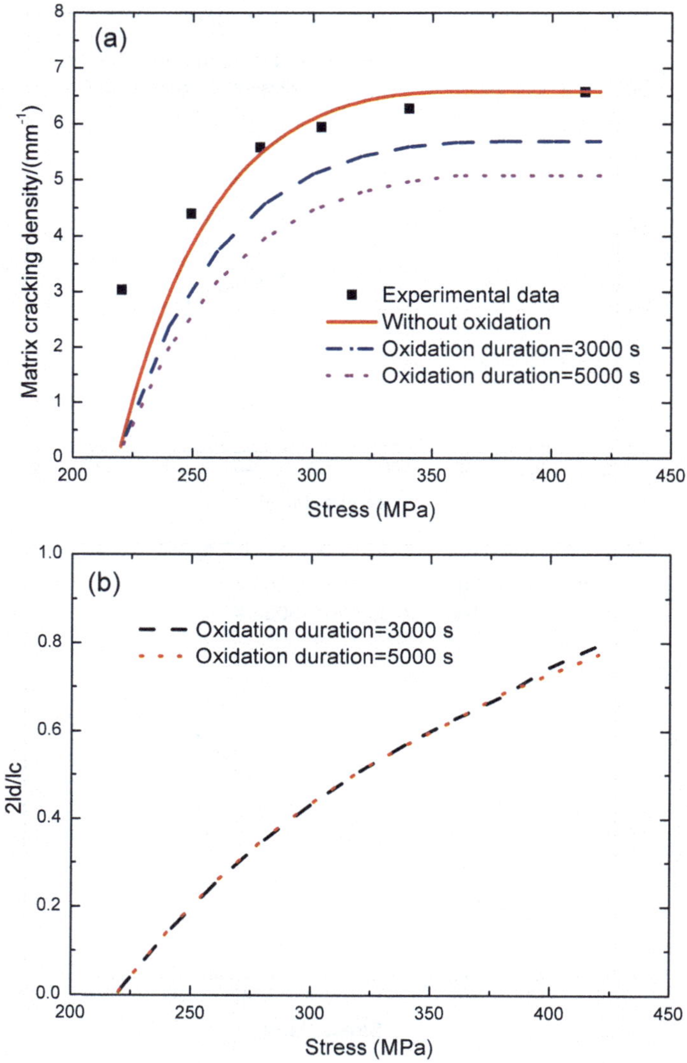

Fig. 2.33 **a** Experimental and predicted time-dependent matrix multicracking density versus applied stress curves; **b** the time-dependent fiber/matrix interface debonding ratio versus applied stress curves; **c** the time-dependent fiber/matrix interface oxidation ratio versus applied stress curves; and **d** the time-dependent broken fiber fraction versus applied stress curve of unidirectional SiC/borosilicate composite

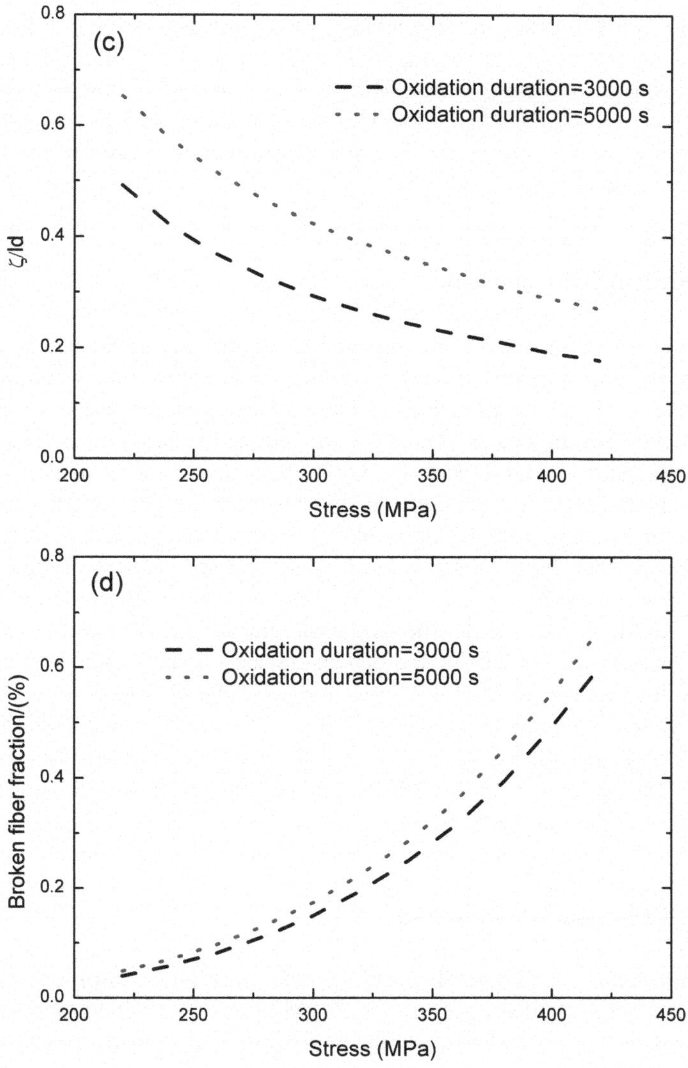

Fig. 2.33 (continued)

stress of $\sigma_{sat} = 360$ MPa, and the matrix cracking density increased from $\varphi = 0.26$ mm to the saturation value of $\varphi = 7.1$ mm. At the oxidation temperature of $T = 800$ °C for the oxidation duration of $t = 3000$ s, the matrix cracking density increased from $\varphi = 0.22$ mm at the first matrix cracking stress of $\sigma_{mc} = 160$ MPa to $\varphi = 6.4$ mm at the saturation matrix cracking stress of $\sigma_{sat} = 360$ MPa, the interface debonding ratio increased from $2l_d/l_c = 0.7$ to $2l_d/l_c = 75.6\%$, the interface oxidation ratio decreased from $\zeta/l_d = 75.4$ to $\zeta/l_d = 21\%$, and the broken fiber fraction increased from $P = 0.008$ to $P = 0.22\%$. At the oxidation temperature of $T = 800$ °C for the oxidation duration of $t = 5000$ s, the matrix cracking density increased from $\varphi = 0.17$ mm at the first matrix cracking stress of $\sigma_{\text{mc}} = 160$ MPa to $\varphi = 5.6$ mm at the saturation matrix cracking stress of $\sigma_{\text{sat}} = 360$ MPa, the interface debonding ratio increased from $2l_d/l_c = 0.8$ to $2l_d/l_c = 73.7\%$, the interface oxidation ratio decreased from $\zeta/l_d = 89.6$ to $\zeta/l_d = 31.5\%$, and the broken fiber fraction increased from $P = 0.01$ to $P = 0.26\%$, as shown in Fig. 2.31.

2.4.5.5 SiC/CAS-II Composite

For the unidirectional SiC/CAS-II composite, the matrix multiple fracture evolution began from the applied stress of $\sigma_{\text{mc}} = 260$ MPa and approached saturation at the applied stress of $\sigma_{\text{sat}} = 380$ MPa, and the matrix cracking density increased from $\varphi = 0.26$ mm to the saturation value of $\varphi = 6.3$ mm. At the oxidation temperature of $T = 800$ °C for the oxidation duration of $t = 3000$ s, the matrix cracking density increased from $\varphi = 0.26$ mm at the first matrix cracking stress of $\sigma_{\text{mc}} = 260$ MPa to $\varphi = 6.3$ mm at the saturation matrix cracking stress of $\sigma_{\text{sat}} = 380$ MPa, the interface debonding ratio increased from $2l_d/l_c = 0.7$ to $2l_d/l_c = 42.1\%$, the interface oxidation ratio decreased from $\zeta/l_d = 86.1$ to $\zeta/l_d = 37.5\%$, and the broken fiber fraction increased from $P = 0.02$ to $P = 0.1\%$. At the oxidation temperature of $T = 800$ °C for the oxidation duration of $t = 5000$ s, the matrix cracking density increased from $\varphi = 0.19$ mm at the first matrix cracking stress of $\sigma_{\text{mc}} = 260$ MPa to $\varphi = 5.4$ mm at the saturation matrix cracking stress of $\sigma_{\text{sat}} = 380$ MPa, the interface debonding ratio increased from $2l_d/l_c = 0.8$ to $2l_d/l_c = 42.7\%$, the interface oxidation ratio decreased from $\zeta/l_d = 98.3$ to $\zeta/l_d = 52.1\%$, and the broken fiber fraction increased from $P = 0.02$ to $P = 0.1\%$, as shown in Fig. 2.32.

2.4.5.6 SiC/Borosilicate Composite

For the unidirectional SiC/borosilicate composite, the matrix multiple fracture evolution began at an applied stress of $\sigma_{\text{mc}} = 220$ MPa and approached saturation at the applied stress of $\sigma_{\text{sat}} = 420$ MPa, and the matrix cracking density increased from $\varphi = 0.2$ mm to the saturation value of $\varphi = 6.5$ mm. At the oxidation temperature of $T = 800$ °C for the oxidation duration of $t = 3000$ s, the matrix cracking density increased from $\varphi = 0.16$ mm at the first matrix cracking stress of $\sigma_{\text{mc}} = 220$ MPa to $\varphi = 5.6$ mm at the saturation matrix cracking stress of $\sigma_{\text{sat}} = 420$ MPa, the interface debonding ratio increased from $2l_d/l_c = 0.8$ to $2l_d/l_c = 79.4\%$, the interface oxidation ratio decreased from $\zeta/l_d = 49.1$ to $\zeta/l_d = 17.7\%$, and the broken fiber fraction increased from $P = 0.04$ to $P = 0.6\%$. At the oxidation temperature of $T = 800$ °C for the oxidation duration of $t = 5000$ s, the matrix cracking density increased from $\varphi = 0.13$ mm at the first matrix cracking stress of $\sigma_{\text{mc}} = 220$ MPa to $\varphi = 5.0$ mm at the saturation matrix cracking stress of $\sigma_{\text{sat}} = 420$ MPa, the interface debonding ratio increased from $2l_d/l_c = 0.85$ to $2l_d/l_c = 77.4\%$, the interface oxidation ratio decreased from $\zeta/l_d = 65.3$ to $\zeta/l_d = 27.1\%$, and the broken fiber fraction increased from $P = 0.05$ to $P = 0.6\%$, as shown in Fig. 2.33.

2.5 Cyclic-Dependent Matrix Multicracking Evolution of Fiber-Reinforced Ceramic-Matrix Composites Considering Interface Wear and Fiber Fracture

In this section, the synergistic effects of fiber/matrix interface wear and fiber fracture on matrix multiple cracking in fiber-reinforced CMCs are investigated using the critical matrix strain energy criterion. The shear-lag model combined with the fiber/matrix interface wear model, fiber fracture model, and the fiber/matrix interface fracture mechanics debonding criterion is adopted to analyze the fiber and matrix axial stress distribution inside of the damaged composite. The relationships between the matrix multiple cracking, fatigue peak stress, applied cyclic number, fiber/matrix interface wear and debonding, and fiber failure are established. The effects of the fiber volume, fiber/matrix interface shear stress, fiber/matrix interface debonded energy, cycle number, fatigue peak stress, and fibers strength on cyclic-dependent matrix multiple cracking evolution are discussed. Comparisons of matrix multiple cracking with/without cyclic fatigue loading are analyzed. The experimental multiple matrix cracking of unidirectional SiC/CAS, SiC/CAS-II, SiC/borosilicate, and mini-SiC/SiC composites with/without cyclic fatigue loading is predicted.

2.5.1 *Cyclic-Dependent Stress Analysis*

Upon first loading to fatigue peak stress $\sigma_{\max 1}$, matrix cracking and fiber/matrix interface debonding occur. After experiencing N applied cycles, the fiber/matrix interface shear stress in the interface debonded region degrades from the initial value τ_i to τ_f, and the fibers strength also degrades due to the fiber/matrix interface wear. Upon increasing applied stress, matrix cracks propagate along the fiber/matrix interface. To analyze stress distributions in fibers and the matrix, a unit cell is extracted from the ceramic composite system. The unit cell contains a single fiber surrounded by a hollow cylinder of matrix. The fiber radius is r_f, and the matrix radius is R ($R = r_f/V_f^{1/2}$). The length of the unit cell is $l_c/2$, which is just the half matrix crack space. The fiber/matrix interface debonded length l_d can be divided into two regions, i.e., the interface debonded region with the interface shear stress of τ_f ($x \in [0, \zeta]$) and the interface debonded region with the interface shear stress of τ_i ($x \in [\zeta, l_d]$), in which ζ denotes the interface debonded length at fatigue peak stress $\sigma_{\max 1}$. Considering fiber fracture, the fiber and matrix axial stress distributions in the interface wear region, interface debonded region, and interface bonded region can be described using the following equations.

$$\sigma_{\mathrm{f}}(z)=\begin{cases}\Phi-\dfrac{2\tau_{\mathrm{f}}}{r_{\mathrm{f}}}z,\ z\in[0,\zeta]\\ \Phi-\dfrac{2\tau_{\mathrm{f}}}{r_{\mathrm{f}}}\zeta-\dfrac{2\tau_{\mathrm{i}}}{r_{\mathrm{f}}}(z-\zeta),\ z\in[\zeta,l_{\mathrm{d}}]\\ \sigma_{\mathrm{fo}}+\left[\Phi-\sigma_{\mathrm{fo}}-\dfrac{2\tau_{\mathrm{f}}}{r_{\mathrm{f}}}\zeta-\dfrac{2\tau_{\mathrm{i}}}{r_{\mathrm{f}}}(l_{\mathrm{d}}-\zeta)\right]\exp\left(-\rho\dfrac{z-l_{\mathrm{d}}}{r_{\mathrm{f}}}\right),\ z\in\left[l_{\mathrm{d}},\dfrac{l_{\mathrm{c}}}{2}\right]\end{cases}\tag{2.70}$$

$$\sigma_{\mathrm{m}}(z)=\begin{cases}2\dfrac{V_{\mathrm{f}}}{V_{\mathrm{m}}}\dfrac{\tau_{\mathrm{f}}}{r_{\mathrm{f}}}z,\ z\in[0,\zeta]\\ 2\dfrac{V_{\mathrm{f}}}{V_{\mathrm{m}}}\dfrac{\tau_{\mathrm{f}}}{r_{\mathrm{f}}}\zeta+2\dfrac{V_{\mathrm{f}}}{V_{\mathrm{m}}}\dfrac{\tau_{\mathrm{i}}}{r_{\mathrm{f}}}(z-\zeta),\ z\in[\zeta,l_{\mathrm{d}}]\\ \sigma_{\mathrm{mo}}-\dfrac{V_{\mathrm{f}}}{V_{\mathrm{m}}}\left[\Phi-\sigma_{\mathrm{fo}}-\dfrac{2\tau_{\mathrm{f}}}{r_{\mathrm{f}}}\zeta-\dfrac{2\tau_{\mathrm{i}}}{r_{\mathrm{f}}}(l_{\mathrm{d}}-\zeta)\right]\exp\left(-\rho\dfrac{z-l_{\mathrm{d}}}{r_{\mathrm{f}}}\right),\ z\in\left[l_{\mathrm{d}},\dfrac{l_{\mathrm{c}}}{2}\right]\end{cases}\tag{2.71}$$

$$\tau_{\mathrm{i}}(z)=\begin{cases}\tau_{\mathrm{f}},\ z\in[0,\zeta]\\ \tau_{\mathrm{i}},\ z\in[\zeta,l_{\mathrm{d}}]\\ \dfrac{\rho}{2}\left[\Phi-\sigma_{\mathrm{fo}}-2\dfrac{\tau_{\mathrm{f}}}{r_{\mathrm{f}}}\zeta-\dfrac{2\tau_{\mathrm{i}}}{r_{\mathrm{f}}}(l_{\mathrm{d}}-\zeta)\right]\exp\left(-\rho\dfrac{z-l_{\mathrm{d}}}{r_{\mathrm{f}}}\right),\ z\in\left[l_{\mathrm{d}},\dfrac{l_{\mathrm{c}}}{2}\right]\end{cases}\tag{2.72}$$

where V_{f} and V_{m} denote the fiber and matrix volume fraction, respectively; T denotes the intact fiber stress; ρ denotes the shear-lag model parameter (Budiansky et al. 1986); and σ_{fo} and σ_{mo} denote the fiber and matrix axial stress in the interface bonded region, respectively.

$$\sigma_{\mathrm{fo}}=\frac{E_{\mathrm{f}}}{E_{\mathrm{c}}}\sigma+E_{\mathrm{f}}(\alpha_{\mathrm{c}}-\alpha_{\mathrm{f}})\Delta T\tag{2.73}$$

$$\sigma_{\mathrm{mo}}=\frac{E_{\mathrm{m}}}{E_{\mathrm{c}}}\sigma+E_{\mathrm{m}}(\alpha_{\mathrm{c}}-\alpha_{\mathrm{m}})\Delta T\tag{2.74}$$

where E_{f}, E_{m}, and E_{c} denote the fiber, matrix, and composite elastic modulus, respectively; α_{f}, α_{m}, and α_{c} denote the fiber, matrix, and composite thermal expansion coefficient, respectively; and ΔT denotes the temperature difference between fabricated temperature T_0 and testing temperature T_1 ($\Delta T = T_1 - T_0$).

2.5.2 Cyclic Dependent Interface Debonding

The fiber/matrix interface debonding length is determined using the fracture mechanics approach (Gao et al. 1988).

$$\xi_{\mathrm{d}} = -\frac{F}{4\pi r_{\mathrm{f}}}\frac{\partial w_{\mathrm{f}}(0)}{\partial l_{\mathrm{d}}} - \frac{1}{2}\int_{0}^{l_{\mathrm{d}}} \tau_{\mathrm{i}}\frac{\partial v(z)}{\partial l_{\mathrm{d}}}\mathrm{d}z \tag{2.75}$$

where $F(=\pi r_{\mathrm{f}}^2\sigma/V_{\mathrm{f}})$ denotes the fiber load at the matrix cracking plane; $w_{\mathrm{f}}(0)$ denotes the fiber axial displacement at the matrix cracking plane; and $v(z)$ denotes the relative displacement between fibers and the matrix.

The fiber and matrix axial displacements (i.e., $w_{\mathrm{f}}(z)$ and $w_{\mathrm{m}}(z)$) are described using the following equations.

$$\begin{aligned} w_{\mathrm{f}}(z) &= \int_{x}^{\frac{l_{\mathrm{c}}}{2}} \frac{\sigma_{\mathrm{f}}(z)}{E_{\mathrm{f}}}\mathrm{d}z \\ &= \frac{\Phi}{E_{\mathrm{f}}}(l_{\mathrm{d}} - z) - \frac{\tau_{\mathrm{f}}}{r_{\mathrm{f}}E_{\mathrm{f}}}\left(2\zeta l_{\mathrm{d}} - z^2 - \zeta^2\right) - \frac{\tau_{\mathrm{i}}}{r_{\mathrm{f}}E_{\mathrm{f}}}(l_{\mathrm{d}} - \zeta)^2 + \frac{\sigma_{\mathrm{fo}}}{E_{\mathrm{f}}}\left(\frac{l_{\mathrm{c}}}{2} - l_{\mathrm{d}}\right) \\ &\quad + \frac{r_{\mathrm{f}}}{\rho E_{\mathrm{f}}}\left[T - \sigma_{\mathrm{fo}} - \frac{2\tau_{\mathrm{f}}}{r_{\mathrm{f}}}\zeta - \frac{2\tau_{\mathrm{i}}}{r_{\mathrm{f}}}(l_{\mathrm{d}} - \zeta)\right]\left[1 - \exp\left(-\rho\frac{l_{\mathrm{c}}/2 - l_{\mathrm{d}}}{r_{\mathrm{f}}}\right)\right] \end{aligned} \tag{2.76}$$

$$\begin{aligned} w_{\mathrm{m}}(z) &= \int_{x}^{\frac{l_{\mathrm{c}}}{2}} \frac{\sigma_{\mathrm{m}}(z)}{E_{\mathrm{m}}}\mathrm{d}z \\ &= \frac{V_{\mathrm{f}}\tau_{\mathrm{f}}}{r_{\mathrm{f}}V_{\mathrm{m}}E_{\mathrm{m}}}\left(2\zeta l_{\mathrm{d}} - \zeta^2 - z^2\right) + \frac{V_{\mathrm{f}}\tau_{\mathrm{i}}}{r_{\mathrm{f}}V_{\mathrm{m}}E_{\mathrm{m}}}(l_{\mathrm{d}} - \zeta)^2 + \frac{\sigma_{\mathrm{mo}}}{E_{\mathrm{m}}}\left(\frac{l_{\mathrm{c}}}{2} - l_{\mathrm{d}}\right) \\ &\quad - \frac{r_{\mathrm{f}}V_{\mathrm{f}}}{\rho V_{\mathrm{m}}E_{\mathrm{m}}}\left[T - \sigma_{\mathrm{fo}} - \frac{2\tau_{\mathrm{f}}}{r_{\mathrm{f}}}\zeta - \frac{2\tau_{\mathrm{i}}}{r_{\mathrm{f}}}(l_{\mathrm{d}} - \zeta)\right]\left[1 - \exp\left(-\rho\frac{l_{\mathrm{c}}/2 - l_{\mathrm{d}}}{r_{\mathrm{f}}}\right)\right] \end{aligned} \tag{2.77}$$

The relative displacement $v(z)$ between the fiber and the matrix is described using the following equation.

$$\begin{aligned} v(z) &= |w_{\mathrm{f}}(z) - w_{\mathrm{m}}(z)| \\ &= \frac{\Phi}{E_{\mathrm{f}}}(l_{\mathrm{d}} - z) - \frac{E_{\mathrm{c}}\tau_{\mathrm{f}}}{r_{\mathrm{f}}V_{\mathrm{m}}E_{\mathrm{f}}E_{\mathrm{m}}}\left(2\zeta l_{\mathrm{d}} - \zeta^2 - z^2\right) - \frac{E_{\mathrm{c}}\tau_{\mathrm{i}}}{r_{\mathrm{f}}V_{\mathrm{m}}E_{\mathrm{f}}E_{\mathrm{m}}}(l_{\mathrm{d}} - \zeta)^2 \\ &\quad + \frac{r_{\mathrm{f}}}{\rho E_{\mathrm{f}}}\left[\Phi - \sigma_{\mathrm{fo}} - \frac{2\tau_{\mathrm{f}}}{r_{\mathrm{f}}}\zeta - \frac{2\tau_{\mathrm{i}}}{r_{\mathrm{f}}}(l_{\mathrm{d}} - \zeta)\right]\left[1 - \exp\left(-\rho\frac{l_{\mathrm{c}}/2 - l_{\mathrm{d}}}{r_{\mathrm{f}}}\right)\right] \\ &\quad + \frac{r_{\mathrm{f}}V_{\mathrm{f}}}{\rho V_{\mathrm{m}}E_{\mathrm{m}}}\left[\Phi - \sigma_{\mathrm{fo}} - \frac{2\tau_{\mathrm{f}}}{r_{\mathrm{f}}}\zeta - \frac{2\tau_{\mathrm{i}}}{r_{\mathrm{f}}}(l_{\mathrm{d}} - \zeta)\right]\left[1 - \exp\left(-\rho\frac{l_{\mathrm{c}}/2 - l_{\mathrm{d}}}{r_{\mathrm{f}}}\right)\right] \end{aligned} \tag{2.78}$$

Substituting $w_{\mathrm{f}}(z = 0)$ and $v(z)$ into Eq. (2.75), it leads to the following equation.

$$\frac{r_f\Phi^2}{4E_f} - \frac{\Phi\tau_f}{E_f}\zeta - \frac{\Phi\tau_i}{E_f}(l_d - \zeta) - \frac{r_f\sigma\Phi}{4E_c} - \frac{r_f\Phi\tau_i}{2\rho E_f} + \frac{E_c\tau_f^2}{r_f V_m E_f E_m}\zeta^2 + \frac{2E_c\tau_f\tau_i}{r_f V_m E_f E_m}\zeta(l_d - \zeta) + \frac{E_c\tau_f\tau_i}{\rho V_m E_f E_m}\zeta + \frac{E_c\tau_i^2}{r_f V_m E_f E_m}(l_d - \zeta)^2 + \frac{E_c\tau_i^2}{\rho V_m E_f E_m}(l_d - \zeta) - \xi_d = 0 \quad (2.79)$$

Solving Eq. (2.79), the fiber/matrix interface debonding length is determined using the following equation.

$$l_d = \left(1 - \frac{\tau_f}{\tau_i}\right)\zeta + \frac{r_f}{2}\left(\frac{V_m E_m T}{E_c\tau_i} - \frac{1}{\rho}\right) - \left[\left(\frac{r_f}{2\rho}\right)^2 + \frac{r_f^2 V_f V_m E_f E_m T^2}{4E_c^2\tau_i^2}\left(\frac{\sigma}{V_f T} - 1\right) + \frac{r_f V_m E_f E_m}{E_c\tau_i^2}\xi_d\right]^{\frac{1}{2}} \quad (2.80)$$

2.5.3 Cyclic-Dependent Interface Wear

When a CMC is subjected to a cyclic loading between a peak stress and a valley stress, damage mechanisms of matrix cracking, fiber/matrix interface debonding, and frictional sliding occur. Under subsequent applied cycles, the fiber/matrix interface shear stress degrades with increasing applied cycle numbers due to the fiber/matrix interface wear. Evans et al. (1995) developed an approach to evaluate the fiber/matrix interface shear stress by analyzing the parabolic regions of stress-strain hysteresis loops based on the Vagaggini's hysteresis loop models (Vagaggini et al. 1995). The initial fiber/matrix interface shear stress of unidirectional SiC/CAS composite was approximately 20 MPa and degraded to 5 MPa at the 30th cycle. The degradation of fiber/matrix interface shear stress with increasing applied cycles can be described using the following equation (Evans et al. 1995).

$$\tau_i(N) = \tau_{io} + \left[1 - \exp\left(-\omega N^{\lambda}\right)\right]\left(\tau_{i\min} - \tau_{io}\right) \quad (2.81)$$

where τ_{io} denotes the initial fiber/matrix interface shear stress; τ_{imin} denotes the steady-state fiber/matrix interface shear stress; and ω and λ are empirical constants.

Lee and Stinchcomb (1994) performed fiber fracture mirror experiments of CMCs under scanning electron microscope and developed the fiber strength degradation model under cyclic loading.

$$\sigma_0(N) = \sigma_0\left[1 - p_1(\log N)^{p_2}\right] \quad (2.82)$$

where σ_0 denotes the fiber strength at the first cycle; $\sigma_0(N)$ denotes the fiber strength at the Nth cycle; and p_1 and p_2 are empirical parameters.

2.5.4 Cyclic-Dependent Fiber Failure

As fibers begin to break, the stress dropped by broken fibers must be transferred to intact fibers at the cross section. Two dominant failure criterions are present in the literatures for modeling fibers failure, i.e., the global load sharing (GLS) and local load sharing (LLS) criterions. The GLS criterion assumes that the stress from any one fiber is transferred equally to all other intact fibers at the same cross-sectional plane. The GLS assumption neglects any local stress concentrations in the neighborhood of existing breaks and is expected to be accurate when the interface shear stress is sufficiently low. The LLS assumes that the load from broken fibers is transferred to the neighborhood intact fibers and is expected to be accurate when the interface shear stress is sufficiently high. The two-parameter Weibull model is adopted to describe the fiber strength distribution. The fiber fracture probability $P(\Phi)$ can be described using the following equation (Curtin 1991).

$$P(\Phi) = 1 - \exp\left(-\int_{L_0} \frac{1}{l_0}\left[\frac{\sigma_f(z)}{\sigma_0}\right]^m \mathrm{d}z\right) \tag{2.83}$$

where σ_0 denotes the fiber strength at the tested gauge length of l_0; m denotes the fiber Weibull modulus; and L_0 denotes the integral length.

The fiber fracture probabilities in the fiber/matrix interface wear region, interface debonded region, and interface bonded region of $P_a(\Phi)$, $P_b(\Phi)$, and $P_c(\Phi)$ are given by the following equations.

$$P_a(\Phi) = 1 - \exp\left\{-\frac{1}{\eta\kappa^m(m+1)}\left[1 - \left(1 - \frac{\eta\zeta}{l_s}\right)^{m+1}\right]\left(\frac{\Phi}{\sigma_c}\right)^{m+1}\right\} \tag{2.84}$$

$$P_b(\Phi) = 1 - \exp\left\{-\frac{1}{m+1}\left(\frac{\Phi}{\sigma_c}\right)^{m+1}\left[\left(1 - \frac{\eta\zeta}{l_s}\right)^{m+1} - \left(1 - \frac{\eta\zeta}{l_s} - \frac{l_d - \zeta}{l_s}\right)^{m+1}\right]\right\} \tag{2.85}$$

$$P_c(\Phi) = 1 - \exp\left\{-\frac{1}{\rho(m+1)\left(1 - \frac{\sigma_{fo}}{T} - \frac{\eta\zeta}{l_s} - \frac{l_d-\zeta}{l_s}\right)}\left(\frac{r_f}{l_s}\right)\left(\frac{\Phi}{\sigma_c}\right)^{m+1}\right.$$

$$\left[\left(1 - \frac{\eta\zeta}{l_s} - \frac{l_d - \zeta}{l_s}\right)^{m+1} - \left(1 - \frac{\eta\zeta}{l_s} - \frac{l_d - \zeta}{l_s}\right.\right.$$

$$-\rho\frac{l_c/2-l_d}{r_f}\left(1-\frac{\sigma_{fo}}{\Phi}-\frac{\eta\zeta}{l_s}-\frac{l_d-\zeta}{l_s}\right)\Bigg)^{m+1}\Bigg]\Bigg\} \quad (2.86)$$

where κ denotes the fiber strength degradation ratio.

$$\kappa=\frac{\sigma_0(N)}{\sigma_0} \quad (2.87)$$

and l_f denotes the slip length over which the fiber stress would decay to zero if not interrupted by the far-field equilibrium stresses.

$$l_f=(1-\eta)\zeta+l_s \quad (2.88)$$

and

$$l_s=\frac{r_f\Phi}{2\tau_i} \quad (2.89)$$

The average load per fiber σ/V_f must be equal to the product of the stress carried by the unbroken fibers Φ, and the fraction $(1-P(\Phi))$ of unbroken fibers, plus a contribution due to the residual stress $<\Phi_b>$ carried by broken fibers which do not break exactly at the matrix crack plane. The equilibrium relationship at the matrix crack plane can be described using the following equation.

$$\frac{\sigma}{V_f}=(1-P(\Phi))\Phi+P(\Phi)\langle\Phi_b\rangle \quad (2.90)$$

where

$$P(\Phi)=P_a(\Phi)+P_b(\Phi)+P_c(\Phi) \quad (2.91)$$

When a fiber breaks, the load carried by the fiber drops to zero at the position of break. Similar to the case of matrix cracking, the fiber/matrix interface debonds and the stress builds up in the fiber through the interface shear stress. During the process of loading, the stress in a broken fiber Φ_b as a function of the distance x from the break can be described using the following equation.

$$\Phi_b(z)=\begin{cases}\frac{2\tau_f}{r_f}z,\ z\in[0,\zeta]\\ \frac{2\tau_f}{r_f}\zeta+\frac{2\tau_i}{r_f}(z-\zeta),\ z\in[\zeta,l_f]\end{cases} \quad (2.92)$$

In order to calculate the average stress carried by broken fibers $<\Phi_b>$, it is necessary to construct the probability distribution $f(z)$ of the distance z of a fiber break from the reference matrix crack plane, provided that a break occurs within a distance

$\pm l_f$. For this conditional probability distribution, Phoenix and Raj (1992) deduced the following from based on the Weibull statistics.

$$f(z) = \frac{1}{P(\Phi)l_f}\left(\frac{\Phi}{\sigma_c}\right)^{m+1} \exp\left[-\left(\frac{z}{l_f}\right)\left(\frac{\Phi}{\sigma_c}\right)^{m+1}\right], z \in [0, l_f] \tag{2.93}$$

where σ_c denotes the characteristic strength of fibers in the characteristic length scale of δ_c.

$$\sigma_c = \left(\frac{l_0\sigma_0^m\tau_i}{r_f}\right)^{\frac{1}{m+1}}, \delta_c = \left(\frac{\sigma_0 r_f l_0^{1/m}}{\tau_i}\right)^{\frac{m}{m+1}} \tag{2.94}$$

The averaging stress carried by broken fibers $<\Phi_b>$ during the process of loading using Eqs. (2.92) and (2.93) leads to the following equation.

$$\begin{aligned}\langle\Phi_b\rangle = {} & \frac{2(1-\eta)l_f\tau_i}{r_f P(\Phi)}\left(\frac{\sigma_c}{\Phi}\right)^{m+1} \exp\left[-\left(\frac{\zeta}{l_f}\right)\left(\frac{\Phi}{\sigma_c}\right)^{m+1}\right] \\ & - \frac{2(\zeta + l_s)\tau_i}{r_f P(\Phi)} \exp\left[-\left(\frac{\Phi}{\sigma_c}\right)^{m+1}\right] + \frac{2l_f\tau_f}{r_f P(\Phi)}\left(\frac{\sigma_c}{\Phi}\right)^{m+1}\end{aligned} \tag{2.95}$$

Substituting Eqs. (2.91) and (2.95) into Eq. (2.90), the relationship between the applied stress σ and the peak stress Φ on the intact fibers can be obtained.

2.5.5 Cyclic-Dependent Matrix Multicracking

Solti et al. (1995) developed the critical matrix strain energy (CMSE) criterion to predict multiple matrix cracking in fiber-reinforced CMCs. The concept of critical matrix strain energy presupposes the existence of an ultimate or critical strain energy. Beyond the critical value of matrix strain energy, as more energy is entered into the composite with increasing applied stress, matrix cannot support the extra stress and continues to fail. The failure is assumed to consist of the formation of new matrix cracks and fiber/matrix interface debonding, to make the total energy within the matrix remain constant and equal to its critical value.

The matrix strain energy is determined using the following equation.

$$U_m = \frac{1}{2E_m}\int_{A_m}\int_0^{l_c} \sigma_m^2(x)\mathrm{d}x\mathrm{d}A_m \tag{2.96}$$

where A_{m} is the cross-sectional area of matrix in the unit cell. Substituting the matrix axial stress in Eq. (2.71) into Eq. (2.96), the matrix strain energy considering matrix cracking and fiber/matrix interface partially debonding is described using the following equation.

$$U_{\mathrm{m}}=\frac{A_{\mathrm{m}}}{E_{\mathrm{m}}}\left\{\frac{4}{3}\left(\frac{V_{\mathrm{f}}}{V_{\mathrm{m}}}\frac{\tau_{\mathrm{f}}}{r_{\mathrm{f}}}\right)^{2}\zeta^{3}+4\left(\frac{V_{\mathrm{f}}}{V_{\mathrm{m}}}\frac{\tau_{\mathrm{f}}}{r_{\mathrm{f}}}\right)^{2}\zeta^{2}(l_{\mathrm{d}}-\zeta)+4\left(\frac{V_{\mathrm{f}}}{r_{\mathrm{f}}V_{\mathrm{m}}}\right)^{2}\tau_{\mathrm{f}}\tau_{\mathrm{i}}\zeta(l_{\mathrm{d}}-\zeta)^{2}\right.$$
$$+\frac{4}{3}\left(\frac{V_{\mathrm{f}}}{V_{\mathrm{m}}}\frac{\tau_{\mathrm{i}}}{r_{\mathrm{f}}}\right)^{2}(l_{\mathrm{d}}-\zeta)^{3}+\sigma_{\mathrm{mo}}^{2}\left(\frac{l_{\mathrm{c}}}{2}-l_{\mathrm{d}}\right)-2\sigma_{\mathrm{mo}}\frac{V_{\mathrm{f}}}{V_{\mathrm{m}}}\frac{r_{\mathrm{f}}}{\rho}\left[T-\sigma_{\mathrm{fo}}-\frac{2\tau_{\mathrm{f}}}{r_{\mathrm{f}}}\zeta\right.$$
$$\left.-\frac{2\tau_{\mathrm{i}}}{r_{\mathrm{f}}}(l_{\mathrm{d}}-\zeta)\right]\left[1-\exp\left(-\rho\frac{l_{\mathrm{c}}/2-l_{\mathrm{d}}}{r_{\mathrm{f}}}\right)\right]+\frac{r_{\mathrm{f}}}{2\rho}\left(\frac{V_{\mathrm{f}}}{V_{\mathrm{m}}}\right)^{2}$$
$$\left.\left[T-\sigma_{\mathrm{fo}}-\frac{2\tau_{\mathrm{f}}}{r_{\mathrm{f}}}\zeta-\frac{2\tau_{\mathrm{i}}}{r_{\mathrm{f}}}(l_{\mathrm{d}}-\zeta)\right]^{2}\left[1-\exp\left(-2\rho\frac{l_{\mathrm{c}}/2-l_{\mathrm{d}}}{r_{\mathrm{f}}}\right)\right]\right\} \quad (2.97)$$

When the fiber/matrix interface completely debonds, the matrix strain energy is described using the following equation.

$$U_{\mathrm{m}}(\sigma,l_{\mathrm{c}},l_{\mathrm{d}}=l_{\mathrm{c}}/2)=\frac{A_{\mathrm{m}}}{E_{\mathrm{m}}}\left[\frac{4}{3}\left(\frac{V_{\mathrm{f}}}{V_{\mathrm{m}}}\frac{\tau_{\mathrm{f}}}{r_{\mathrm{f}}}\right)^{2}\zeta^{3}+4\left(\frac{V_{\mathrm{f}}}{V_{\mathrm{m}}}\frac{\tau_{\mathrm{f}}}{r_{\mathrm{f}}}\right)^{2}\zeta^{2}(l_{\mathrm{d}}-\zeta)\right.$$
$$\left.+4\left(\frac{V_{\mathrm{f}}}{r_{\mathrm{f}}V_{\mathrm{m}}}\right)^{2}\tau_{\mathrm{f}}\tau_{\mathrm{i}}\zeta(l_{\mathrm{d}}-\zeta)^{2}+\frac{4}{3}\left(\frac{V_{\mathrm{f}}}{V_{\mathrm{m}}}\frac{\tau_{\mathrm{i}}}{r_{\mathrm{f}}}\right)^{2}(l_{\mathrm{d}}-\zeta)^{3}\right] \quad (2.98)$$

Evaluating the matrix strain energy at a critical stress of σ_{cr}, the critical matrix strain energy of U_{crm} can be obtained. The critical matrix strain energy is described using the following equation.

$$U_{\mathrm{crm}}=\frac{1}{2}kA_{\mathrm{m}}l_{0}\frac{\sigma_{\mathrm{mocr}}^{2}}{E_{\mathrm{m}}} \quad (2.99)$$

where k ($k\in[0,1]$) is the critical matrix strain energy parameter; and l_0 is the initial matrix crack spacing, and σ_{mocr} is determined using the following equation.

$$\sigma_{\mathrm{mocr}}=\frac{E_{\mathrm{m}}}{E_{\mathrm{c}}}\sigma_{\mathrm{cr}}+E_{\mathrm{m}}(\alpha_{\mathrm{c}}-\alpha_{\mathrm{m}})\Delta T \quad (2.100)$$

where σ_{cr} is the critical stress corresponding to composite's proportional limit stress, i.e., the stress at which the stress-strain curve starts to deviate from linearity due to damage accumulation of matrix cracks (Li 2017). The critical stress is defined to be ACK matrix cracking stress and is determined using the following equation (Aveston et al. 1971).

$$\sigma_{\mathrm{cr}} = \left(\frac{6V_{\mathrm{f}}^2 E_{\mathrm{f}} E_{\mathrm{c}}^2 \tau_{\mathrm{i}} \xi_{\mathrm{m}}}{r_{\mathrm{f}} V_{\mathrm{m}} E_{\mathrm{m}}^2} \right)^{\frac{1}{3}} - E_{\mathrm{c}}(\alpha_{\mathrm{c}} - \alpha_{\mathrm{m}})\Delta T \tag{2.101}$$

where ξ_{m} denotes the matrix fracture energy.

The energy balance relationship to evaluate multiple matrix cracking is determined using the following equation.

$$U_{\mathrm{m}}(\sigma > \sigma_{\mathrm{cr}}, l_{\mathrm{c}}, l_{\mathrm{d}}) = U_{\mathrm{crm}}(\sigma_{\mathrm{cr}}, l_0) \tag{2.102}$$

The multiple matrix cracking versus applied stress can be solved by Eq. (2.102) when the critical matrix cracking stress σ_{cr} and the fiber/matrix interface debonded length l_{d} are determined by Eqs. (2.80) and (2.101).

2.5.6 *Results and Discussion*

The effects of the fiber volume, interface shear stress, interface debonding energy, applied cycle number, fatigue peak stress, and fiber strength on matrix multiple fracture, interface debonding, and fiber failure are analyzed. The ceramic composite system of SiC/CAS is used for the case study, and its material properties are given by: $V_{\mathrm{f}} = 30\%$, $E_{\mathrm{f}} = 200$ GPa, $E_{\mathrm{m}} = 97$ GPa, $r_{\mathrm{f}} = 7.5$ μm, $\xi_{\mathrm{m}} = 6$ J/m^2, $\xi_{\mathrm{d}} = 0.8$ J/m^2, $\tau_{\mathrm{i}} = 25$ MPa, $\alpha_{\mathrm{f}} = 4 \times 10^{-5}$ °C, $\alpha_{\mathrm{m}} = 5 \times 10^{-5}$ °C, $\Delta T = -1000$ °C, $\sigma_0 = 1.2$ GPa, and $m = 5$.

2.5.6.1 Effect of Fiber Volume on Cyclic-Dependent Matrix Multiple Fracture, Interface Debonding, and Fiber Failure

The cyclic-dependent matrix multicracking density, fiber/matrix interface debonding length, interface debonding ratio, interface wear ratio, and the broken fiber fraction corresponding to different fiber volume (i.e., $V_{\mathrm{f}} = 30$ and 35%) are shown in Fig. 2.34. With increasing fiber volume, the first matrix cracking stress, matrix saturation cracking stress, and matrix multicracking density increase; the fiber/matrix interface debonding length, interface debonding ratio, interface wear ratio, and the broken fiber fraction decrease.

When the fiber volume is $V_{\mathrm{f}} = 30\%$, the cyclic-dependent matrix multicracking density increases from $\varphi = 0.035$ mm at the first matrix cracking stress of $\sigma_{\mathrm{mc}} = 201$ MPa to $\varphi = 1.55$ mm at the matrix saturation stress of $\sigma_{\mathrm{sat}} = 360$ MPa; the cyclic-dependent fiber/matrix interface debonding length increases from $l_{\mathrm{d}}/r_{\mathrm{f}} = 18.6$ to 28; the cyclic-dependent interface debonding ratio increases from $2l_{\mathrm{d}}/l_{\mathrm{c}} = 0.9$ to 65.6%; the cyclic-dependent interface wear ratio decreases from $\zeta/l_{\mathrm{d}} = 1.0$ to 67%; and the broken fiber increases from $P = 0.3$ to 11%.

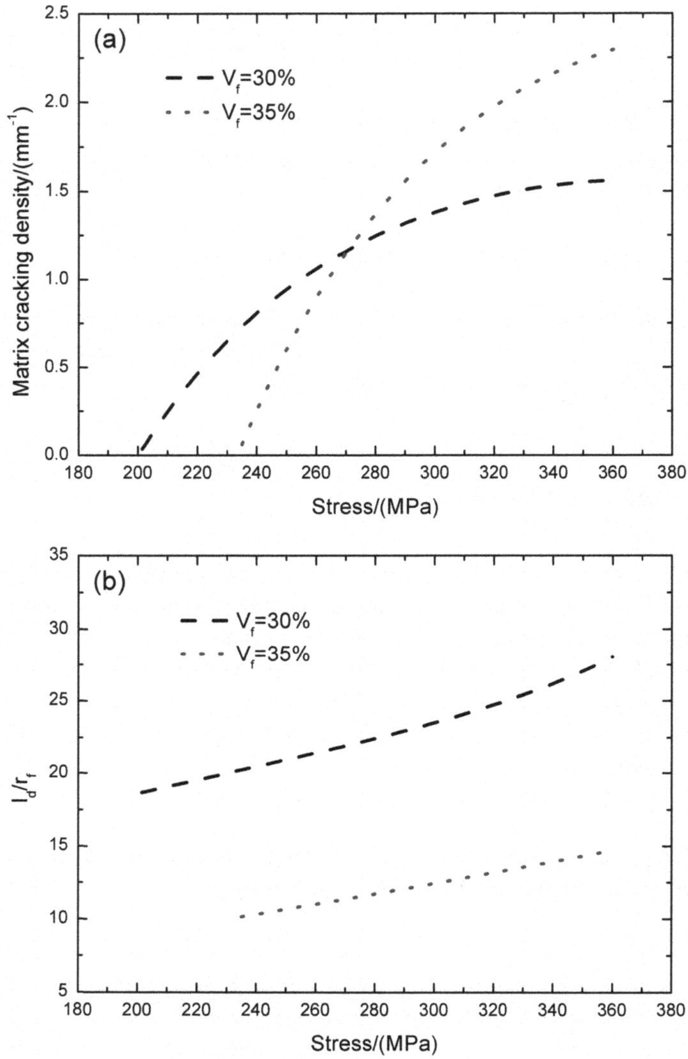

Fig. 2.34 Effect of fiber volume (i.e., $V_f = 30$ and 35%) on **a** the cyclic-dependent matrix multicracking density versus applied stress curves; **b** the cyclic-dependent fiber/matrix interface debonding length versus applied stress curves; **c** the cyclic-dependent fiber/matrix interface debonding ratio versus applied stress curves; **d** the cyclic-dependent fiber/matrix interface wear ratio versus applied stress curves; and **e** the cyclic-dependent broken fiber fraction versus applied stress curves

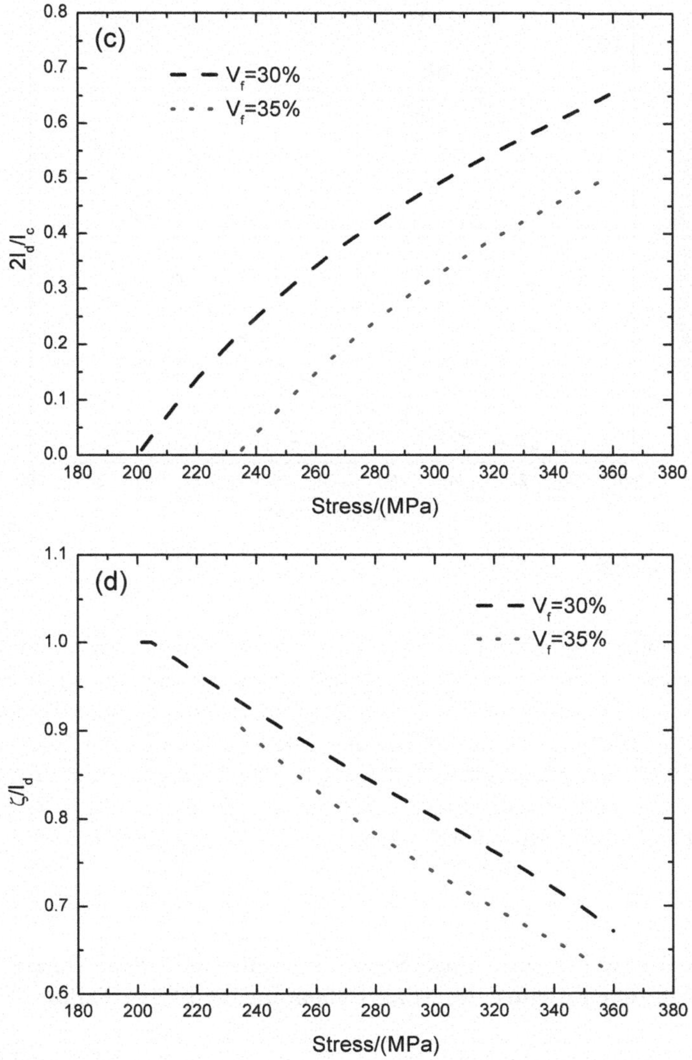

Fig. 2.34 (continued)

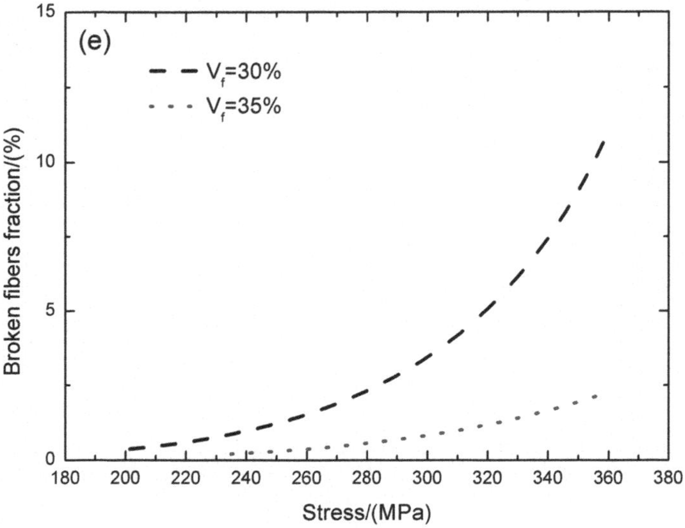

Fig. 2.34 (continued)

When the fiber volume is $V_f = 35\%$, the cyclic-dependent matrix multicracking density increases from $\varphi = 0.06$ mm at the first matrix cracking stress of $\sigma_{mc} = 235$ MPa to $\varphi = 2.3$ mm at the matrix saturation stress of $\sigma_{sat} = 360$ MPa; the cyclic-dependent fiber/matrix interface debonding length increases from $l_d/r_f = 10.1$ to 14.7; the cyclic-dependent interface debonding ratio increases from $2l_d/l_c = 0.95$ to 50.7%; the cyclic-dependent interface wear ratio decreases from $\zeta/l_d = 90.3$ to 62.4%; and the cyclic-dependent broken fiber fraction increases from $P = 0.22$ to 2.2%.

2.5.6.2 Effect of Interface Shear Stress on Cyclic-Dependent Matrix Multiple Fracture, Interface Debonding, and Fiber Failure

The cyclic-dependent matrix multicracking density, fiber/matrix interface debonding length, interface debonding ratio, interface wear ratio, and broken fiber fraction corresponding to different fiber/matrix interface shear stress (i.e., $\tau_i = 25$ and 30 MPa) are shown in Fig. 2.35. With increasing fiber/matrix interface shear stress, the first matrix cracking stress, saturation matrix cracking stress, and matrix cracking density increase; the fiber/matrix interface debonding length and interface debonding ratio decrease; the interface wear ratio increases; and the broken fiber fraction decreases.

When the interface shear stress is $\tau_i = 25$ MPa, the cyclic-dependent matrix multicracking density increases from $\varphi = 0.04$ mm at the first matrix cracking stress of $\sigma_{mc} = 221$ MPa to $\varphi = 1.7$ mm at the matrix saturation stress of $\sigma_{sat} = 360$ MPa; the cyclic-dependent fiber/matrix interface debonding length increases from $l_d/r_f =$ 16.3 to 22.6; the cyclic-dependent interface debonding ratio increases from $2l_d/l_c =$

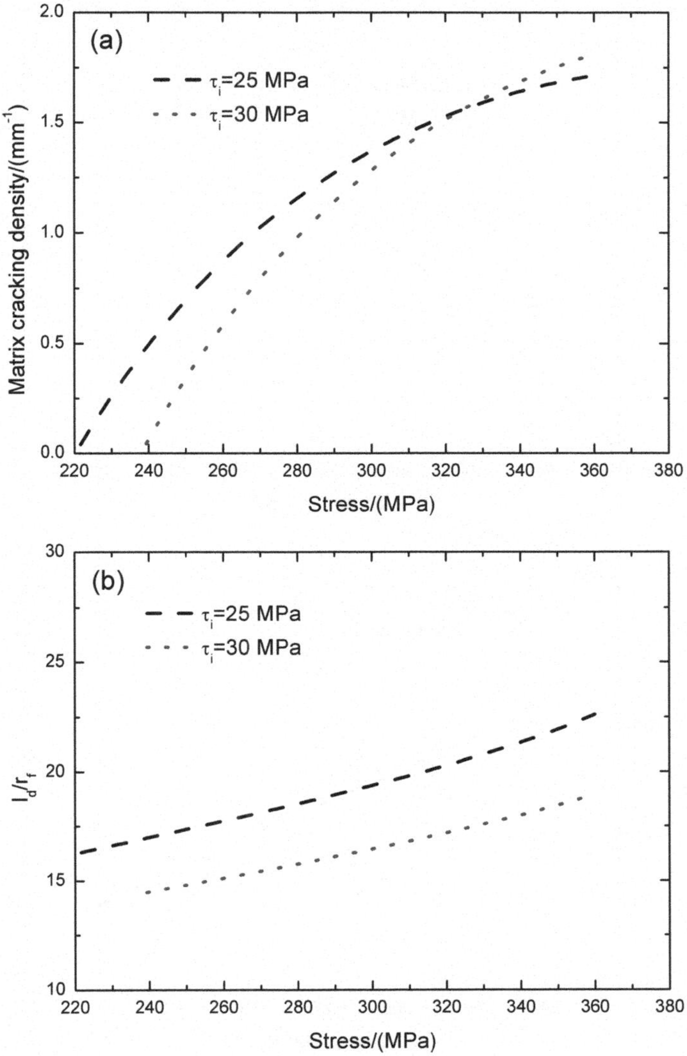

Fig. 2.35 Effect of fiber/matrix interface shear stress (i.e., $\tau_i = 25$ and 30 MPa) on **a** the cyclic-dependent matrix cracking density versus applied stress curves; **b** the cyclic-dependent fiber/matrix interface debonded length versus applied stress curves; **c** the cyclic-dependent fiber/matrix interface debonding ratio versus applied stress curves; **d** the cyclic-dependent fiber/matrix interface wear ratio versus applied stress curves; and **e** the cyclic-dependent broken fiber fraction versus applied stress curves

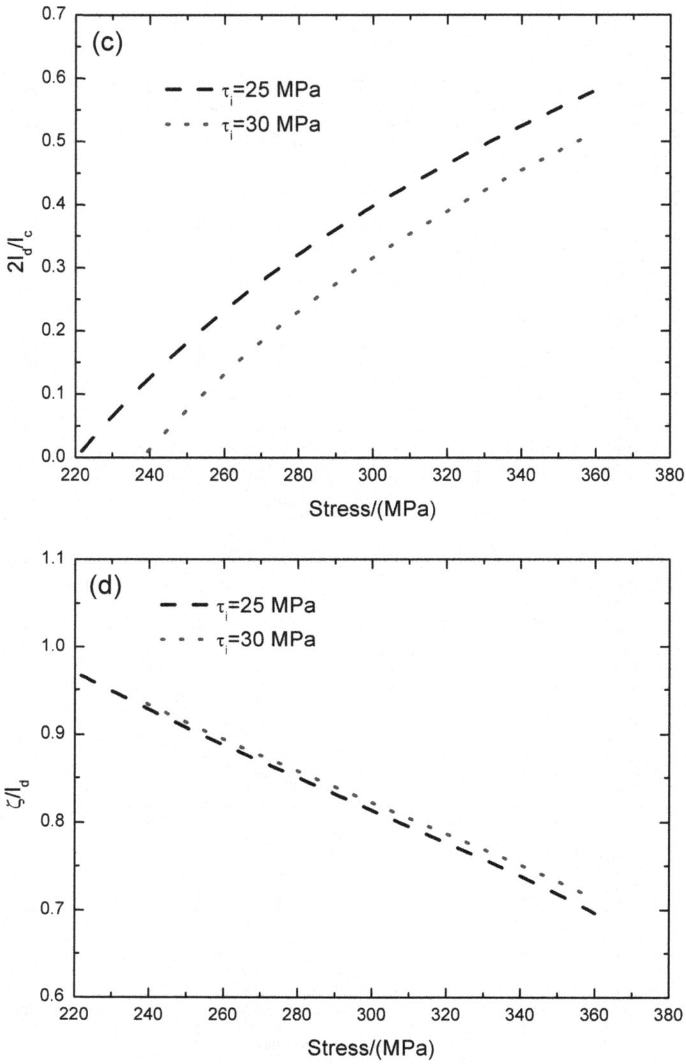

Fig. 2.35 (continued)

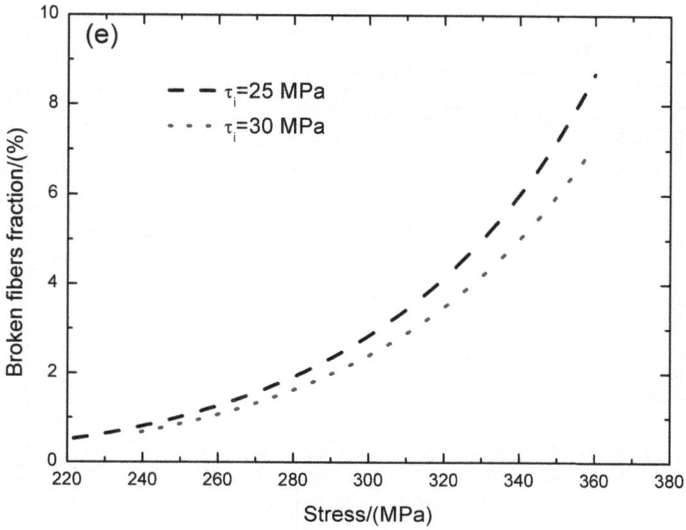

Fig. 2.35 (continued)

0.98 to 58.1%; the cyclic-dependent interface wear ratio decreases from $\zeta/l_d = 96.6$ to 69.6%; and the cyclic-dependent broken fiber fraction increases from $P = 0.5$ to 8.7%.

When the interface shear stress is $\tau_i = 30$ MPa, the cyclic-dependent matrix multicracking density increases from $\varphi = 0.045$ mm at the first matrix cracking stress of $\sigma_{mc} = 239$ MPa to $\varphi = 1.8$ mm at the matrix saturation stress of $\sigma_{sat} = 360$ MPa; the cyclic-dependent fiber/matrix interface debonding length increases from $l_d/r_f = 14.4$ to 18.9; the cyclic-dependent interface debonding ratio increases from $2l_d/l_c = 0.98$ to 51.5%; the cyclic-dependent interface wear ratio decreases from $\zeta/l_d = 93.4$ to 71.3%; and the cyclic-dependent broken fiber fraction increases from $P = 0.68$ to 7.1%.

2.5.6.3 Effect of Interface Debonding Energy on Cyclic-Dependent Matrix Multiple Fracture, Interface Debonding, and Fiber Failure

The cyclic-dependent matrix multicracking density, fiber/matrix interface debonding length, interface debonding ratio, interface wear ratio, and broken fiber fraction corresponding to different fiber/matrix interface debonding energy (i.e., $\xi_d = 1.0$ and 1.5 J/m^2) are shown in Fig. 2.36. With increasing interface debonding energy, the matrix multicracking density increases; the interface debonding length, interface debonding ratio, interface wear ratio, and the broken fiber fraction decrease.

When the fiber/matrix interface debonded energy is $\xi_d = 1.0$ J/m^2, the cyclic-dependent matrix multicracking density increases from $\varphi = 0.04$ mm at the first

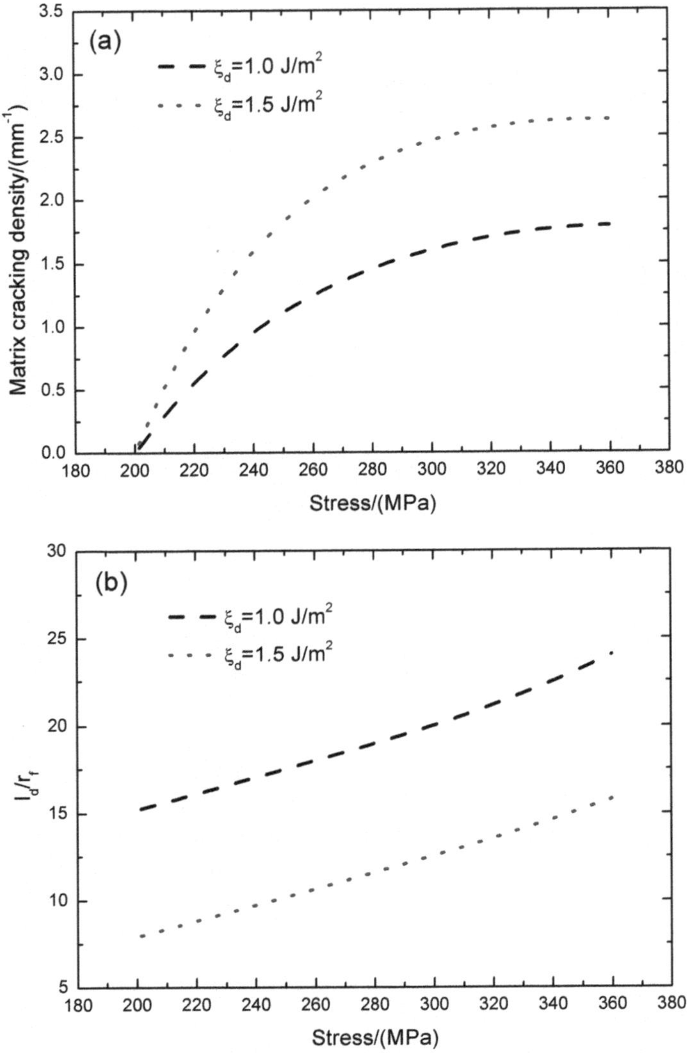

Fig. 2.36 Effect of fiber/matrix interface debonding energy (i.e., $\xi_d = 1.0$ and 1.5 J/m^2) on **a** the cyclic-dependent matrix multicracking density versus applied stress curves; **b** the cyclic-dependent fiber/matrix interface debonded length versus applied stress curves; **c** the cyclic-dependent fiber/matrix interface debonding ratio versus applied stress curves; **d** the cyclic-dependent fiber/matrix interface wear ratio versus applied stress curves; and **e** the cyclic-dependent broken fiber fraction versus applied stress curves

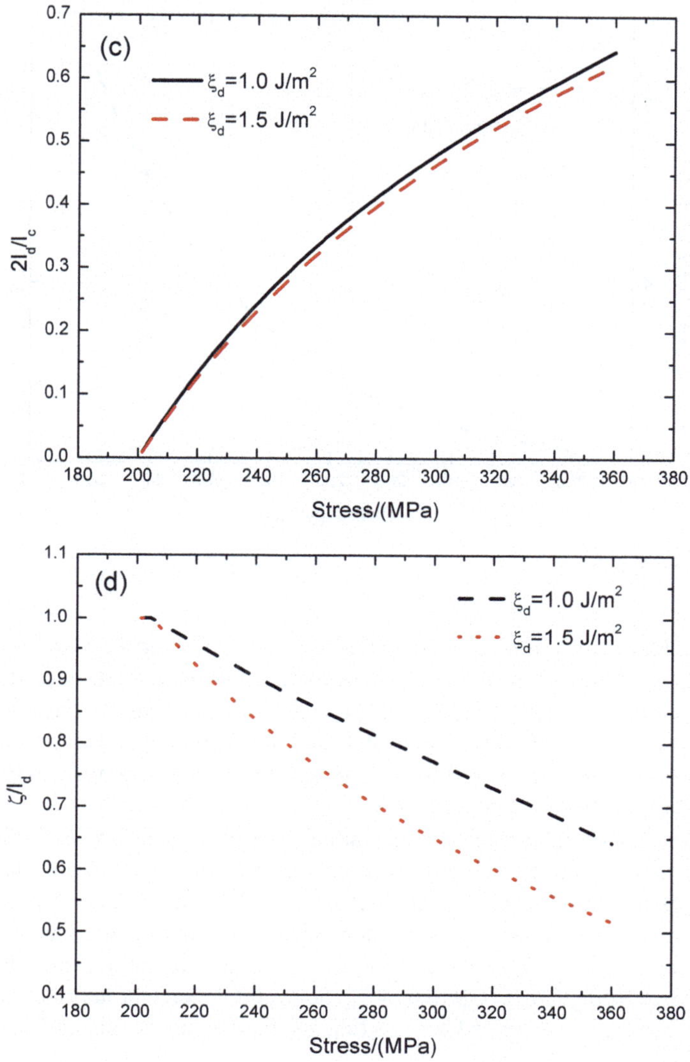

Fig. 2.36 (continued)

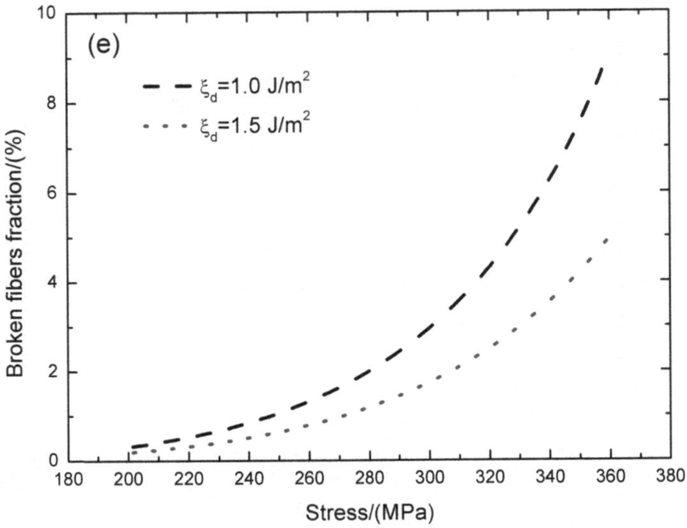

Fig. 2.36 (continued)

matrix cracking stress of σ_{mc} = 201 MPa to φ = 1.7 mm at the matrix saturation stress of σ_{sat} = 360 MPa; the cyclic-dependent fiber/matrix interface debonding length increases from l_d/r_f = 15.2 to 24; the cyclic-dependent interface debonding ratio increases from $2l_d/l_c$ = 0.97 to 64.4%; and the cyclic-dependent interface wear ratio decreases from ζ/l_d = 1.0 to 64.2%; and the cyclic-dependent broken fiber fraction increases from P = 0.3 to 9.0%.

When the fiber/matrix interface debonding energy is ξ_d = 1.5 J/m^2, the cyclic-dependent matrix multicracking density increases from φ = 0.07 mm at the first matrix cracking stress of σ_{mc} = 201 MPa to φ = 2.6 mm at the matrix saturation stress of σ_{sat} = 360 MPa; the cyclic-dependent fiber/matrix interface debonding length increases from l_d/r_f = 8.0 to 15.8; the cyclic-dependent interface debonding ratio increases from $2l_d/l_c$ = 0.9 to 62.4%; the cyclic-dependent interface wear ratio decreases from ζ/l_d = 1.0 to 51.6%; and the cyclic-dependent broken fiber fraction increases from P = 0.2 to 4.9%.

2.5.6.4 Effect of Applied Cycle Number on Cyclic-Dependent Matrix Multiple Fracture, Interface Debonding, and Fiber Failure

The cyclic-dependent matrix multicracking density, fiber/matrix interface debonding length, interface debonding ratio, interface wear ratio, and broken fiber fraction corresponding to different applied cycle numbers of N = 100 and 1000 are shown in Fig. 2.37. With increasing applied cycle number, the matrix multicracking density decreases; and the fiber/matrix interface debonding length, interface wear ratio, and the broken fiber fraction increase.

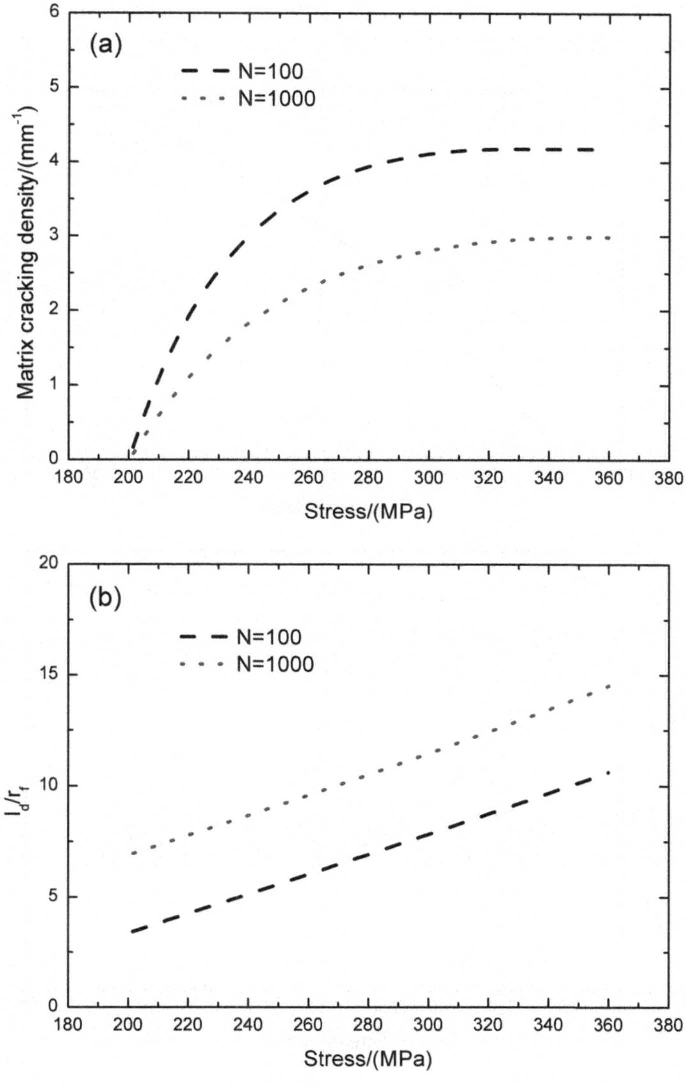

Fig. 2.37 Effect of applied cycle number (i.e., N = 100 and 1000) on **a** the cyclic-dependent matrix multicracking density versus applied stress curves; **b** the cyclic-dependent fiber/matrix interface debonding length versus applied stress curves; **c** the cyclic-dependent fiber/matrix interface debonding ratio versus applied stress curves; **d** the cyclic-dependent fiber/matrix interface wear ratio versus applied stress curves; and **e** the cyclic-dependent broken fiber fraction versus applied stress curves

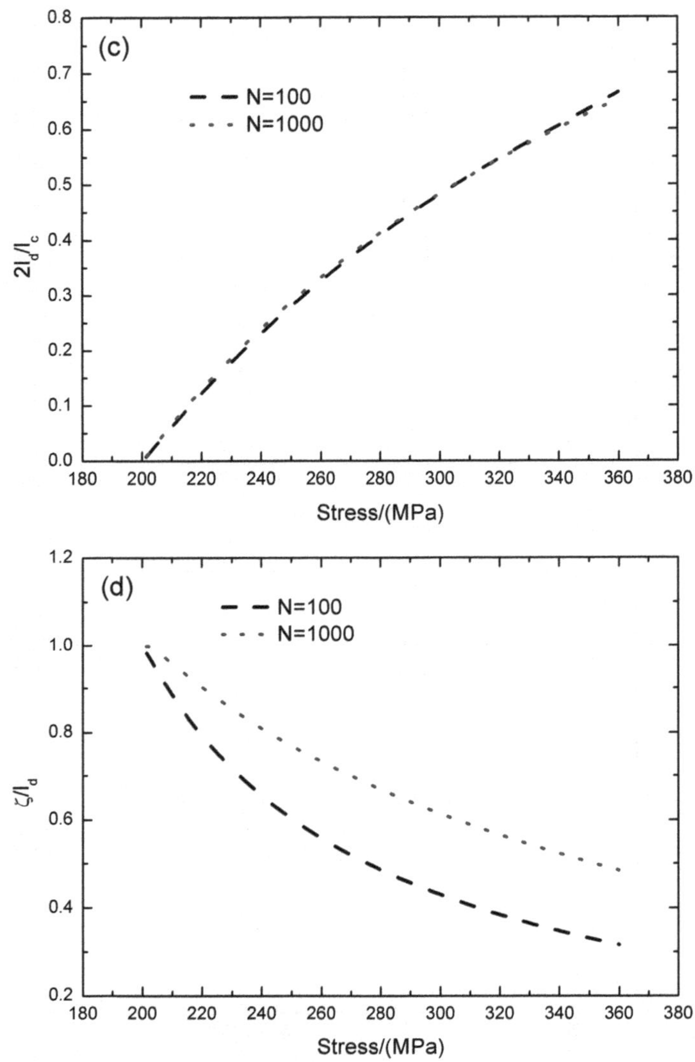

Fig. 2.37 (continued)

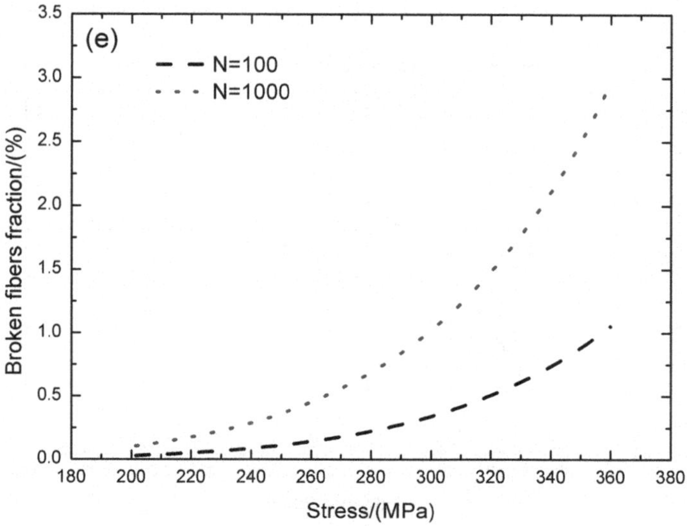

Fig. 2.37 (continued)

When the cycle number is $N = 100$, the cyclic-dependent matrix multicracking density increases from $\varphi = 0.16$ mm at the first matrix cracking stress of $\sigma_{mc} =$ 201 MPa to $\varphi = 4.1$ mm at the saturation matrix cracking stress of $\sigma_{sat} = 320$ MPa; the cyclic-dependent fiber/matrix interface debonding length increases from $l_d/r_f =$ 3.4 to 10.6; the cyclic-dependent interface debonding ratio increases from $2l_d/l_c =$ 0.85 to 66.5%; the cyclic-dependent interface wear ratio decreases from $\zeta/l_d = 98.3$ to 31.6%; and the cyclic-dependent broken fiber fraction increases from $P = 0.02$ to 1%.

When the applied cycle number is $N = 1000$, the cyclic-dependent matrix multicracking density increases from $\varphi = 0.09$ mm at the first matrix cracking stress of $\sigma_{mc} = 201$ MPa to $\varphi = 3.0$ mm at the saturation matrix cracking stress of σ_{sat} $= 360$ MPa; the cyclic-dependent fiber/matrix interface debonding length increases from $l_d/r_f = 6.9$ to 14.5; the interface debonding ratio increases from $2l_d/l_c = 0.9$ to 65.3%; the cyclic-dependent interface wear ratio decreases from $\zeta/l_d = 1.0$ to 48.5%; and the cyclic-dependent broken fiber fraction increases from $P = 0.1$ to 2.9%.

2.5.6.5 Effect of Fatigue Peak Stress on Cyclic-Dependent Matrix Multiple Fracture, Interface Debonding, and Fiber Failure

The cyclic-dependent matrix multicracking density, fiber/matrix interface debonding length, interface debonding ratio, interface wear ratio, and the broken fiber fraction corresponding to different fatigue peak stresses (i.e., $\sigma_{max1} = 150$ and 180 MPa) are shown in Fig. 2.38. With increasing fatigue peak stress, the matrix multicracking

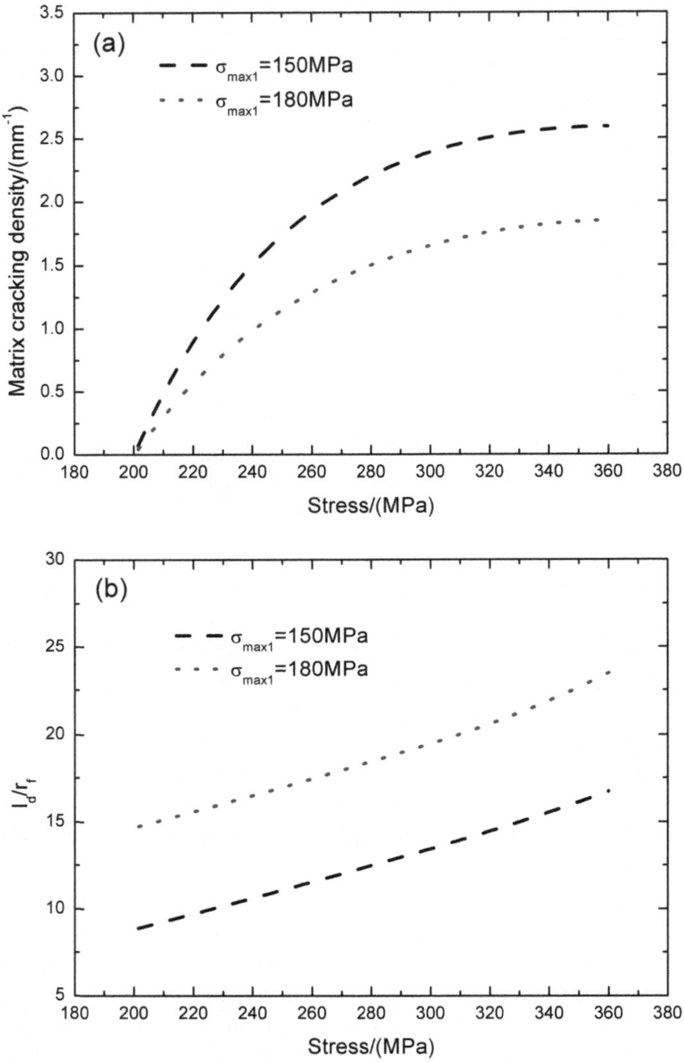

Fig. 2.38 Effect of fatigue peak stress (i.e., $\sigma_{\max 1} = 150$ and 180 MPa) on **a** the cyclic-dependent matrix multicracking density versus applied stress curves; **b** the cyclic-dependent interface debonding length versus applied stress curves; **c** the cyclic-dependent interface debonding ratio versus applied stress curves; **d** the cyclic-dependent interface wear ratio versus applied stress curves; and **e** the cyclic-dependent broken fiber fraction versus applied stress curves

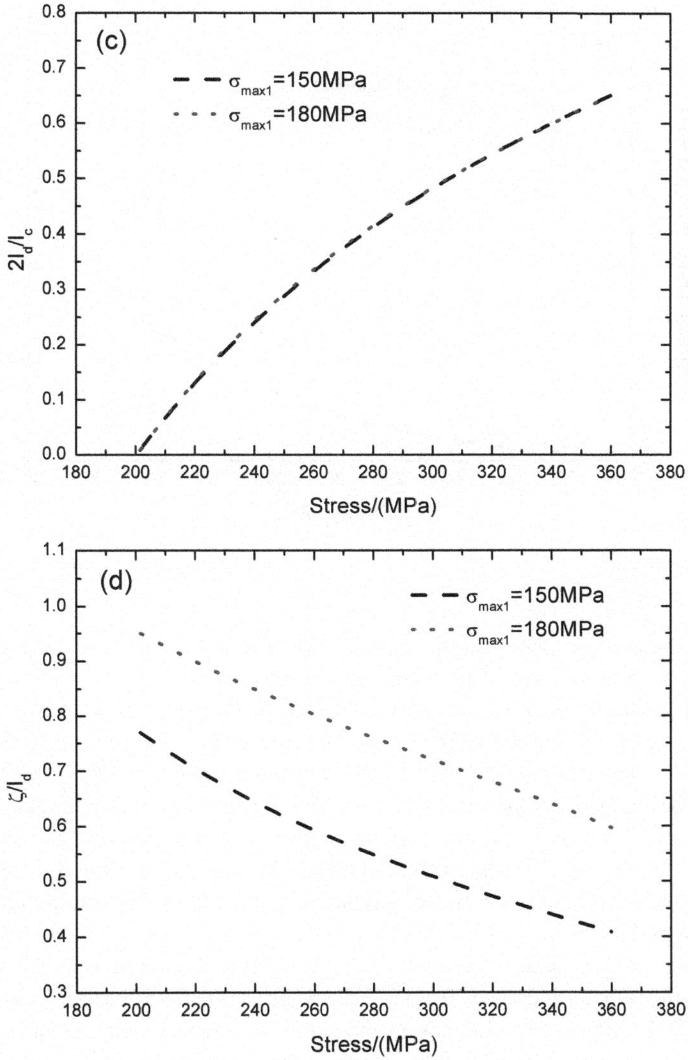

Fig. 2.38 (continued)

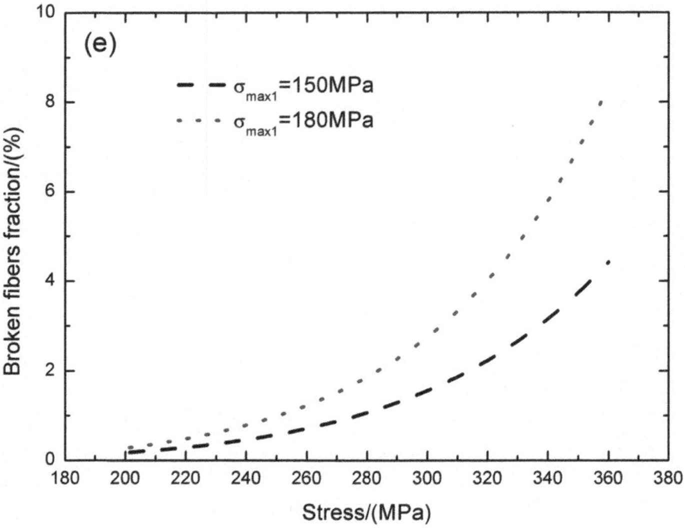

Fig. 2.38 (continued)

density decreases; the fiber/matrix interface debonding length and interface wear ratio increase; and the broken fiber fraction increases.

When the fatigue peak stress is $\sigma_{\max 1}$ = 150 MPa, the cyclic-dependent matrix multicracking density increases from $\varphi = 0.07$ mm at the first matrix cracking stress of σ_{mc} = 201 MPa to φ = 2.6 mm at the saturation matrix cracking stress of σ_{sat} = 360 MPa; the cyclic-dependent fiber/matrix interface debonding length increases from $l_{\mathrm{d}}/r_{\mathrm{f}}$ = 8.8 to 16.7; the cyclic-dependent interface debonding ratio increases from $2l_{\mathrm{d}}/l_{\mathrm{c}}$ = 0.9 to 65.1%; the cyclic-dependent interface wear ratio decreases from ζ/l_{d} = 76.9 to 40.8%; and the cyclic-dependent broken fiber fraction increases from P = 0.18 to 4.4%.

When the fatigue peak stress is $\sigma_{\max 1}$ = 180 MPa, the cyclic-dependent matrix multicracking density increases from $\varphi = 0.04$ mm at the first matrix cracking stress of σ_{mc} = 201 MPa to φ = 1.8 mm at the saturation matrix cracking stress of σ_{sat} = 360 MPa; the cyclic-dependent fiber/matrix interface debonding length increases from $l_{\mathrm{d}}/r_{\mathrm{f}}$ = 14.7 to 23.4; the cyclic-dependent interface debonding ratio increases from $2l_{\mathrm{d}}/l_{\mathrm{c}}$ = 0.9 to 65.2%; the cyclic-dependent interface wear ratio decreases from ζ/l_{d} = 95 to 59.7%; and the cyclic-dependent broken fiber fraction increases from P = 0.3 to 8.3%.

2.5.6.6 Effect of Fiber Strength on Cyclic-Dependent Matrix Multiple Fracture, Interface Debonding, and Fiber Failure

The cyclic-dependent matrix multicracking density, interface debonding length, interface debonding ratio, interface wear ratio, and the broken fiber fraction corresponding to different fiber strength (i.e., σ_0 = 1.0 and 1.5 GPa) are shown in Fig. 2.39. With increasing fiber strength, the matrix multicracking density increases; the fiber/matrix interface debonding length and interface debonding ratio decrease; the interface wear ratio increases; and the broken fiber fraction decreases.

When the fiber strength is $\sigma_0 = 1.0$ GPa, the cyclic-dependent matrix multicracking density increases from $\varphi = 0.03$ mm at the first matrix cracking stress of $\sigma_{mc} =$ 201 MPa to $\varphi = 1.38$ mm at the saturation matrix cracking stress of $\sigma_{sat} = 320$ MPa; the cyclic-dependent fiber/matrix interface debonding length increases from $l_d/r_f =$ 18.7 to 27.2; the cyclic-dependent interface debonding ratio increases from $2l_d/l_c =$ 0.9 to 56.6%; the cyclic-dependent interface wear ratio decreases from $\zeta/l_d = 1.0$ to 69.2%; and the broken fiber fraction increases from $P = 0.9$ to 16.9%.

When the fiber strength is $\sigma_0 = 1.5$ GPa, the cyclic-dependent matrix multicracking density increases from $\varphi = 0.03$ mm at the first matrix cracking stress of $\sigma_{mc} =$ 201 MPa to $\varphi = 1.5$ mm at the saturation matrix cracking stress of $\sigma_{sat} = 320$ MPa; the cyclic-dependent fiber/matrix interface debonding length increases from $l_d/r_f =$ 18.6 to 24.1; the cyclic-dependent interface debonding ratio increases from $2l_d/l_c =$ 0.9 to 54.2%; the cyclic-dependent interface wear ratio decreases from $\zeta/l_d = 1.0$ to 78%; and the cyclic-dependent broken fiber fraction increases from $P = 0.12$ to 1.5%.

2.5.6.7 Comparisons of Matrix Multicracking Evolution With/Without Interface Wear

Comparisons of matrix multicracking density, fiber/matrix interface debonding length, interface debonding ratio, interface wear ratio, and broken fiber fraction with/without interface wear (i.e., σ_{max1} = 200 MPa and N = 2000) are shown in Fig. 2.40. Considering fiber/matrix interface wear, the matrix multicracking density decreases; the fiber/matrix interface debonding length increases, and the interface debonding ratio decreases.

Without considering fiber/matrix interface wear, the matrix multicracking density increases from $\varphi = 0.13$ mm at the first matrix cracking stress of $\sigma_{mc} = 201$ MPa to $\varphi = 3.9$ mm at the saturation matrix cracking stress of $\sigma_{sat} = 360$ MPa; the fiber/matrix interface debonding length increases from $l_d/r_f = 4.5$ to 11.5; and the interface debonding ratio increases from $2l_d/l_c = 0.92$ to 68%.

Considering interface wear for σ_{max1} = 200 MPa and N = 2000, the cyclic-dependent matrix multicracking density increases from $\varphi = 0.03$ mm at the first matrix cracking stress of σ_{mc} = 201 MPa to $\varphi = 1.5$ mm at the matrix cracking stress of σ_{sat} = 360 MPa; the cyclic-dependent fiber/matrix interface debonding length increases from $l_d/r_f = 18.6$ to 28; the cyclic-dependent interface debonding

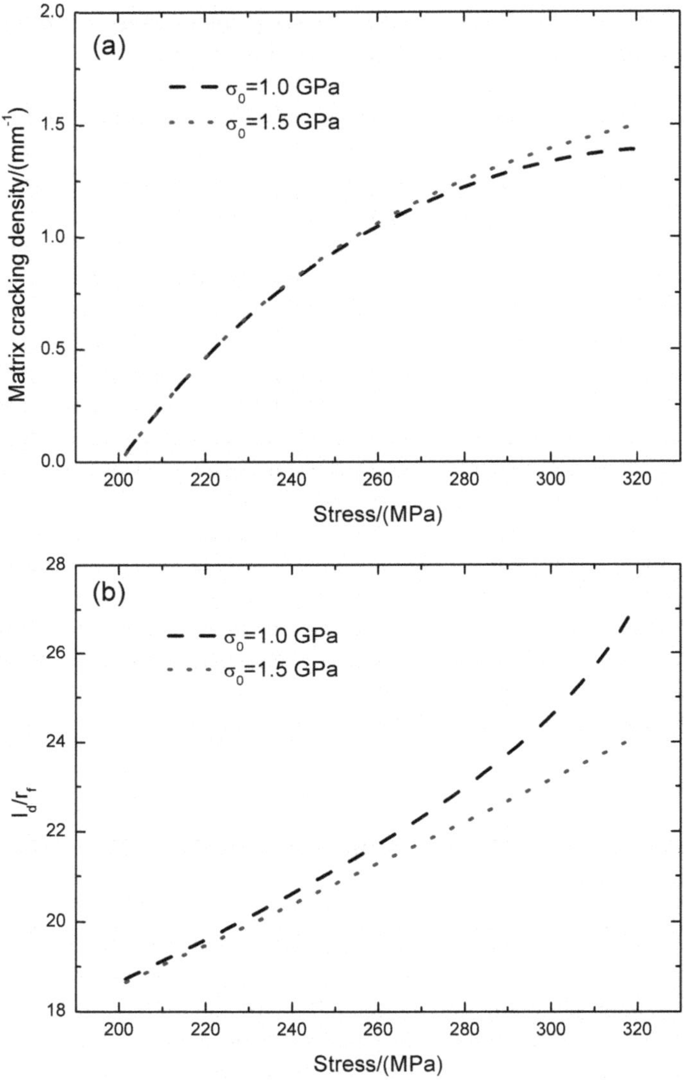

Fig. 2.39 Effect of fiber strength (i.e., $\sigma_0 = 1.0$ and 1.5 GPa) on **a** the cyclic-dependent matrix multicracking density versus applied stress curves; **b** the cyclic-dependent fiber/matrix interface debonding length versus applied stress curves; **c** the cyclic-dependent fiber/matrix interface debonding ratio versus applied stress curves; **d** the cyclic-dependent fiber/matrix interface wear ratio versus applied stress curves; and **e** the cyclic-dependent broken fiber fraction versus applied stress curves

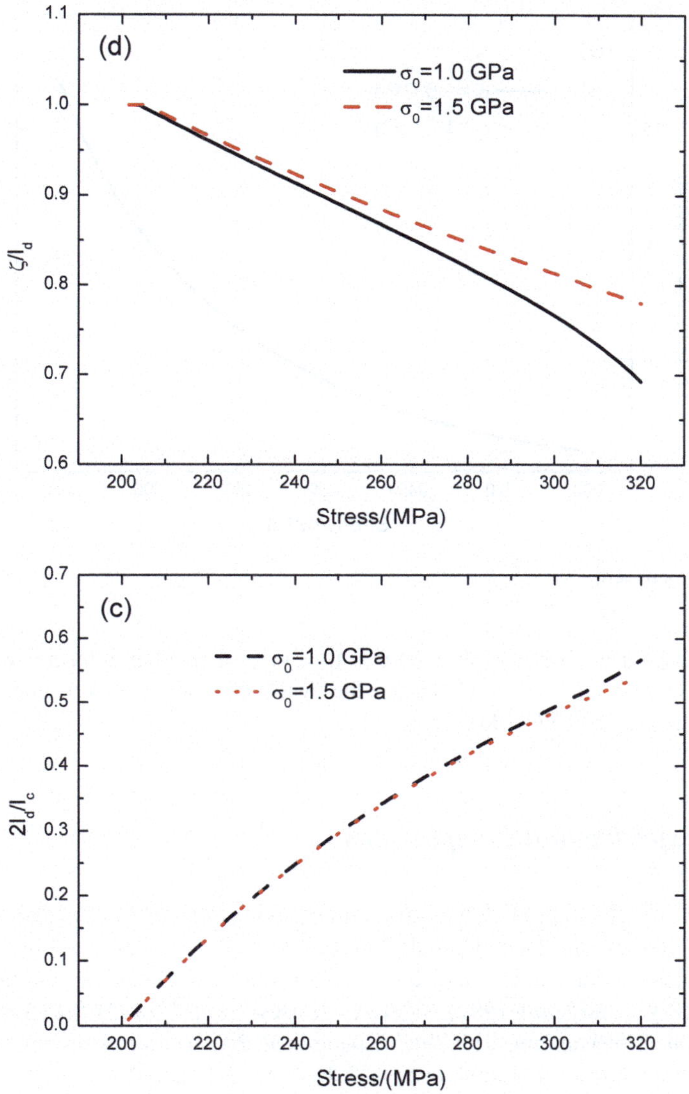

Fig. 2.39 (continued)

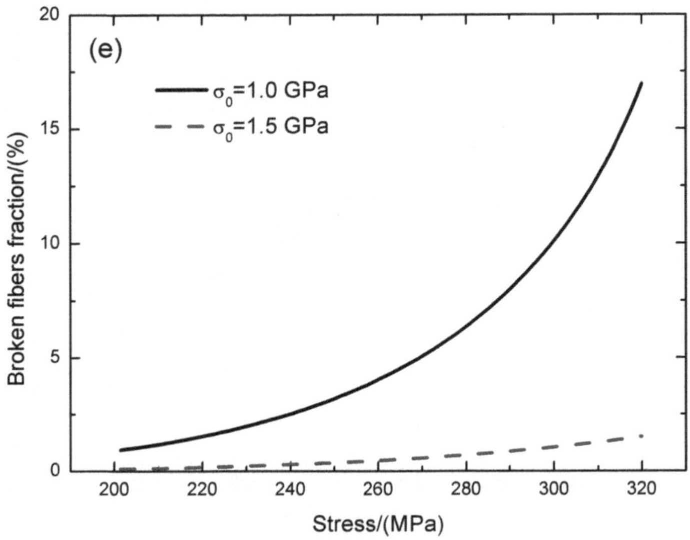

Fig. 2.39 (continued)

ratio increases from $2l_d/l_c = 0.9$ to 65.6%; the cyclic-dependent interface wear ratio decreases from $\zeta/l_d = 1.0$ to 67%; and the cyclic-dependent broken fiber fraction increases from $P = 0.37$ to 11%.

2.5.7 *Experimental Comparisons*

The experimental and predicted matrix multicracking density, fiber/matrix interface debonding length, interface debonding ratio, interface wear ratio, and broken fiber fraction versus applied stress for different fiber-reinforced CMCs, i.e., unidirectional SiC/CAS (Pryce and Smith 1992), SiC/CAS-II (Beyerle et al. 1992), SiC/Borosilicate (Okabe et al. 1999) and mini-SiC/SiC (Zhang et al. 2016) composites, are predicted using present analysis, as shown in Figs. 2.41, 2.42, 2.43 and 2.44.

2.5.7.1 SiC/CAS Composite

For the SiC/CAS composite without cyclic fatigue loading, the matrix multicracking starts at the applied stress of $\sigma_{mc} = 160$ MPa and approaches saturation at the applied stress of $\sigma_{sat} = 288$ MPa; the matrix multicracking density increases from $\varphi = 0.2$ mm to the saturation value of $\varphi = 7.0$ mm. With cyclic loading of $\sigma_{max1} = 150$ MPa and $N = 1500$, the cyclic-dependent matrix multicracking density increases from $\varphi = 0.19$ mm at $\sigma_{mc} = 160$ MPa to $\varphi = 5.8$ mm at 290 MPa; the cyclic-dependent interface debonding length increases from $l_d/r_f = 2.6$ to 8.8; the cyclic-dependent interface

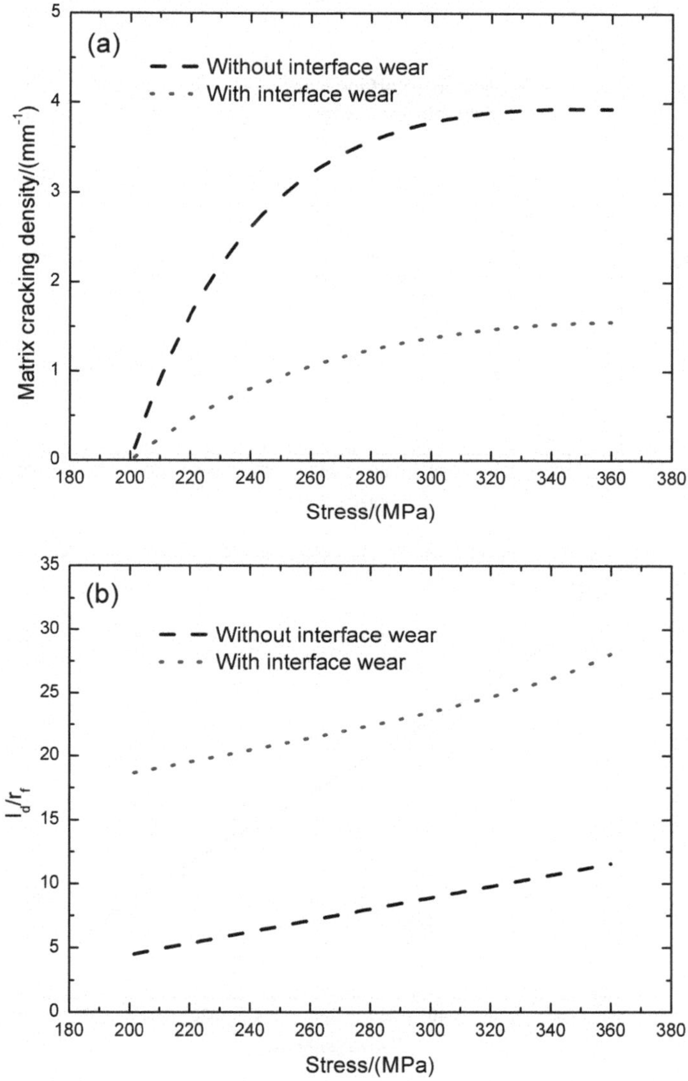

Fig. 2.40 Comparisons of with/without interface wear for **a** the matrix cracking density versus applied stress curves; **b** the fiber/matrix interface debonding length versus applied stress curves; **c** the fiber/matrix interface debonding ratio versus applied stress curves; **d** the fiber/matrix interface wear ratio versus applied stress curves; and **e** the broken fiber fraction versus applied stress curves

debonding ratio increases from $2l_d/l_c = 0.7$ to 76.7%; the cyclic-dependent interface wear ratio decreases from ζ/l_d = 96.6 to 28.7%; and the cyclic-dependent broken fiber fraction increases from $P = 0.09$ to 4%. With cyclic loading of $\sigma_{\max 1} = 150$ MPa and $N = 2000$, the cyclic-dependent matrix multicracking density increases from $\varphi =$ 0.12 mm at $\sigma_{mc} = 160$ MPa to $\varphi = 4.5$ mm at $\sigma_{sat} = 310$ MPa; the cyclic-dependent

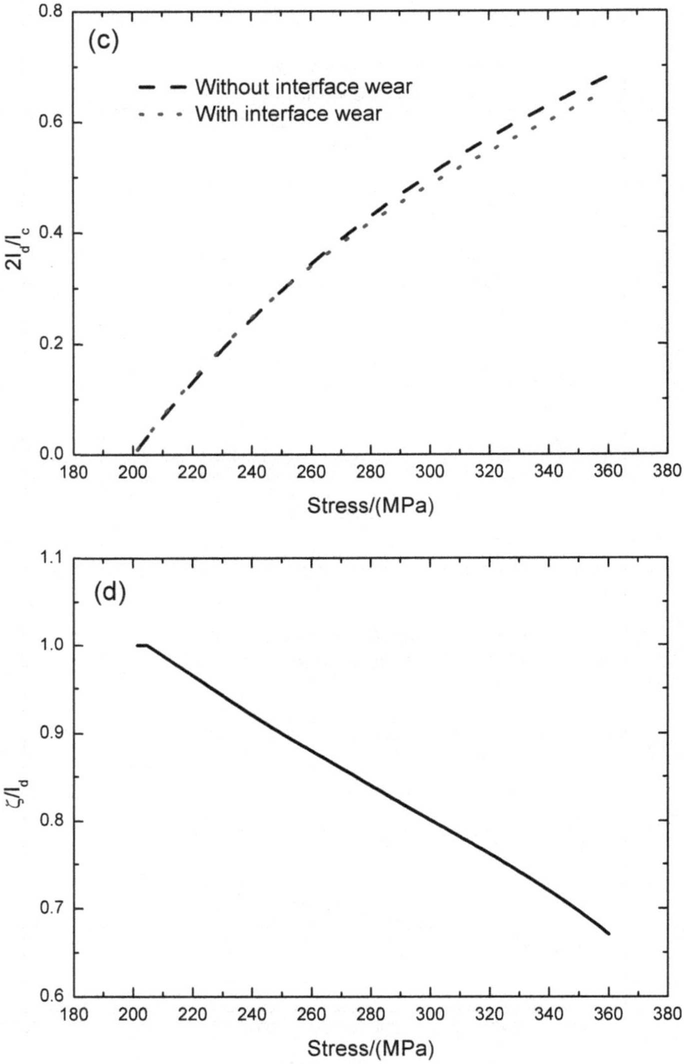

Fig. 2.40 (continued)

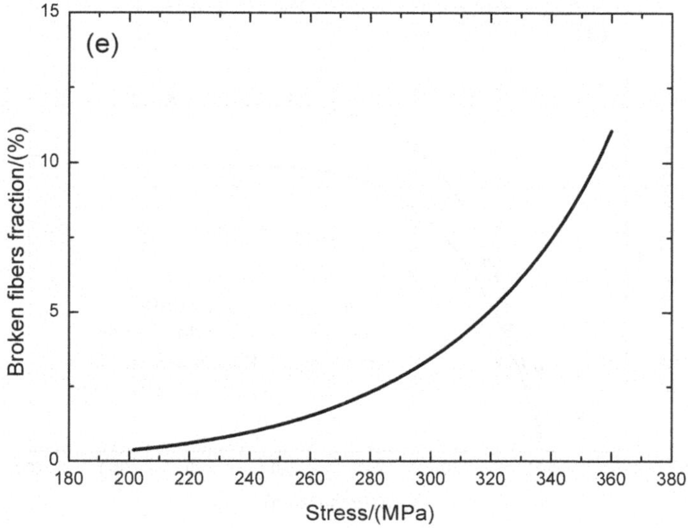

Fig. 2.40 (continued)

interface debonding length increases from $l_d/r_f = 4.4$ to 11; the cyclic-dependent interface debonding ratio increases from $2l_d/l_c = 0.8$ to 75.5%; the cyclic-dependent interface wear ratio decreases from $\zeta/l_d = 98.6$ to 39.5%; and the cyclic-dependent broken fiber fraction increases from $P = 0.1$ to 6.7%, as shown in Fig. 2.41.

2.5.7.2 SiC/CAS-II Composite

For the SiC/CAS-II composite without cyclic fatigue loading, the matrix multicracking starts at the applied stress of $\sigma_{mc} = 260$ MPa and approaches saturation at the applied stress of $\sigma_{sat} = 360$ MPa; and the matrix multicracking density increases from $\varphi = 1.9$ mm to the saturation value of $\varphi = 9.2$ mm. With cyclic fatigue loading of $\sigma_{max1} = 240$ MPa and $N = 1000$, the cyclic-dependent matrix multicracking density increases from $\varphi = 0.2$ mm at $\sigma_{mc} = 260$ MPa to $\varphi = 5.4$ mm at $\sigma_{sat} = 380$ MPa; the cyclic-dependent interface debonding length increases from $l_d/r_f = 2.7$ to 5.4; the interface debonding ratio increases from $2l_d/l_c = 0.8$ to 44.4%; the cyclic-dependent interface wear ratio decreases from $\zeta/l_d = 88.5$ to 44.8%; and the cyclic-dependent broken fiber fraction increases from $P = 0.29$ to 1.7%. With cyclic fatigue loading of $\sigma_{max1} = 240$ MPa and $N = 1500$, the cyclic-dependent matrix multicracking density increases from $\varphi = 0.135$ mm at $\sigma_{mc} = 260$ MPa to $\varphi = 4.1$ mm at $\sigma_{sat} = 380$ MPa; the cyclic-dependent fiber/matrix interface debonding length increases from $l_d/r_f =$ 4.4 to 7.2; the cyclic-dependent interface debonding ratio increases from $2l_d/l_c =$ 0.89 to 44.9%; the cyclic-dependent interface wear ratio decreases from $\zeta/l_d = 93.4$ to 57.1%; and the cyclic-dependent broken fiber fraction increases from $P = 0.5$ to 3.1%, as shown in Fig. 2.42.

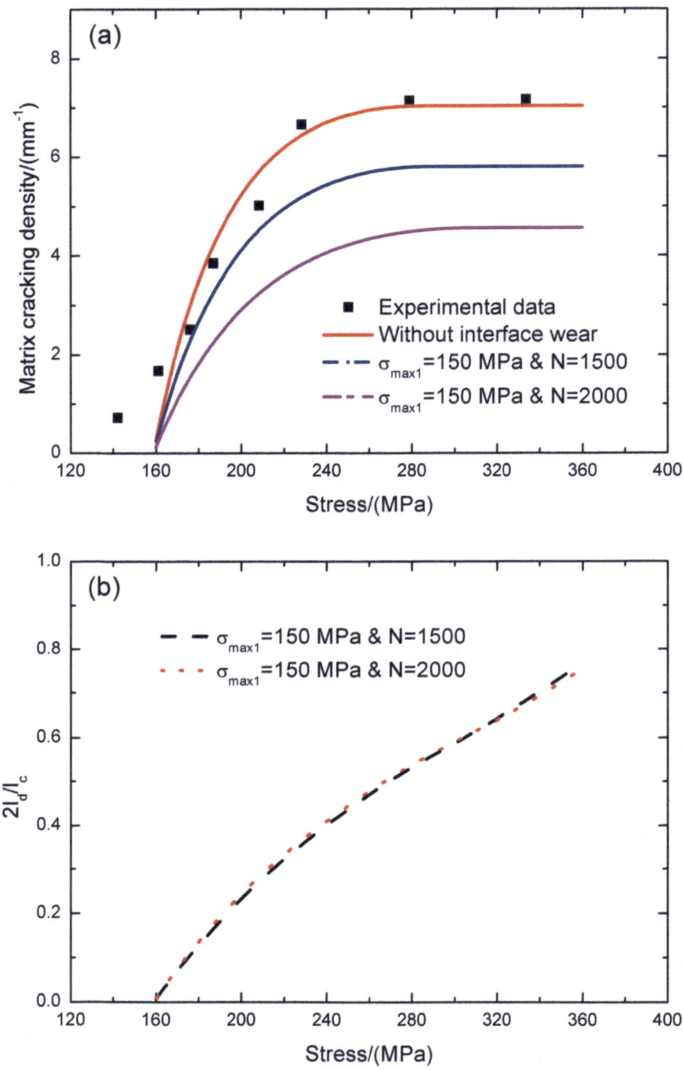

Fig. 2.41 **a** Experimental and predicted matrix multicracking density versus applied stress curves; **b** the fiber/matrix interface debonding ratio versus applied stress curves; **c** the fiber/matrix interface debonding length versus applied stress curves; **d** the fiber/matrix interface wear ratio versus of applied stress curves; and **e** the broken fiber fraction versus applied stress curves of unidirectional SiC/CAS composite with/without interface wear

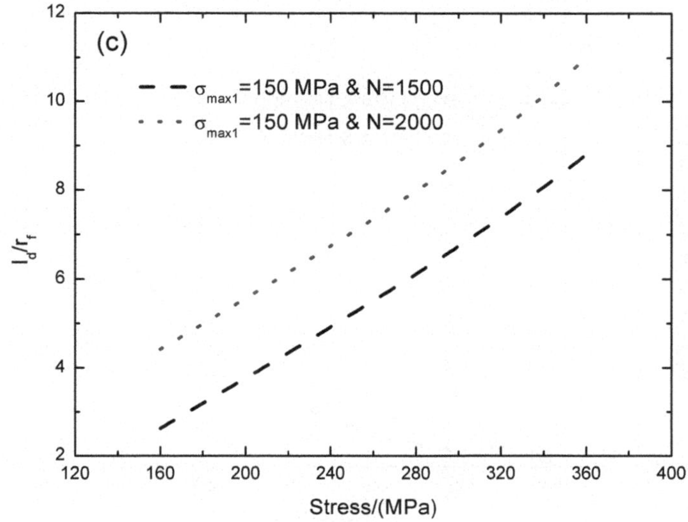

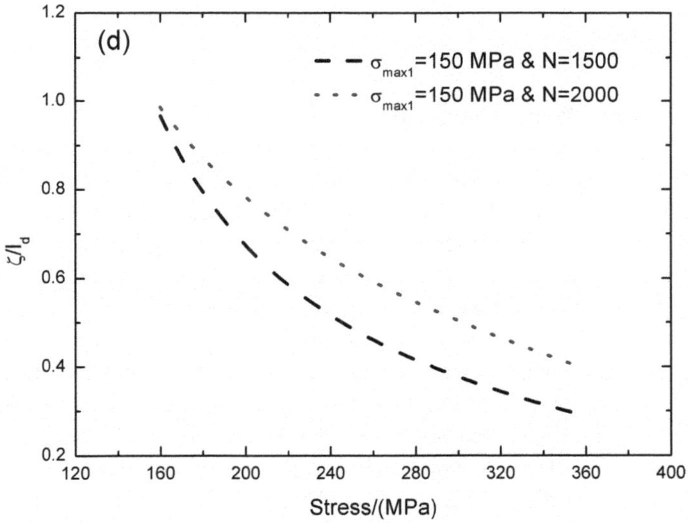

Fig. 2.41 (continued)

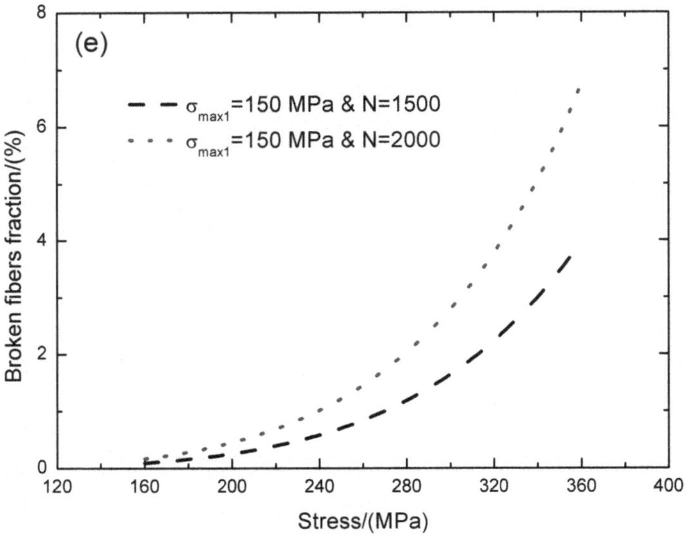

Fig. 2.41 (continued)

2.5.7.3 SiC/Borosilicate Composite

For the SiC/borosilicate composite without cyclic fatigue loading, the matrix multicracking starts at the applied stress of $\sigma_{mc} = 220$ MPa and approaches saturation at the applied stress of $\sigma_{sat} = 360$ MPa; the matrix multicracking density increases from $\varphi = 0.2$ mm to the saturation value of $\varphi = 6.5$ mm. With cyclic fatigue loading of $\sigma_{max1} = 200$ MPa and $N = 1000$, the cyclic-dependent matrix multicracking density increases from $\varphi = 0.11$ mm at $\sigma_{mc} = 220$ MPa to $\varphi = 4.5$ mm at $\sigma_{sat} = 374$ MPa; the cyclic-dependent fiber/matrix interface debonding length increases from $l_d/r_f =$ 4.8 to 12; the cyclic-dependent interface debonding ratio increases from $2l_d/l_c = 0.9$ to 86.8%; the cyclic-dependent interface wear ratio decreases from $\zeta/l_d = 92.4$ to 36.9%; and the cyclic-dependent broken fiber fraction increases from $P = 0.3$ to 6.8%. With cyclic fatigue loading of $\sigma_{max1} = 200$ MPa and $N = 1200$, the cyclic-dependent matrix multicracking density increases from $\varphi = 0.09$ mm at $\sigma_{mc} = 220$ MPa to $\varphi =$ 13.8 mm at $\sigma_{sat} = 420$ MPa; the cyclic-dependent fiber/matrix interface debonding length increases from $l_d/r_f = 5.7$ to 13.8; the cyclic-dependent interface debonding ratio increases from $2l_d/l_c = 0.9$ to 89.2%; and the cyclic-dependent interface wear ratio decreases from $\zeta/l_d = 94$ to 39.3%; and the cyclic-dependent broken fiber fraction increases from $P = 0.4$ to 8.7%, as shown in Fig. 2.43.

2.5.7.4 SiC/SiC Composite

For the mini-SiC/SiC composite without cyclic fatigue loading, the matrix multicracking starts at the applied stress of $\sigma_{mc} = 135$ MPa and approaches saturation at the applied stress of $\sigma_{sat} = 250$ MPa; the matrix multicracking density increases

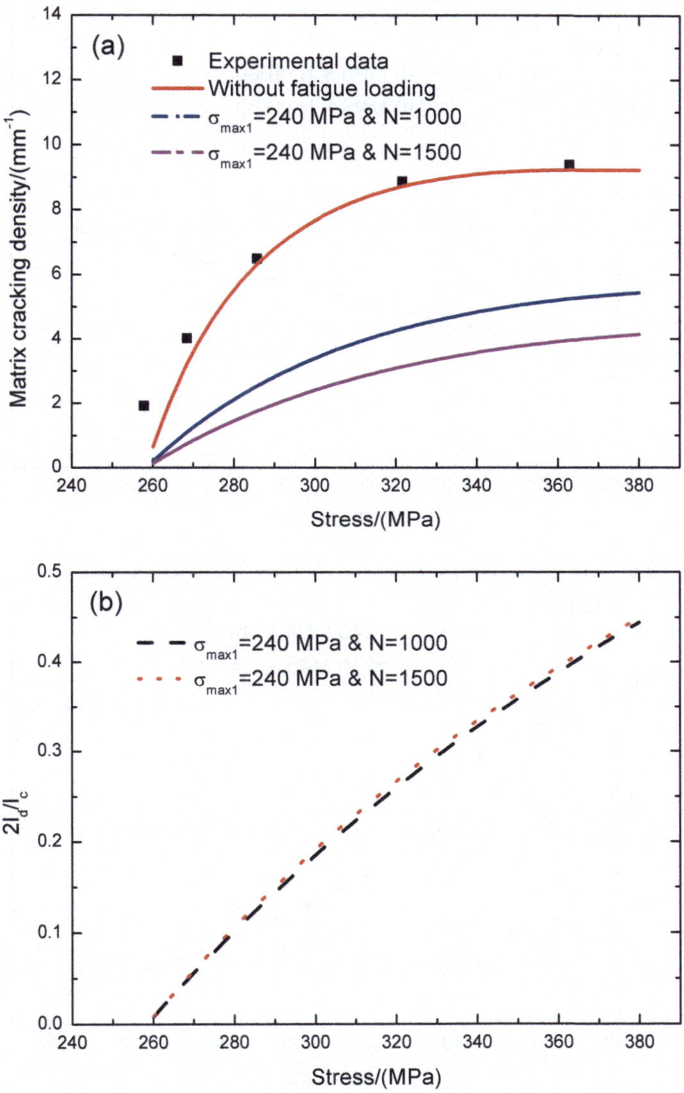

Fig. 2.42 **a** Experimental and predicted matrix multicracking density versus applied stress curves; **b** the fiber/matrix interface debonding ratio versus applied stress curves; **c** the fiber/matrix interface debonding length versus applied stress curves; **d** the fiber/matrix interface wear ratio versus of applied stress curves; and **e** the broken fiber fraction versus applied stress curves of unidirectional SiC/CAS-II composite with/without interface wear

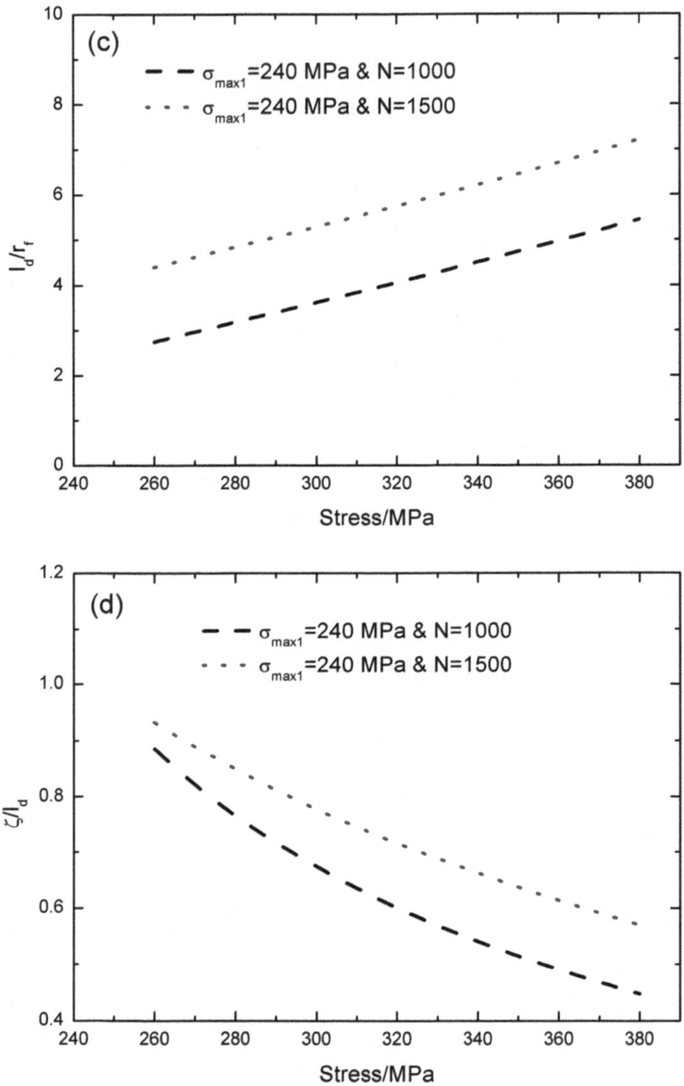

Fig. 2.42 (continued)

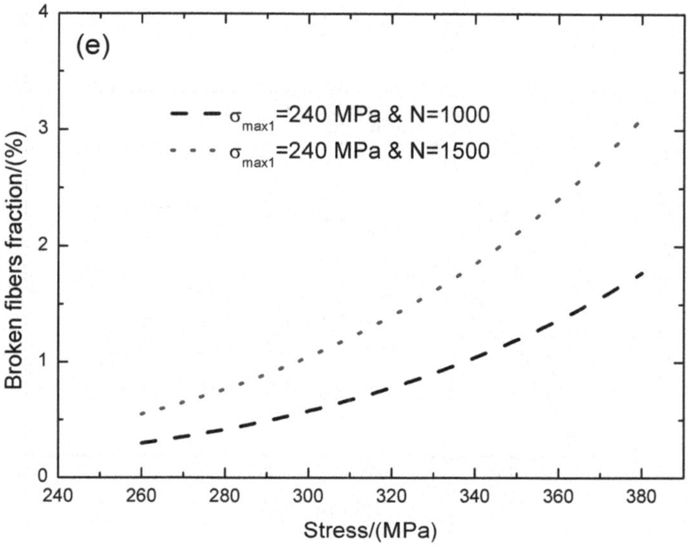

Fig. 2.42 (continued)

from $\varphi = 0.4$ mm to the saturation value of $\varphi = 2.4$ mm. With cyclic loading of $\sigma_{\max 1} = 120$ MPa and $N = 1000$, the cyclic-dependent matrix multicracking density increases from $\varphi = 0.05$ mm at $\sigma_{\rm mc} = 135$ MPa to $\varphi = 1.9$ mm at $\sigma_{\rm sat} = 260$ MPa; the cyclic-dependent fiber/matrix interface debonding length increases from $l_{\rm d}/r_{\rm f} = 14.7$ to 39.2; the cyclic-dependent fiber/matrix interface debonding ratio increases from $2l_{\rm d}/l_{\rm c} = 1.0$ to 96.5%; the cyclic-dependent fiber/matrix interface wear ratio decreases from $\zeta/l_{\rm d} = 90.2$ to 34%; and the cyclic-dependent broken fiber fraction increases from $P = 0.07$ to 14.4%. With cyclic loading of $\sigma_{\max 1} = 120$ MPa and $N = 1500$, the cyclic-dependent matrix multicracking density increases from $\varphi = 0.03$ mm at $\sigma_{\rm mc} = 135$ MPa to $\varphi = 1.4$ mm at $\sigma_{\rm sat} = 270$ MPa; the cyclic-dependent fiber/matrix interface debonding length increases from $l_{\rm d}/r_{\rm f} = 22.6$ to 46.1; the cyclic-dependent fiber/matrix interface debonding ratio increases from $2l_{\rm d}/l_{\rm c} = 1.0$ to 86.5%; the cyclic-dependent fiber/matrix interface wear ratio decreases from $\zeta/l_{\rm d} = 93.8$ to 46%; and the cyclic-dependent broken fiber fraction increases from $P = 0.12$ to 17.9%, as shown in Fig. 2.44.

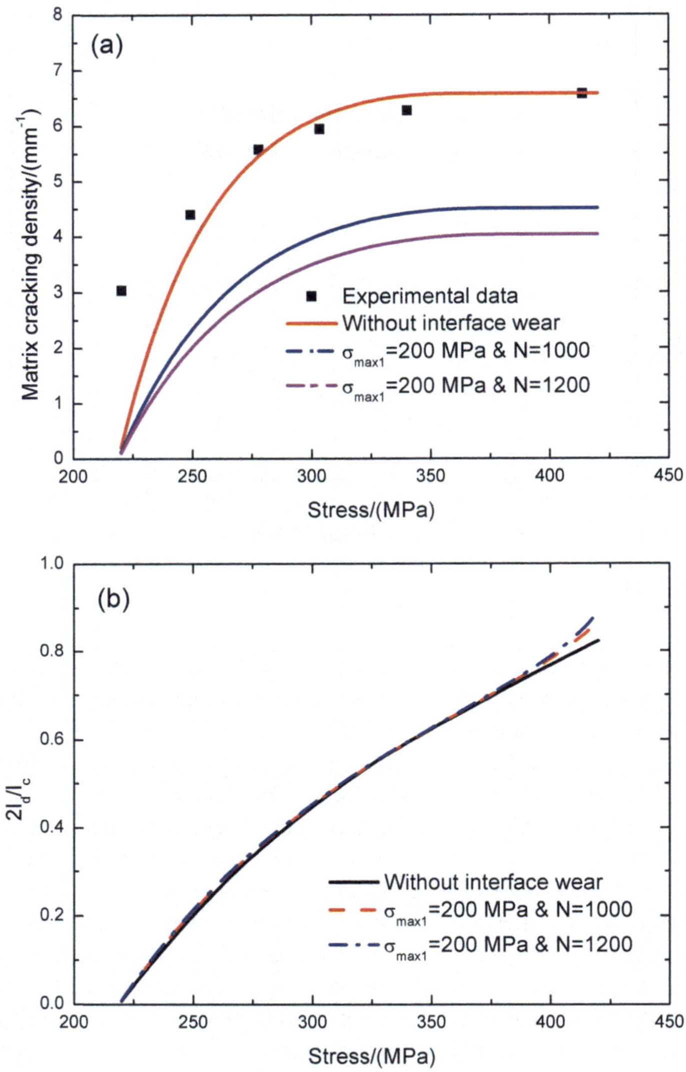

Fig. 2.43 **a** Experimental and predicted matrix multicracking density versus applied stress curves; **b** the fiber/matrix interface debonding ratio versus applied stress curves; **c** the fiber/matrix interface debonding length versus applied stress curves; **d** the fiber/matrix interface wear ratio versus of applied stress curves; and **e** the broken fiber fraction versus applied stress curves of unidirectional SiC/borosilicate composite with/without interface wear

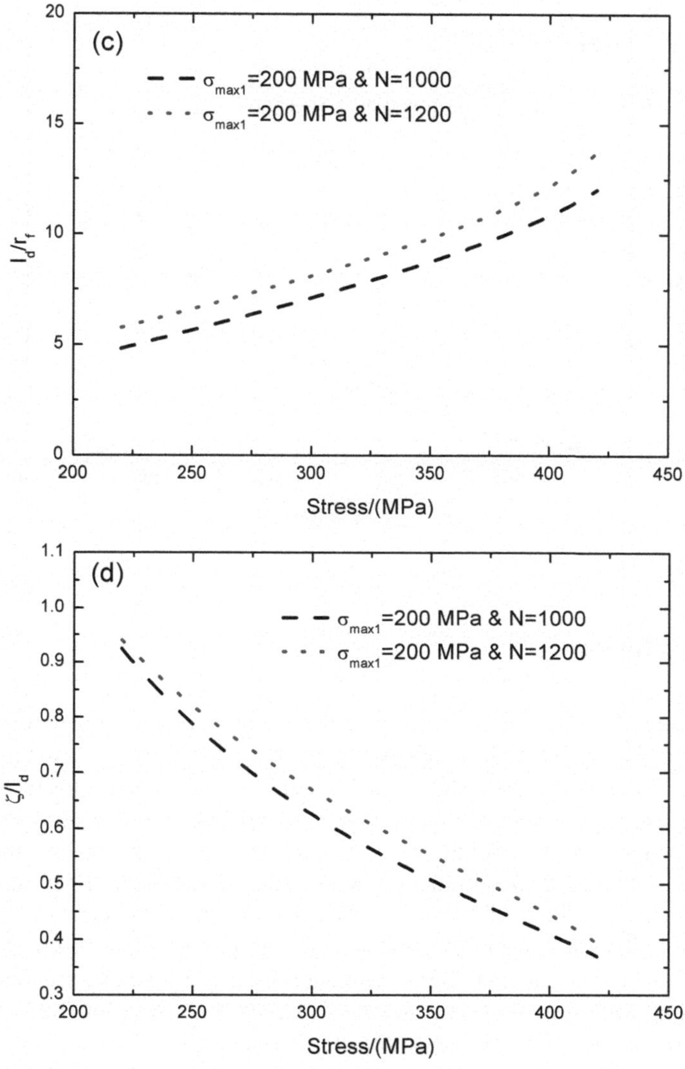

Fig. 2.43 (continued)

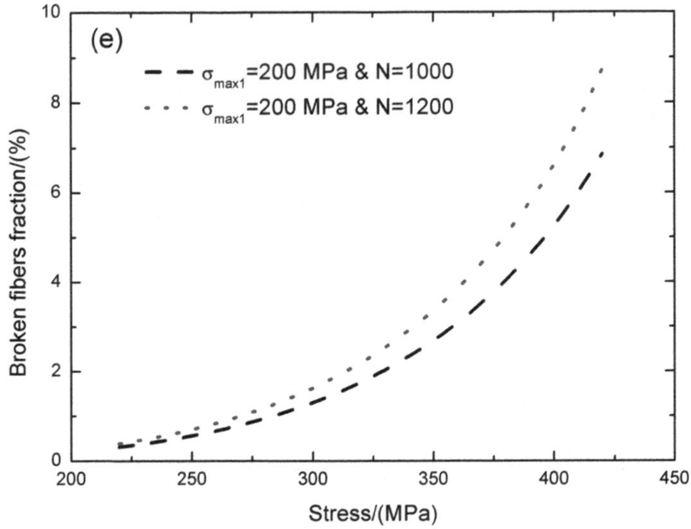

Fig. 2.43 (continued)

2.6 Conclusion

In this chapter, the time-, stress-, and cycle-dependent matrix multicracking of fiber-reinforced CMCs with interface debonding, interface wear, interface oxidation, and fiber fracture is investigated. The shear-lag model is combined with the interface debonding, interface wear, interface oxidation, and fiber fracture models, and the fiber/matrix interface debonding criterion is adopted to determine the microstress field of the damaged fiber-reinforced CMCs. The effects of the fiber volume and interface shear stress in the debonding and oxidation region, the interface debonding energy, the oxidation temperature, and time on the matrix multicracking, interface debonding and oxidation, and fiber fracture are discussed. The experimental matrix multicracking evolution of unidirectional C/SiC, SiC/SiC, mini-SiC/SiC, SiC/CAS, SiC/CAS-II, and SiC/borosilicate composites is predicted.

(1) When the fiber volume and interface shear stress in the debonding region increased, the matrix first cracking stress and saturation cracking stress increased, the saturation matrix cracking space decreased, the matrix multiple fracture evolution rate increased, the interface debonding ratio and the broken fiber fraction decreased, and the interface oxidation ratio increased.
(2) When the interface debonding energy and the interface shear stress in the oxidation region increased, the matrix multicracking density increased, the interface debonding ratio and broken fiber fraction decreased, and the fiber/matrix interface oxidation ratio increased.

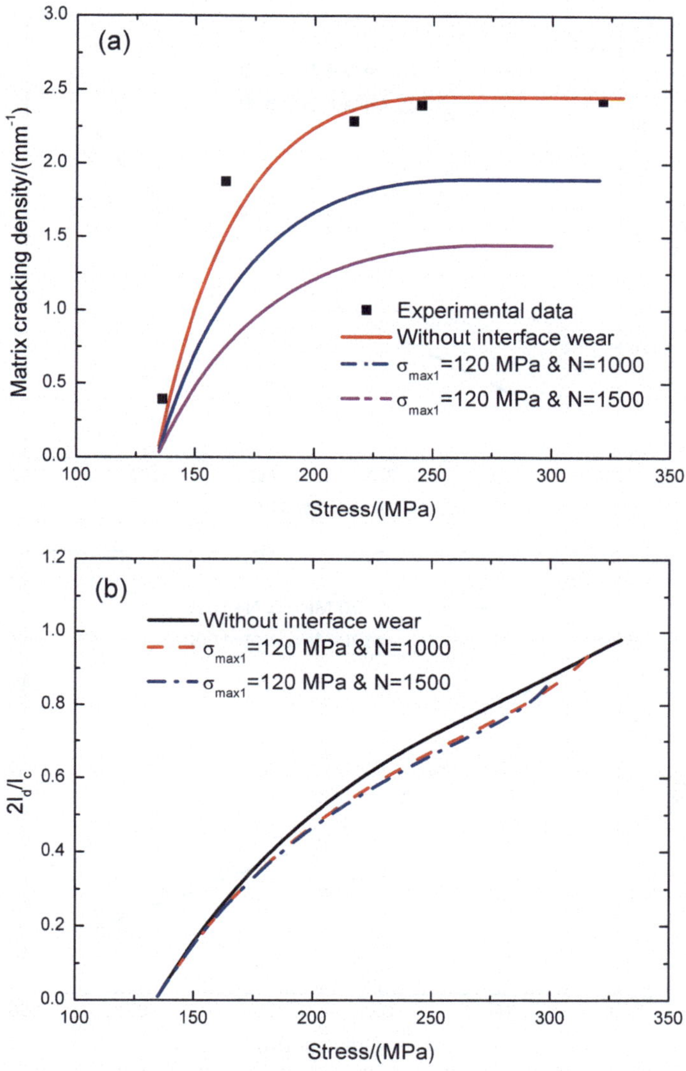

Fig. 2.44 **a** Experimental and predicted matrix multicracking density versus applied stress curves; **b** the fiber/matrix interface debonding ratio versus applied stress curves; **c** the fiber/matrix interface debonding length versus applied stress curves; **d** the fiber/matrix interface wear ratio versus of applied stress curves; and **e** the broken fiber fraction versus applied stress curves of mini-SiC/SiC composite with/without interface wear

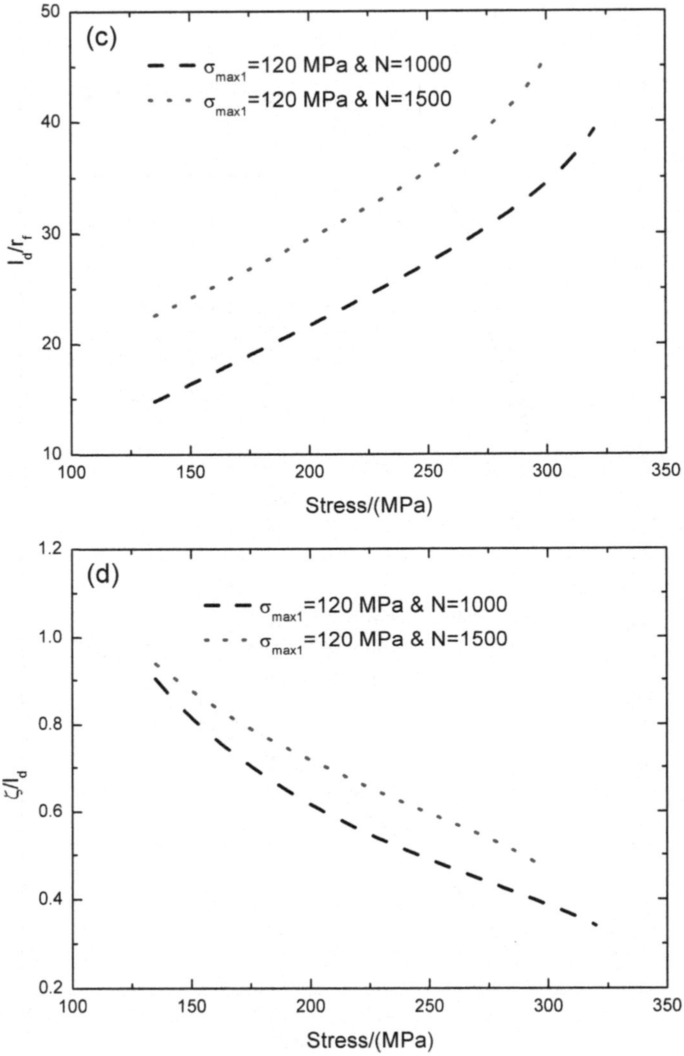

Fig. 2.44 (continued)

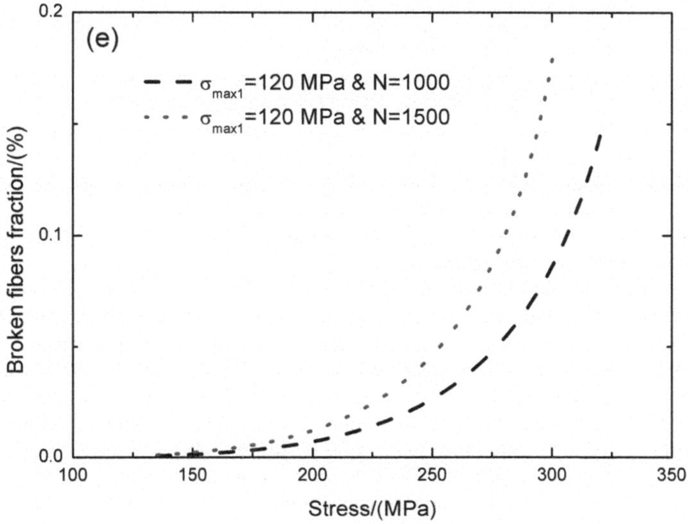

Fig. 2.44 (continued)

(3) When the oxidation temperature and oxidation duration increased, the matrix multicracking density decreased, and the interface debonding ratio, oxidation ratio, and broken fiber fraction increased.
(4) Considering the fiber/matrix interface wear, the matrix multicracking density decreases; the fiber/matrix interface debonding length increases, and the interface debonding ratio decreases.
(5) With increasing applied cycle number, the matrix multicracking density decreases; and the fiber/matrix interface debonded length, interface wear ratio, and the broken fiber fraction increase.

References

Aveston J, Cooper GA, Kelly A (1971) In: The properties of fiber composites, conference on proceedings. National Physical Laboratory. IPC Science and Technology Press, Guildford, pp 15–26

Beyerle DS, Spearing SM, Zok FW, Evans AG (1992) Damage and failure in unidirectional ceramic-matrix composites. J Am Ceram Soc 75:2719–2725. https://doi.org/10.1111/j.1151-2916.1992.tb05495.x

Beyerle DS, Spearing SM, Zok FW, Evans AG (1993) Damage and failure in unidirectional ceramic-matrix composites. J Am Ceram Soc 76:560. https://doi.org/10.1111/j.1151-2916.1993.tb03828.x

Budiansky B, Hutchinson JW, Evans AG (1986) Matrix fracture in fiber-reinforced ceramics. J Mech Phys Solids 34:167–189. https://doi.org/10.1016/0022-5096(86)90035-9

Casas L, Martinez-Esnaola JM (2003) Modelling the effect of oxidation on the creep behavior of fiber-reinforced ceramic matrix composites. Acta Mater 51:3745–3757. https://doi.org/10.1016/S1359-6454(03)00189-7

Chiang YC (2000) On crack-wake debonding in fiber-reinforced ceramics. Eng Fract Mech 65:15–28. https://doi.org/10.1016/S0013-7944(99)00130-7

Chiang YC (2001) On fiber debonding and matrix cracking in fiber-reinforced ceramics. Compos Sci Technol 61:1743–1756. https://doi.org/10.1016/S0266-3538(01)00078-1

Curtin WA (1991) Theory of mechanical properties of ceramic-matrix composites. J Am Ceram Soc 74:2837–2845. https://doi.org/10.1111/j.1151-2916.1991.tb06852.x

Evans AG, Zok FW, McMeeking RM (1995) Fatigue of ceramic matrix composites. Acta Metall Mater 43:859–875. https://doi.org/10.1016/0956-7151(94)00304-Z

Gao YC, Mai Y, Cotterell B (1988) Fracture of fiber-reinforced materials. J Appl Math Phys 39:550–572. https://doi.org/10.1007/BF00948962

Gowayed Y, Ojard G, Santhosh U, Jefferson G (2015) Modeling of crack density in ceramic matrix composites. J Compos Mater 49:2285–2294. https://doi.org/10.1177/0021998314545188

Halbig MC, McGuffin-Cawley JD, Eckel AD, Brewer DN (2008) Oxidation kinetics and stress effects for the oxidation of continuous carbon fibers within a microcracked C/SiC ceramic matrix composite. J Am Ceram Soc 91:519–526. https://doi.org/10.1111/j.1551-2916.2007.02170.x

Hsueh CH (1996) Crack-wake interfacial debonding criteria for fiber-reinforced ceramic composites. Acta Mater 44:2211–2216. https://doi.org/10.1016/1359-6454(95)00369-X

Lara-Curzio E (1999) Analysis of oxidation-assisted stress-rupture of continuous fiber-reinforced ceramic matrix composites at intermediate temperatures. Compos Part A 30:549–554. https://doi.org/10.1016/S1359-835X(98)00148-1

Lee SS, Stinchcomb WW (1994) Damage mechanisms of cross-ply Nicalon/CAS-II laminate under cyclic tension. Ceram Eng Sci Proc 15:40–48. https://doi.org/10.1002/9780470314500.ch5

Li L (2017) Synergistic effects of fiber debonding and fracture on matrix cracking in fiber-reinforced ceramic-matrix composites. Mater Sci Eng A 682:482–490. https://doi.org/10.1016/j.msea.2016.11.077

Li L (2018a) Damage, fracture and fatigue of ceramic-matrix composites. Springer Nature Singapore Pte Ltd. ISBN 978-981-13-1782-8. https://doi.org/10.1007/978-981-13-1783-5

Li L (2018b) Effect of interface oxidation on matrix multi-cracking evolution of fiber-reinforced ceramic-matrix composites at elevated temperature. J Ceram Sci Technol 9(4):397–410. https://www.ceramic-science.com/articles/all-articles.html?article_id=100682

Li L (2018c) Modeling strength degradation of fiber-reinforced ceramic-matrix composites subjected to cyclic loading at elevated temperatures in oxidative environment. Appl Compos Mater 15:1–19. https://doi.org/10.1007/s10443-017-9609-5

Li L (2019a) Thermomechanical fatigue of ceramic-matrix composites. Wiley-VCH. ISBN 978-3-527-34637-0. https://onlinelibrary.wiley.com/doi/book/10.1002/9783527822614

Li L (2019b) Modeling matrix multicracking development of fiber-reinforced ceramic-matrix composites considering fiber debonding. Int J Appl Ceram Technol 16:97–107. https://onlinelibrary.wiley.com/doi/epdf/10.1111/ijac.13068

Li L (2019c) Synergistic effects of fiber/matrix interface wear and fibers fracture on matrix multiple cracking in fiber-reinforced ceramic-matrix composites. Compos Interfaces 26(3):193–219. https://www.tandfonline.com/doi/full/10.1080/09276440.2018.1488490

Li L, Song Y, Sun Y (2014) Modeling the tensile behavior of unidirectional C/SiC ceramic-matrix composites. Mech Compos Mater 49:659–672. https://doi.org/10.1007/s11029-013-9382-y

Li L, Song Y, Sun Y (2015) Modeling the tensile behavior of cross-ply C/SiC ceramic-matrix composites. Mech Compos Mater 51:359–376. https://doi.org/10.1007/s11029-015-9507-6

Meng Q, Wang Z (2015a) Theoretical analysis of interfacial debonding and fiber pull-out in fiber-reinforced polymer-matrix composites. Arch Appl Mech 85:745–759. https://doi.org/10.1007/s00419-015-0987-6

Meng Q, Wang Z (2015b) Theoretical model of fiber debonding and pull-out in unidirectional hybrid-fiber-reinforced brittle-matrix composites. J Compos Mater 49:1739–1751. https://doi.org/10.1177/0021998314540191

Meng Q, Wang Z (2015c) Prediction of interfacial strength and failure mechanisms in particle-reinforced metal-matrix composites based on a micromechanical model. Eng Fract Mech 142:170–183. https://doi.org/10.1016/j.engfracmech.2015.06.001
Meng Q, Li B, Li T, Feng XQ (2017) A multiscale crack-bridging model of cellulose nanopaper. J Mech Phys Solids 103:22–39. https://doi.org/10.1016/j.jmps.2017.03.004
Meng Q, Li B, Li T, Feng XQ (2018) Effects of nanofiber orientations on the fracture toughness of cellulose nanopaper. Eng Fract Mech 194:350–361. https://doi.org/10.1016/j.engfracmech.2018.03.034
Morscher GN, Yun HM, DiCarlo JA (2005) Matrix cracking in 3D orthogonal melt-infiltrated SiC/SiC composites with various Z-fiber types. J Am Ceram Soc 88:146–153. https://doi.org/10.1111/j.1551-2916.2004.00029.x
Morscher GN, Singh M, Kiser D, Freedman M, Bhatt R (2007) Modeling stress-dependent matrix cracking and stress-strain behavior in 2D woven SiC fiber reinforced CVI SiC composites. Compos Sci Technol 67:1009–1017. https://doi.org/10.1016/j.compscitech.2006.06.007
Naslain R (2004) Design, preparation and properties of non-oxide CMCs for application in engines and nuclear reactors: an overview. Compos Sci Technol 64:155–170. https://doi.org/10.1023/B:JMSC.0000048745.18938.d5
Ogasawara T, Ishikawa T, Ito H, Watanabe N, Davies IJ (2001) Multiple cracking and tensile behavior for an orthogonal 3-D woven Si-Ti-C-O fiber/Si-Ti-C-O matrix composite. J Am Ceram Soc 84:1565–1574. https://doi.org/10.1111/j.1151-2916.2001.tb00878.x
Okabe T, Komotori J, Shimizu M, Takeda N (1999) Mechanical behavior of SiC fiber reinforced brittle-matrix composites. J Mater Sci 34:3405–3412. https://doi.org/10.1023/A:1004637300310
Olivier C (1998) Élaboration et éude du comportement mécanique de composites unidirectionnels C/Si_3N_4 et SiC/Si_3N_4. PhD thesis, INSA Lyon, Lyon
Parthasarathy TA, Cox B, Surde O, Przybyla C, Cinibulk MK (2018) Modeling environmentally induced property degradation of SiC/BN/SiC ceramic matrix composites. J Am Ceram Soc 101:973–997. https://doi.org/10.1111/jace.15325
Phoenix SL, Raj R (1992) Scalings in fracture probabilities for a brittle matrix fiber composite. Acta Metall Mater 40:2813–2828. https://doi.org/10.1016/0956-7151(92)90447-M
Pryce AW, Smith PA (1992) Behavior of unidirectional and crossply ceramic matrix composites under quasi-static tensile loading. J Mater Sci 27:2695–2704. https://doi.org/10.1007/BF00540692
Rajan VP, Zok FW (2014) Matrix cracking of fiber-reinforced ceramic composites in shear. J Mech Phys Solids 73:3–21. https://doi.org/10.1016/j.jmps.2014.08.007
Schmidt S, Beyer S, Knabe H, Immich H, Meistring R, Gessler A (2004) Advanced ceramic matrix composite materials for current and future propulsion system applications. Acta Astronaut 55:409–420. https://doi.org/10.1016/j.actaastro.2004.05.052
Sevener KM, Tracy JM, Chen Z, Kiser JD, Daly S (2017) Crack opening behavior in ceramic matrix composites. J Am Ceram Soc 100:4734–4747. https://doi.org/10.1111/jace.14976
Solti JP, Mall S, Robertson DD (1995) Modeling damage in unidirectional ceramic-matrix composites. Compos Sci Technol 54:55–66. https://doi.org/10.1016/0266-3538(95)00041-0
Stang H, Shah S (1986) Failure of fiber-reinforced composites by pull-out fracture. J Mater Sci 21:953–957. https://doi.org/10.1007/BF01117378
Sun YJ, Singh RN (1998) The generation of multiple matrix cracking and fiber-matrix interfacial debonding in a glass composite. Acta Mater 46(5):1657–1667. https://doi.org/10.1016/S1359-6454(97)00347-9
Vagaggini E, Domergue JM, Evans AG (1995) Relationship between hysteresis measurements and the constituent properties of ceramic matrix composites: I, Theory. J Am Ceram Soc 78:2709–2720. https://doi.org/10.1111/j.1151-2916.1995.tb08046.x

Verrilli MJ, Opila EJ, Calomino A, Kiser JD (2004) Effect of environment on the stress-rupture behavior of a carbon-fiber-reinforced silicon carbide ceramic matrix composite. J Am Ceram Soc 87:1536–1542. https://doi.org/10.1111/j.1551-2916.2004.01536.x
Zhang S, Gao X, Chen J, Dong H, Song Y (2016) Strength model of the matrix element in SiC/SiC composites. Mater Des 101:66–71. https://doi.org/10.1016/j.matdes.2016.03.166

Chapter 3
Time-, Stress-, and Cycle-Dependent Tensile Strength of Fiber-Reinforced Ceramic-Matrix Composites

Abstract In this chapter, the strength degradation of non-oxide and oxide/oxide fiber-reinforced ceramic-matrix composites (CMCs) subjected to multiple fatigue loading at room temperature, oxidation environment at elevated temperature, and cyclic loading at elevated temperatures in oxidative environments is investigated. Considering damage mechanisms of matrix cracking, interface debonding, interface wear, interface oxidation, and fiber fracture, the residual strength model of CMCs is established by combining the microstress field of the damaged composites, the damage models, and the fracture criterion. The relationships between the composite residual strength, fatigue peak stress, interface debonding, fiber failure, oxidation time and temperature, and applied cycle number are established. The effects of the peak stress level, initial and steady-state interface shear stress, fiber Weibull modulus, fiber strength, oxidation temperature and time on the degradation of composite strength and fiber failure are investigated. The evolution of residual strength versus oxidation temperature and time and applied cycle number curves of non-oxide and oxide/oxide CMCs is predicted.

Keywords Ceramic-matrix composites (CMCs) · Residual strength · Matrix cracking · Interface debonding · Interface oxidation · Fiber failure

3.1 Introduction

Ceramic-matrix composites (CMCs) possess high strength-to-weight ratio at elevated temperatures and are being designed in the hot-section components of commercial aeroengine, i.e., CFM 56-5B and LEAP turbofan engine (CFM International, Cincinnati, OH USA) (Gonczy 2015; Li 2018a, 2019). The Civil Aviation Administration of China (CAAC) issues and enforces regulations and minimum standards covering the safe manufacture, and operation and maintenance of civil aircraft. As new materials, these ceramic composite components will have to meet the certification regulations of the CAAC for airworthiness. The CMC producer and user will have to show the CAAC the component is well designed for operation, safety, and durability, and the failure risk is within the accepted level. However, during cyclic loading at

L. Li, *Time-Dependent Mechanical Behavior of Ceramic-Matrix Composites at Elevated Temperatures*, Advanced Ceramics and Composites 1,
https://doi.org/10.1007/978-981-15-3274-0_3

elevated temperatures, the composite strength degrades with flight hours or cycles, which increases the failure risk level.

Under cyclic fatigue loading, the material performance of CMCs degrades with applied cycles due to damage mechanisms of matrix cracking, interface debonding, interface wear, and fiber fracture. Lee et al. (1998) investigated the tension–tension fatigue behavior of 2D SiC/[Si–N–C] composite at room and elevated temperature. At room temperature, the fatigue limit was about 75% tensile strength; at 1000 °C in air atmosphere, the fatigue run-out stress level of 110 MPa was 50 MPa lower than that at room temperature, however 35 MPa higher above the proportional limit. The area of the stress–strain loops was calculated to generate hysteresis energy density values. The fatigue hysteresis energy density decreases with applied cycles. Bertrand et al. (2015) investigated the tension–tension fatigue behavior of Sylramic-iBN/BN/SiC composite at elevated temperature in a simulated combustion condition. After prior fatigue for 90,000 cycles, the tensile strength reduced by 70% under high-peak stress level of 125 MPa for 1250 °C and 35% under low-peak stress level of 88 MPa for 1350 °C. The strength degradation depends not only on the test temperature but also on the applied stress level. Ruggles-Wrenn et al. (2008) investigated tension–tension fatigue behavior of 2D Nextel™ 720/alumina composite at 1200 °C in air and in steam environment under the loading frequency of 0.1 and 10 Hz. In air environment, the fatigue life appears to be independent of the loading frequency; in steam atmosphere, the fatigue limit, number of applied cycles to failure, and failure time all decrease as the loading frequency decreases, and the strength and stiffness degradation increase with decreasing frequency of prior fatigue. Mehrman et al. (2007) investigated the effect of hold times on the tension–tension fatigue behavior of 2D Nextel™ 720/alumina composite at 1200 °C in air and in steam conditions. In air environment, the fatigue lives with hold times were shorter than those obtained in fatigue; in steam atmosphere, the fatigue lives with hold times reduced significantly and were less than those obtained in air. Ruggles-Wrenn and Lanser (2016) investigated the tension–compression fatigue behavior of 2D Nextel™ 720/alumina composite at 1200 °C in air and in steam condition. The tensile strength degraded approximately 40% after prior fatigue in steam condition, and the tension–compression cycling is much more damaging than tension–tension fatigue, due to fiber microbuckling during compression portion of the cycle. For the non-oxide CMCs, the degradation under cyclic loading involves interface wear at room temperature and oxidation of interphase and fibers at elevated temperature, and is accelerated by the presence of moisture. However, for the oxide/oxide CMCs, the composite is inherently resistant to oxidation at elevated temperature, but the formation of matrix cracks and degradation of fiber strength also affect the fatigue performance. Whitworth (2000) evaluated the residual strength degradation in graphite/epoxy composite laminates subjected to cyclic loading, and the effects of peak stress and applied cycles on the degradation of residual strength have been analyzed. Keiji (2011) developed the residual tensile strength model of fiber-reinforced CMCs after fatigue loading, and the effects of fiber strength and the interfacial shear stress on the fatigue life and residual strength of the composite have been investigated. Shah et al. (2000) developed a probabilistic modeling approach to quantify the scatter in the first matrix cracking strength and the ultimate tensile strength in CMCs. Murthy et al. (2008) investigated the probabilistic analysis and reliability assessment of the turbine vane withstanding a maximum temperature of 1315 °C within the substrate and the hot surface temperature of 1482 °C, considering the random variables of the material properties, strength, and pressure loading on the vane. Li (2016a) developed approaches to model the tensile strength of CMCs subjected to multiple fatigue loading. It was found that the loading sequence, peak stress, and cycle number affect the tensile strength of the composite. Li (2015, 2018b) investigated the tensile strength degradation after non-stress oxidation at elevated temperature. It was found that the oxidation temperature and oxidation time affect the damage and fracture in CMCs. However, under cyclic loading at elevated temperature in oxidative environments, the composite strength significantly degrades compared with that under non-stress oxidation at elevated temperature, due to serious oxidation occurred as matrix cracks open under stress, and the wide range of interface wear/interface oxidation/fiber fracture.

In this chapter, the strength degradation of non-oxide and oxide/oxide fiber-reinforced CMCs subjected to multiple fatigue loading at room temperature, oxidation environment at elevated temperature, and cyclic loading at elevated temperatures in oxidative environments is investigated. Considering damage mechanisms of matrix cracking, interface debonding, interface wear, interface oxidation, and fiber fracture, the composite residual strength model is established by combining the microstress field of the damaged composites, the damage models, and the fracture criterion. The relationships between the composite residual strength, fatigue peak stress, interface debonding, fiber failure, oxidation time and temperature, and applied cycle number are established. The effects of the peak stress level, initial and steady-state interface shear stress, fiber Weibull modulus, fiber strength, oxidation temperature and time on the degradation of composite strength and fiber failure are investigated. The evolution of residual strength versus oxidation temperature and time and applied cycle number curves of non-oxide and oxide/oxide CMCs is predicted.

3.2 Cyclic-Dependent Tensile Strength of Fiber-Reinforced CMCs Under Multiple Fatigue Loading at Room Temperature

In this section, an analytical method is developed to investigate the effect of interface wear on the tensile strength of fiber-reinforced CMCs subjected to multiple fatigue loading. The shear-lag model is used to describe the microstress field of the damaged composite considering fiber failure and the difference existed in the new and original interface debonded region. The statistical matrix multicracking model and fracture mechanics interface debonding criterion are used to determine the matrix crack spacing and interface debonding length. The interface shear stress degradation model and fiber strength degradation model are adopted to analyze the interface wear effect on the tensile strength of the composite subjected to multiple fatigue loading. Under tensile loading, the fiber failure probabilities are determined by combining the interface wear model and fiber failure model based on the assumption that the fiber strength is subjected to two-parameter Weibull distribution and the loads carried by broken and intact fibers satisfy the global load sharing criterion. The composite can no longer support the applied load when the total loads supported by the broken and intact fibers approach its peak value. The conditions of a single matrix crack and matrix multicrackings for tensile strength corresponding to multiple fatigue peak stress levels and different cycle numbers are analyzed.

3.2.1 Cyclic-Dependent Stress Analysis

Upon first loading to fatigue peak stress $\sigma_{\max\_1}$, it is assumed that matrix multicracking and fiber/matrix interface debonding occur. After experiencing N applied cycles, the fiber/matrix interface shear stress in the interface debonding region decreases from initial value τ_i to τ_f due to the interface wear. When the fatigue peak stress increases from $\sigma_{\max\_1}$ to $\sigma_{\max\_2}$, the matrix cracks propagate along the fiber/matrix interface. To analyze the stress distributions in the fiber and the matrix, a unit cell is extracted from the ceramic composite system. The unit cell contains a single fiber surrounded by a hollow cylinder of matrix. The fiber radius is r_f, and the matrix radius is $R(R = r_f/V_f^{1/2})$. The length of the unit cell is $l_c/2$, which is just the half matrix crack space. The interface debonded length is l_d, which can be divided into two regions, i.e., the interface debonding region with the interface shear stress of τ_f ($z\in[0, \zeta]$) and interface debonding region with the interface shear stress of τ_i ($z \in [\zeta, l_d]$), in which ζ denotes the interface debonding length at the fatigue peak stress $\sigma_{\max\_1}$. On the matrix crack plane, fibers carry all the loads of σ/V_f, in which σ denotes the far-field applied stress and V_f denotes the fiber volume. The shear-lag model adopted by Budiansky et al. (1986) is applied to perform the stress and strain calculations in the interface debonding region ($z \in [0, l_d]$) and interface bonding region ($z \in [l_d, l_c/2]$).

$$\sigma_f(z) = \begin{cases} \dfrac{\sigma}{V_f} - \dfrac{2\tau_f}{r_f}z, z \in [0, \zeta] \\ \dfrac{\sigma}{V_f} - \dfrac{2\tau_f}{r_f}\zeta - \dfrac{2\tau_i}{r_f}(z - \zeta), z \in [\zeta, l_d] \\ \sigma_{fo} + \left[\dfrac{V_m}{V_f}\sigma_{mo} - 2\dfrac{\tau_f}{r_f}\zeta - 2\dfrac{\tau_i}{r_f}(l_d - \zeta)\right] \exp\left(-\rho\dfrac{z - l_d}{r_f}\right), z \in \left[l_d, \dfrac{l_c}{2}\right] \end{cases} \tag{3.1}$$

$$\sigma_{\rm m}(z)=\begin{cases}2\dfrac{V_{\rm f}}{V_{\rm m}}\dfrac{\tau_{\rm f}}{r_{\rm f}}z,\ z\in[0,\zeta]\\ 2\dfrac{V_{\rm f}}{V_{\rm m}}\dfrac{\tau_{\rm f}}{r_{\rm f}}\zeta+2\dfrac{V_{\rm f}}{V_{\rm m}}\dfrac{\tau_{\rm i}}{r_{\rm f}}(x-\zeta),\ z\in[\zeta,l_{\rm d}]\\ \sigma_{\rm mo}-\left[\sigma_{\rm mo}-2\dfrac{V_{\rm f}}{V_{\rm m}}\dfrac{\tau_{\rm f}}{r_{\rm f}}\zeta-2\dfrac{V_{\rm f}}{V_{\rm m}}\dfrac{\tau_{\rm i}}{r_{\rm f}}(l_{\rm d}-\zeta)\right]\exp\left(-\rho\dfrac{z-l_{\rm d}}{r_{\rm f}}\right),\ z\in\left[l_{\rm d},\dfrac{l_{\rm c}}{2}\right]\end{cases}\tag{3.2}$$

$$\tau_{\rm i}(z)=\begin{cases}\tau_{\rm f},\ z\in[0,\zeta]\\ \tau_{\rm i},\ z\in[\zeta,l_{\rm d}]\\ \dfrac{\rho}{2}\left[\dfrac{V_{\rm m}}{V_{\rm f}}\sigma_{\rm mo}-2\dfrac{\tau_{\rm f}}{r_{\rm f}}\zeta-\dfrac{2\tau_{\rm i}}{r_{\rm f}}(l_{\rm d}-\zeta)\right]\exp\left(-\rho\dfrac{z-l_{\rm d}}{r_{\rm f}}\right),\ z\in\left[l_{\rm d},\dfrac{l_{\rm c}}{2}\right]\end{cases}\tag{3.3}$$

where $V_{\rm m}$ denotes the matrix volume fraction and ρ denotes the shear-lag model parameter (Budiansky et al. 1986).

$$\rho^2=\frac{4E_{\rm c}G_{\rm m}}{V_{\rm m}E_{\rm m}E_{\rm f}\phi}\tag{3.4}$$

in which $G_{\rm m}$ denotes the matrix shear modulus, and

$$\phi=-\frac{2\ln V_{\rm f}+V_{\rm m}(3-V_{\rm f})}{2V_{\rm m}^2}\tag{3.5}$$

$\sigma_{\rm fo}$ and $\sigma_{\rm mo}$ denote the fiber and matrix axial stress in the interface bonded region, respectively.

$$\sigma_{\rm fo}=\frac{E_{\rm f}}{E_{\rm c}}\sigma+E_{\rm f}(\alpha_{\rm c}-\alpha_{\rm f})\Delta T\tag{3.6}$$

$$\sigma_{\rm mo}=\frac{E_{\rm m}}{E_{\rm c}}\sigma+E_{\rm m}(\alpha_{\rm c}-\alpha_{\rm m})\Delta T\tag{3.7}$$

where $E_{\rm f}$, $E_{\rm m}$, and $E_{\rm c}$ denote the fiber, matrix, and composite elastic modulus, respectively; $\alpha_{\rm f}$, $\alpha_{\rm m}$, and $\alpha_{\rm c}$ denote the fiber, matrix, and composite thermal expansion coefficient, respectively; and ΔT denotes the temperature difference between fabricated temperature T_0 and room temperature $T_1(\Delta T=T_1-T_0)$.

Under cyclic loading, repeated forward and reverse slip can occur at the fiber/matrix interface. Macroscopically, the interface slip results in hysteresis in the stress–strain behavior and a temperature rise of the specimens. At the microscale, cyclic slip may result in interfacial wear, lowering the interfacial frictional stress and fiber strength. As a result of this, both the fiber axial stress distribution and their failure probability would change, because longer portions of the fibers are subjected to peak stress Φ. After cyclic fatigue loading, the unit cell can be divided into three regions, i.e., the interface wear region ($z\in[0,\zeta]$), interface debonding region (z

$\in [\zeta, l_d]$), and interface bonding region ($z \in [l_d, l_c/2]$). When fiber fractures, the fiber axial stress distributions in the interface wear region, interface debonding region, and interface bonding region are

$$\sigma_f(z) = \begin{cases} \Phi - \dfrac{2\tau_f}{r_f} z, z \in [0, \zeta] \\ \Phi - \dfrac{2\tau_f}{r_f}\zeta - \dfrac{2\tau_i}{r_f}(z - \zeta), z \in [\zeta, l_d] \\ \sigma_{fo} + \left[\Phi - \sigma_{fo} - \dfrac{2\tau_f}{r_f}\zeta - \dfrac{2\tau_i}{r_f}(l_d - \zeta)\right]\exp\left(-\rho\dfrac{z - l_d}{r_f}\right), z \in \left[l_d, \dfrac{l_c}{2}\right] \end{cases} \tag{3.8}$$

3.2.2 Matrix Multicracking

Upon loading of fiber-reinforced CMCs, cracks typically initiate within composite matrix since the strain-to-failure of matrix is usually less than that of fiber. The matrix crack spacing decreases with the increases in stress above matrix initial cracking stress σ_{mc} and may eventually approach saturation at stress σ_{sat}. Four dominant failure criterions are present in the literature for modeling matrix multicracking evolution in unidirectional CMCs, i.e., the maximum stress criterion, energy balance approach, critical matrix strain energy criterion, and statistical failure approach. The maximum stress criterion (Daniel and Lee 1993) assumes that a new matrix crack will form whenever the matrix stress exceeds the ultimate strength of matrix, which is assumed to be single-valued and a known material property. The energy balance failure criteria involve calculation of the energy balance relationship before and after the formation of a single dominant crack as originally proposed by Aveston et al. (1971). The progression of matrix cracking as determined by energy criterion is dependent upon matrix strain energy release rate. The energy criterion is represented by Zok and Spearing (1992) and Zhu and Weitsman (1994). The concept of a critical matrix strain energy criterion (Solti et al. 1995) presupposes the existence of an ultimate or critical strain energy limit beyond which the matrix fails. Beyond this, as more energy is placed into the composite, the matrix, unable to support the additional load, continues to fail. As more energy is placed into the system, the matrix fails such that all the additional energy is transferred to fibers. Failure may consist of the formation of matrix cracks, propagation of existing cracks, or interface debonding. Statistical failure approach (Curtin 1993) assumes that matrix cracking is governed by statistical relations, which relate the size and spatial distribution of matrix flaws to their relative propagation stress. The brittle nature of matrix material and the possible formation of initial crack distribution throughout microstructure suggest that a statistical approach to matrix multicracking evolution is warranted in fiber-reinforced CMCs. The tensile strength of the brittle matrix is assumed to be described by two-parameter Weibull distribution where the probability of matrix failure P_{m} is (Curtin 1993)

$$P_{\mathrm{m}} = 1 - \exp\left\{ -\left[\frac{\sigma - (\sigma_{\mathrm{mc}} - \sigma_{\mathrm{th}})}{(\sigma_{\mathrm{R}} - \sigma_{\mathrm{th}}) - (\sigma_{\mathrm{mc}} - \sigma_{\mathrm{th}})} \right]^{\beta} \right\} \quad (3.9)$$

in which σ_{R} denotes the matrix characteristic strength; σ_{mc} denotes the matrix initial cracking stress; σ_{th} denotes the matrix thermal residual stress; and β denotes the matrix Weibull modulus.

As applied stress increases, the number of matrix cracks increases and matrix crack space decreases. To estimate the instantaneous matrix crack space with increase in applied stress, it leads to the form of

$$P_{\mathrm{m}} = l_{\mathrm{sat}} / l_{\mathrm{c}} \quad (3.10)$$

where

$$l_{\mathrm{sat}} = \Lambda(\sigma_{\mathrm{mc}}/\sigma_{\mathrm{R}}, \sigma_{\mathrm{th}}/\sigma_{\mathrm{R}})\delta_{\mathrm{R}} \quad (3.11)$$

where Λ denotes the final nominal crack space, which is a pure number and depends only upon the micromechanical and statistical quantities characterizing the cracking. δ_{R} denotes characteristic interface sliding length.

$$\delta_{\mathrm{R}} = r_{\mathrm{f}} \frac{V_{\mathrm{m}} E_{\mathrm{m}}}{V_{\mathrm{f}} E_{\mathrm{c}}} \frac{\sigma_{\mathrm{R}}}{2\tau_{\mathrm{i}}} \quad (3.12)$$

Using Eqs. (3.10) and (3.11), the instantaneous matrix crack space is derived by (Curtin 1993)

$$l_{\mathrm{c}} = r_{\mathrm{f}} \frac{V_{\mathrm{m}} E_{\mathrm{m}}}{V_{\mathrm{f}} E_{\mathrm{c}}} \frac{\sigma_{\mathrm{R}}}{2\tau_{\mathrm{i}}} \Lambda \left\{ 1 - \exp\left[-\left(\frac{\sigma - (\sigma_{\mathrm{mc}} - \sigma_{\mathrm{th}})}{(\sigma_{\mathrm{R}} - \sigma_{\mathrm{th}}) - (\sigma_{\mathrm{mc}} - \sigma_{\mathrm{th}})} \right)^{\beta} \right] \right\}^{-1} \quad (3.13)$$

3.2.3 *Interface Debonding*

When matrix crack propagates to the fiber/matrix interface, it deflects along the interface. There are two approaches to the problem of interface debonding, i.e., the shear strength approach and the fracture mechanics approach. The shear strength approach is based on a maximum shear stress criterion in which interface debonding occurs as the shear stress reaches the interface shear strength (Hsueh 1996). On the other hand, the fracture mechanics approach treats the interface debonding as a particular crack propagation problem in which interface debonding occurs as the strain energy release rate of the interface achieves the debonded toughness (Gao et al. 1988). It has been proved that the fracture mechanics approach is preferred to the shear strength approach for interface debonding (Sun and Singh 1998). The fracture mechanics approach is adopted in the present analysis. The interface debonding criterion is (Gao et al. 1988)

$$\xi_{\mathrm{d}} = -\frac{F}{4\pi r_{\mathrm{f}}}\frac{\partial w_{\mathrm{f}}(0)}{\partial l_{\mathrm{d}}} - \frac{1}{2}\int_{0}^{l_{\mathrm{d}}} \tau_{\mathrm{i}}\frac{\partial v(z)}{\partial l_{\mathrm{d}}}\mathrm{d}z \tag{3.14}$$

in which $F(=\pi r_{\mathrm{f}}^2\sigma/V_{\mathrm{f}})$ denotes the fiber load at the matrix cracking plane; $w_{\mathrm{f}}(0)$ denotes the fiber axial displacement on the matrix cracking plane; and $v(z)$ denotes the relative displacement between the fiber and the matrix.

The axial displacements of the fiber and matrix, i.e., $w_{\mathrm{f}}(z)$ and $w_{\mathrm{m}}(z)$, are

$$\begin{aligned} w_{\mathrm{f}}(z) &= \int_{z}^{\frac{l_{\mathrm{c}}}{2}} \frac{\sigma_{\mathrm{f}}(z)}{E_{\mathrm{f}}}\mathrm{d}z \\ &= \frac{\Phi}{E_{\mathrm{f}}}(l_{\mathrm{d}} - z) - \frac{\tau_{\mathrm{f}}}{r_{\mathrm{f}}E_{\mathrm{f}}}\left(2\zeta l_{\mathrm{d}} - z^2 - \zeta^2\right) - \frac{\tau_{\mathrm{i}}}{r_{\mathrm{f}}E_{\mathrm{f}}}(l_{\mathrm{d}} - \zeta)^2 + \frac{\sigma_{\mathrm{fo}}}{E_{\mathrm{f}}}\left(\frac{l_{\mathrm{c}}}{2} - l_{\mathrm{d}}\right) \\ &\quad + \frac{r_{\mathrm{f}}}{\rho E_{\mathrm{f}}}\left[\Phi - \sigma_{\mathrm{fo}} - \frac{2\tau_{\mathrm{f}}}{r_{\mathrm{f}}}\zeta - \frac{2\tau_{\mathrm{i}}}{r_{\mathrm{f}}}(l_{\mathrm{d}} - \zeta)\right]\left[1 - \exp\left(-\rho\frac{l_{\mathrm{c}}/2 - l_{\mathrm{d}}}{r_{\mathrm{f}}}\right)\right] \end{aligned} \tag{3.15}$$

$$\begin{aligned} w_{\mathrm{m}}(z) &= \int_{z}^{\frac{l_{\mathrm{c}}}{2}} \frac{\sigma_{\mathrm{m}}(z)}{E_{\mathrm{m}}}\mathrm{d}z \\ &= \frac{V_{\mathrm{f}}\tau_{\mathrm{f}}}{r_{\mathrm{f}}V_{\mathrm{m}}E_{\mathrm{m}}}\left(2\zeta l_{\mathrm{d}} - \zeta^2 - z^2\right) + \frac{V_{\mathrm{f}}\tau_{\mathrm{i}}}{r_{\mathrm{f}}V_{\mathrm{m}}E_{\mathrm{m}}}(l_{\mathrm{d}} - \zeta)^2 + \frac{\sigma_{\mathrm{mo}}}{E_{\mathrm{m}}}\left(\frac{l_{\mathrm{c}}}{2} - l_{\mathrm{d}}\right) \\ &\quad - \frac{r_{\mathrm{f}}}{\rho E_{\mathrm{m}}}\left[\sigma_{\mathrm{mo}} - \frac{2V_{\mathrm{f}}\tau_{\mathrm{f}}}{r_{\mathrm{f}}V_{\mathrm{m}}}\zeta - \frac{2V_{\mathrm{f}}\tau_{\mathrm{i}}}{r_{\mathrm{f}}V_{\mathrm{m}}}(l_{\mathrm{d}} - \zeta)\right]\left[1 - \exp\left(-\rho\frac{l_{\mathrm{c}}/2 - l_{\mathrm{d}}}{r_{\mathrm{f}}}\right)\right] \end{aligned} \tag{3.16}$$

Using Eqs. (3.15) and (3.16), the relative displacement between the fiber and the matrix, i.e., $v(z)$, is

$$\begin{aligned} v(z) &= |w_{\mathrm{f}}(z) - w_{\mathrm{m}}(z)| \\ &= \frac{\Phi}{E_{\mathrm{f}}}(l_{\mathrm{d}} - z) - \frac{E_{\mathrm{c}}\tau_{\mathrm{f}}}{r_{\mathrm{f}}V_{\mathrm{m}}E_{\mathrm{f}}E_{\mathrm{m}}}\left(2\zeta l_{\mathrm{d}} - \zeta^2 - z^2\right) - \frac{E_{\mathrm{c}}\tau_{\mathrm{i}}}{r_{\mathrm{f}}V_{\mathrm{m}}E_{\mathrm{f}}E_{\mathrm{m}}}(l_{\mathrm{d}} - \zeta)^2 \\ &\quad + \frac{r_{\mathrm{f}}}{\rho E_{\mathrm{f}}}\left[\Phi - \sigma_{\mathrm{fo}} - \frac{2\tau_{\mathrm{f}}}{r_{\mathrm{f}}}\zeta - \frac{2\tau_{\mathrm{i}}}{r_{\mathrm{f}}}(l_{\mathrm{d}} - \zeta)\right]\left[1 - \exp\left(-\rho\frac{l_{\mathrm{c}}/2 - l_{\mathrm{d}}}{r_{\mathrm{f}}}\right)\right] \end{aligned}$$

$$+\frac{r_f}{\rho E_m}\left[\sigma_{mo}-\frac{2V_f\tau_f}{r_fV_m}\zeta-\frac{2V_f\tau_i}{r_fV_m}(l_d-\zeta)\right]\left[1-\exp\left(-\rho\frac{l_c/2-l_d}{r_f}\right)\right] \tag{3.17}$$

Substituting $w_f(x = 0)$ and $v(z)$ into Eq. (3.14), it leads to the form of

$$\begin{aligned}&\frac{r_f\Phi^2}{4E_f}-\frac{\Phi\tau_f}{E_f}\zeta-\frac{\Phi\tau_i}{E_f}(l_d-\zeta)-\frac{r_f\sigma\Phi}{4E_c}-\frac{r_f\Phi\tau_i}{2\rho E_f}+\frac{E_c\tau_f^2}{r_fV_mE_fE_m}\zeta^2\\&+\frac{2E_c\tau_f\tau_i}{r_fV_mE_fE_m}\zeta(l_d-\zeta)+\frac{E_c\tau_f\tau_i}{\rho V_mE_fE_m}\zeta\\&+\frac{E_c\tau_i^2}{r_fV_mE_fE_m}(l_d-\zeta)^2+\frac{E_c\tau_i^2}{\rho V_mE_fE_m}(l_d-\zeta)-\xi_d=0\end{aligned} \tag{3.18}$$

To solve Eq. (3.18), the interface debonding length l_d is

$$\begin{aligned}l_d=&\left(1-\frac{\tau_f}{\tau_i}\right)\zeta+\frac{r_f}{2}\left(\frac{V_mE_m\Phi}{E_c\tau_i}-\frac{1}{\rho}\right)\\&-\left[\left(\frac{r_f}{2\rho}\right)^2+\frac{r_f^2V_fV_mE_fE_m\Phi^2}{4E_c^2\tau_i^2}\left(\frac{\sigma}{V_f\Phi}-1\right)+\frac{r_fV_mE_fE_m}{E_c\tau_i^2}\xi_d\right]^{\frac{1}{2}}\end{aligned} \tag{3.19}$$

3.2.4 Interface Wear

When a CMC is subjected to a cyclic loading between a peak stress and a valley stress, upon first loading to the peak stress, the damage mechanisms of matrix multicracking, interface debonding, and frictional slipping would occur. Under subsequent cycles, the interface shear stress degrades with the increase of cycle number due to interface wear at room temperature (Li and Song 2010; Rouby and Reynaud 1993; Evans et al. 1995; Reynaud 1996; Holmes and Cho 1992; Zhu et al. 1999; Fantozzi and Reynaud 2009; Cho et al. 1991). Evidences of interface wear that a reduction in the height of asperities occurs along the fiber coating for a different thermal misfit, surface roughness, and frictional sliding velocity have been presented by push-out and push-back tests on a ceramic composite system (Cherouali et al. 1998). The interface wear process can be facilitated by temperature rising that occurs along the fiber/matrix interface, as frictional dissipation proceeds (Holmes and Cho 1992), i.e., the temperature rising exceeded 100 K under fatigue loading at 75 Hz between stress levels of 220 and 10 MPa in unidirectional SiC/CAS-II composite. Cho et al. (1991) first developed an approach to estimate the interface shear stress from frictional heating measurement. By analyzing the frictional heating data, Holmes and Cho (1992) found that the interfacial shear stress in unidirectional SiC/CAS-II composite undergoes a rapid decrease at the initial stage of fatigue loading, i.e., from an initial value of over 20 MPa to approximately 5 MPa after 25,000 cycles. Evans et al. (1995) developed an approach to evaluate the interface shear stress by analyzing the parabolic regions of stress-strain hysteresis loops based on Vagaggini's hysteresis loop models (Vagaggini et al. 1995). The initial interface shear stress of unidirectional SiC/CAS composite was about 20 MPa, and degraded to about 5 MPa at the 30th cycle.

This variation in interface shear stress, $\tau_i(N)$, as provided by Evans et al. (1995) is given by the following equation.

$$\tau_i(N) = \tau_{io} + \left(1 - \exp\left(-\omega N^{\lambda}\right)\right)(\tau_{imin} - \tau_{io}) \tag{3.20}$$

where τ_{io} denotes the initial interface shear stress, i.e., $\tau_i(N)$ at $N = 1$, before fatigue loading; τ_{imin} denotes the steady-state interface shear stress during cycling; and ω and λ are empirical constants.

Lee and Stinchcomb (1994) performed fiber fracture mirror experiments of fiber-reinforced CMCs under scanning electron microscope and found that the fiber strength degrades with cycles increasing under cyclic loading.

$$\sigma_0(N) = \sigma_0\left[1 - p_1(\log N)^{p_2}\right] \tag{3.21}$$

in which σ_0 denotes the fiber strength at the first cycle; $\sigma_0(N)$ denotes the fiber strength at the Nth cycle; and p_1 and p_2 are empirical parameters.

3.2.5 Cyclic-Dependent Fiber Failure

There are relatively fewer models for fiber failure of fiber-reinforced CMCs compared to analyses for damage mechanisms, i.e., matrix multicracking and interface debonding. As fibers begin to break, the loads dropped by broken fibers must be transferred to intact fibers at the cross section. Two dominant failure criterions are present in the literatures for modeling fiber failure, i.e., the global load sharing (GLS) criterion and local load sharing (LLS) criterion. The GLS criterion assumes that the load from any one fiber is transferred equally to all other intact fibers at the same cross-sectional plane. The GLS assumption neglects any local stress concentrations in the neighborhood of existing breaks and is expected to be accurate when interface shear stress is sufficiently low. Models that include GLS explicitly have been developed, i.e., Thouless and Evans (1988), Cao and Thouless (1990), Sutcu (1989), Schwietert and Steif (1990), Curtin (1991), Weitsman and Zhu (1993), Hild et al. (1994), Zhu and Weitsman (1994), Curtin et al. (1998), Paar et al. (1998), Liao and Reifsnider (2000), and so on. The LLS assumes that the load from broken fibers is transferred to neighborhood intact fibers and is expected to be accurate when interface shear stress is sufficiently high. Models that include LLS explicitly have been developed, i.e., Zhou and Curtin (1995), Dutton et al. (2000), Xia and Curtin (2000), and so on. The ceramic reinforcing fibers are brittle materials and must also be described statistically by a flaw distribution. The two-parameter Weibull model is adopted to describe the fiber strength distribution. The fiber fracture probability P is (Curtin et al. 1998)

$$P(\sigma)= 1-\exp\left(-\int_{L_0}\frac{1}{l_0}\left[\frac{\sigma_f(z)}{\sigma_0}\right]^m dz\right) \tag{3.22}$$

where σ_0 denotes the fiber strength at the tested gauge length of l_0; m denotes the fiber Weibull modulus; and L_0 denotes the integral length.

The fiber fracture probabilities in the interface wear region, interface debonding region, and interface bonding region of $P_a(\Phi)$, $P_b(\Phi)$, and $P_c(\Phi)$ are

$$P_a(\Phi) = 1-\exp\left\{-\frac{1}{\eta\kappa^m(m+1)}\left[1-\left(1-\frac{\eta\zeta}{l_s}\right)^{m+1}\right]\left(\frac{\Phi}{\sigma_c}\right)^{m+1}\right\} \tag{3.23}$$

$$P_b(\Phi) = 1-\exp\left\{-\frac{1}{m+1}\left(\frac{\Phi}{\sigma_c}\right)^{m+1}\left[\left(1-\frac{\eta\zeta}{l_s}\right)^{m+1}-\left(1-\frac{\eta\zeta}{l_s}-\frac{l_d-\zeta}{l_s}\right)^{m+1}\right]\right\} \tag{3.24}$$

$$P_c(\Phi) = 1-\exp\left\{-\frac{1}{\rho(m+1)\left(1-\frac{\sigma_{fo}}{\Phi}-\frac{\eta\zeta}{l_s}-\frac{l_d-\zeta}{l_s}\right)}\left(\frac{r_f}{l_s}\right)\left(\frac{\Phi}{\sigma_c}\right)^{m+1}\right.$$
$$\left.\left[\left(1-\frac{\eta\zeta}{l_s}-\frac{l_d-\zeta}{l_s}\right)^{m+1}-\left(1-\frac{\eta\zeta}{l_s}-\frac{l_d-\zeta}{l_s}-\rho\frac{l_c/2-l_d}{r_f}\left(1-\frac{\sigma_{fo}}{\Phi}-\frac{\eta\zeta}{l_s}-\frac{l_d-\zeta}{l_s}\right)\right)^{m+1}\right]\right\} \tag{3.25}$$

where κ denotes the fiber strength degradation ratio,

$$\kappa = \frac{\sigma_0(N)}{\sigma_0} \tag{3.26}$$

$$\eta = \frac{\tau_f}{\tau_i} \tag{3.27}$$

and l_f denotes the slip length over which the fiber stress would decay to zero if not interrupted by the far-field equilibrium stresses.

$$l_f = (1-\eta)\zeta + l_s \tag{3.28}$$

and

$$l_s = \frac{r_f\Phi}{2\tau_i} \tag{3.29}$$

For a single matrix crack, the global load sharing criterion was used to determine the loads carried by the intact and broken fibers. The average load per fiber σ/V_f

must equal the product of the stress carried by the unbroken fibers Φ and the fraction $(1 - P(\Phi))$ of unbroken fibers, plus a contribution due to the residual stress $\langle\Phi_b\rangle$ carried by the broken fiber which do not break exactly at the matrix crack plane. The mechanical equilibrium at the matrix crack plane is

$$\frac{\sigma}{V_f} = (1 - P(\Phi))\Phi + P(\Phi)\langle\Phi_b\rangle \tag{3.30}$$

where

$$P(\Phi) = P_a(\Phi) + P_b(\Phi) + P_c(\Phi) \tag{3.31}$$

When a fiber breaks, the load carried by the fiber drops to zero at the position of break. Similar to the case of matrix cracking, the fiber/matrix interface debonds and the stress builds up in the fiber through the interface shear stress. During the process of loading, the stress in a broken fiber S_b as a function of the distance x from the break can be written as

$$\Phi_b(z) = \begin{cases} \frac{2\tau_f}{r_f}z, \ z \in [0, \zeta] \\ \frac{2\tau_f}{r_f}\zeta + \frac{2\tau_i}{r_f}(z - \zeta), \ z \in [\zeta, l_f] \end{cases} \tag{3.32}$$

In order to calculate the average stress carried by broken fibers $\langle\Phi_b\rangle$, it is necessary to construct the probability distribution $f(z)$ of the distance x of a fiber break from the reference matrix crack plane, provided that a break occurs within a distance $\pm l_f$. For this conditional probability distribution, Phoenix and Raj (1992) deduced the following based on Weibull statistics, as

$$f(z) = \frac{1}{P(\Phi)l_f}\left(\frac{\Phi}{\sigma_c}\right)^{m+1} \exp\left[-\left(\frac{z}{l_f}\right)\left(\frac{\Phi}{\sigma_c}\right)^{m+1}\right], z \in [0, l_f] \tag{3.33}$$

in which σ_c denotes the characteristic strength of fibers in the characteristic length scale of δ_c.

$$\sigma_c = \left(\frac{l_0\sigma_0^m\tau_i}{r_f}\right)^{\frac{1}{m+1}}, \delta_c = \left(\frac{\sigma_0 r_f l_0^{1/m}}{\tau_i}\right)^{\frac{m}{m+1}} \tag{3.34}$$

The averaging stress carried by broken fibers $\langle\Phi_b\rangle$ during the process of loading using Eqs. (3.32) and (3.33) leads to

$$\langle\Phi_b\rangle = \frac{2(1-\eta)l_f\tau_i}{r_f P(\Phi)}\left(\frac{\sigma_c}{\Phi}\right)^{m+1} \exp\left[-\left(\frac{\zeta}{l_f}\right)\left(\frac{\Phi}{\sigma_c}\right)^{m+1}\right]$$

$$- \frac{2(\zeta + l_s)\tau_i}{r_f P(\Phi)} \exp\left[-\left(\frac{\Phi}{\sigma_c}\right)^{m+1}\right] + \frac{2 l_f \tau_f}{r_f P(\Phi)} \left(\frac{\sigma_c}{\Phi}\right)^{m+1} \quad (3.35)$$

By substituting Eqs. (3.31) and (3.35) into Eq. (3.30), the relationship between the applied stress σ and the peak stress Φ on the intact fibers can be obtained. The maximum value of σ versus Φ gives the ultimate strength of the composite considering the effect of interface wear under multiple fatigue loading.

When there is more than one matrix crack, a balance of forces at the matrix crack plane requires that the applied force per fiber equals the force carried by the unbroken fibers plus the pullout force carried by those fibers broken away from the matrix crack plane and can be expressed as

$$\frac{\sigma}{V_f} = \left[1 - P(\Phi)\left(1 + \frac{2l_f}{l_c}\right)\right]\Phi + P(\Phi)\frac{2l_f}{l_c}\langle\Phi_b\rangle \quad (3.36)$$

where

$$P(\Phi) = \lambda P_a(\Phi) + P_b(\Phi) \quad (3.37)$$

where λ denotes the fraction of interface wear region in the multiple matrix cracks.

$$\lambda = \frac{l_{sat}}{l_f - 2\zeta} \quad (3.38)$$

3.2.6 Results and Discussion

The ceramic composite system of C/SiC is used for the case study, and its basic material properties are given by: $V_f = 30\%$, $E_f = 230$ GPa, $E_m = 350$ GPa, $r_f = 3.5$ μm, $\tau_{io} = 50$ MPa, $\tau_{imin} = 1$ MPa, $\omega = 0.0001$, $\lambda = 1.3$, $p_1 = 0.02$, $p_2 = 1.4$, $m = 5$, $l_0 = 25 \times 10^{-3}$ m, $\sigma_0 = 0.52$ GPa.

The interface debonding length of $2l_d/l_c$ versus applied cycle number curves under single fatigue peak stress of $\sigma_{max} = 120$ MPa and multiple fatigue peak stresses of $\sigma_{max\_1} = 120$ MPa and $\sigma_{max\_2} = 180$ MPa considering interface wear is shown in Fig. 3.1.

Under single fatigue peak stress $\sigma_{max} = 120$ MPa, the interface debonding length increases from $2l_d/l_c = 0$ under $\tau_i = 50$ MPa at the first cycle to $2l_d/l_c = 0.38$ under $\tau_i = 1$ MPa at the 10,000th cycle and then increases slowly to $2l_d/l_c = 0.42$ at the 15000th cycle.

Under multiple fatigue peak stresses of $\sigma_{max\_1} = 120$ MPa and $\sigma_{max\_2} = 180$ MPa, the interface debonding length increases from $2l_d/l_c = 0$ under $\tau_i = 50$ MPa at the first cycle to $2l_d/l_c = 0.38$ under $\tau_i = 1$ MPa at the 10,000th cycle; with increasing fatigue peak stress from $\sigma_{max\_1} = 120$ MPa to $\sigma_{max\_2} = 180$ MPa, the interface

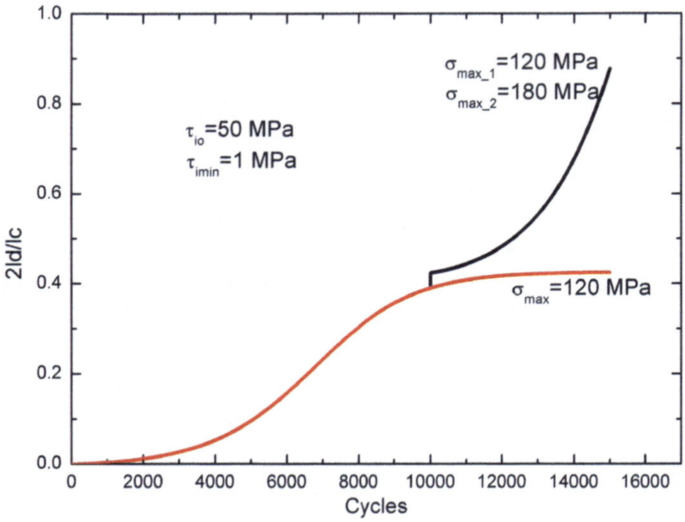

Fig. 3.1 Interface debonding length versus applied cycle number curves under single fatigue peak stress of $\sigma_{\max} = 120$ MPa and multiple fatigue peak stresses of $\sigma_{\max\_1} = 120$ MPa and $\sigma_{\max\_2} = 180$ MPa considering the interface wear

debonding length increases from $2l_d/l_c = 0.38$ to $2l_d/l_c = 0.42$ at the 10,000th cycle and then increases rapidly to $2l_d/l_c = 0.87$, due to the increase of fatigue peak stress and interface wear in the new debonding region.

The effects of applied cycle number and fatigue peak stress on fiber failure and composite tensile strength corresponding to the conditions of a single matrix crack and multiple matrix cracks are investigated.

3.2.6.1 Case 1: A Single Matrix Crack

The effect of applied cycle number, i.e., $N = 1$, 3000, 5000, and 10,000, on fiber fracture, composite tensile strength, and intact fiber stress under fatigue peak stress $\sigma_{\max\_1} = 120$ MPa, is shown in Fig. 3.2.

When the applied cycle number is $N = 1$, the failure probability of the composite under an applied stress of $\sigma = 381$ MPa is $P = 22.6\%$; however, after experiencing $N = 3000$ cycles, the failure probability of the composite under an applied stress of $\sigma = 323$ MPa is $P = 32.5\%$; after experiencing $N = 5000$ cycles, the failure probability of the composite under an applied stress of $\sigma = 257$ MPa is $P = 28.9\%$; after experiencing $N = 10{,}000$ cycles, the failure probability of the composite under an applied stress of $\sigma = 184$ MPa is $P = 27.4\%$. The stress carried by intact fibers increases as the broken fiber fraction increases, in which the intact fiber stress approaches 1.12, 1.19, 1.17, and 1.16 of fiber average stress σ/V_f, corresponding to different cycle

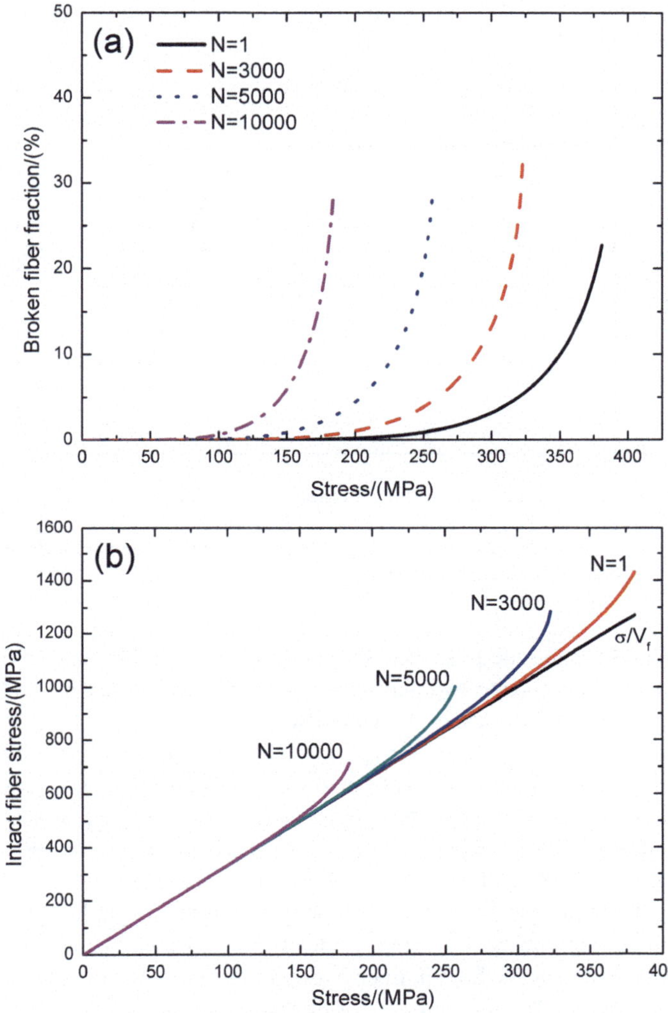

Fig. 3.2 **a** Broken fiber fraction versus applied stress and **b** the intact fiber stress versus applied stress corresponding to different cycle numbers, i.e., $N = 1$, 3000, 5000, and 10,000 cycles under fatigue peak stress of $\sigma_{\max} = 120$ MPa for conditions of a single matrix crack

numbers of $N = 1$, 3000, 5000, and 10,000 under fatigue peak stress of $\sigma_{\max\_1} = 120$ MPa.

The residual tensile strength versus applied cycle number curve under fatigue peak stress $\sigma_{\max\_1} = 120$ MPa is shown in Fig. 3.3, in which the tensile strength decreases from about $\sigma_{\mathrm{UTS}} = 381$ MPa at $N = 1$ to about $\sigma_{\mathrm{UTS}} = 184$ MPa at $N = 10{,}000$ due to the degradation of the interface shear stress and fiber strength.

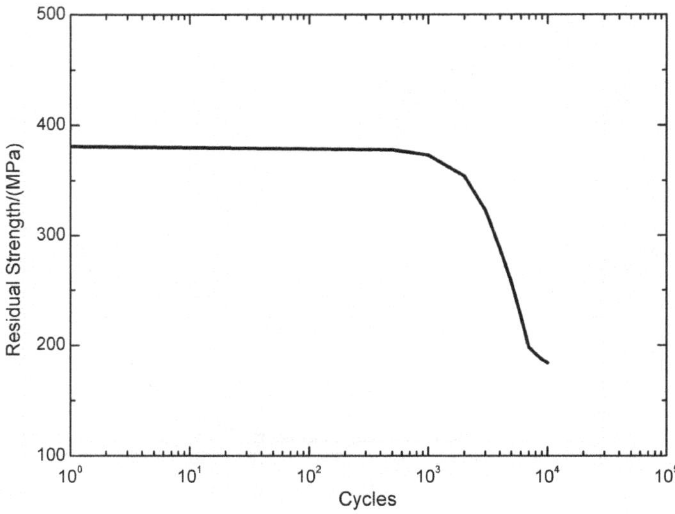

Fig. 3.3 Residual tensile strength versus cycle number curve under fatigue peak stress of σ_{max} = 120 MPa for conditions of a single matrix crack

The effect of the fatigue peak stress, i.e., σ_{max_1} = 120, 160, and 200 MPa, on the fiber fracture, composite residual tensile strength, and intact fiber stress after experiencing N = 5000 cycles, is shown in Fig. 3.4.

When the fatigue peak stress is σ_{max_1} = 120 MPa, the failure probability of the composite under an applied stress of σ = 257 MPa after experiencing N = 5000 cycles is P = 28.9%; when the fatigue peak stress is σ_{max_1} = 160 MPa, the failure probability of the composite under an applied stress of σ = 196 MPa after experiencing N = 5000 cycles is P = 17%; when the fatigue peak stress is σ_{max_1} = 200 MPa, the failure probability of the composite under an applied stress of σ = 189 MPa after experiencing N = 5000 cycles is P = 21.6%. The stress carried by intact fibers increases as the broken fiber fraction increases, in which the intact fiber stress approaches to 1.17, 1.1, and 1.12 of fiber average stress σ/V_f, corresponding to different fatigue peak stresses of σ_{max_1} = 120, 160, and 200 MPa.

3.2.6.2 Case 2: Matrix Multicracking

The effect of applied cycle number, i.e., N = 1, 3000, 5000, and 10,000, on the fiber fracture, composite residual tensile strength, and intact fiber stress under the fatigue peak stress of σ_{max_1} = 120 MPa, is shown in Fig. 3.5.

When the applied cycle number is N = 1, the failure probability of the composite under an applied stress of σ = 337 MPa is P = 2.7%; however, after experiencing N = 3000 cycles, the failure probability of the composite under an applied stress of σ = 301 MPa is P = 5.4%; after experiencing N = 5000 cycles, the failure probability of

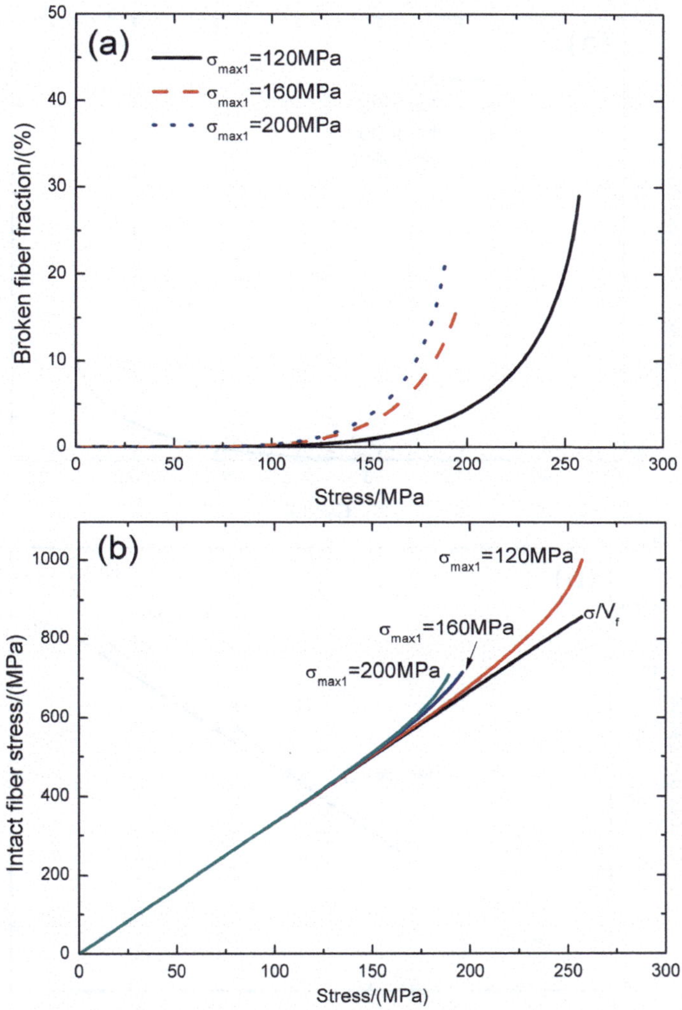

Fig. 3.4 **a** Broken fiber fraction versus applied stress and **b** the intact fiber stress versus applied stress corresponding to different fatigue peak stresses, i.e., $\sigma_{\text{max\_1}} = 120$, 160, and 200 MPa after experiencing $N = 5000$ cycles for conditions of a single matrix crack

the composite under an applied stress of $\sigma = 237$ MPa is $P = 4.5\%$; after experiencing $N = 10000$ cycles, the failure probability of the composite under an applied stress of $\sigma = 168$ MPa is $P = 5\%$. The stress carried by intact fibers increases as the broken fiber fraction increases, in which the intact fiber stress approaches to 1.17, 1.14, 1.1, and 1.11 of fiber average stress σ/V_{f}, corresponding to different cycle numbers of $N = 1$, 3000, 5000, and 10000.

The residual tensile strength versus applied cycle number curve under fatigue peak stress $\sigma_{\text{max\_1}} = 120$ MPa is shown in Fig. 3.6, in which the tensile strength

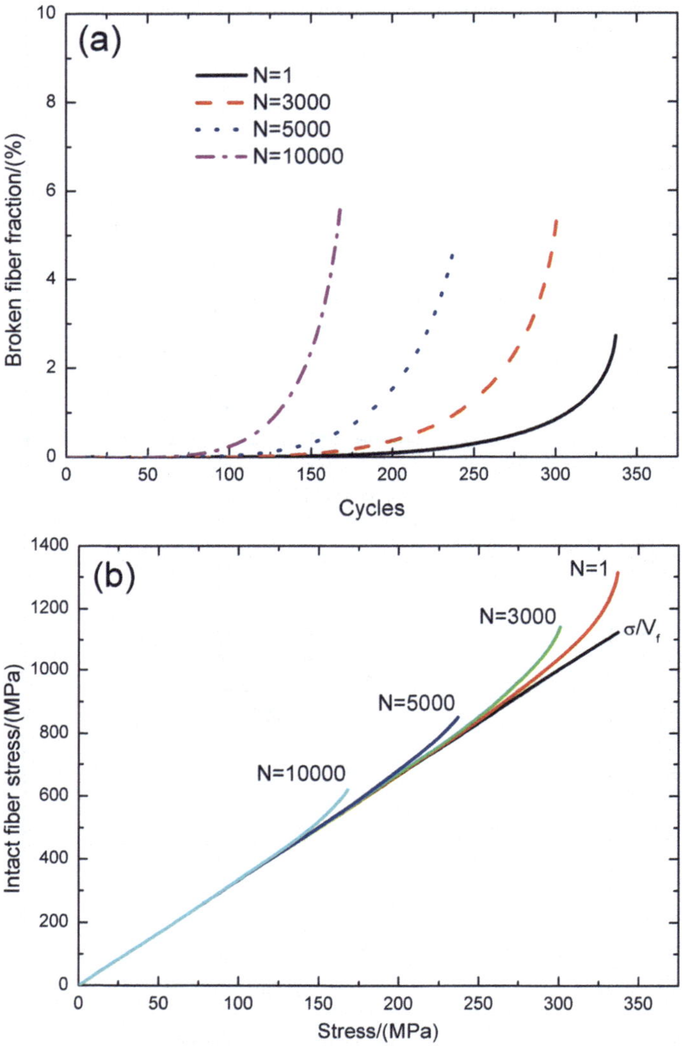

Fig. 3.5 **a** Broken fiber fraction versus applied stress and **b** the intact fiber stress versus applied stress corresponding to different cycle numbers, i.e., $N = 1$, 3000, 5000, and 10,000 cycles under fatigue peak stress of $\sigma_{max} = 120$ MPa for conditions of multiple matrix cracks

decreases from about $\sigma_{UTS} = 337$ MPa at $N = 1$ to about $\sigma_{UTS} = 168$ MPa at $N =$ 10000 due to the degradation of the interface shear stress and fiber strength.

The effect of the fatigue peak stress level, i.e., $\sigma_{max\_1} = 120$, 160, and 200 MPa, on fiber fracture, composite residual tensile strength, and intact fiber stress after experiencing $N = 1000$ cycles, is shown in Fig. 3.7.

When the fatigue peak stress is $\sigma_{max\_1} = 120$ MPa, the failure probability of the composite under an applied stress of $\sigma = 332$ MPa after experiencing $N =$

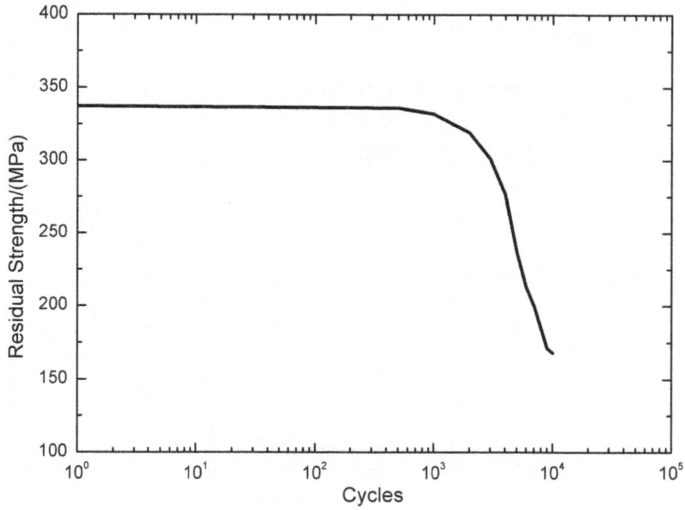

Fig. 3.6 Residual tensile strength versus applied cycle number curve under fatigue peak stress of $\sigma_{\max} = 120$ MPa for conditions of multiple matrix cracks

1000 cycles is $P = 3.1\%$; when the fatigue peak stress is $\sigma_{\max\_1} = 160$ MPa, the failure probability of the composite under an applied stress of $\sigma = 306$ MPa after experiencing $N = 1000$ cycles is $P = 4.8\%$; when the fatigue peak stress is $\sigma_{\max\_1} = 200$ MPa, the failure probability of the composite under an applied stress of $\sigma = 295$ MPa after experiencing $N = 1000$ cycles is $P = 6.4\%$. The stress carried by intact fibers increases as the broken fiber fraction increases, in which the intact fiber stress approaches to 1.15, 1.13, and 1.15 of fiber average stress σ/V_f, corresponding to different fatigue peak stresses of $\sigma_{\max\_1} = 120$, 160, and 200 MPa.

3.2.7 *Experimental Comparisons*

Sørensen et al. (2000) investigated the rate of strength decrease of fiber-reinforced ceramic composite during fatigue loading. The tensile strength of virgin and pre-fatigued specimens is determined experimentally. Four specimens were loaded in monotonic uniaxial tension under load control of 100 MPa/s to establish the strength of the virgin material. Four other specimens were cycled under fatigue loading with the fatigue peak stress of $\sigma_{\max} = 240$ MPa and valley stress of $\sigma_{\min} = 10$ MPa for different cycle numbers without fatigue failure; then, these specimens were loaded in monotonic tension to measure the residual tensile strength.

The experimental tensile stress-strain curves of unidirectional SiC/CAS-II composite corresponding to the conditions of virgin specimen and fatigued specimen after experiencing $N = 100{,}000$ under $\sigma_{\max} = 240$ MPa are shown in Fig. 3.8, in which

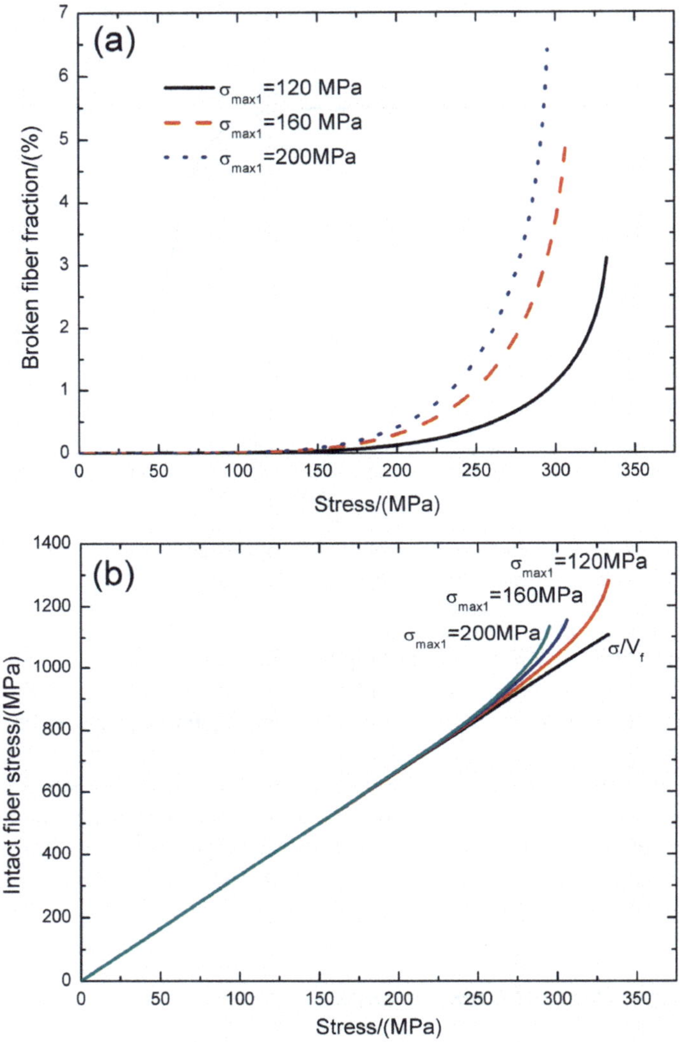

Fig. 3.7 **a** Broken fiber fraction versus applied stress and **b** the intact fiber stress versus applied stress corresponding to different fatigue peak stresses, i.e., $\sigma_{\max\_1} = 120$, 160, and 200 MPa after experiencing $N = 1000$ cycles for conditions of multiple matrix cracks

the tensile stress-strain curve of the virgin specimen exhibits linear-elastic behavior at low applied stress; however, the tensile stress-strain curve of the pre-fatigued specimen is nonlinear even at low applied stress due to the distributed matrix multicracking and interface debonding under fatigue loading. The composite residual tensile strength decreases from $\sigma_{\mathrm{UTS}} = 504$ MPa without pre-fatigue to $\sigma_{\mathrm{UTS}} = 470$ MPa after experiencing $N = 100{,}000$ and $\sigma_{\mathrm{UTS}} = 239$ MPa after experiencing $N = 350{,}000$, as shown in Fig. 3.9. The basic material properties of unidirectional

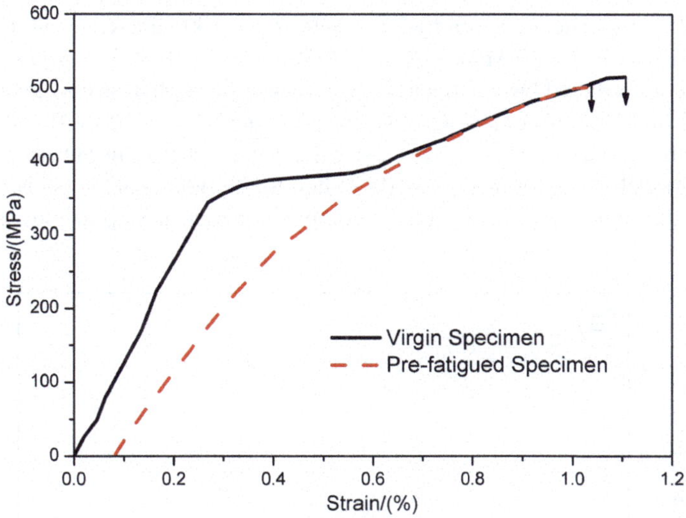

Fig. 3.8 Tensile stress-strain curves of unidirectional SiC/CAS-II composite corresponding to virgin specimen and pre-fatigued specimen under $\sigma_{\max} = 240$ MPa experiencing $N = 100{,}000$ cycles

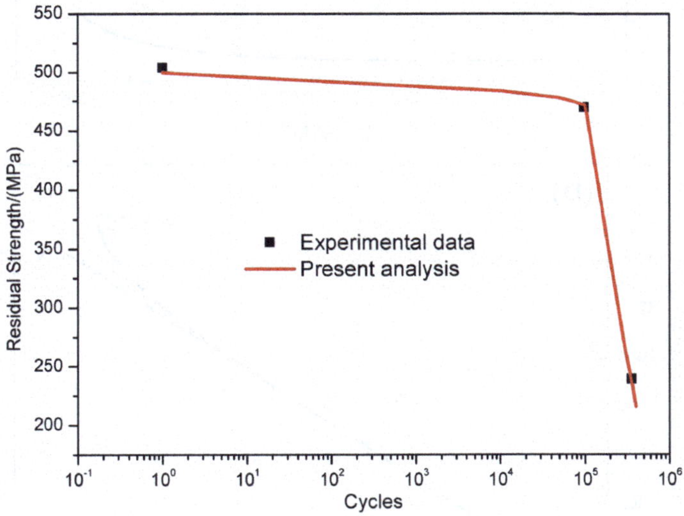

Fig. 3.9 Residual tensile strength of unidirectional SiC/CAS-II composites as a function of applied cycles under fatigue peak stress of $\sigma_{\max} = 240$ MPa

SiC/CAS-II composite are given by $V_f = 35\%$, $E_f = 190$ GPa, $E_m = 90$ GPa, $r_f = 7.5$ μm, $m_f = 3$, $\tau_{io} = 20$ MPa, $\tau_{imin} = 6$ MPa, $\omega = 1\times10^{-6}$, $\lambda = 1.1$, $p_1 = 0.01$, $p_2 = 1.2$. The predicted broken fiber fraction versus the applied stress curves of unidirectional SiC/CAS-II composite after experiencing $N = 1$, 100,000 and 300,000 cycles at fatigue peak stress of $\sigma_{max} = 240$ MPa using matrix multiple crack strength predicted model is shown in Fig. 3.10a. When the fatigue peak stress is $\sigma_{max\_1} = 200$ MPa, the failure probability of the composite under an applied stress of $\sigma =$

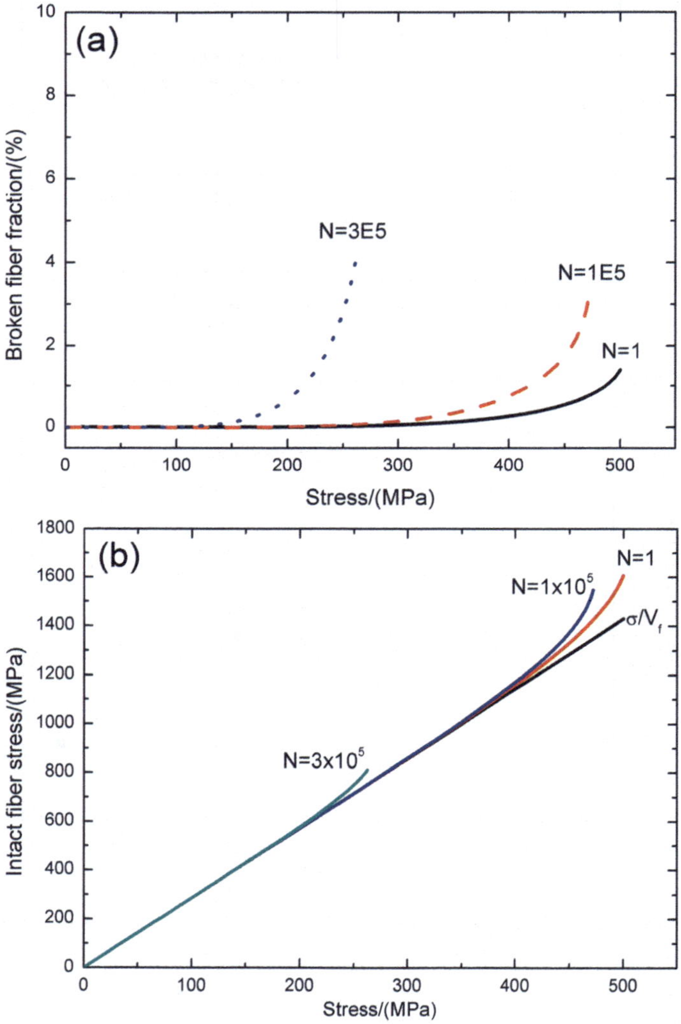

Fig. 3.10 **a** Broken fiber fraction versus applied stress and **b** the intact fiber stress versus applied stress of unidirectional SiC/CAS-II composite under fatigue peak stress of $\sigma_{max} = 240$ MPa using multiple matrix cracking strength predicted model

500 MPa upon first loading is $P = 1.4\%$; after experiencing $N = 100{,}000$ cycles, the failure probability of the composite under an applied stress of $\sigma = 472$ MPa is $P = 3.2\%$; after experiencing $N = 300{,}000$ cycles, the failure probability of the composite under an applied stress of $\sigma = 263$ MPa is $P = 4.3\%$, as shown in Fig. 3.10a. The stress carried by intact fibers increases as the broken fiber fraction increases, as shown in Fig. 3.10b, in which the intact fiber stress approaches to 1.12, 1.14, and 1.1 of fiber average stress σ/V_f, corresponding to different cycle numbers of $N =$ 1, 100,000, and 300,000 cycles. It can be found that the composite residual tensile strength decreases with the increase of cycle number. The predicted residual strength corresponding to different cycle numbers using matrix multiple crack strength model agreed with experimental data, as shown in Fig. 3.9.

3.3 Time-Dependent Tensile Residual Strength of Fiber-Reinforced CMCs Considering Interface Oxidation at Elevated Temperature

In this section, an analytical method is developed to investigate the effect of oxidation on the tensile residual strength of carbon fiber-reinforced CMCs. The shear-lag model is used to describe the microstress field of the damaged composite considering fiber failure. The statistical matrix multicracking model and fracture mechanics interface debonding criterion are used to determine the matrix crack spacing and interface debonded length. The fiber strength degradation model and oxidation region propagation model are adopted to analyze the oxidation effect on tensile strength of the composite, which is controlled by diffusion of oxygen gas through matrix cracks. Under tensile loading, the fiber failure probabilities are determined by combining the oxidation model and fiber statistical failure model based on the assumption that fiber strength is subjected to two-parameter Weibull distribution and the loads carried by broken and intact fibers satisfy the global load sharing criterion. The composite can no longer support the applied load when the total loads supported by broken and intact fibers approach its maximum value. The conditions of a single matrix crack and matrix multicrackings for tensile strength considering oxidation time and temperature are analyzed.

3.3.1 Time-Dependent Residual Strength Model

Matrix cracks will serve as avenues for the ingress of the environment atmosphere into the composite. The oxidation of fiber is assumed to be controlled by diffusion of oxygen gas through the matrix cracks. When the oxidizing gas ingresses into the composite, a sequence of events is triggered starting first with the oxidation of the fiber. For simplicity, it is assumed that both the Weibull and Young moduli of the fibers remain constant and that the only effect of oxidation is to decrease the strength of the fibers. The time-dependent strength of the fibers will be controlled by surface defects resulting from oxidation, with the thickness of the oxide layer representing the size of the average strength-controlling flaw (Naslain et al. 2004).

According to linear-elastic fracture mechanics, the relationship between strength and flaw size is given by (Lara-Curzio 1999)

$$K_{\mathrm{IC}} = Y\sigma_0\sqrt{a} \tag{3.39}$$

where K_{IC} denotes the critical stress intensity factor; Y is a geometric parameter; σ_0 is the fiber strength; and a is the size of the strength-controlling flaw.

Considering that the oxidation of fibers is controlled by diffusion of oxygen through oxide layer, the oxide layer will grow on fiber's surface according to (Lara-Curzio 1999)

$$\alpha = \sqrt{kt} \tag{3.40}$$

where α is the thickness of the oxide layer at time t and k is the parabolic rate constant.

By assuming the fracture toughness of the fibers remains constant and that the fiber strength σ_0 is related to the mean oxide layer thickness according to Eq. (3.40), i.e., $a = \alpha$, then the time dependence of the fiber strength will be given by (Lara-Curzio 1999)

$$\sigma_0(t) = \sigma_0, t \le \frac{1}{k}\left(\frac{K_{\mathrm{IC}}}{Y\sigma_0}\right)^4 \tag{3.41}$$

$$\sigma_0(t) = \frac{K_{IC}}{Y\sqrt[4]{kt}}, t > \frac{1}{k}\left(\frac{K_{\mathrm{IC}}}{Y\sigma_0}\right)^4 \tag{3.42}$$

Equations (3.41) and (3.42) indicate that there exists an incubation period equal to the time required to grow an oxide layer as thick as the size of the average critical flaw in the virgin fibers. Also note that afterward, the characteristic fiber strength changes with time as $\approx t^{-1/4}$.

Filipuzzi and Naslain (1994) have measured and modeled the change in the interface oxidation length ζ of the carbon interface that oxidation occurs according to

$$\mathrm{C} + \mathrm{O_2} \rightarrow \mathrm{CO_2} \tag{3.43}$$

The oxidation region length of ζ is (Filipuzzi and Naslain 1994)

$$\zeta = \varphi_1\left(1 - \mathrm{e}^{-\varphi_2 t}\right) \tag{3.44}$$

where φ_1 and φ_2 are fitting parameters dependent on temperature. Casas and Martinez-Esnaola (2003) performed the thermodynamic calculations and found that the deceleration of the oxidation phenomena, as a consequence of the reduced oxygen activity due to the diffusion through the glassy phases, can represent several orders of magnitude in the oxidation timescale. This effect has been incorporated into the model using a delay factor b in Eq. (3.44), which becomes

$$\zeta = \varphi_1\left(1 - e^{-\frac{\varphi_2 t}{b}}\right) \tag{3.45}$$

The two-parameter Weibull model is adopted to describe fiber strength distribution. The fiber fracture probability P is (Curtin et al. 1998)

$$P = 1 - \exp\left(-\int_{L_0} \frac{1}{l_0}\left[\frac{\sigma_f(z)}{\sigma_0}\right]^m dz\right) \tag{3.46}$$

where σ_0 denotes the fiber strength at tested gauge length of l_0; m denotes the fiber Weibull modulus; and L_0 denotes the integral length.

When the fiber-reinforced CMCs are subjected to oxidation, the notch would form at the fiber surface leading to the degradation of fiber strength and the increase of fiber stress concentration and fracture probability. The fracture probabilities of oxidized fibers in the oxidation region, unoxidized fibers in the oxidation region, fibers in the interface debonded region, and interface bonded region of P_a, P_b, P_c, and P_d are

$$P_a(\Phi) = 1 - \exp\left\{-2\frac{\zeta}{l_0}\left[\frac{\Phi}{\sigma_0(t)}\right]^m\right\} \tag{3.47}$$

$$P_b(\Phi) = 1 - \exp\left\{-2\frac{\zeta}{l_0}\left(\frac{\Phi}{\sigma_0}\right)^{m_f}\right\} \tag{3.48}$$

$$P_c(\Phi) = 1 - \exp\left\{-\frac{r_f\Phi^{m+1}}{l_0\sigma_0^m\tau_i(m+1)}\left[1 - \left(1 - \frac{l_d}{l_s}\right)^{m+1}\right]\right\} \tag{3.49}$$

$$P_d(\Phi) = 1 - \exp\left\{-\frac{2r_f\Phi^m}{\rho l_0\sigma_0^m(m+1)\left(1 - \frac{\sigma_{fo}}{\Phi} - \frac{l_d}{l_s}\right)} \times\left[\left(1 - \frac{l_d}{l_s} - \left(1 - \frac{\sigma_{fo}}{\Phi} - \frac{l_d}{l_s}\right)\frac{\rho l_d}{r_f}\right)^{m+1} - \left(1 - \frac{l_d}{l_s} - \left(1 - \frac{\sigma_{fo}}{\Phi} - \frac{l_d}{l_s}\right)\frac{\rho l_c}{2r_f}\right)^{m+1}\right]\right\} \tag{3.50}$$

where $\sigma_0(t)$ denotes the oxidized fiber strength; t denotes the oxidation time; and L_s denotes the slip length over which the fiber stress would decay to zero if not interrupted by the far-field equilibrium stresses.

$$l_s = \frac{r_f\Phi}{2\tau_i} \tag{3.51}$$

For a single matrix crack, the global load sharing criterion was used to determine the loads carried by the intact and broken fibers. The average load per fiber σ/V_{f} must equal the product of the stress carried by the unbroken fibers Φ and the fraction $(1-P)$ of unbroken fibers, plus a contribution due to the residual stress $\langle\Phi_{\mathrm{b}}\rangle$ carried by the broken fibers which do not break exactly at the matrix crack plane. The mechanical equilibrium at the matrix crack plane is

$$\frac{\sigma}{V_{\mathrm{f}}}=(1-P(\Phi))\Phi+P(\Phi)\langle\Phi_{\mathrm{b}}\rangle \tag{3.52}$$

where

$$P(\Phi)=\theta P_a(\Phi)+(1-\theta)P_b(\Phi)+P_c(\Phi)+P_d(\Phi) \tag{3.53}$$

where θ denotes the oxidation fiber fraction in the oxidized region.

When a fiber breaks, the load carried by the fiber drops to zero at the position of break. Similar to the case of matrix cracking, the fiber/matrix interface debonds and the stress builds up in the fiber through the interface shear stress. During the process of loading, the stress in a broken fiber Φ_{b} as a function of the distance x from the break can be written as

$$\Phi_b(z)=\frac{2\tau_{\mathrm{i}}}{r_{\mathrm{f}}}z \tag{3.54}$$

In order to calculate the average stress carried by broken fibers $\langle\Phi_{\mathrm{b}}\rangle$, it is necessary to construct the probability distribution $f(z)$ of the distance z of a fiber break from the reference matrix crack plane, provided that a break occurs within a distance $\pm l_{\mathrm{s}}$. For this conditional probability distribution, Phoenix and Raj (1992) deduced the following based on Weibull statistics, as

$$f(z)=\frac{1}{P(\Phi)l_{\mathrm{s}}}\left(\frac{\Phi}{\sigma_{\mathrm{c}}}\right)^{m+1}\exp\left[-\left(\frac{z}{l_{\mathrm{s}}}\right)\left(\frac{\Phi}{\sigma_{\mathrm{c}}}\right)^{m+1}\right],\ z\in[0,l_{\mathrm{s}}] \tag{3.55}$$

where σ_{c} denotes the characteristic strength of fibers in the characteristic length scale of δ_{c}.

$$\sigma_{\mathrm{c}}=\left(\frac{l_0\sigma_0^{m_{\mathrm{f}}}\tau_{\mathrm{i}}}{r_{\mathrm{f}}}\right)^{\frac{1}{m+1}},\ \delta_{\mathrm{c}}=\left(\frac{\sigma_0 r_{\mathrm{f}} l_0^{1/m}}{\tau_{\mathrm{i}}}\right)^{\frac{m}{m+1}} \tag{3.56}$$

The averaging stress carried by broken fibers $\langle\Phi_{\mathrm{b}}\rangle$ during the process of loading using Eqs. (3.54) and (3.55) leads to

$$\langle\Phi_{\mathrm{b}}\rangle=\Phi\left[\left(\frac{\sigma_{\mathrm{c}}}{\Phi}\right)^{m+1}-\frac{1-P(\Phi)}{P(\Phi)}\right] \tag{3.57}$$

When there is more than one matrix crack, a balance of forces at the matrix crack plane requires that the applied force per fiber equals the force carried by the unbroken fibers plus the pullout force carried by those fibers broken away from the matrix crack plane and can be expressed as

$$\frac{\sigma}{V_f} = \left[1 - P(\Phi)\left(1 + \frac{2l_s}{l_c}\right)\right]\Phi + P(\Phi)\frac{2l_s}{l_c}\langle\Phi_b\rangle \tag{3.58}$$

where

$$P(\Phi) = \gamma[\eta P_a(\Phi) + (1-\eta)P_b(\Phi)] + P_c(\Phi) \tag{3.59}$$

3.3.2 Results and Discussion

The ceramic composite system of C/SiC is used for the case study, and its basic material properties are given by: $V_f = 20\%$, $E_f = 230$ GPa, $E_m = 350$ GPa, $r_f = 3.5$ μm, $k = \exp[11.383 - (8716/T)]\times10^{-18}/60$ m^2/s, $K_{IC} = 0.5$ MPa/m$^{1/2}$, $Y = 1$, $\varphi_1 = 7.021\times10^{-3}\times\exp(8231/T)$, $\varphi_2 = 227.1\times\exp(-17{,}090/T)$, $m = 5$, $l_0 = 25\times10^{-3}$ m, $\sigma_0 = 0.52$ GPa.

The fiber strength versus oxidation time curves corresponding to the elevated temperatures of 700, 800, and 900 °C is shown in Fig. 3.11a, in which the no. of degradation time of $t_0 = \frac{1}{k}\left(\frac{K_{IC}}{Y\sigma_0}\right)^4$ decreases with the increase of temperature, i.e., from 2 h at the temperature of 700 °C to 0.4 h at the temperature of 900 °C. The fiber strength σ_0(t) decreases with the increase of oxidation time and temperature; i.e., at the oxidation time of 10 h, the carbon fiber strength decreases to 0.67, 0.54, and 0.45 of original strength. The oxidation region length ζ versus oxidation time curves corresponding to the elevated temperatures of 700, 800, and 900 °C is shown in Fig. 3.11b, in which the oxidation length increases with the increase of oxidation time and temperature; i.e., the oxidation length increases to $3.6r_f$, $8.5r_f$, and $17r_f$ after 10 h oxidation corresponding to the temperature of 700, 800, and 900 °C, respectively. The degradation of fiber strength and increase of oxidation region length would lead to fiber fracture under low applied stress and decrease of fiber strength.

The effects of the oxidation time, oxidation temperature, and matrix crack spacing on fiber failure and composite tensile residual strength corresponding to the conditions of a single matrix crack and multiple matrix cracks are investigated.

3.3.2.1 Case 1: A Single Matrix Crack

The effect of oxidation temperature (i.e., $T = 700$ and 900 °C) on fiber fracture, composite tensile strength, and intact fiber stress is shown in Fig. 3.12. At room tem-

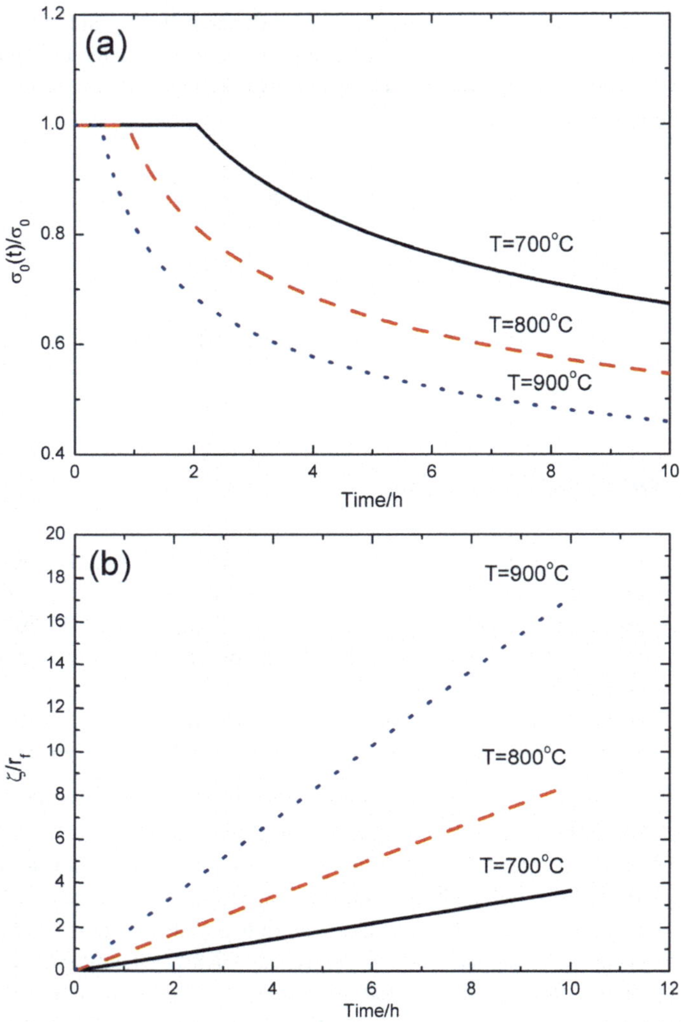

Fig. 3.11 **a** Fiber strength versus the oxidation time and **b** the oxidation region length versus oxidation time corresponding to different temperatures of $T = 700$, 800, and 900 °C

perature without oxidation, the composite tensile strength is about $\sigma_{\text{UTS}} = 146$ MPa with the fiber broken fraction of $P = 21.3\%$; however, after oxidizing 4 h at elevated temperature of $T = 700$ °C, the composite tensile strength decreases to $\sigma_{\text{UTS}} = 115$ MPa with the fiber broken fraction of $P = 14.1\%$; after oxidizing 4 h at elevated temperature of $T = 900$ °C, the composite tensile strength decreases to $\sigma_{\text{UTS}} = 96$ MPa with the fiber broken fraction of $P = 9.5\%$, as shown in Fig. 3.12a. The stress carried by intact fibers increases as the broken fiber fraction increases, as shown in Fig. 3.12b, in which the intact fiber stress approaches to 1.13, 1.16, and 1.1 of fiber

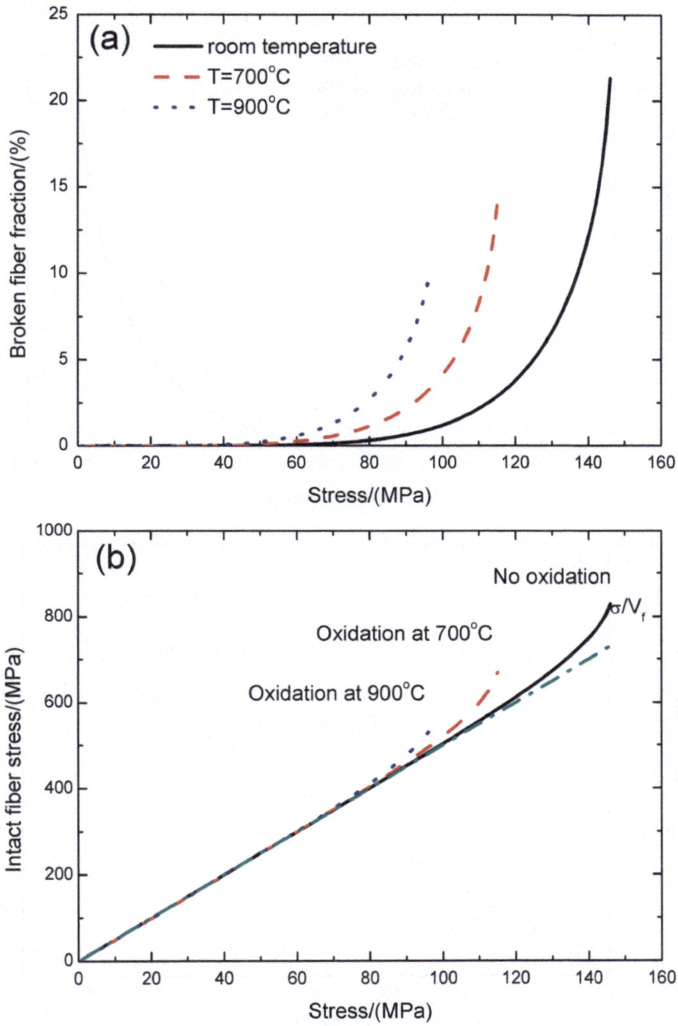

Fig. 3.12 **a** Broken fiber fraction versus applied stress and **b** the intact fiber stress versus applied stress for conditions of no oxidation and after oxidation of 4 h at temperatures of $T = 700$ and 900 °C when the interface partially debonding

average stress σ/V_f, corresponding to the conditions of room temperature without oxidation and the oxidation temperatures of $T = 700$ and 900 °C.

The effect of oxidation time (i.e., $t = 0$, 2, and 6 h) at elevated temperature of $T = 800$ °C, on fiber fracture, composite tensile strength, and intact fiber stress, is shown in Fig. 3.13. With the increase of oxidation time, the composite tensile strength decreases and the fiber broken fraction at low applied stress increases; i.e., after oxidizing 2 h, the composite tensile strength decreases to $\sigma_{UTS} = 126$ MPa with the broken fiber fraction of $P = 11.5\%$; after oxidizing 6 h, the composite tensile

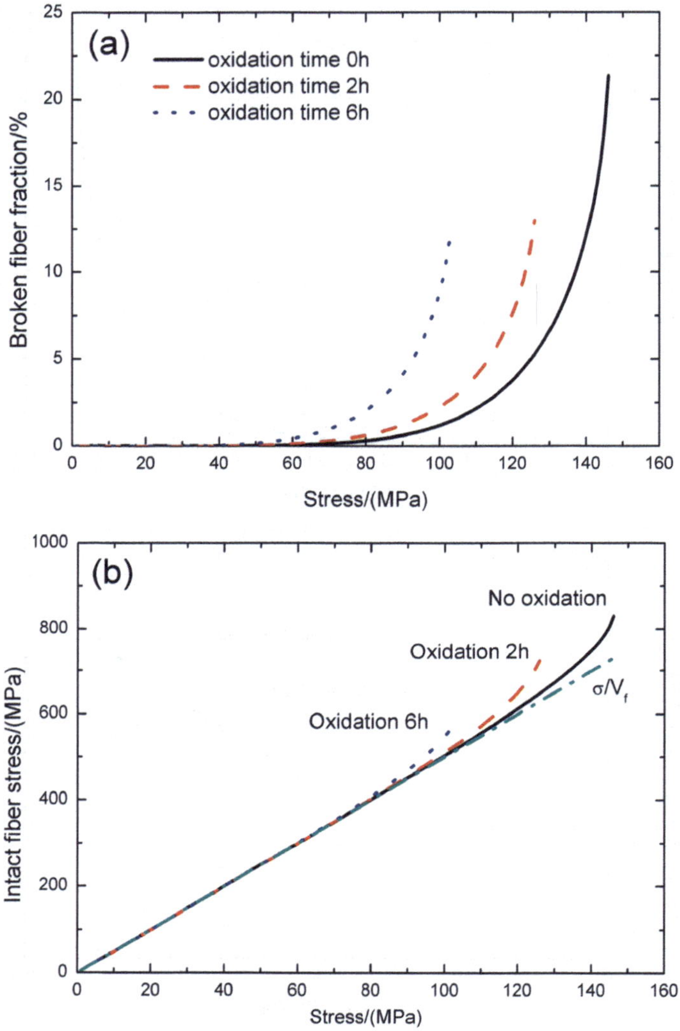

Fig. 3.13 **a** Broken fiber fraction versus the applied stress and **b** the intact fiber stress versus the applied stress for conditions of no oxidation and after oxidation of $t = 2$ and 6 h at temperature of $T = 800$ °C when the interface partially debonding

strength decreases to $\sigma_{\mathrm{UTS}} = 103$ MPa with the broken fiber fraction of $P = 12.1\%$, as shown in Fig. 3.13a. The stress carried by the intact fiber increases as the broken fiber fraction increases, as shown in Fig. 3.13b, in which the intact fiber stress approaches 1.15 and 1.14 of fiber average stress σ/V_{f}, corresponding to the oxidation time of $t = 2$ and 6 h.

3.3.2.2 Case 2: Matrix Multicracking

The effect of oxidation temperature (i.e., $T = 700$, 800, and 900 °C) on fiber fracture, composite tensile strength, and intact fiber stress is shown in Fig. 3.14. After oxidizing $t = 4$ h at elevated temperature of $T = 700$ °C, the composite tensile strength decreases to $\sigma_{\mathrm{UTS}} = 101$ MPa with the fiber broken fraction of $P = 6.7\%$; after oxidizing $t =$

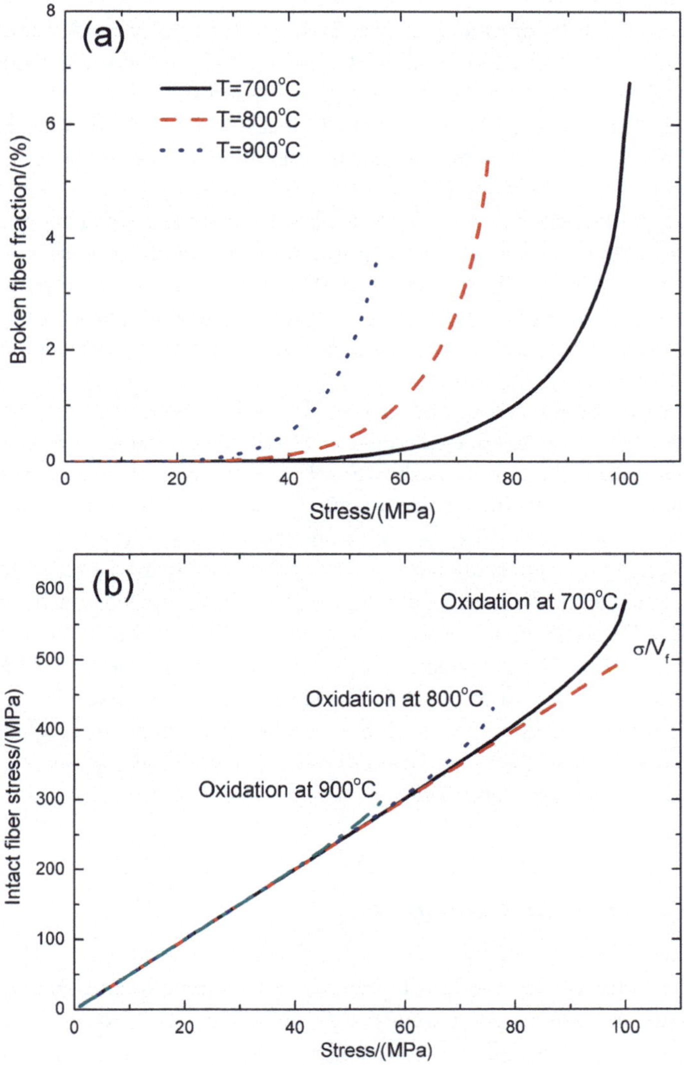

Fig. 3.14 **a** Broken fiber fraction versus applied stress and **b** the intact fiber stress versus applied stress for conditions of no oxidation and after oxidation of $t = 4$ h at temperatures of $T = 700$, 800, and 900 °C when the interface complete debonding

4 h at elevated temperature of T = 800 °C, the composite tensile strength decreases to σ_{UTS} = 76 MPa with the fiber broken fraction of P = 5.6%; after oxidizing t = 4 h at elevated temperature of T = 900 °C, the composite tensile strength decreases to σ_{UTS} = 56 MPa with the fiber broken fraction of P = 3.7%, as shown in Fig. 3.14a. The stress carried by the intact fiber increases as the broken fiber fraction increases, as shown in Fig. 3.14b, in which the intact fiber stress approaches to 1.16, 1.13, and 1.07 of fiber average stress σ/V_f, corresponding to the oxidation temperatures of T = 700, 800, and 900 °C.

The effect of oxidation time (i.e., t = 2, 4, and 6 h) at elevated temperature of T = 750 °C, on fiber fracture, composite tensile strength, and intact fiber stress, is shown in Fig. 3.15. With the increase of oxidation time, the composite tensile strength decreases and the fiber broken fraction at low applied stress increases; i.e., after oxidizing t = 2 h, the composite tensile strength decreases to σ_{UTS} = 106 MPa with the broken fiber fraction of P = 4.4%; after oxidizing t = 4 h, the composite tensile strength decreases to σ_{UTS} = 87 MPa with the broken fiber fraction of P = 5.3%; after oxidizing t = 6 h, the composite tensile strength decreases to σ_{UTS} = 76 MPa with the broken fiber fraction of P = 5.9%. The stress carried by intact fibers increases as the broken fiber fraction increases, in which the intact fiber stress approaches to 1.12, 1.13, and 1.14 of fiber average stress σ/V_f, corresponding to the oxidation time of t = 2, 4, and 6 h.

The effect of matrix crack spacing (i.e., l_c = $30r_f$, $50r_f$, and $70r_f$) at elevated temperature of T = 800 °C for oxidizing t = 4 h, on fiber fracture, composite tensile strength, and intact fiber stress, is shown in Fig. 3.16. With the decrease of matrix crack spacing, the composite tensile strength decreases and the broken fiber fraction at low applied stress increases; i.e., at the matrix crack spacing of l_c = $30r_f$, the composite tensile strength is about σ_{UTS} = 74 MPa with the broken fiber fraction of P = 5.2%; at the matrix crack spacing of l_c = $50r_f$, the composite tensile strength is about σ_{UTS} = 77 MPa with the broken fiber fraction of P = 5.6%; at the matrix crack spacing of l_c = $70r_f$, the composite tensile strength is about σ_{UTS} = 79 MPa with the broken fiber fraction of P = 6.2%. The stress carried by intact fibers increases as the broken fiber fraction increases, in which the intact fiber stress approaches to 1.14, 1.12, and 1.11 of fiber average stress σ/V_f, corresponding to the matrix crack spacing of l_c = $30r_f$, $50r_f$, and $70r_f$.

3.3.3 *Experimental Comparisons*

Wang (2010) investigated the tensile behavior of unidirectional C/SiC composite after oxidizing at elevated temperatures of T = 650 and 800 °C for t = 2, 4, and 6 h. The specimens were tensiled to fracture under displacement control of 0.06 mm/min conducted on an MTS model 809 servo-hydraulic load frame. After tensile tests, direct observations of matrix multicracking, fiber pullout, and oxidation were made using optical microscope.

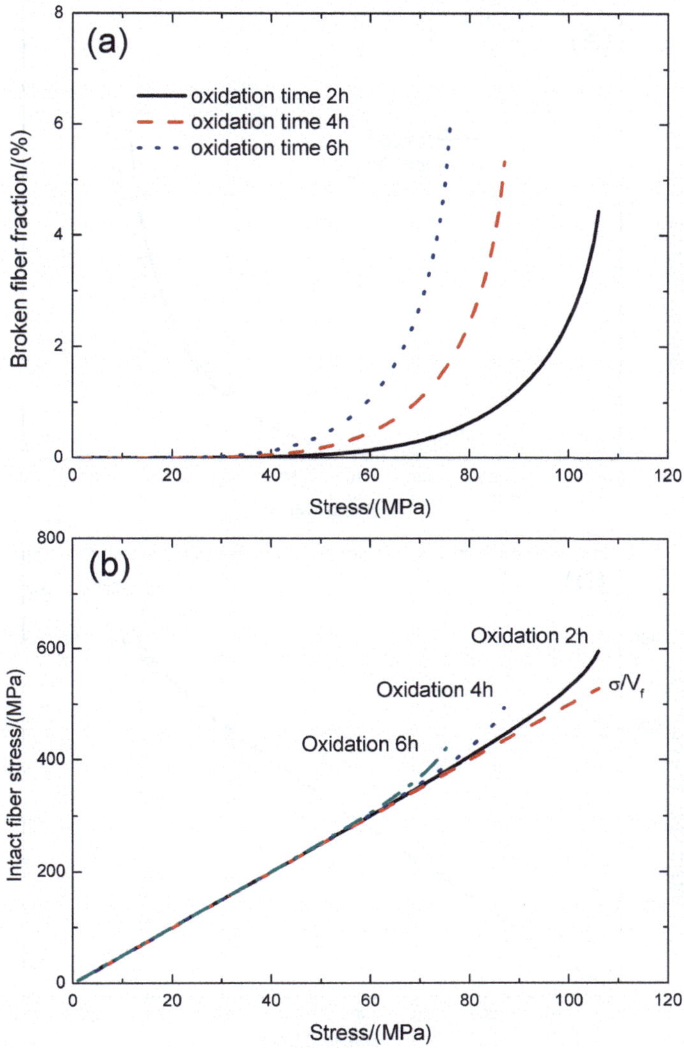

Fig. 3.15 **a** Broken fiber fraction versus applied stress and **b** the intact fiber stress versus applied stress for conditions after oxidation of t = 2, 4, and 6 h at temperature of T = 750 °C when the interface complete debonding

The experimental tensile stress-strain curves of unidirectional C/SiC composite corresponding to the conditions of no oxidation (Li et al. 2014) and oxidation at T = 650 °C for t = 4 h (Wang 2010) are shown in Fig. 3.17, in which the composite elastic modulus and tensile strength decrease, and the failure strain increases. The composite tensile strength decreases from σ_{UTS} = 230 MPa without oxidation to σ_{UTS} = 224, 137, and 102 MPa after t = 2 h, 4 h, and 6 h oxidizing at an elevated temperature of T = 650 °C and σ_{UTS} = 92, 85 and 54 MPa after t = 2, 4, and 6 h oxidizing at an

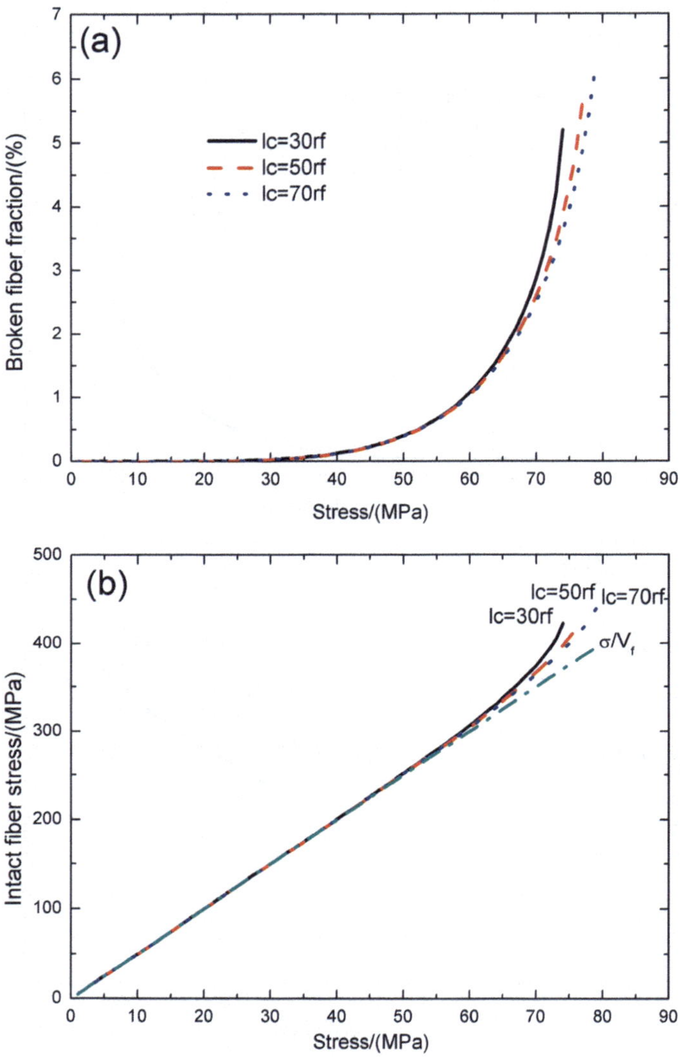

Fig. 3.16 **a** Broken fiber fraction versus applied stress and **b** the intact fiber stress versus applied stress for conditions after oxidation of t = 4 h at temperature of T = 800 °C corresponding to a different matrix crack spacing of $l_c = 30r_f$, $50r_f$, and $70r_f$

elevated temperature of T = 800 °C. The basic material properties of unidirectional C/SiC composite are given by: V_f = 42%, E_f = 230 GPa, E_m = 350 GPa, r_f = 3.5 μm, $k = \exp[11.383 - (8716/T)] \times 10^{-18}/60$ m²/s, K_{IC} = 0.5 MPa/m$^{1/2}$, Y = 1, $\varphi_1 = 7.021 \times 10^{-3} \times \exp(8231/T)$, $\varphi_2 = 227.1 \times \exp(-17{,}090/T)$, m_f = 5, $l_0 = 25 \times 10^{-3}$ m, σ_c = 0.787 GPa. The predicted broken fiber fraction versus applied stress curves of unidirectional C/SiC composite after oxidizing t = 2 h, 4 h, and 6 h at temperatures of T = 650 and 800 °C using a single matrix crack and matrix multiple

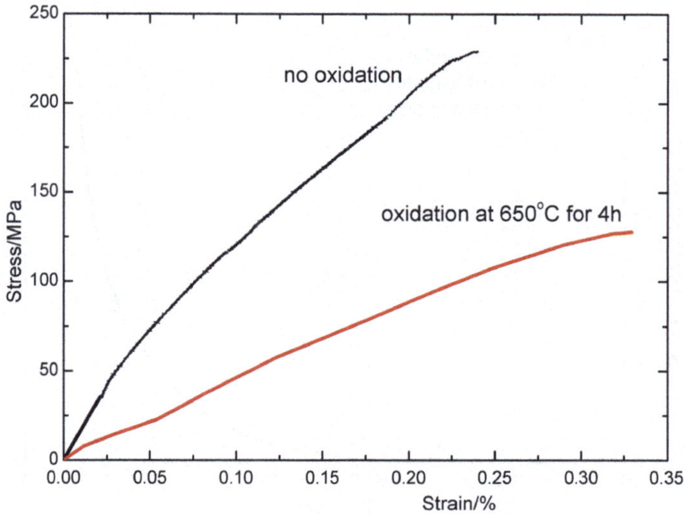

Fig. 3.17 Tensile stress-strain curves of unidirectional C/SiC composite corresponding to the conditions of no oxidation and oxidation at $T = 650$ °C for $t = 4$ h

crack strength predicted model is shown in Figs. 3.18 and 3.19. It can be found that the composite tensile strength decreases with the increase of oxidation time and temperature. The predicted composite strength corresponding to different oxidation times and temperatures is given in Table 3.1, in which the predicted results using matrix multiple crack strength model agreed better with experimental data.

3.4 Time-Dependent Tensile Residual Strength of Fiber-Reinforced CMCs Under Cyclic Loading at Elevated Temperature

In this section, the strength degradation of non-oxide and oxide/oxide fiber-reinforced CMCs subjected to cyclic loading at elevated temperatures in oxidative environments is investigated. Considering the damage mechanisms of matrix cracking, interface debonding, interface wear, interface oxidation, and fiber fracture, the composite residual strength model is established by combining the microstress field of the damaged composites, the damage models, and the fracture criterion. The relationships between the composite residual strength, fatigue peak stress, interface debonding, fiber failure, and cycle number are established. The effects of the peak stress level, initial and steady-state interface shear stress, fiber Weibull modulus and fiber strength, and testing temperature on the degradation of composite strength and fiber failure are investigated. The evolution of residual strength versus cycle number curves of

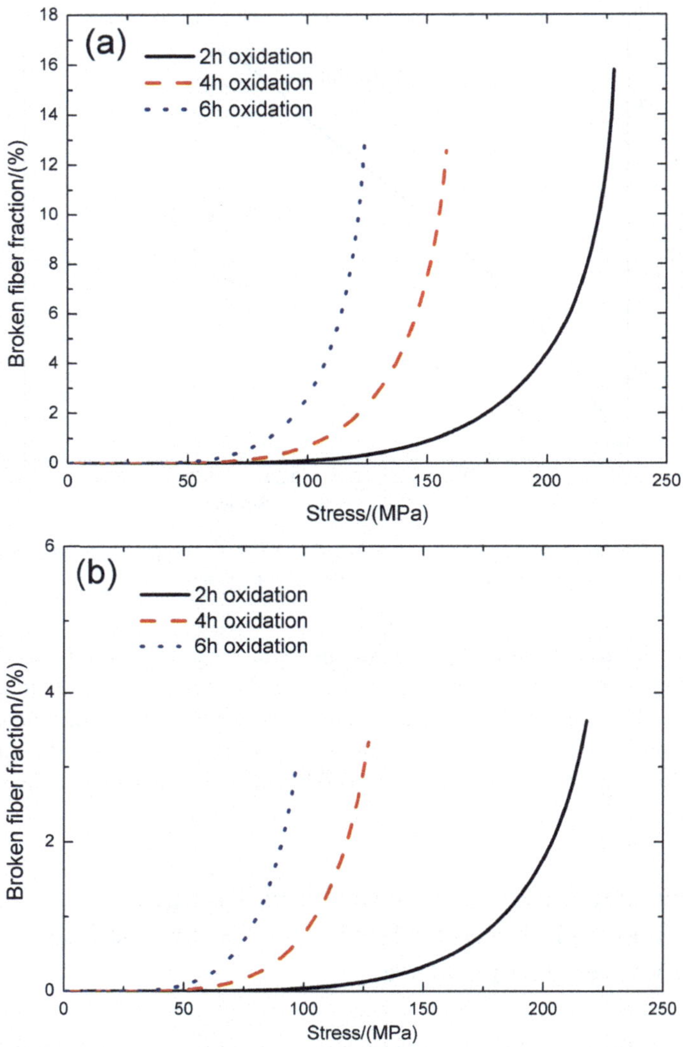

Fig. 3.18 Broken fiber fraction versus applied stress of unidirectional C/SiC composite after oxidizing for $t = 2$, 4, and 6 h at $T = 650$ °C using **a** single matrix cracking strength predicted model and **b** multiple matrix cracking strength predicted model

non-oxide and oxide/oxide CMCs under cyclic loading at elevated temperatures in oxidative environments is predicted.

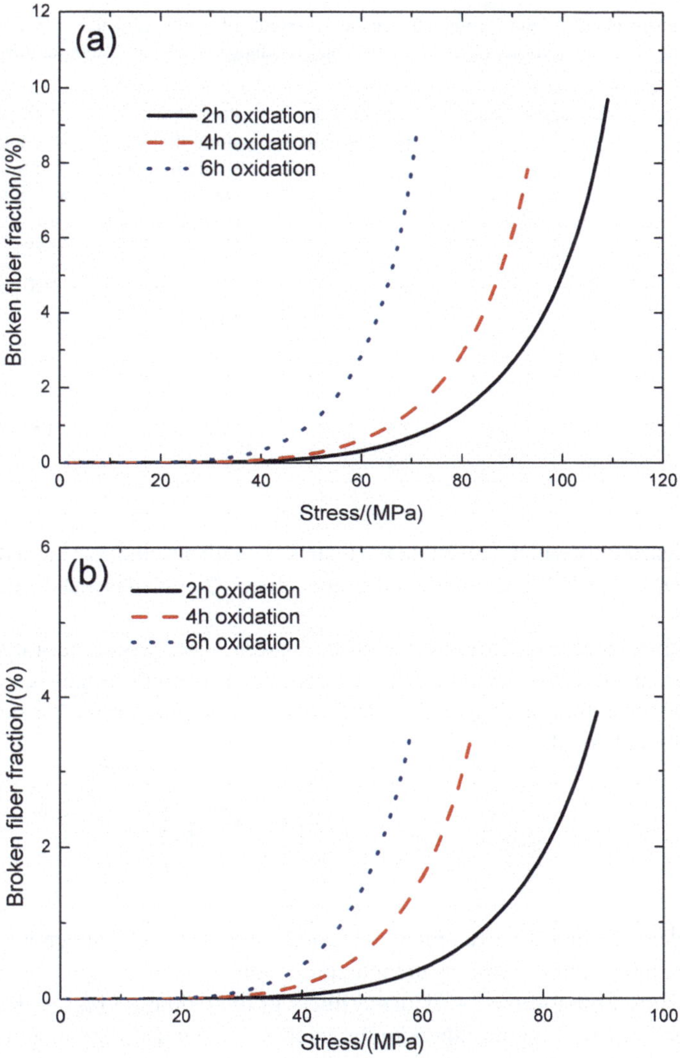

Fig. 3.19 Broken fiber fraction versus applied stress of unidirectional C/SiC composite after oxidizing for $t = 2$, 4, and 6 h at $T = 800$ °C using **a** single matrix cracking strength predicted model and **b** multiple matrix cracking strength predicted model

3.4.1 Cyclic-Dependent Residual Strength Model

Under cyclic fatigue loading, the interface shear stress and fiber strength degrade with increasing applied cycles due to the interface wear and interface oxidation (Holmes 1991; McNulty and Zok 1999; Li 2014; Fantozzi and Reynaud 2014; Ruggles-Wrenn and Lee 2016). A unit cell is extracted from the ceramic composite system to analyze

Table 3.1 Experimental and predicted tensile strength of unidirectional C/SiC composite corresponding to the conditions of no oxidation and oxidation at elevated temperatures of T = 650 and 800 °C

Temperature (°C)	Oxidation time (h)	Composite strength (MPa)	Predicted strength using single matrix cracking strength model (MPa)	Predicted strength using multiple matrix cracking strength model (MPa)
650	2	224	228	218
	4	137	158	127
	6	102	124	97
800	2	92	109	89
	4	85	93	68
	6	54	71	58

the stress distributions in the fiber and the matrix. The fiber radius is r_f, and the matrix radius is $R(R = r_f/V_f^{1/2})$. The length of the unit cell is $l_c/2$, which is just the half matrix crack space.

When fiber breaks, the stress dropped by broken fibers would be transferred to intact fibers at the cross section. The two-parameter Weibull model is adopted to describe the fiber strength distribution. The fiber fracture probability P is determined by the following equation.

$$P = 1 - \exp\left(- \int_{L_0} \frac{1}{l_0} \left[\frac{\sigma_f(z)}{\sigma_0} \right]^m \mathrm{d}z \right) \tag{3.60}$$

where σ_0 denotes the fiber strength at the tested gauge length l_0; m denotes the fiber Weibull modulus; and L_0 denotes the integral length.

The fracture probabilities of oxidized fibers in the oxidation region, unoxidized fibers in the oxidation region, fibers in the interface debonded region, and interface bonded region of P_a, P_b, P_c, and P_d are determined by the following equations.

$$P_a = 1 - \exp\left\{ -2\frac{l_t}{l_0} \left[\frac{\Phi}{\sigma_0(t)} \right]^m \right\} \tag{3.61}$$

$$P_b = 1 - \exp\left\{ -2\frac{l_t}{l_0} \left(\frac{\Phi}{\sigma_0} \right)^m \right\} \tag{3.62}$$

$$P_c = 1 - \exp\left\{ -\frac{r_f \Phi^{m+1}}{l_0(\sigma_0(N))^m \tau_i(N)(m+1)} \left[1 - \left(1 - \frac{l_d(N)}{l_f(N)} \right)^{m+1} \right] \right\} \tag{3.63}$$

$$P_d = 1 - \exp\left\{ -\frac{2r_f \Phi^m}{\rho l_0 (\sigma_0(N))^m (m+1)\left(1 - \frac{\sigma_{fo}}{\Phi} - \frac{l_d(N)}{l_s(N)}\right)} \times \left[\left(1 - \frac{l_d(N)}{l_f(N)}\right.\right.\right.$$
$$\left.\left.\left. - \left(1 - \frac{\sigma_{fo}}{\Phi} - \frac{l_d(N)}{l_f(N)}\right)\frac{\rho l_d(N)}{r_f}\right)^{m+1} - \left(1 - \frac{l_d(N)}{l_f(N)} - \left(1 - \frac{\sigma_{fo}}{\Phi} - \frac{l_d(N)}{l_f(N)}\right)\frac{\rho l_c}{2r_f}\right)^{m+1}\right]\right\} \tag{3.64}$$

where l_d denotes the interface debonded length; σ_{fo} denotes the fiber stress at the interface bonded region; l_f denotes the interface slip length; l_t denotes the interface oxidation region length; and σ_0(t) denotes the time dependence of fiber strength.

$$\tau_i(N) = \tau_{io} + \left[1 - \exp\left(-\omega N^\lambda\right)\right](\tau_{imin} - \tau_{io}) \tag{3.65}$$

$$\sigma_o(N) = \sigma_o\left[1 - p_1(\log N)^{p_2}\right] \tag{3.66}$$

$$\sigma_0(t) = \sigma_0,\ t \le \frac{1}{k}\left(\frac{K_{IC}}{Y\sigma_0}\right)^4 \tag{3.67}$$

$$\sigma_0(t) = \frac{K_{IC}}{Y\sqrt[4]{kt}},\ t > \frac{1}{k}\left(\frac{K_{IC}}{Y\sigma_0}\right)^4 \tag{3.68}$$

where τ_{io} denotes the initial interface shear stress; τ_{imin} denotes the steady-state interface shear stress; ω and λ are empirical constants; p_1 and p_2 are empirical parameters; K_{IC} denotes the critical stress intensity factor; Y is a geometric parameter; and k is the parabolic rate constant.

The relationship between the applied stress, the stress carried by intact fibers, and broken fibers is determined by the following equation:

$$\frac{\sigma}{V_f} = \left[1 - P_f\left(1 + \frac{2l_f}{l_c}\right)\right]\Phi + P_r\frac{2l_f}{l_c}\langle\Phi_b\rangle \tag{3.69}$$

where

$$P_f = \gamma[\theta P_a + (1-\theta)P_b] + P_c + P_d \tag{3.70}$$

$$P_r = P_c + P_d \tag{3.71}$$

where θ denotes the oxidation fiber fraction in the oxidized region and γ denotes the fraction of oxidation in the multiple matrix cracks.

The stress carried by broken fibers is determined by the following equation:

$$\langle\Phi_b\rangle = \frac{\Phi}{P_r}\left(\frac{\sigma_c}{\Phi}\right)^{m+1}\left(\frac{\sigma_o(N)}{\sigma_o}\right)^m\frac{\tau_i(N)}{\tau_i}$$

$$\left\{1 - \exp\left[-\left(\frac{\Phi}{\sigma_c}\right)^{m+1}\left(\frac{\sigma_o}{\sigma_o(N)}\right)^m \frac{\tau_i}{\tau_i(N)}\right]\right\}$$
$$-\frac{\Phi}{P_r}\exp\left\{-\left(\frac{\Phi}{\sigma_c}\right)^{m+1}\left(\frac{\sigma_o}{\sigma_o(N)}\right)^m \frac{\tau_i}{\tau_i(N)}\right\} \tag{3.72}$$

3.4.2 Results and Discussion

The ceramic composite system of SiC/SiC is used for the case study, and its basic material properties are given by Li (2016b): $V_f = 30\%$, $E_f = 230$ GPa, $E_m = 350$ GPa, $r_f = 7.5$ μm, $m = 3$, $l_0 = 25\times10^{-3}$ m, and $\sigma_0 = 0.5$ GPa; the interface shear stress degradation model parameters are given by: $\tau_{io} = 50$ MPa, $\tau_{imin} = 1.0$ MPa, $\omega = 0.0001$, and $\lambda = 1.2$; the fiber strength degradation model parameters are given by: $p_1 = 0.02$ and $p_2 = 1.4$. The effects of fatigue peak stress level, interface shear stress, fiber Weibull modulus, fiber strength, and testing temperature on the composite residual strength versus applied cycles are analyzed.

3.4.2.1 Effect of Peak Stress Level on Cyclic-Dependent Residual Strength and Fiber Failure

The effects of peak stress level, i.e., $\sigma_{max} = 180$ and 200 MPa, on the evolution of cyclic-dependent residual strength versus applied cycles, and the fiber failure versus applied stress are shown in Fig. 3.20.

When the fatigue peak stress is $\sigma_{max} = 180$ MPa, the composite residual strength decreases with increasing applied cycles, i.e., $\sigma_{UTS} = 364$, 343, 314, and 194 MPa corresponding to the cycle number of $N = 1$, 100, 1000, and 3000; the broken fiber fraction versus applied stress curves after experiencing $N = 1$, 100, 1000, and 3000 applied cycles under $\sigma_{max} = 180$ MPa is shown in Fig. 3.20b, in which the residual strength decreases with increasing cycle number.

When the fatigue peak stress is $\sigma_{max} = 200$ MPa, the composite residual strength decreases with increasing applied cycles, i.e., $\sigma_{UTS} = 364$, 338, 310, and 166 MPa corresponding to the cycle number of $N = 1$, 100, 1000, and 3000, as shown in Fig. 3.20a; the broken fiber fraction versus applied stress curves after experiencing $N = 1$, 100, 1000, and 3000 applied cycles under $\sigma_{max} = 200$ MPa is shown in Fig. 3.20c, in which the residual strength decreases with applied cycle number.

With increasing peak stress level, the composite residual strength degrades due to the increasing interface wear and interface debonding region.

3.4.2.2 Effect of Interface Shear Stress on Cyclic-Dependent Residual Strength and Fiber Failure

The effects of interface shear stress (i.e., τ_i = 20 and 40 MPa) on the evolution of composite residual strength versus applied cycles, and the broken fiber fraction

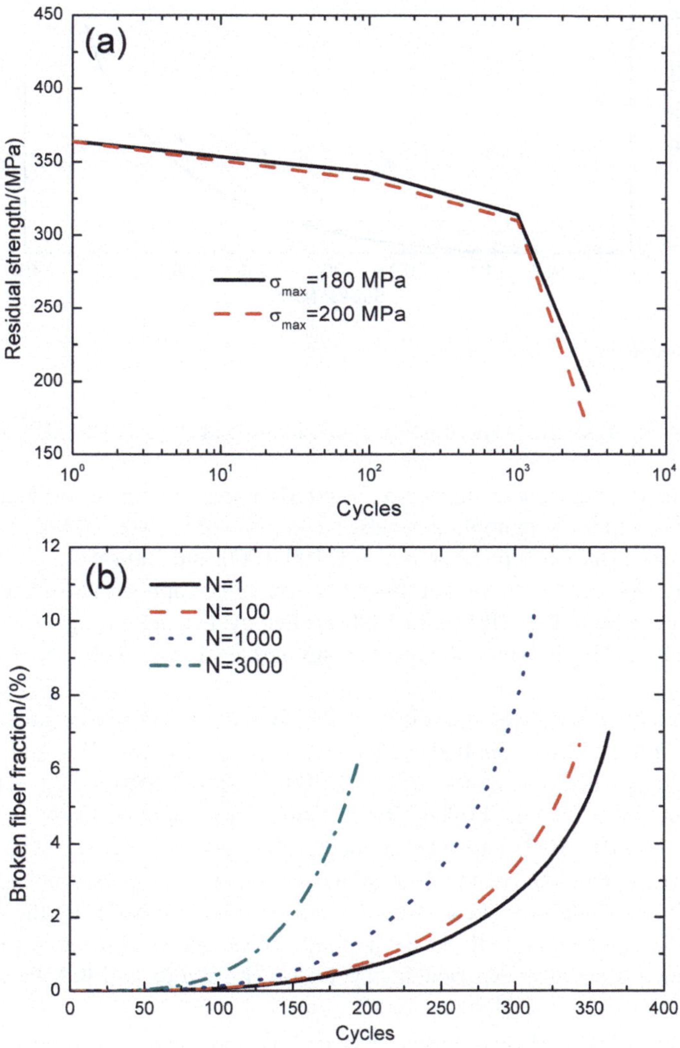

Fig. 3.20 **a** Residual strength versus applied cycle number curves under σ_{max} = 180 and 200 MPa; **b** the broken fiber fraction versus applied stress curves after experiencing different applied cycles under σ_{max} = 180 MPa; and **c** the broken fiber fraction versus applied stress curves after experiencing different applied cycles under σ_{max} = 200 MPa

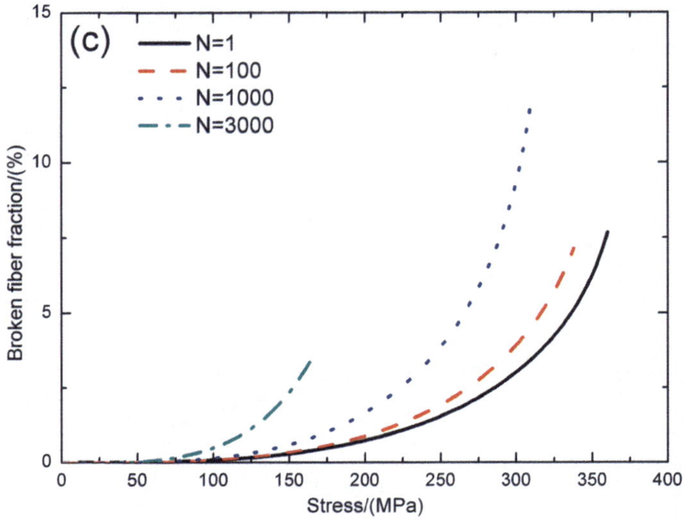

Fig. 3.20 (continued)

versus applied stress after experiencing cyclic loading at $\sigma_{max} = 180$ MPa are shown in Fig. 3.21.

When the interface shear stress is $\tau_i = 40$ MPa, the composite residual strength decreases with increasing applied cycles, i.e., $\sigma_{UTS} = 338$, 318, 294, and 160 MPa corresponding to the cycle number of $N = 1$, 100, 1000, and 3000 at $\sigma_{max} = 180$ MPa, as shown in Fig. 3.21a; the broken fiber fraction versus applied stress curves after experiencing $N = 1$, 100, 1000, and 3000 applied cycles under $\sigma_{max} = 180$ MPa is shown in Fig. 3.21b, in which the residual strength decreases with increasing cycle number.

When the interface shear stress is $\tau_i = 20$ MPa, the composite residual strength decreases with increasing applied cycles, i.e., $\sigma_{UTS} = 270$, 261, 236, and 175 MPa corresponding to the cycle number of $N = 1$, 100, 1000, and 3000 at $\sigma_{max} = 180$ MPa, as shown in Fig. 3.21a; the broken fiber fraction versus applied stress curves after experiencing 1, 100, 1000, and 3000 applied cycles under $\sigma_{max} = 180$ MPa is shown in Fig. 3.21**c**, in which the residual strength decreases with increasing cycle number.

The effects of interface shear stress (i.e., $\tau_f = 5$ and 10 MPa) on the evolution of composite residual strength versus applied cycles, and the broken fiber fraction versus applied stress after experiencing cyclic loading at $\sigma_{max} = 180$ MPa are shown in Fig. 3.22.

When the interface shear stress is $\tau_f = 10$ MPa, the composite residual strength decreases with increasing applied cycles, i.e., $\sigma_{UTS} = 292$, 277, 259, and 228 MPa corresponding to the cycle number of $N = 1$, 100, 1000, and 3000 at $\sigma_{max} = 180$ MPa, as shown in Fig. 3.22a; the broken fiber fraction versus applied stress curves after experiencing 1, 100, 1000, and 3000 applied cycles under $\sigma_{max} = 180$ MPa is shown in Fig. 3.22b, in which the residual strength decreases with increasing cycle number.

When the interface shear stress is $\tau_f = 5$ MPa, the composite residual strength decreases with increasing applied cycles, i.e., $\sigma_{UTS} = 292$, 276, 256, and 198 MPa corresponding to the cycle number of $N = 1$, 100, 1000, and 3000 at $\sigma_{max} = 180$ MPa, as shown in Fig. 3.22a; the broken fiber fraction versus applied stress curves after experiencing $N = 1$, 100, 1000, and 3000 applied cycles under $\sigma_{max} = 180$ MPa is

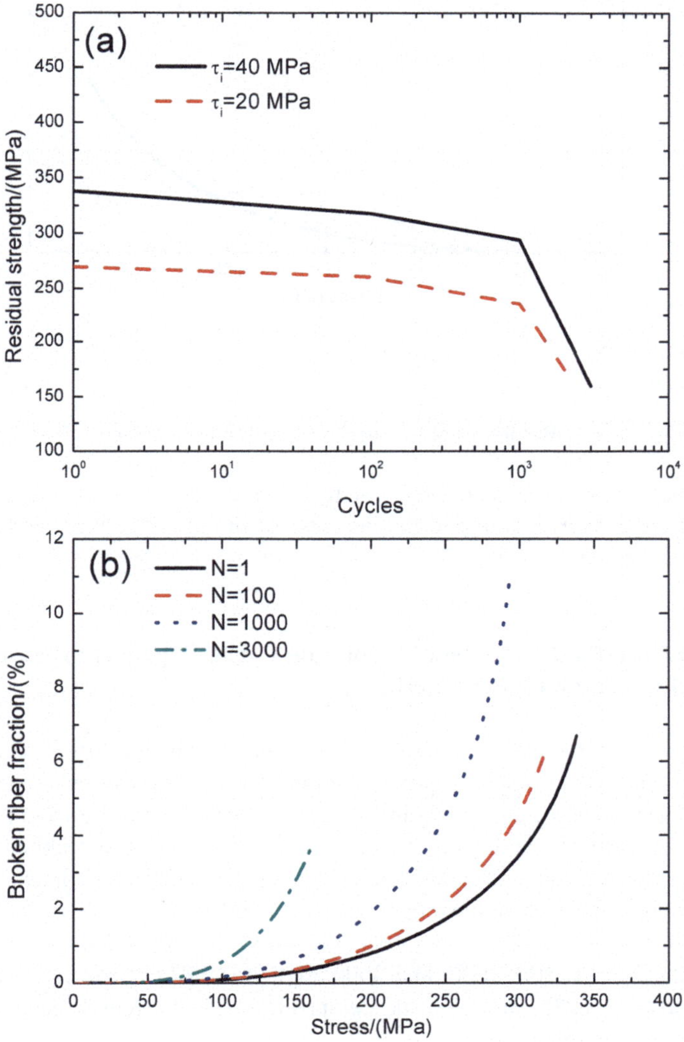

Fig. 3.21 **a** Residual strength versus applied cycle number curves corresponding to $\tau_i = 20$ and 40 MPa under $\sigma_{max} = 180$ MPa; **b** the broken fiber fraction versus applied stress curves after experiencing different applied cycles corresponding to $\tau_i = 40$ MPa under $\sigma_{max} = 180$ MPa; and **c** the broken fiber fraction versus applied stress curves after experiencing different applied cycles corresponding to $\tau_i = 20$ MPa under $\sigma_{max} = 180$ MPa

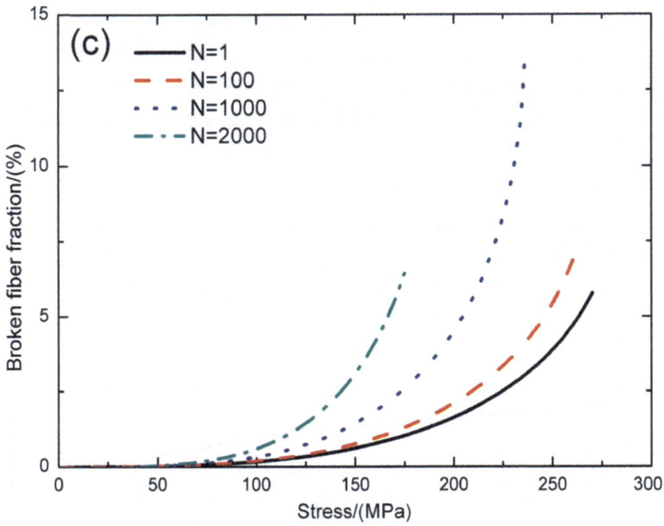

Fig. 3.21 (continued)

shown in Fig. 3.22c, in which the residual strength decreases with increasing cycle number.

With increasing initial and steady-state interface shear stress, the composite residual strength increases due to the decrease of the interface wear and interface debonding range.

3.4.2.3 Effect of Fiber Weibull Modulus on Cyclic-Dependent Residual Strength and Fiber Failure

The effects of fiber Weibull modulus (i.e., $m = 3$ and 5) on the evolution of composite residual strength versus applied cycles, and the broken fiber fraction versus applied stress after experiencing cyclic loading at $\sigma_{\max} = 180$ MPa are shown in Fig. 3.23.

When the fiber Weibull modulus is $m = 3$, the composite residual strength decreases with increasing applied cycles, i.e., $\sigma_{\text{UTS}} = 360$, 338, 310, and 166 MPa corresponding to the cycle number of $N = 1$, 100, 1000, and 3000 at $\sigma_{\max} = 180$ MPa, as shown in Fig. 3.23a; the broken fiber fraction versus applied stress curves after experiencing $N = 1$, 100, 1000, and 3000 applied cycles under $\sigma_{\max} = 180$ MPa is shown in Fig. 3.23b, in which the composite residual strength decreases with increasing cycle number.

When the fiber Weibull modulus is $m = 5$, the composite residual strength decreases with increasing applied cycles, i.e., $\sigma_{\text{UTS}} = 242$, 230, 202, and 147 MPa corresponding to the cycle number of $N = 1$, 100, 1000, and 3000 at $\sigma_{\max} = 180$ MPa, as shown in Fig. 3.23a; the broken fiber fraction versus applied stress curves after experiencing 1, 100, 1000, and 3000 applied cycles under $\sigma_{\max} = 180$ MPa is shown in Fig. 3.23c, in which the composite residual strength decreases with increasing cycle number.

With increasing fiber Weibull modulus, the composite residual strength after experiencing cyclic loading decreases.

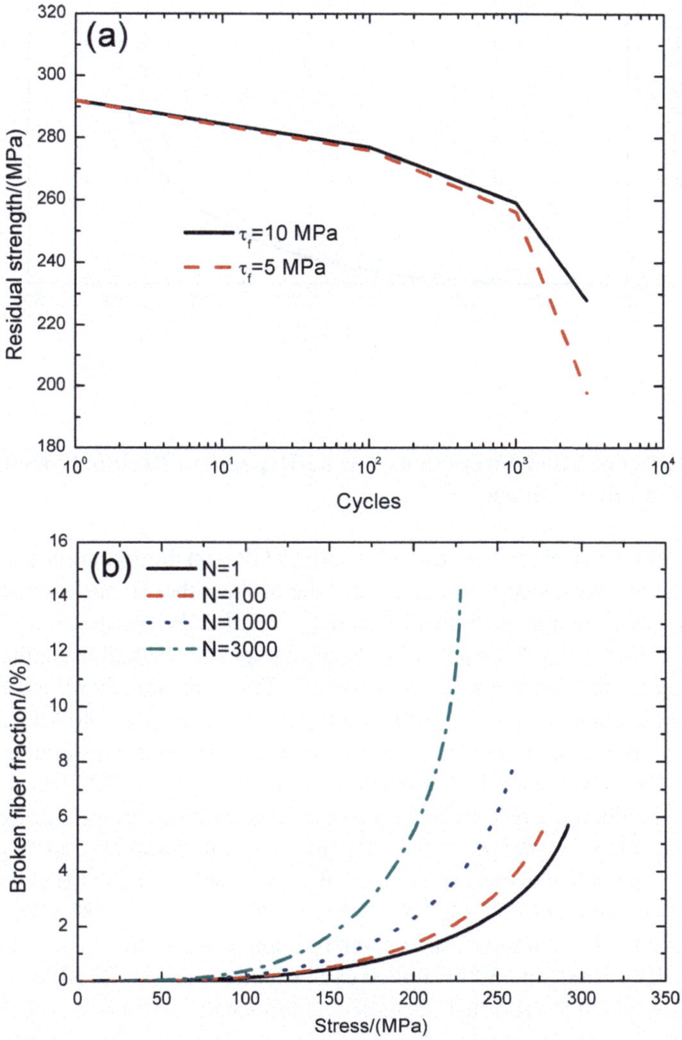

Fig. 3.22 **a** Residual strength versus applied cycle number curves corresponding to $\tau_f = 5$ and 10 MPa under $\sigma_{max} = 180$ MPa; **b** the broken fiber fraction versus applied stress curves after experiencing different applied cycles corresponding to $\tau_f = 10$ MPa under $\sigma_{max} = 180$ MPa; and **c** the broken fiber fraction versus applied stress curves after experiencing different applied cycles corresponding to $\tau_f = 5$ MPa under $\sigma_{max} = 180$ MPa

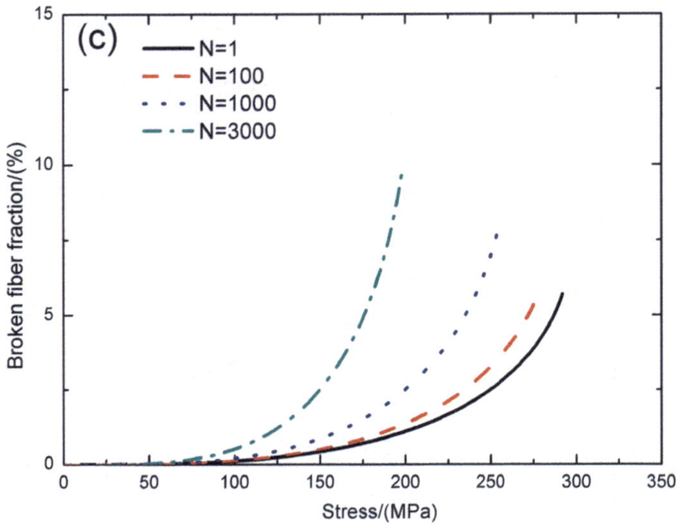

Fig. 3.22 (continued)

3.4.2.4 Effect of Fiber Strength on Cyclic-Dependent Residual Strength and Fiber Failure

The effects of fiber strength (i.e., $\sigma_0 = 0.3$ and 0.5 GPa) on the evolution of composite residual strength versus applied cycles, and the broken fiber fraction versus applied stress after experiencing cyclic loading at σ_{max} = 180 MPa are shown in Fig. 3.24.

When the fiber strength is $\sigma_0 = 0.3$ GPa, the composite residual strength decreases with increasing applied cycles, i.e., σ_{UTS} = 303, 289, 258, and 179 MPa corresponding to the cycle number of N = 1, 100, 1000, and 3000 at σ_{max} = 180 MPa, as shown in Fig. 3.24a; the broken fiber fraction versus applied stress curves after experiencing N = 1, 100, 1000, and 3000 applied cycles under σ_{max} = 180 MPa is shown in Fig. 3.24b, in which the residual strength decreases with increasing cycle number.

When the fiber strength is $\sigma_0 = 0.5$ GPa, the composite residual strength decreases with increasing applied cycles, i.e., σ_{UTS} = 427, 410, 380, and 285 MPa corresponding to the cycle number of N = 1, 100, 1000, and 3000 at σ_{max} = 180 MPa, as shown in Fig. 3.24a; the broken fiber fraction versus applied stress curves after experiencing N = 1, 100, 1000, and 3000 applied cycles under σ_{max} = 180 MPa is shown in Fig. 3.24c, in which the residual strength decreases with increasing cycle number.

With increasing fiber strength, the composite residual strength after experiencing cyclic loading increases.

3.4.2.5 Effect of Testing Temperature on Cyclic-Dependent Residual Strength and Fiber Failure

The effects of testing temperature (i.e., T = 600 and 800 °C) on the evolution of composite residual strength versus applied cycles, and the broken fiber fraction versus

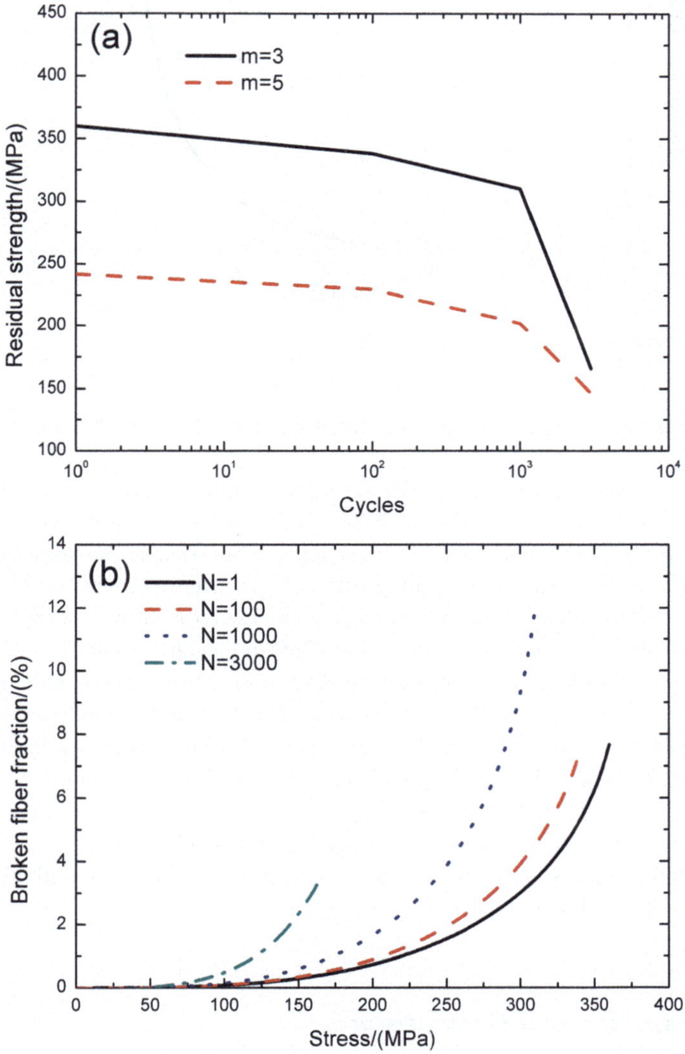

Fig. 3.23 **a** Residual strength versus applied cycle number curves corresponding to $m = 3$ and 5 under $\sigma_{\max} = 180$ MPa; **b** the broken fiber fraction versus applied stress curves after experiencing different applied cycles corresponding to $m = 3$ under $\sigma_{\max} = 180$ MPa; and **c** the broken fiber fraction versus applied stress curves after experiencing different applied cycles corresponding to $m = 5$ under $\sigma_{\max} = 180$ MPa

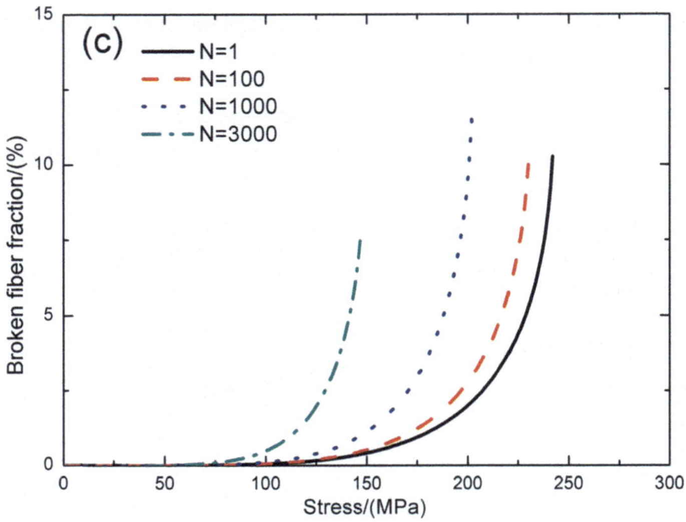

Fig. 3.23 (continued)

applied stress after experiencing cyclic loading at $\sigma_{max} = 180$ MPa are shown in Fig. 3.25.

When $T = 600$ °C, the composite residual strength decreases with increasing applied cycles, i.e., $\sigma_{UTS} = 363$, 353, 326, and 260 MPa corresponding to the cycle number of $N = 1$, 100, 1000, and 3000 at $\sigma_{max} = 180$ MPa, as shown in Fig. 3.25a; the broken fiber fraction versus applied stress curves after experiencing $N = 1$, 100, 1000, and 3000 applied cycles under $\sigma_{max} = 180$ MPa is shown in Fig. 3.25b, in which the residual strength decreases with increasing cycle number.

When $T = 800$ °C, the composite residual strength decreases with increasing applied cycles, i.e., $\sigma_{UTS} = 363$, 341, 316, and 214 MPa corresponding to the cycle number of $N = 1$, 100, 1000, and 3000 at $\sigma_{max} = 180$ MPa, as shown in Fig. 3.25a; the broken fiber fraction versus applied stress curves after experiencing $N = 1$, 100, 1000, and 3000 applied cycles under $\sigma_{max} = 180$ MPa is shown in Fig. 3.25c, in which the residual strength decreases with increasing cycle number.

With increasing testing temperature, the composite residual strength after experiencing cyclic loading decreases.

3.4.3 *Experimental Comparisons*

3.4.3.1 Non-oxide Composite

Lee et al. (1998) investigated the tension–tension fatigue behavior of SiC/Si–N–C composite at room temperature and 1000 °C. At room temperature, the composite tensile strength was about 197 MPa, after experiencing 1,000,000 applied cycles under the fatigue peak stress of σ_{max} = 125 MPa, the tensile strength degrades to approximately σ_{UTS} = 167 MPa, and the strength decreases about 15%. The experimental and theoretical predicted composite residual strength versus applied cycle curves is shown in Fig. 3.26a. The residual strength decreases with increasing

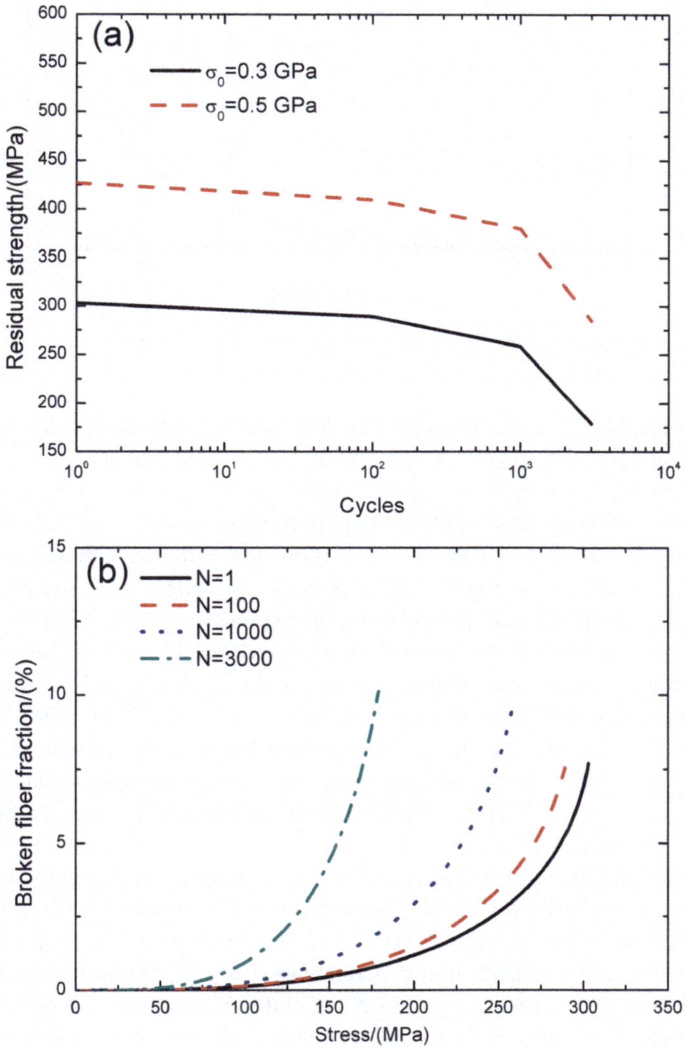

Fig. 3.24 **a** Residual strength versus applied cycle number curves corresponding to $\sigma_0 = 0.3$ and 0.5 GPa under σ_{max} = 180 MPa; **b** the broken fiber fraction versus applied stress curves after experiencing different applied cycles corresponding to $\sigma_0 = 0.3$ GPa under σ_{max} = 180 MPa; and **c** the broken fiber fraction versus applied stress curves after experiencing different applied cycles corresponding to $\sigma_0 = 0.5$ GPa under σ_{max} = 180 MPa

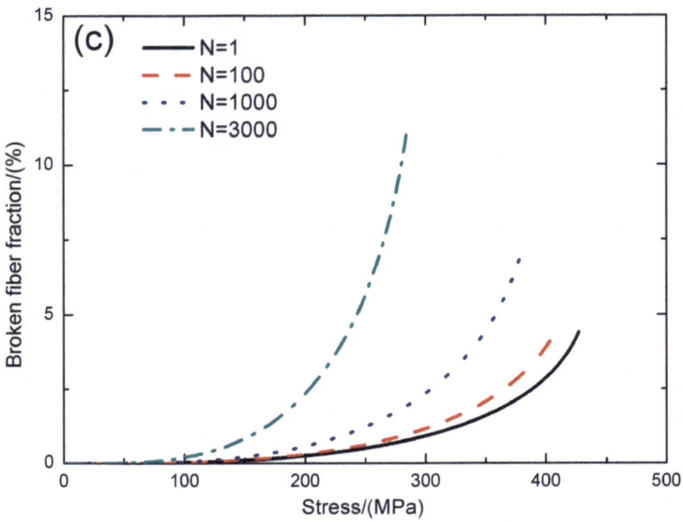

Fig. 3.24 (continued)

applied cycles, i.e., σ_{UTS} = 197, 192, 186, 179, and 170 MPa corresponding to the cycle number of N = 1, 1000, 10,000, 100,000, and 1,000,000 at σ_{max} = 125 MPa. The broken fiber fraction versus applied stress curves after experiencing N = 1, 1000, 10,000, 100,000, and 1,000,000 applied cycles under σ_{max} = 125 MPa is shown in Fig. 3.26b. At T = 1000 °C, the composite tensile strength was about σ_{UTS} = 214 MPa, after experiencing N = 100,000 under the fatigue peak stress of σ_{max} = 100 MPa, the tensile strength degrades to approximately σ_{UTS} = 157 MPa, and the strength decreases about 26%. The experimental and theoretical predicted composite residual strength versus applied cycle curves is shown in Fig. 3.27a. The residual strength decreases with increasing applied cycles, i.e., σ_{UTS} = 214, 206, 193, 113, and 99 MPa corresponding to the cycle number of N = 1, 1000, 10,000, 50,000, and 100,000 at σ_{max} = 100 MPa. The broken fiber fraction versus applied stress curves after experiencing N = 1, 1000, 10,000, and 100,000 applied cycles under σ_{max} = 100 MPa is shown in Fig. 3.27b.

Bertrand et al. (2015) investigated the tension–tension fatigue behavior of SiC/SiC composite at T = 1250 and 1350 °C. The composite tensile strength was about σ_{UTS} = 274 MPa at elevated temperature. At T = 1250 °C, the composite residual strength decreases to σ_{UTS} = 116 MPa after experiencing N = 100,000 applied cycles under the fatigue peak stress of σ_{max} = 125 MPa, and the composite strength degrades approximately 57%. The experimental and predicted composite residual strength versus applied cycle curves is shown in Fig. 3.28a. The residual strength decreases with increasing applied cycles, i.e., σ_{UTS} = 274, 264, 251, 140, and 117 MPa corresponding to the cycle number of N = 1, 1000, 10,000, 50,000, and 100,000 at σ_{max} = 125 MPa. The broken fiber fraction versus applied stress curves after experiencing N = 1, 1000, 10,000, and 100,000 applied cycles under σ_{max} = 125 MPa is shown in Fig. 3.28b. At $T = 1350$ °C, the composite residual strength decreases to $\sigma_{UTS} = 247$ MPa after experiencing $N = 100{,}000$ applied cycles under the fatigue peak stress of $\sigma_{max} = 90$ MPa, and the composite strength degrades approximately 10%. The experimental and predicted composite residual strength versus applied cycle curves is shown in Fig. 3.29a. The residual strength decreases with increasing

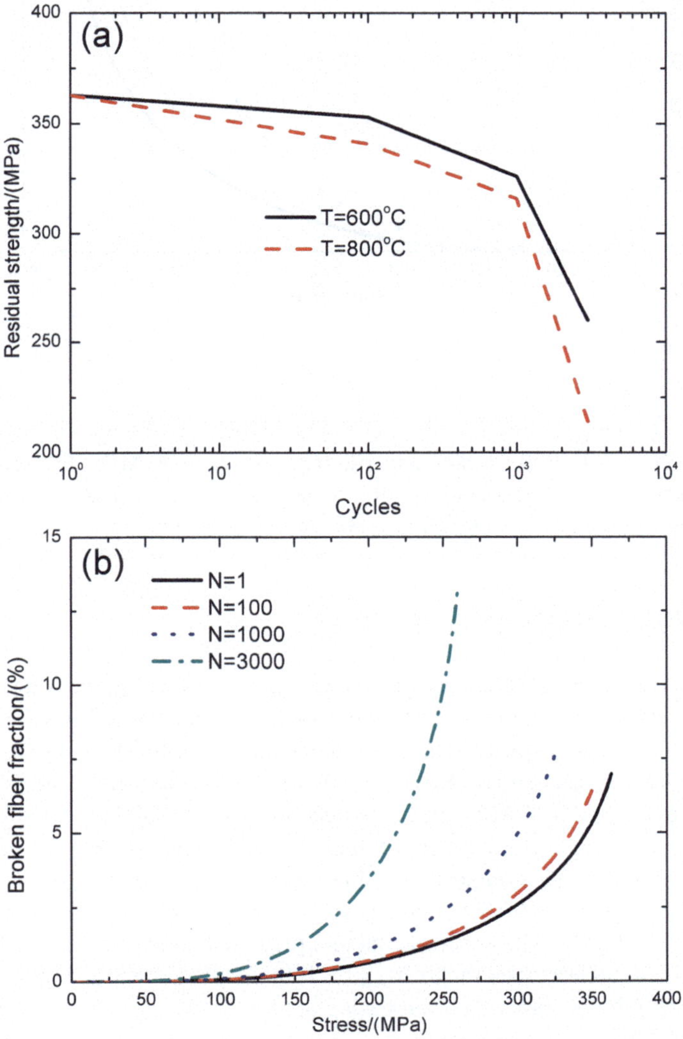

Fig. 3.25 **a** Residual strength versus applied cycle number curves corresponding to $T = 600$ and 800 °C under $\sigma_{max} = 180$ MPa; **b** the broken fiber fraction versus applied stress curves after experiencing different applied cycles corresponding to $T = 600$ °C under $\sigma_{max} = 180$ MPa; and **c** the broken fiber fraction versus applied stress curves after experiencing different applied cycles corresponding to $T = 800$ °C under $\sigma_{max} = 180$ MPa

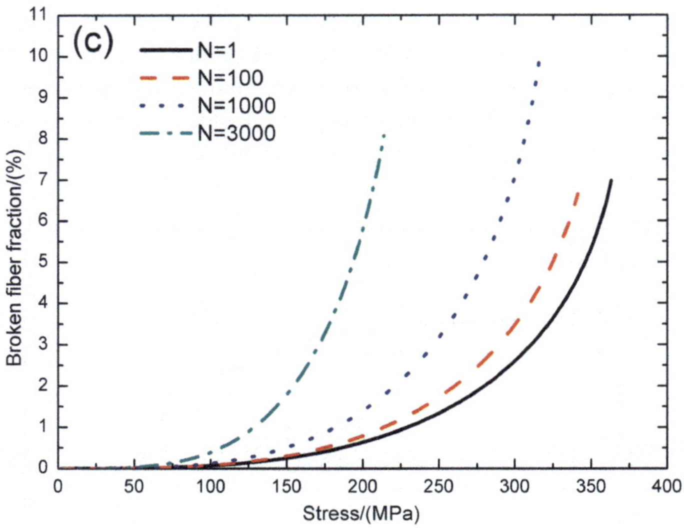

Fig. 3.25 (continued)

applied cycles, i.e., $\sigma_{UTS} = 274$, 262, 248, and 247 MPa corresponding to the cycle number of $N = 1$, 10,000, 50,000, and 100,000 at $\sigma_{max} = 90$ MPa. The broken fiber fraction versus applied stress curves after experiencing $N = 1$, 10,000, and 100,000 applied cycles under $\sigma_{max} = 90$ MPa is shown in Fig. 3.29b.

3.4.3.2 Oxide–Oxide Composite

Ruggles-Wrenn et al. (2008) investigated the tension–tension fatigue behavior of 2D Nextel^{TM} 720/alumina composite under the loading frequency of $f = 1.0$ Hz at $T = 1200$ °C in steam atmosphere. The ultimate tensile strength was $\sigma_{UTS} = 190$ MPa. Under the fatigue peak stress of $\sigma_{max} = 100$ MPa, the composite residual strength decreases to $\sigma_{UTS} = 174$ MPa after experiencing $N = 100{,}000$ applied cycles, and the composite strength degrades approximately 8.4%; under the fatigue peak stress of $\sigma_{max} = 125$ MPa, the composite residual strength decreases to $\sigma_{UTS} = 168$ MPa after experiencing $N = 100{,}000$ applied cycles, and the composite strength degrades approximately 11.5%. The experimental and predicted composite residual strength versus applied cycle curves under $\sigma_{max} = 100$ MPa is shown in Fig. 3.30a. The residual strength decreases with increasing applied cycles, i.e., $\sigma_{UTS} = 190$, 188, 185, and 173 MPa corresponding to the cycle number of $N = 1$, 1000, 10,000, and 100,000 at $\sigma_{max} = 100$ MPa. The broken fiber fraction versus applied stress curves after experiencing $N = 1$, 1000, 10,000, and 100,000 applied cycles under $\sigma_{max} = 100$ MPa is shown in Fig. 3.30b. The experimental and predicted composite residual strength versus applied cycle curves under $\sigma_{max} = 125$ MPa is shown in Fig. 3.31a. The residual strength decreases with increasing applied cycles, i.e., $\sigma_{UTS} = 190$, 183,

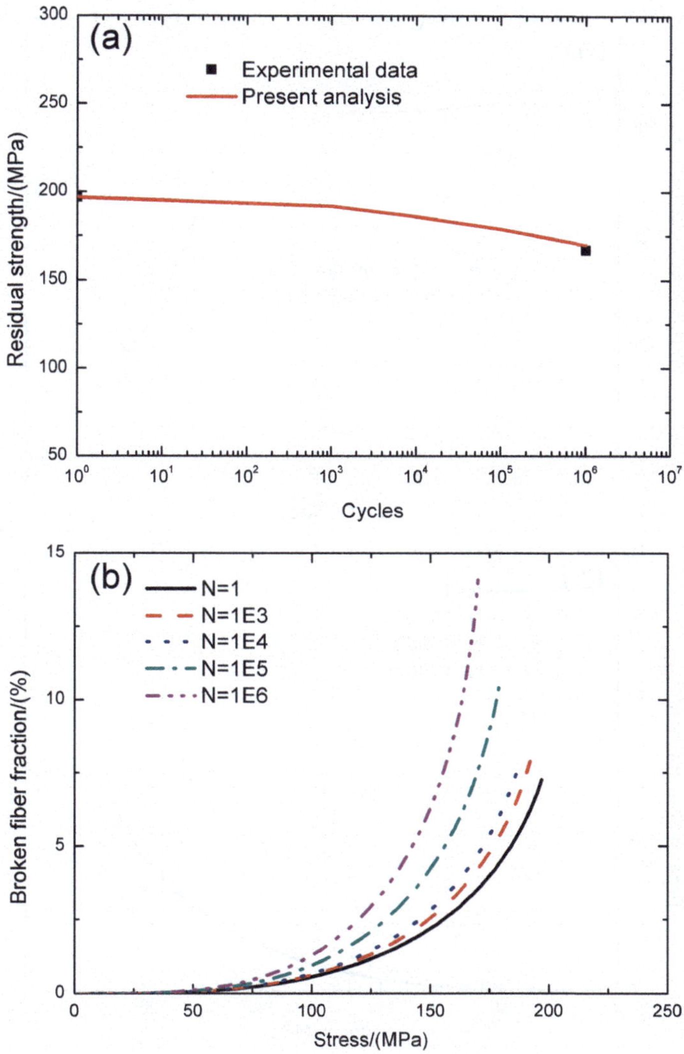

Fig. 3.26 **a** Residual strength versus applied cycles and **b** the broken fiber versus applied cycles of SiC/Si–N–C composites under fatigue peak stress of $\sigma_{max} = 125$ MPa at room temperature

182, and 167 MPa corresponding to the cycle number of $N = 1$, 1000, 10,000, and 100,000 at $\sigma_{max} = 125$ MPa. The broken fiber fraction versus applied stress curves after experiencing $N = 1$, 1000, 10,000, and 100,000 applied cycles under $\sigma_{max} =$ 125 MPa is shown in Fig. 3.31b.

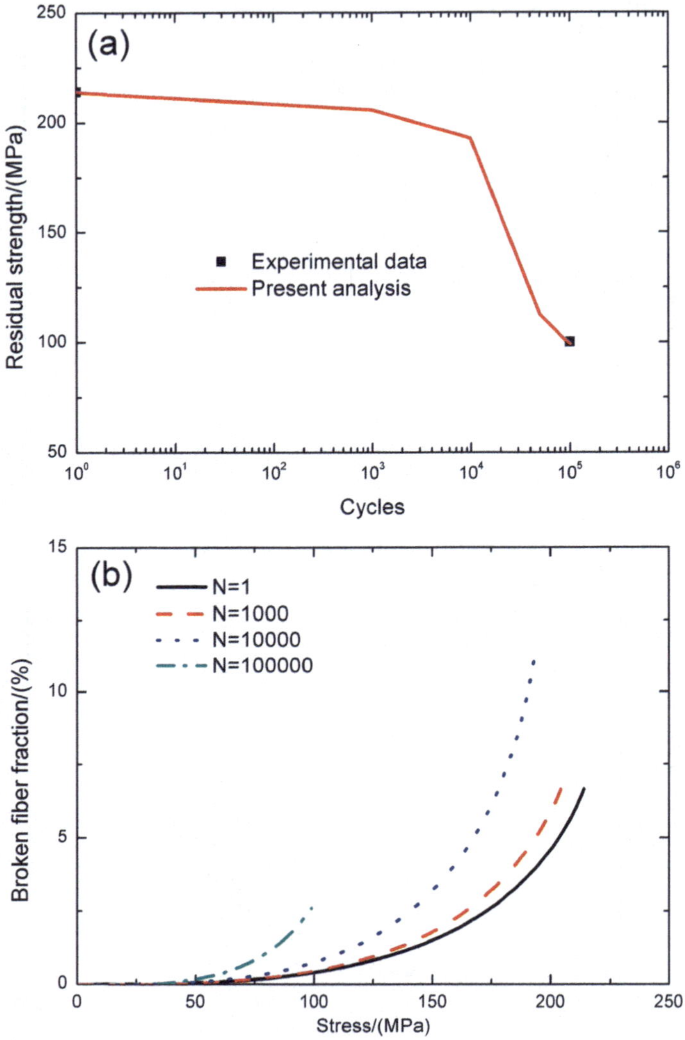

Fig. 3.27 **a** Residual strength versus applied cycles and **b** the broken fiber versus applied cycles of SiC/Si–N–C composites under fatigue peak stress of $\sigma_{max} = 100$ MPa at $T = 1000$ °C

3.5 Conclusion

In this chapter, the cyclic- and time-dependent strength degradation of non-oxide and oxide/oxide fiber-reinforced CMCs subjected to multiple fatigue loading at room temperature, oxidation environment at elevated temperature, and cyclic loading at elevated temperatures in oxidative environments is investigated. Considering the damage mechanisms of matrix cracking, interface debonding, interface wear, interface oxidation, and fiber fracture, the composite residual strength model is established

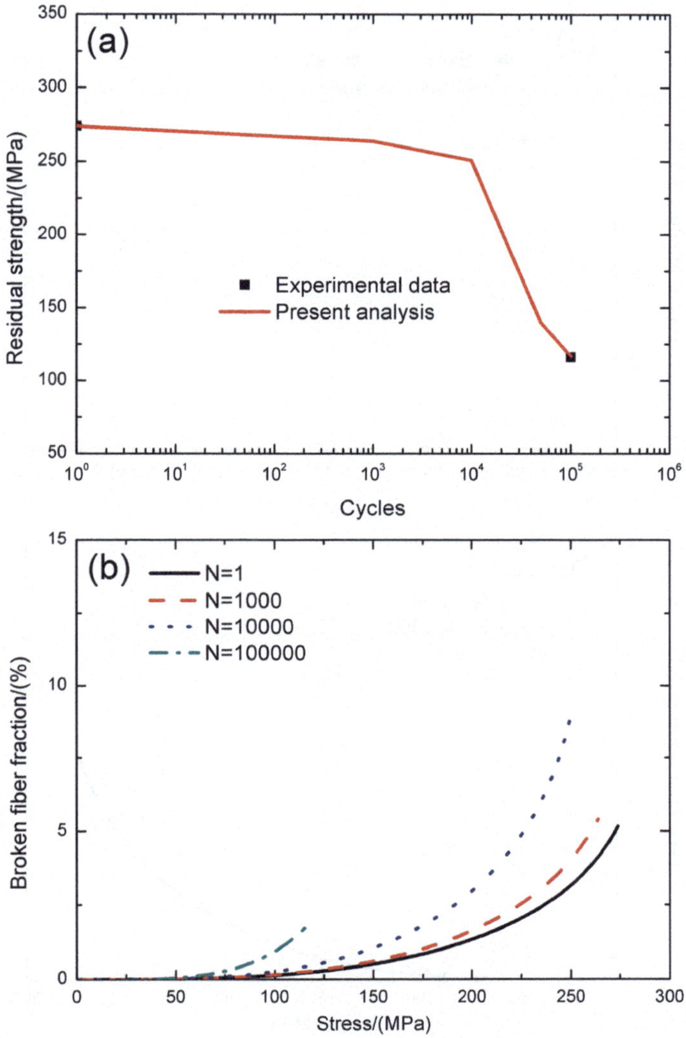

Fig. 3.28 **a** Residual strength versus applied cycles and **b** the broken fiber fraction versus applied cycles of SiC/Si–N–C composites under fatigue peak stress of σ_{max} = 125 MPa at T = 1250 °C

by combining the microstress field of the damaged composites, the damage models, and the fracture criterion. The relationships between the composite residual strength, fatigue peak stress, interface debonding, fiber failure, oxidation time and temperature, and applied cycle number are established. The effects of the peak stress level, initial and steady-state interface shear stress, fiber Weibull modulus, fiber strength, oxidation temperature and time on the degradation of composite strength and fiber

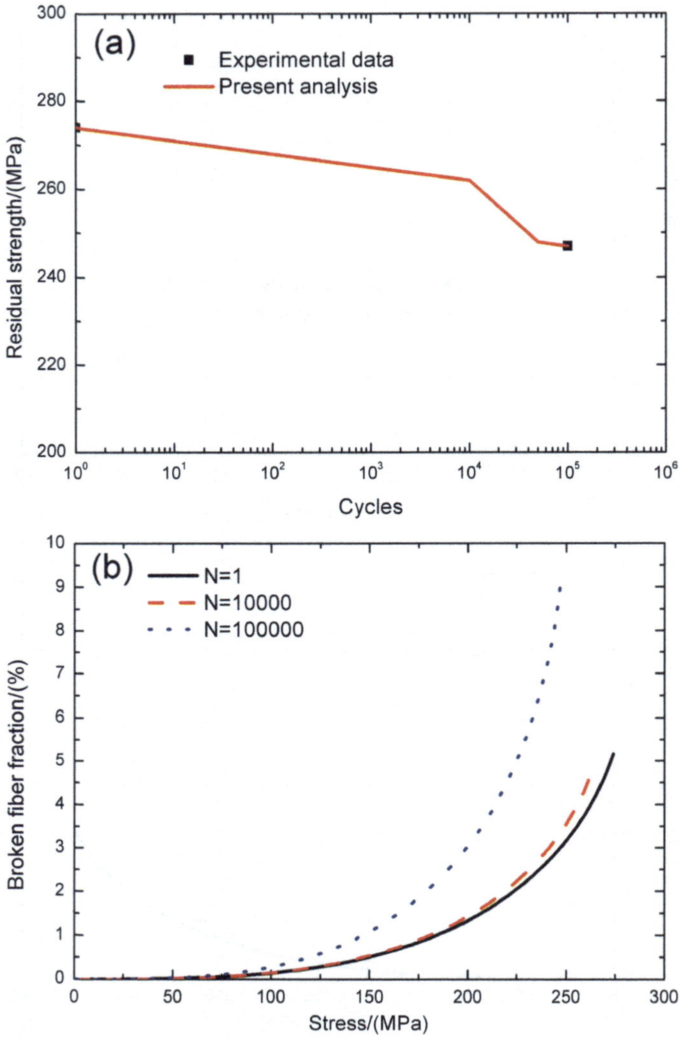

Fig. 3.29 **a** Residual strength versus applied cycles and **b** the broken fiber versus applied cycles of SiC/Si–N–C composites under fatigue peak stress of $\sigma_{\max} = 90$ MPa at $T = 1350$ °C

failure are investigated. The evolution of residual strength versus oxidation temperature and time and applied cycle number curves of non-oxide and oxide/oxide CMCs is predicted.

1. With increasing peak stress level, the composite residual strength after experiencing cyclic loading decreases, due to the increasing range of interface debonding and interface wear.

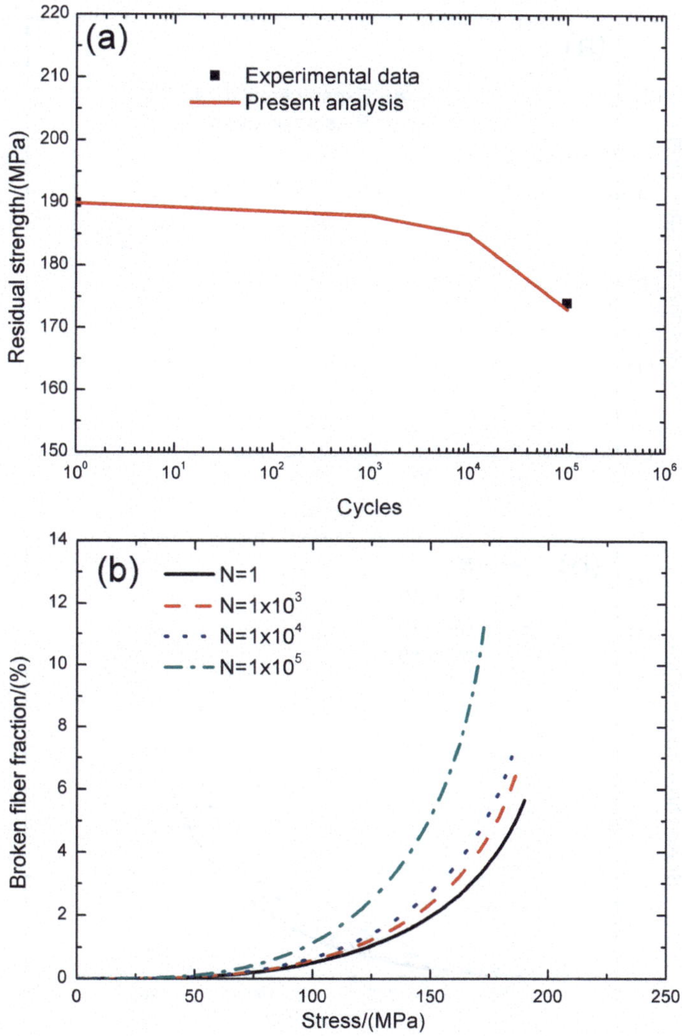

Fig. 3.30 **a** Residual strength versus applied cycles and **b** the broken fiber versus applied cycles of Nextel^{TM} 720/alumina composites under fatigue peak stress of $\sigma_{max} = 100$ MPa at $T = 1200$ °C

2. With increasing initial and steady-state interface shear stress, the composite residual strength after experiencing cyclic loading increases, due to the decrease of interface debonding and interface wear range.
3. With increasing fiber Weibull modulus, the composite residual strength after experiencing cyclic loading decreases.
4. With increasing fiber strength, the composite residual strength after experiencing cyclic loading increases.

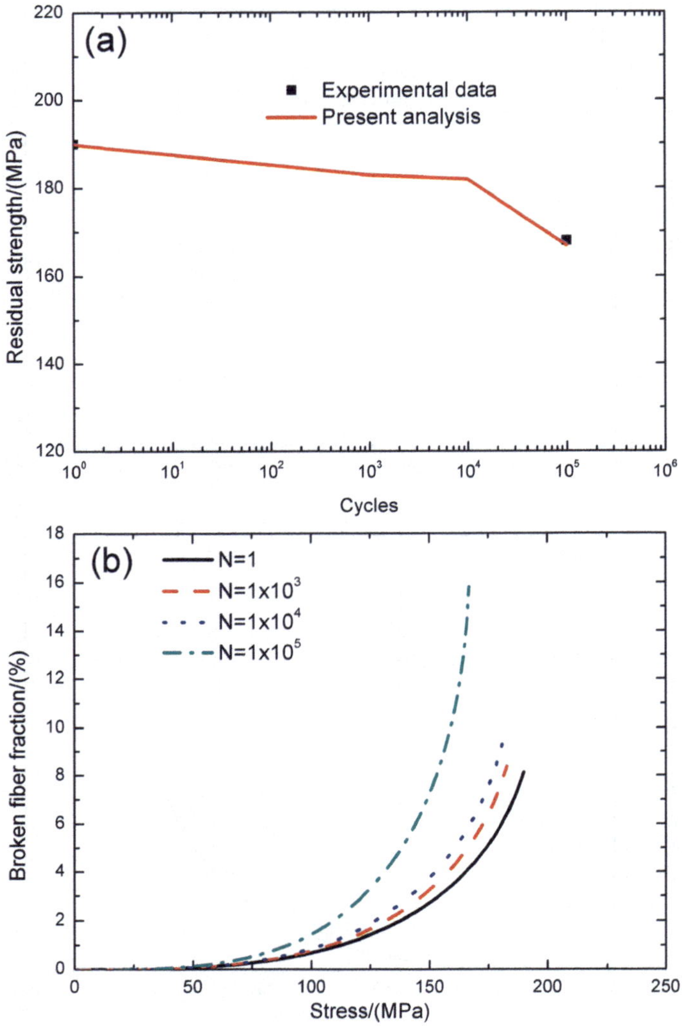

Fig. 3.31 **a** Residual strength versus applied cycles and **b** the broken fiber fraction versus applied cycles of Nextel^{TM} 720/alumina composites under fatigue peak stress of $\sigma_{max} = 125$ MPa at $T = 1200$ °C

5. With increasing testing temperature, the composite residual strength after experiencing cyclic loading decreases due to the interface oxidation and fiber strength degradation.

References

Aveston J, Cooper GA, Kelly A (1971) Single and multiple fracture. Properties of fiber composites: conference on proceedings. National Physical Laboratory, IPC, England, pp 15–26

Bertrand DJ, Sabelkin V, Zawada L, Mall S (2015) Fatigue behavior of Sylramic-iBN/BN/CVI SiC ceramic matrix composite in combustion environment. J Mater Sci 50:7437–7447. https://doi.org/10.1007/s10853-015-9302-8

Budiansky B, Hutchinson JW, Evans AG (1986) Matrix fracture in fiber-reinforced ceramics. J Mech Phys Solids 34:167–189. https://doi.org/10.1016/0022-5096(86)90035-9

Cao HC, Thouless MD (1990) Tensile tests of ceramic-matrix composites: theory and experiment. J Am Ceram Soc 73:2091–2094. https://doi.org/10.1111/j.1151-2916.1990.tb05273.x

Casas L, Martinez-Esnaola JM (2003) Modeling the effect of oxidation on the creep behavior of fiber-reinforced ceramic matrix composites. Acta Mater 51:3745–3757. https://doi.org/10.1016/j.jeurceramsoc.2015.11.005

Cherouali H, Fantozzi G, Reynaud P, Rouby D (1998) Analysis of interfacial sliding in brittle-matrix composites during push-out and push-back tests. Mater Sci Eng A 250:169–177. https://doi.org/10.1016/S0921-5093(98)00589-9

Cho CD, Holmes JW, Barber JR (1991) Estimation of interfacial shear in ceramic composites from frictional heating measurements. J Am Ceram Soc 74:2802–2808. https://doi.org/10.1111/j.1151-2916.1991.tb06846.x

Curtin WA (1991) Theory of mechanical properties of ceramic-matrix composites. J Am Ceram Soc 74:2837–2845. https://doi.org/10.1111/j.1151-2916.1991.tb06852.x

Curtin WA (1993) Multiple matrix cracking in brittle matrix composites. Acta Metall Mater 41:1369–1377. https://doi.org/10.1016/0956-7151(93)90246-O

Curtin WA, Ahn BK, Takeda N (1998) Modeling brittle and tough stress-strain behavior in unidirectional ceramic matrix composites. Acta Mater 46:3409–3420. https://doi.org/10.1016/S1359-6454(98)00041-X

Daniel IM, Lee JW (1993) The behavior of ceramic matrix fiber composites under longitudinal loading. Compos Sci Technol 46:105–113. https://doi.org/10.1016/0266-3538(93)90166-E

Dutton RE, Pagano NJ, Kim RY (2000) Modeling the ultimate tensile strength of unidirectional glass-matrix composites. J Am Ceram Soc 83:166–174. https://doi.org/10.1111/j.1151-2916.2000.tb01166.x

Evans AG, Zok FW, McMeeking RM (1995) Fatigue of ceramic matrix composites. Acta Metall Mater 43:859–875. https://doi.org/10.1016/0956-7151(94)00304-Z

Fantozzi G, Reynaud P (2009) Mechanical hysteresis in ceramic matrix composites. Mater Sci Eng A 521–522:18–23. https://doi.org/10.1016/j.msea.2008.09.128

Fantozzi G, Reynaud P (2014) Mechanical behavior of SiC fiber-reinforced ceramic matrix composites. Comprehen Hard Mat 345–366. https://doi.org/10.1016/B978-0-08-096527-7.00031-3

Filipuzzi L, Naslain R (1994) Oxidation mechanisms and kinetics of 1D-SiC/C/SiC composite materials: II, Modelling. J Am Ceram Soc 77:467–480. https://doi.org/10.1111/j.1151-2916.1994.tb07016.x

Gao Y, Mai Y, Cotterell B (1988) Fracture of fiber-reinforced materials. J Appl Math Phys 39:550–572. https://doi.org/10.1007/BF00948962

Gonczy ST (2015) Federal Aviation Administration (FAA) airworthiness certification for ceramic matrix composite components in civil aircraft systems. MATEC Web Conf 29:00002. https://doi.org/10.1051/matecconf/20152900002

Hild F, Domergue JM, Leckie FA, Evans AG (1994) Tensile and flexural ultimate strength of fiber-reinforced ceramic-matrix composites. Int J Solids Strut 31:1035–1045. https://doi.org/10.1016/0020-7683(94)90010-8

Holmes JW (1991) Influence of stress ratio on the elevated-temperature fatigue of a silicon carbide fiber-reinforced silicon nitride composite. J Am Ceram Soc 74:1639–1645. https://doi.org/10.1111/j.1151-2916.1991.tb07152.x

Holmes JW, Cho CD (1992) Experimental observation of frictional heating in fiber-reinforced ceramics. J Am Ceram Soc 75:929–938. https://doi.org/10.1111/j.1151-2916.1992.tb04162.x

Hsueh CH (1996) Crack-wake interface debonding criterion for fiber-reinforced ceramic composites. Acta Mater 44:2211–2216. https://doi.org/10.1016/1359-6454(95)00369-X

Keiji O (2011) Prediction of residual tensile strength after fatigue in unidirectional brittle fiber-reinforced ceramic composites. J Solid Mech Mater Eng 5:64–74. https://doi.org/10.1299/jmmp.5.64

Lara-Curzio E (1999) Analysis of oxidation-assisted stress-rupture of continuous fiber-reinforced ceramic matrix composites at intermediate temperatures. Compos. Part A 30:549–554. https://doi.org/10.1016/S1359-835X(98)00148-1

Lee SS, Stinchcomb WW (1994) Damage mechanisms of cross-ply Nicalon/CAS-II laminate under cyclic tension. Ceram Eng Sci Proc 15:40–48. https://doi.org/10.1002/9780470314500.ch5

Lee SS, Zawada LP, Staehler JM, Folsom GA (1998) Mechanical behavior and high-temperature performance of a woven Nicalon^{TM}/Si–N–C ceramic-matrix composite. J Am Ceram Soc 81:1797–1811. https://doi.org/10.1111/j.1151-2916.1998.tb02550.x

Li L (2014) Assessment of the interfacial properties from fatigue hysteresis loss energy in ceramic-matrix composites with different fiber preforms at room and elevated temperatures. Mater Sci Eng, A 613:17–36. https://doi.org/10.1016/j.msea.2014.06.092

Li L (2015) Modeling the effect of oxidation on tensile strength of carbon fiber-reinforced ceramic-matrix composites. Appl Compos Mater 22:921–943. https://doi.org/10.1007/s10443-015-9443-6

Li L (2016a) Modeling the tensile strength of carbon fiber-reinforced ceramic-matrix composites under multiple fatigue loading. Appl Compos Mater 23:313–336. https://doi.org/10.1007/s10443-015-9462-3

Li L (2016b) Damage development in fiber-reinforced ceramic-matrix composites under cyclic fatigue loading using hysteresis loops at room and elevated temperatures. Int J Fract 199:39–58. https://doi.org/10.1007/s10704-016-0085-y

Li L (2018a) Damage, fracture and fatigue of ceramic-matrix composites. Springer. ISBN 978-981-13-1782-8. https://doi.org/10.1007/978-981-13-1783-5

Li L (2018b) Modeling strength degradation of fiber-reinforced ceramic-matrix composites subjected to cyclic loading at elevated temperatures in oxidative environments. Appl Compos Mater 25(1):1–19. https://doi.org/10.1007/s10443-017-9609-5

Li L (2019) Thermomechanical fatigue of ceramic-matrix composites. Wiley. ISBN 978-3-527-34637-0. https://onlinelibrary.wiley.com/doi/book/10.1002/9783527822614

Li L, Song YD (2010) An approach to estimate interface shear stress of ceramic matrix composites from hysteresis loops. Appl Compos Mater 17:309–328. https://doi.org/10.1007/s10443-009-9122-6

Li L, Song Y, Sun Y (2014) Modeling the tensile behavior of unidirectional C/SiC ceramic-matrix composites. Mech Compos Mater 49:659–672. https://doi.org/10.1007/s11029-013-9382-y

Liao K, Reifsnider KL (2000) A tensile strength model for unidirectional fiber-reinforced brittle matrix composite. Int J Fract 106:95–115. https://doi.org/10.1023/A:1007645817753

McNulty JC, Zok FW (1999) Low-cycle fatigue of Nicalon-fiber-reinforced ceramic composites. Compos Sci Technol 59:1597–1607. https://doi.org/10.1016/S0266-3538(99)00019-6

Mehrman JM, Ruggles-Wrenn MB, Baek SS (2007) Influence of hold times on the elevated-temperature fatigue behavior of an oxide-oxide ceramic composite in air and in steam environment. Compos Sci Technol 67:1425–1438. https://doi.org/10.1016/j.compscitech.2006.09.005

Murthy PLN, Nemeth NN, Brewer DN, Mital S (2008) Probabilistic analysis of a SiC/SiC ceramic matrix composite turbine vane. Compos B 39:694–703. https://doi.org/10.1016/j.compositesb.2007.05.006

Naslain R, Guette A, Rebillat F, Gallet S, Lamouroux F, Filipuzzi L, Louchet C (2004) Oxidation mechanisms and kinetics of SiC-matrix composites and their constituents. J Mater Sci 39:7303–7316. https://doi.org/10.1023/B:JMSC.0000048745.18938.d5

Paar R, Valles J-L, Danzer R (1998) Influence of fiber properties on the mechanical behavior of unidirectionally-reinforced ceramic matrix composites. Mat Sci Eng A 250:209–216. https://doi.org/10.1016/S0921-5093(98)00593-0

Phoenix SL, Raj R (1992) Scalings in fracture probabilities for a brittle matrix fiber composite. Acta Metall Mater 40:2813–2828. https://doi.org/10.1016/0956-7151(92)90447-M

Reynaud P (1996) Cyclic fatigue of ceramic-matrix composites at ambient and elevated temperatures. Compos Sci Technol 56:809–814. https://doi.org/10.1016/0266-3538(96)00025-5

Rouby D, Reynaud P (1993) Fatigue behavior related to interface modification during load cycling in ceramic-matrix fiber composites. Comps Sci Technol 48:109–118. https://doi.org/10.1016/0266-3538(93)90126-2

Ruggles-Wrenn MB, Lanser RL (2016) Tension-compression fatigue of an oxide/oxide ceramic composite at elevated temperature. Mater Sci Eng A 659:270–277. https://doi.org/10.1016/j.msea.2016.02.057

Ruggles-Wrenn MB, Lee MD (2016) Fatigue behavior of an advanced SiC/SiC ceramic composite with a self-healing matrix at 1300 °C in air and in steam. Mater Sci Eng A 677:438–445. https://doi.org/10.1016/j.msea.2016.09.076

Ruggles-Wrenn MB, Hetrick G, Baek SS (2008) Effects of frequency and environment on fatigue behavior of an oxide-oxide ceramic composite at 1200 °C. Int J Fatigue 30:502–516. https://doi.org/10.1016/j.ijfatigue.2007.04.004

Schwietert HR, Steif PS (1990) A theory for the ultimate strength of a brittle-matrix composite. J Mech Phys Solids 38:325–343. https://doi.org/10.1016/0022-5096(90)90002-L

Shah AR, Murthy PLN, Mital SK, Bhatt RT (2000) Probabilistic modeling of ceramic matrix composite strength. J Compos Mater 34:670–688. https://doi.org/10.1177/002199830003400803

Solti JP, Mall S, Robertson DD (1995) Modeling damage in unidirectional ceramic-matrix composites. Compos Sci Technol 54:55–66. https://doi.org/10.1016/0266-3538(95)00041-0

Sørensen BF, Holmes JW, Vanswijgenhoven EL (2000) Rate of strength decrease of fiber-reinforced ceramic-matrix composites during fatigue. J Am Ceram Soc 83:1469–1475. https://doi.org/10.1111/j.1151-2916.2000.tb01412.x

Sun YJ, Singh RN (1998) The generation of multiple matrix cracking and fiber-matrix interfacial debonding in a glass composite. Acta Mater 46:1657–1667. https://doi.org/10.1016/S1359-6454(97)00347-9

Sutcu M (1989) Weibull statistics applied to fiber failure in ceramic composites and work of fracture. Acta Metall Mater 37:651–661. https://doi.org/10.1016/0001-6160(89)90249-6

Thouless MD, Evans AG (1988) Effects of pull-out on the mechanical properties of ceramic matrix composites. Acta Metall Mater 36:517–522. https://doi.org/10.1016/0001-6160(88)90083-1

Vagaggini E, Domergue JM, Evans AG (1995) Relationship between hysteresis measurements and the constituent properties of ceramic matrix composites: I, theory. J Am Ceram Soc 78:2709–2720. https://doi.org/10.1111/j.1151-2916.1995.tb08046.x

Wang ZJ (2010) Research on mechanical properties of ceramic matrix composites in oxidation environment. Master thesis, Nanjing University of Aeronautics and Astronautics

Weitsman Y, Zhu H (1993) Multiple-fracture of ceramic composites. J Mech Phys Solids 41:351–388. https://doi.org/10.1016/0022-5096(93)90012-5

Whitworth HA (2000) Evaluation of the residual strength degradation in composite laminates under fatigue loading. Compos Struct 48:261–264. https://doi.org/10.1016/S0263-8223(99)00113-0

Xia Z, Curtin WA (2000) Toughness-to-brittle transitions in ceramic-matrix composites with increasing interfacial shear stress. Acta Mater 48:4879–4892. https://doi.org/10.1016/S1359-6454(00)00291-3

Zhou SJ, Curtin WA (1995) Failure of fiber composites: a lattice green function model. Acta Metall Mater 43:3093–3104. https://doi.org/10.1016/0956-7151(95)00003-E

Zhu H, Weitsman Y (1994) The progression of failure mechanisms in unidirectional reinforced ceramic composites. J Mech Phys Solids 42:1601–1632. https://doi.org/10.1016/0022-5096(94)90089-2

Zhu SJ, Mizuno M, Kagawa Y, Mutoh Y (1999) Monotonic tension, fatigue and creep behavior of SiC-fiber-reinforced SiC-matrix composites: a review. Compos Sci Technol 59:833–851. https://doi.org/10.1016/S0266-3538(99)00014-7
Zok FW, Spearing SM (1992) Matrix crack spacing in brittle matrix composites. Acta Metall Mater 40:2033–2043. https://doi.org/10.1016/0956-7151(92)90189-L

Chapter 4
Time-Dependent Tensile Behavior of Fiber-Reinforced Ceramic-Matrix Composites

Abstract In this chapter, the time-dependent tensile damage and fracture of fiber-reinforced ceramic-matrix composites (CMCs) subjected to pre-exposure at elevated temperatures and thermal fatigue are investigated. The damage mechanisms of the interface oxidation and fiber failure are considered in the stress analysis, matrix multicracking, interface debonding, and fiber failure. Combining the stress analysis and damage models, the tensile stress-strain curves of fiber-reinforced CMCs for different damage stages can be obtained. The effects of the pre-exposure temperature and time, thermal fatigue temperature, thermal cyclic number, the interface shear stress, fiber strength, and fiber Weibull modulus on tensile damage and fracture processes are analyzed. The experimental tensile damage and fracture process of fiber-reinforced CMCs with different fiber preforms are predicted for a different pre-exposure temperature and time.

Keywords Ceramic-matrix composites (CMCs) · Tensile · Matrix multicracking · Interface debonding · Interface oxidation · Fiber failure

4.1 Introduction

With the rapid development of aerothermodynamics, structural mechanics, and material science, turbofan engines with large bypass ratio are developing toward high efficiency, such as low fuel consumption, low emission, low noise, easy maintenance, high reliability, and long life. Without changing the existing layout of turbofan engines, relying on innovative materials and novel configurations becomes a fundamental solution. Over the past half century, the thrust-to-weight ratio of commercial aeroengine technology, especially the combustion chamber technology, has been significantly improved. At present, the requirements of high thrust and high thrust-to-weight ratio of the engine are more and more stringent to reduce the emission of NO_x and CO. As a result, the turbocharging ratio, turbine inlet temperature, combustion chamber temperature, and rotational speed of the engine must also be continuously increased. As far as materials are concerned, the current high-heat gas of high-efficiency aeroengine has already reached the limit of operation temperature of traditional titanium alloy and nickel-based superalloy. The existing alloy material

L. Li, *Time-Dependent Mechanical Behavior of Ceramic-Matrix Composites at Elevated Temperatures*, Advanced Ceramics and Composites 1,
https://doi.org/10.1007/978-981-15-3274-0_4

cannot meet the heat-resistant requirements of the next generation of advanced engine design. The high-temperature components have to be protected by air-cooling and thermal barrier coating. However, the application of air-cooling reduces the combustion efficiency of the engine. In addition, it complicates the structure of components, which not only increases the difficulty of processing, but also increases the cost of development and maintenance. The high-performance aeroengine pursues to continuously raise the turbine inlet temperature. The high-temperature strength, corrosion resistance, and anti-oxidation performance of hot-section components are increasingly demanded. The turbine inlet temperature of engine with thrust–weight ratio of 15–20 will reach 1927 °C. Ceramic-matrix composites (CMCs) with excellent properties such as high temperature resistance, low density, metal-like fracture behavior, insensitivity to cracks, and no catastrophic damage can replace superalloys to meet the needs of hot-section components in higher-temperature environments of aeroengine. It not only is beneficial to greatly reduce weight, but also can save cooling air or even need no cooling, thus increasing the total pressure ratio, and further increase working temperature about 400–500 °C and structure weight loss 50–70% compared with traditional superalloy (Ohnabe et al. 1999; Bouillon et al. 2005; Halbig et al. 2013; Li 2018a, b, 2019a, b, c).

Under tensile loading of fiber-reinforced CMCs, the tensile stress-strain curve can be divided into four stages, i.e., (1) the linear-elastic stage, and the strain increases in direct proportion to the stress; (2) the damage stage with matrix cracking and fiber debonding at the interface, which makes the CMCs appear the characteristics of pseudo-plastic fracture and high toughness; (3) the damage stage with the saturation of matrix cracking and complete interface debonding; and (4) the fiber failure stage. Many researchers investigated the tensile behavior of fiber-reinforced CMCs. Li et al. (2013, 2015; Li 2018c) investigated the tensile behavior of unidirectional, cross-ply, 2D, and 2.5D CMCs at room temperature. The damage models of matrix cracking, interface debonding, and fiber failure were considered and combined with shear-lag model to predict the tensile stress-strain curves of fiber-reinforced CMCs with different fiber preforms. Wang et al. (2013) compared the tensile behavior of 1D (unidirectional), 2D (plain woven), and 3D (braided) C/SiC composites at room temperature. The tensile stress-strain curves of C/SiC depended on the fiber volume along the loading direction and the interface properties. Zhang et al. (2016) investigated the strength degradation of 2.5D C/SiC composite after exposure at elevated temperature in air condition. The composite tensile strength and failure strain decrease after exposure at elevated temperature. Mei et al. (2006) investigated the thermal cyclic damage of C/SiC composite in oxidizing environment. Under thermal cyclic between the temperatures of 900 and 1200 °C under constant stress of 62.5 MPa, the damage of matrix cracking, fiber debonding, sliding, and breaking attributed to the strain increase with thermal cyclic number. At elevated temperature, the interface oxidation degrades the interface and fiber properties, which affects the tensile properties of fiber-reinforced CMCs (Halbig and Cawley 1999; Halbig et al. 2008). Hou et al. (2009) investigated the influence of high temperature exposure to air on the damage to 3D C/SiC composites. The composite was exposure in air atmosphere at elevated temperatures of $T = 600$, 900, and 1300 °C for $t = 0$ to 15 h and then was loaded under three-point bend test at room temperature. The matrix microcracks caused by a difference of coefficients of thermal expansion between the matrix and carbon fibers in the cooling process after composite preparation act as oxygen diffuse paths; however, the damage decreases with temperature at the same exposure time. Wallentine (2015) investigated the effect of prior exposure at elevated temperatures on tensile properties and stress-strain behavior of different fiber-reinforced CMCs, i.e., SiC/SiNC, C/SiC, C/SiC-B_4C, and SiC/SiC-B_4C. The CMCs were heat treated in laboratory air for $t = 10$, 20, 40, and 100 h at over-temperature ($T = 1300$ °C or 1400 °C) and for $t = 100$ h at operating temperature ($T = 1200$ °C or 1300 °C), and then tensile loaded to failure at room temperature. The prior exposure at elevated temperature caused a reduction of tensile strength and changed the tensile stress-strain behavior due to oxidation of the interphase and fibers. Gowayed et al. (2015) investigated the effect of oxidation on strain accumulation of SiC/SiNC and SiC/SiC composites under the dwell-fatigue loading. The stress level and cracking density affect the internal damage and accumulation strain. However, the tensile and damage fracture processes of fiber-reinforced CMCs under the effect of pre-exposure at elevated temperatures have not been investigated.

In this chapter, the time-dependent tensile damage and fracture of fiber-reinforced CMCs subjected to the pre-exposure at elevated temperatures and thermal fatigue are investigated. The damage mechanisms of the interface oxidation and fiber failure are considered in the stress analysis, matrix multicracking, interface debonding, and fiber failure. Combining the stress analysis and damage models, the tensile stress-strain curves of fiber-reinforced CMCs for different damage stages can be obtained. The effects of the pre-exposure temperature and time, thermal fatigue temperature and thermal cyclic number, the interface shear stress, fiber strength, and fiber Weibull modulus on tensile damage and fracture processes are analyzed. The experimental tensile damage and fracture process of fiber-reinforced CMCs with different fiber preforms are predicted for a different pre-exposure temperature and time.

4.2 Time-Dependent Tensile Damage and Fracture of Fiber-Reinforced Ceramic-Matrix Composites Subjected to Pre-exposure at Elevated Temperature

In this section, the time-dependent tensile damage and fracture of fiber-reinforced CMCs subjected to pre-exposure at elevated temperature are investigated. The damage mechanisms of the interface oxidation and fiber failure are considered in the stress analysis, matrix multicracking, interface debonding, and fiber failure. Combining the stress analysis and damage models, the tensile stress-strain curves of fiber-reinforced CMCs for different damage stages can be obtained. The effects of the pre-exposure temperature and time, interface shear stress, fiber strength, and fiber Weibull modulus on tensile damage and fracture processes are analyzed. The experimental tensile damage and fracture process of fiber-reinforced CMCs with different fiber preforms are predicted for a different pre-exposure temperature and time.

4.2.1 Stress Analysis Considering Interface Oxidation and Fiber Failure

As the mismatch of the axial thermal expansion coefficient between the carbon fiber and silicon carbide matrix, there are unavoidable microcracks within SiC matrix when the composite is cooled down from high fabricated temperature to ambient temperature. These processing-induced microcracks mainly existed in the surface of the material, which do not propagate through the entire thickness of the composite. However, at elevated temperature, the microcracks serve as avenues for the ingress of the environment atmosphere into the composite. The oxygen reacts with carbon layer along the fiber length at a certain rate of $\mathrm{d}\zeta/\mathrm{d}t$, in which ζ is the length of carbon lost in each side of the crack (Casas and Martinez-Esnaola 2003).

$$\zeta(t) = \varphi_1\left[1 - \exp\left(-\frac{\varphi_2 t}{b}\right)\right] \tag{4.1}$$

where b is a delay factor considering the deceleration of reduced oxygen activity and φ_1 and φ_2 are parameters dependent on temperature and described using the Arrhenius-type laws.

$$\varphi_1 = 7.021 \times 10^{-3} \times \exp\left(\frac{8231}{T}\right) \tag{4.2}$$

$$\varphi_2 = 227.1 \times \exp\left(-\frac{17{,}090}{T}\right) \tag{4.3}$$

where φ_1 is in mm and φ_2 in s^{-1}; φ_1 represents the asymptotic behavior for long times, which decreases with temperature; and the product $\varphi_1\varphi_2$ represents the initial oxidation rate, which is an increasing function of temperature.

When damage of matrix cracking and interface debonding occur in fiber-reinforced CMCs, the shear-lag model can be used to analyze the microstress distributions in the interface oxidation region, interface slip region, and interface bonded region. The distributions of the fiber and matrix axial stress distribution, and the fiber/matrix interface shear stress can be determined using the following equations.

$$\sigma_f(x)=\begin{cases}\frac{\sigma}{V_f}-\frac{2\tau_f}{r_f}x, & x\in[0,\zeta(t)]\\ \frac{\sigma}{V_f}-\frac{2\tau_f}{r_f}\zeta(t)-\frac{2\tau_i}{r_f}[x-\zeta(t)], & x\in[\zeta(t),l_d]\\ \sigma_{fo}+\left\{\frac{\sigma}{V_f}-\frac{2\tau_f}{r_f}\zeta(t)-\frac{2\tau_i}{r_f}[l_d-\zeta(t)]-\sigma_{fo}\right\}\exp\left(-\rho\frac{x-l_d}{r_f}\right), & x\in\left[l_d,\frac{l_c}{2}\right]\end{cases}\tag{4.4}$$

$$\sigma_m(x)=\begin{cases}2\frac{V_f}{V_m}\frac{\tau_f}{r_f}x, & x\in[0,\zeta(t)]\\ 2\frac{V_f}{V_m}\frac{\tau_f}{r_f}\zeta(t)+2\frac{V_f}{V_m}\frac{\tau_i}{r_f}[x-\zeta(t)], & x\in[\zeta(t),l_d]\\ \sigma_{mo}+\left\{2\frac{V_f}{V_m}\frac{\tau_f}{r_f}\zeta(t)+2\frac{V_f}{V_m}\frac{\tau_i}{r_f}[l_d-\zeta(t)]-\sigma_{mo}\right\}\exp\left(-\rho\frac{x-l_d}{r_f}\right), & x\in\left[l_d,\frac{l_c}{2}\right]\end{cases}\tag{4.5}$$

$$\tau_i(x)=\begin{cases}\tau_f, & x\in[0,\zeta(t)]\\ \tau_i, & x\in[\zeta(t),l_d]\\ \frac{\rho}{2}\left\{\frac{\sigma}{V_f}-\frac{2\tau_f}{r_f}\zeta(t)-\frac{2\tau_i}{r_f}[l_d-\zeta(t)]-\sigma_{fo}\right\}\exp\left(-\rho\frac{x-l_d}{r_f}\right), & x\in\left[l_d,\frac{l_c}{2}\right]\end{cases}\tag{4.6}$$

where V_f and V_m denote the fiber and matrix volume, respectively; τ_f and τ_i denote the interface shear stress in the interface oxidation region and interface debonded region, respectively; r_f denotes the fiber radius; l_d and l_c denote the interface debonded length and the matrix crack spacing, respectively; ρ denotes the shear-lag model parameter; and σ_{fo} and σ_{mo} denote the fiber and matrix axial stress in the interface bonded region, respectively.

$$\sigma_{fo}=\frac{E_f}{E_c}\sigma+E_f(\alpha_c-\alpha_f)\Delta T\tag{4.7}$$

$$\sigma_{mo}=\frac{E_m}{E_c}\sigma+E_m(\alpha_c-\alpha_m)\Delta T\tag{4.8}$$

where E_f, E_m, and E_c denote the fiber, matrix, and composite elastic modulus, respectively; α_f, α_m, and α_c denote the fiber, matrix, and composite thermal expansion coefficient; and ΔT denotes the temperature difference between the testing temperature and fabrication temperature.

When the fiber failure occurs, the fiber axial stress in the interface debonded and bonded region can be determined using the following equation.

$$\sigma_f(z)=\begin{cases}\Phi-\frac{2\tau_f}{r_f}z, & z\in[0,\zeta(t)]\\ \Phi-\frac{2\tau_f}{r_f}\zeta(t)-\frac{2\tau_i}{r_f}[z-\zeta(t)], & z\in[\zeta(t),l_d]\\ \sigma_{fo}+\left\{\Phi-\frac{2\tau_f}{r_f}\zeta(t)-\frac{2\tau_i}{r_f}[l_d-\zeta(t)]-\sigma_{fo}\right\}\exp\left(-\rho\frac{z-l_d}{r_f}\right), & z\in\left[l_d,\frac{l_c}{2}\right]\end{cases}\tag{4.9}$$

where Φ denotes the intact fiber stress.

4.2.2 Matrix Multicracking Considering Interface Oxidation

The two-parameter Weibull distribution is used to describe the tensile strength of the matrix, and the failure probability of the matrix can be determined using the following equation (Curtin 1993).

$$P_{\mathrm{m}} = 1 - \exp\left\{ -\left[\frac{\sigma - (\sigma_{\mathrm{mc}} - \sigma_{\mathrm{th}})}{(\sigma_{\mathrm{R}} - \sigma_{\mathrm{th}}) - (\sigma_{\mathrm{mc}} - \sigma_{\mathrm{th}})} \right]^{\beta} \right\} \tag{4.10}$$

where σ_{R} denotes the matrix characteristic strength; σ_{mc} denotes matrix first cracking stress; σ_{th} denotes matrix thermal residual stress; and β denotes matrix Weibull modulus.

When the applied stress increases, the matrix cracking density increases. The matrix failure probability relates to the instantaneous matrix crack space and saturation matrix crack spacing, as follows:

$$P_{\mathrm{m}} = l_{\mathrm{sat}} / l_{\mathrm{c}} \tag{4.11}$$

where

$$l_{\mathrm{sat}} = \Lambda\left(\sigma_{\mathrm{mc}} / \sigma_{\mathrm{R}}, \sigma_{\mathrm{th}} / \sigma_{\mathrm{R}}, \beta\right)\delta_{\mathrm{R}} \tag{4.12}$$

where Λ denotes the final nominal crack space, which is a pure number and depends only on the micromechanical and statistical quantities characterizing the cracking, and δ_{R} denotes the characteristic interface sliding length.

$$\delta_{\mathrm{R}} = \frac{r_{\mathrm{f}}}{2\tau_{\mathrm{i}}} \frac{V_{\mathrm{m}} E_{\mathrm{m}}}{V_{\mathrm{f}} E_{\mathrm{c}}} \sigma_{\mathrm{R}} + \left(1 - \frac{\tau_{\mathrm{f}}}{\tau_{\mathrm{i}}}\right)\zeta(t) \tag{4.13}$$

Using Eqs. (4.10)–(4.13), the instantaneous matrix crack space can be determined using the following equation.

$$l_{\mathrm{c}} = r_{\mathrm{f}} \frac{V_{\mathrm{m}} E_{\mathrm{m}}}{V_{\mathrm{f}} E_{\mathrm{c}}} \frac{\sigma_{\mathrm{R}}}{2\tau_{\mathrm{i}}} \Lambda \left\{ 1 - \exp\left[-\left(\frac{\sigma - (\sigma_{\mathrm{mc}} - \sigma_{\mathrm{th}})}{(\sigma_{\mathrm{R}} - \sigma_{\mathrm{th}}) - (\sigma_{\mathrm{mc}} - \sigma_{\mathrm{th}})} \right)^{\beta} \right] \right\}^{-1} \tag{4.14}$$

4.2.3 Interface Debonding Considering Interface Oxidation

When the matrix cracking propagates to the fiber/matrix interface, the fracture mechanics approach is used to determine the interface debonded length (Gao et al. 1988).

$$\xi_{\mathrm{d}} = -\frac{F}{4\pi r_{\mathrm{f}}}\frac{\partial w_{\mathrm{f}}(z=0)}{\partial l_{\mathrm{d}}} - \frac{1}{2}\int_{0}^{l_{\mathrm{d}}} \tau_{\mathrm{i}}\frac{\partial v(z)}{\partial l_{\mathrm{d}}}\mathrm{d}z \tag{4.15}$$

where ξ_{d} denotes the interface debond energy; $F(=\pi r_{\mathrm{f}}^2\sigma/V_{\mathrm{f}})$ denotes the fiber stress on the matrix cracking plane; $w_{\mathrm{f}}(z=0)$ denotes the fiber axial displacement at the matrix cracking plane; and $v(z)$ denotes the relative displacement between the fiber and the matrix.

$$\begin{aligned} w_{\mathrm{f}}(z) &= \int_{x}^{l_{\mathrm{c}}/2} \frac{\sigma_{\mathrm{f}}(z)}{E_{\mathrm{f}}}\mathrm{d}z \\ &= \frac{\sigma}{V_{\mathrm{f}}E_{\mathrm{f}}}(l_{\mathrm{d}}-z) - \frac{\tau_{\mathrm{f}}}{r_{\mathrm{f}}E_{\mathrm{f}}}\left[2\zeta(t)l_{\mathrm{d}} - \zeta^2 - z^2\right] - \frac{\tau_{\mathrm{i}}}{r_{\mathrm{f}}E_{\mathrm{f}}}[l_{\mathrm{d}} - \zeta(t)]^2 + \frac{\sigma_{\mathrm{fo}}}{E_{\mathrm{f}}}\left(\frac{l_{\mathrm{c}}}{2} - l_{\mathrm{d}}\right) \\ &+ \frac{r_{\mathrm{f}}}{\rho E_{\mathrm{f}}}\left[\frac{V_{\mathrm{m}}}{V_{\mathrm{f}}}\sigma_{\mathrm{mo}} - \frac{2\tau_{\mathrm{f}}}{r_{\mathrm{f}}}\zeta(t) - \frac{2\tau_{\mathrm{i}}}{r_{\mathrm{f}}}(l_{\mathrm{d}} - \zeta(t))\right]\left[1 - \exp\left(-\rho\frac{l_{\mathrm{c}}/2 - l_{\mathrm{d}}}{r_{\mathrm{f}}}\right)\right] \end{aligned} \tag{4.16}$$

$$\begin{aligned} w_{\mathrm{m}}(z) &= \int_{x}^{l_{\mathrm{c}}/2} \frac{\sigma_{\mathrm{m}}(z)}{E_{\mathrm{m}}}\mathrm{d}z \\ &= \frac{V_{\mathrm{f}}\tau_{\mathrm{f}}}{r_{\mathrm{f}}V_{\mathrm{m}}E_{m}}\left(2\zeta(t)l_{\mathrm{d}} - \zeta^2(t) - z^2\right) + \frac{V_{\mathrm{f}}\tau_{\mathrm{i}}}{r_{\mathrm{f}}V_{\mathrm{m}}E_{m}}(l_{\mathrm{d}} - \zeta(t))^2 + \frac{\sigma_{\mathrm{mo}}}{E_{\mathrm{m}}}\left(\frac{l_{\mathrm{c}}}{2} - l_{\mathrm{d}}\right) \\ &- \frac{r_{\mathrm{f}}}{\rho E_{\mathrm{m}}}\left[\sigma_{\mathrm{mo}} - 2\frac{V_{\mathrm{f}}\tau_{\mathrm{f}}}{r_{\mathrm{f}}V_{\mathrm{m}}}\zeta(t) - 2\frac{V_{\mathrm{f}}\tau_{\mathrm{i}}}{r_{\mathrm{f}}V_{\mathrm{m}}}(l_{\mathrm{d}} - \zeta(t))\right] \\ &\left[1 - \exp\left(-\rho\frac{l_{\mathrm{c}}/2 - l_{\mathrm{d}}}{r_{\mathrm{f}}}\right)\right] \end{aligned} \tag{4.17}$$

The relative displacement $v(z)$ between the fiber and the matrix is described using the following equation.

$$\begin{aligned} v(z) &= |w_{\mathrm{f}}(z) - w_{\mathrm{m}}(z)| \\ &= \frac{\sigma}{V_{\mathrm{f}}E_{\mathrm{f}}}(l_{\mathrm{d}} - z) - \frac{E_{\mathrm{c}}\tau_{\mathrm{f}}}{r_{\mathrm{f}}V_{\mathrm{m}}E_{\mathrm{m}}E_{\mathrm{f}}}\left(2\zeta(t)l_{\mathrm{d}} - \zeta^2(t) - z^2\right) \\ &- \frac{E_{\mathrm{c}}\tau_{\mathrm{i}}}{r_{\mathrm{f}}V_{\mathrm{m}}E_{\mathrm{m}}E_{\mathrm{f}}}(l_{\mathrm{d}} - \zeta(t))^2 \\ &+ \frac{r_{\mathrm{f}}E_{\mathrm{c}}}{\rho V_{\mathrm{m}}E_{\mathrm{m}}E_{\mathrm{f}}}\left[\sigma_{\mathrm{mo}} - 2\frac{\tau_{\mathrm{f}}}{r_{\mathrm{f}}}\zeta(t) - 2\frac{\tau_{\mathrm{i}}}{r_{\mathrm{f}}}(l_{\mathrm{d}} - \zeta(t))\right] \\ &quad \left[1 - \exp\left(-\rho\frac{l_{\mathrm{c}}/2 - l_{\mathrm{d}}}{r_{\mathrm{f}}}\right)\right] \end{aligned} \tag{4.18}$$

Substituting $w_{\mathrm{f}}(x=0)$ and $v(z)$ into Eq. (4.15), it leads to the following equation.

$$\frac{E_{\mathrm{c}}\tau_{\mathrm{i}}^{2}}{r_{\mathrm{f}}V_{\mathrm{m}}E_{\mathrm{m}}E_{\mathrm{f}}}(l_{\mathrm{d}}-\zeta(t))^{2}+\frac{E_{\mathrm{c}}\tau_{\mathrm{i}}^{2}}{\rho V_{\mathrm{m}}E_{\mathrm{m}}E_{\mathrm{f}}}(l_{\mathrm{d}}-\zeta(t))-\frac{\tau_{\mathrm{i}}\sigma}{V_{\mathrm{f}}E_{\mathrm{f}}}(l_{\mathrm{d}}-\zeta(t))$$
$$+\frac{2E_{\mathrm{c}}\tau_{\mathrm{f}}\tau_{\mathrm{i}}}{r_{\mathrm{f}}V_{\mathrm{m}}E_{\mathrm{m}}E_{\mathrm{f}}}\zeta(t)(l_{\mathrm{d}}-\zeta(t))$$
$$-\frac{r_{\mathrm{f}}\tau_{\mathrm{i}}\sigma}{2\rho V_{\mathrm{f}}E_{\mathrm{f}}}+\frac{E_{\mathrm{c}}\tau_{\mathrm{f}}^{2}}{r_{\mathrm{f}}V_{\mathrm{m}}E_{\mathrm{m}}E_{\mathrm{f}}}\zeta^{2}(t)+\frac{E_{\mathrm{c}}\tau_{\mathrm{f}}\tau_{\mathrm{i}}}{\rho V_{\mathrm{m}}E_{\mathrm{m}}E_{\mathrm{f}}}\zeta(t)$$
$$-\frac{\tau_{\mathrm{f}}\sigma}{V_{\mathrm{f}}E_{\mathrm{f}}}\zeta(t)+\frac{r_{\mathrm{f}}V_{\mathrm{m}}E_{\mathrm{m}}\sigma^{2}}{4V_{\mathrm{f}}^{2}E_{\mathrm{f}}E_{\mathrm{c}}}-\xi_{d}=0 \tag{4.19}$$

Solving Eq. (4.19), the interface debonding length is determined using the following equation.

$$l_{\mathrm{d}}=\left(1-\frac{\tau_{\mathrm{f}}}{\tau_{\mathrm{i}}}\right)\zeta(t)+\frac{r_{\mathrm{f}}}{2}\left(\frac{V_{\mathrm{m}}E_{\mathrm{m}}\sigma}{V_{\mathrm{f}}E_{\mathrm{c}}\tau_{\mathrm{i}}}-\frac{1}{\rho}\right)-\sqrt{\left(\frac{r_{\mathrm{f}}}{2\rho}\right)^{2}+\frac{r_{\mathrm{f}}V_{\mathrm{m}}E_{\mathrm{m}}E_{\mathrm{f}}}{E_{\mathrm{c}}\tau_{\mathrm{i}}^{2}}\xi_{\mathrm{d}}} \tag{4.20}$$

4.2.4 Fiber Failure Considering Interface and Fiber Oxidation

The two-parameter Weibull model is adopted to describe the fiber strength distribution, and the global load sharing criterion is used to determine the stress distributions between the intact and fracture fibers (Curtin 1991).

$$\frac{\sigma}{V_{\mathrm{f}}}=\Phi(1-P(\Phi))+\frac{2\tau_{\mathrm{f}}}{r_{\mathrm{f}}}\langle L\rangle P(\Phi) \tag{4.21}$$

where $\langle L\rangle$ denotes the average fiber pullout length and $P(\Phi)$ denotes the fiber failure probability.

$$P(\Phi)=1-\exp\left[-\left(\frac{\Phi}{\sigma_{\mathrm{c}}}\right)^{m+1}\right] \tag{4.22}$$

where m denotes the fiber Weibull modulus and σ_{fc} denotes the fiber characteristic strength of a length δ_{c} of fiber.

$$\sigma_{\mathrm{c}}=\left(\frac{l_{\mathrm{o}}\sigma_{\mathrm{o}}^{m}\tau_{\mathrm{i}}}{r_{\mathrm{f}}}\right)^{\frac{1}{m+1}},\quad \delta_{\mathrm{c}}=\left(\frac{\sigma_{\mathrm{o}}r_{\mathrm{f}}l_{\mathrm{o}}^{\frac{1}{m}}}{\tau_{\mathrm{i}}}\right)^{\frac{m}{m+1}} \tag{4.23}$$

The time-dependent fiber strength of $\sigma_{0}(t)$ can be determined using the following equation (Lara-Curzio 1999).

$$\sigma_0(t) = \begin{cases} \sigma_0, & t \le \frac{1}{k}\left(\frac{K_{IC}}{Y\sigma_0}\right)^4 \\ \frac{K_{IC}}{Y\sqrt[4]{kt}}, & t > \frac{1}{k}\left(\frac{K_{IC}}{Y\sigma_0}\right)^4 \end{cases} \tag{4.24}$$

The composite tensile strength is given by the following equation.

$$\sigma_{\mathrm{UTS}} = V_{\mathrm{f}}\sigma_{\mathrm{c}}\left(\frac{2}{m+2}\right)^{\frac{1}{m+1}}\left(\frac{m+1}{m+2}\right) \tag{4.25}$$

4.2.5 *Tensile Stress-Strain Curves Considering Effect of Pre-exposure*

For the fiber-reinforced CMCs without damage, the composite strain can be determined using the following equation.

$$\varepsilon_{\mathrm{c}} = \sigma \big/ E_{\mathrm{c}} \tag{4.26}$$

When the damage forms inside of CMCs, the composite strain can be determined using the following equation.

$$\varepsilon_{\mathrm{c}} = \frac{2}{E_{\mathrm{f}} l_{\mathrm{c}}} \int_{l_{\mathrm{c}}/2} \sigma_{\mathrm{f}}(z)\mathrm{d}x - (\alpha_{\mathrm{c}} - \alpha_{\mathrm{f}})\Delta T \tag{4.27}$$

When matrix cracking and interface debonding occur, the composite strain can be determined using the following equation.

$$\begin{aligned} \varepsilon_{\mathrm{c}} = & \frac{2\sigma}{V_{\mathrm{f}}E_{\mathrm{f}}}\frac{l_{\mathrm{d}}}{l_{\mathrm{c}}} + \frac{2\tau_{\mathrm{f}}}{r_{\mathrm{f}}E_{\mathrm{f}}l_{\mathrm{c}}}\zeta^2(t) - \frac{4\tau_{\mathrm{f}}}{r_{\mathrm{f}}E_{\mathrm{f}}}\frac{l_{\mathrm{d}}}{l_{\mathrm{c}}}\zeta(t) \\ & - \frac{2\tau_{\mathrm{i}}}{r_{\mathrm{f}}E_{\mathrm{f}}l_{\mathrm{c}}}[l_{\mathrm{d}} - \zeta(t)]^2 + \frac{2\sigma_{\mathrm{fo}}}{E_{\mathrm{f}}l_{\mathrm{c}}}\left(\frac{l_{\mathrm{c}}}{2} - l_{\mathrm{d}}\right) \\ & + \frac{2r_{\mathrm{f}}}{\rho E_{\mathrm{f}}l_{\mathrm{c}}}\left\{\frac{\sigma}{V_{\mathrm{f}}} - \frac{2\tau_{\mathrm{f}}}{r_{\mathrm{f}}}\zeta(t) - \frac{2\tau_{\mathrm{i}}}{r_{\mathrm{f}}}[l_{\mathrm{d}} - \zeta(t)] - \sigma_{\mathrm{fo}}\right\} \\ & \times\left[1 - \exp\left(-\rho\frac{l_{\mathrm{c}}/2 - l_{\mathrm{d}}}{r_{\mathrm{f}}}\right)\right] - (\alpha_{\mathrm{c}} - \alpha_{\mathrm{f}})\Delta T \end{aligned} \tag{4.28}$$

When the fiber failure occurs, the composite strain can be determined using the following following equation.

$$\varepsilon_{\mathrm{c}} = \frac{T}{E_{\mathrm{f}}}\frac{2l_{\mathrm{d}}}{l_{\mathrm{c}}} + \frac{2\tau_{\mathrm{f}}}{r_{\mathrm{f}}E_{\mathrm{f}}l_{\mathrm{c}}}\zeta^2(t) - \frac{4\tau_{\mathrm{f}}l_{\mathrm{d}}}{r_{\mathrm{f}}E_{\mathrm{f}}l_{\mathrm{c}}}\zeta(t) - \frac{2\tau_{\mathrm{i}}}{r_{\mathrm{f}}E_{\mathrm{f}}l_{\mathrm{c}}}(l_{\mathrm{d}} - \zeta(t))^2 + \frac{2\sigma_{\mathrm{fo}}}{E_{\mathrm{f}}l_{\mathrm{c}}}\left(\frac{l_{\mathrm{c}}}{2} - l_{\mathrm{d}}\right) + \frac{2r_{\mathrm{f}}}{\rho E_{\mathrm{f}}l_{\mathrm{c}}}\left\{T - \frac{2\tau_{\mathrm{f}}}{r_{\mathrm{f}}}\zeta(t) - \frac{2\tau_{\mathrm{i}}}{r_{\mathrm{f}}}[l_{\mathrm{d}} - \zeta(t)] - \sigma_{\mathrm{fo}}\right\} \times \left[1 - \exp\left(-\rho\frac{l_{\mathrm{c}}/2 - l_{\mathrm{d}}}{r_{\mathrm{f}}}\right)\right] - (\alpha_{\mathrm{c}} - \alpha_{\mathrm{f}})\Delta T \quad (4.29)$$

4.2.6 *Results and Discussion*

The effects of the pre-exposure temperature and time, interface shear stress, fiber strength, and fiber Weibull modulus on tensile damage process are analyzed. The unidirectional C/SiC composite is used in the case analysis, and the material properties are given by: $V_{\mathrm{f}} = 40\%$, $E_{\mathrm{f}} = 230$ GPa, $E_{\mathrm{m}} = 350$ GPa, $r_{\mathrm{f}} = 3.5$ μm, $\beta = 6$, $\sigma_{\mathrm{R}} = 100$ MPa, $l_{\mathrm{sat}} = 150$ μm, $\alpha_{\mathrm{f}} = 0 \times 10^{-6}$/K, $\alpha_{\mathrm{m}} = 4.6 \times 10^{-6}$/K, $\Delta T = -1000$ °C, $\xi_{\mathrm{d}} = 0.1$ J/m^2, $\tau_{\mathrm{i}} = 10$ MPa, $\tau_{\mathrm{f}} = 1$ MPa, $\sigma_{\mathrm{c}} = 1.6$ GPa, and $m = 5$.

4.2.6.1 Effect of Pre-exposure Temperature on Time-Dependent Tensile and Damage

The effect of pre-exposure temperature (i.e., $T = 600$, 700, and 800 °C) on the tensile stress-strain curves, interface debonding and oxidation and fiber failure of C/SiC composite corresponding to pre-exposure time of $t = 20$ h is shown in Fig. 4.1. When the pre-exposure temperature increases, the composite tensile strength and failure strain both decrease, the interface debonded length and the interface oxidation ratio both increase, and the fiber broken fraction increases at low applied stress level.

When the pre-exposure temperature is $T = 600$ °C, the time-dependent composite tensile strength is $\sigma_{\mathrm{UTS}} = 445$ MPa with the failure strain of $\varepsilon_{\mathrm{f}} = 0.57\%$, the time-dependent interface debonding length increases to $2l_{\mathrm{d}}/l_{\mathrm{c}} = 67.5\%$, the time-dependent interface oxidation ratio decreases to $\zeta/l_{\mathrm{d}} = 2.9\%$, and the time-dependent fiber broken fraction increases to $P = 13.3\%$.

When the pre-exposure temperature is $T = 700$ °C, the time-dependent composite tensile strength is $\sigma_{\mathrm{UTS}} = 366$ MPa with the failure strain of $\varepsilon_{\mathrm{f}} = 0.45\%$, the time-dependent interface debonding length increases to $2l_{\mathrm{d}}/l_{\mathrm{c}} = 55\%$, the time-dependent interface oxidation ratio decreases to $\zeta/l_{\mathrm{d}} = 10.3\%$, and the time-dependent fiber broken fraction increases to $P = 12.1\%$.

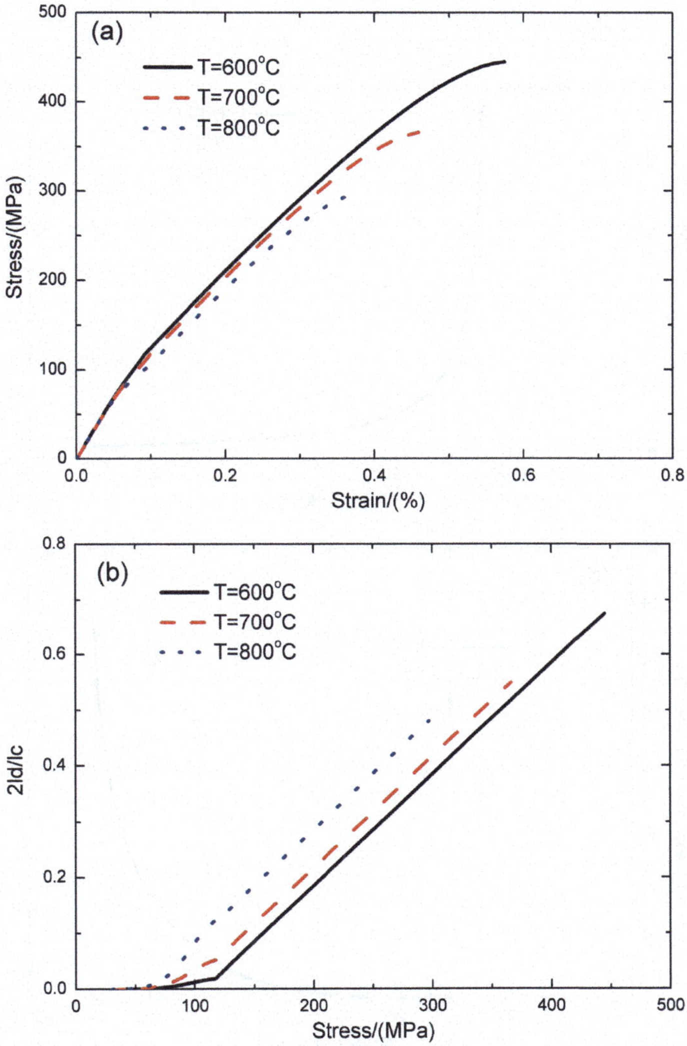

Fig. 4.1 Effect of the pre-exposure temperature on **a** the time-dependent tensile stress-strain curves; **b** the time-dependent interface debonding length versus the applied stress curves; **c** the time-dependent interface oxidation ratio versus the applied stress curves; and **d** the time-dependent broken fiber fraction versus applied stress curves of C/SiC composite

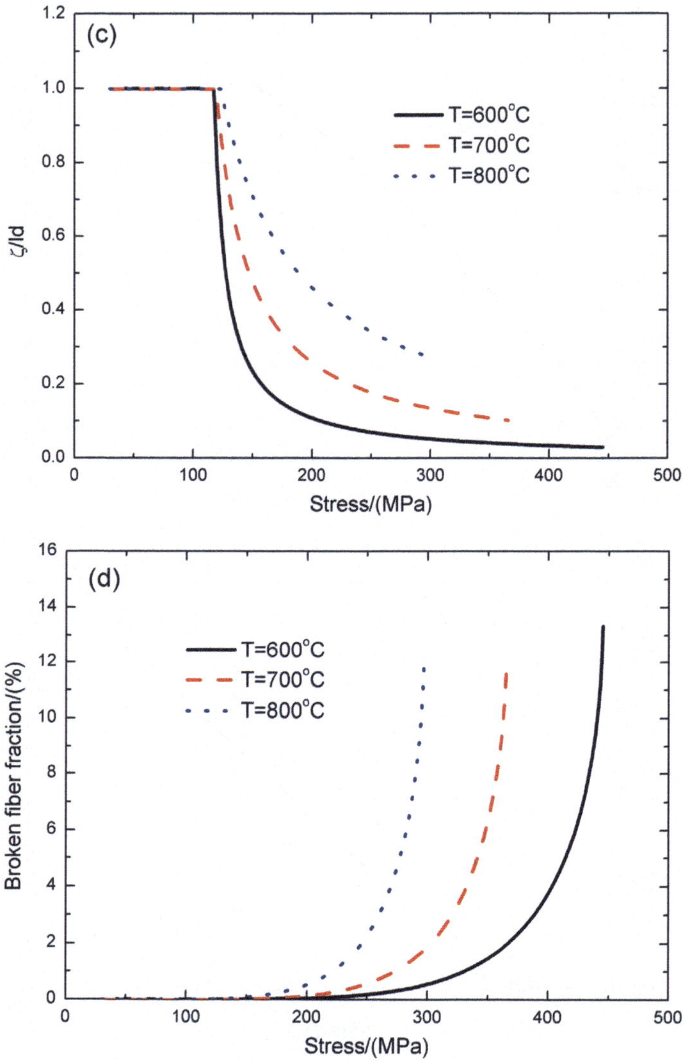

Fig. 4.1 (continued)

When the pre-exposure temperature is $T = 800$ °C, the time-dependent composite tensile strength is $\sigma_{\text{UTS}} = 297$ MPa with the failure strain of $\varepsilon_{\text{f}} = 0.37\%$, the time-dependent interface debonding length increases to $2l_{\text{d}}/l_{\text{c}} = 48\%$, the time-dependent interface oxidation ratio decreases to $\zeta/l_{\text{d}} = 27.5\%$, and the time-dependent fiber broken fraction increases to $P = 12\%$.

4.2.6.2 Effect of Pre-exposure Time on Time-Dependent Tensile and Damage Subjected to Pre-exposure at Elevated Temperature

The effect of pre-exposure time (i.e., t = 10, 20, and 30 h) on the tensile stress-strain curves, interface debonding and oxidation and fiber failure of C/SiC composite corresponding to pre-exposure temperature of $T = 800$ °C is shown in Fig. 4.2. When

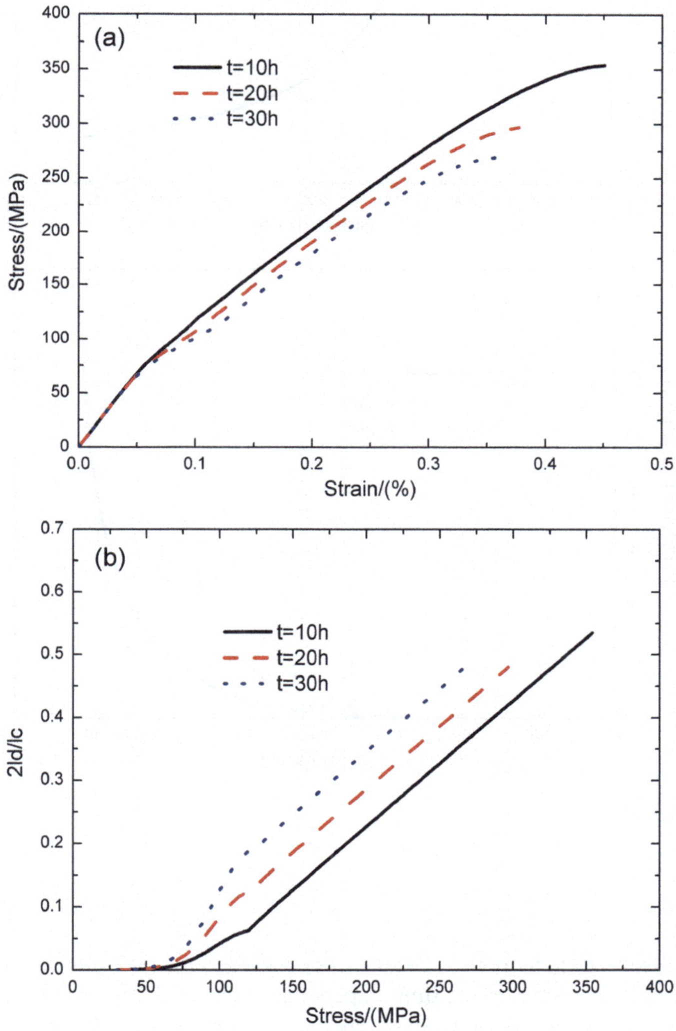

Fig. 4.2 Effect of pre-exposure time on **a** the time-dependent tensile stress-strain curves; **b** the time-dependent interface debonding length versus the applied stress curves; **c** the time-dependent interface oxidation ratio versus the applied stress curves; and **d** the time-dependent broken fiber fraction versus applied stress curves of C/SiC composite

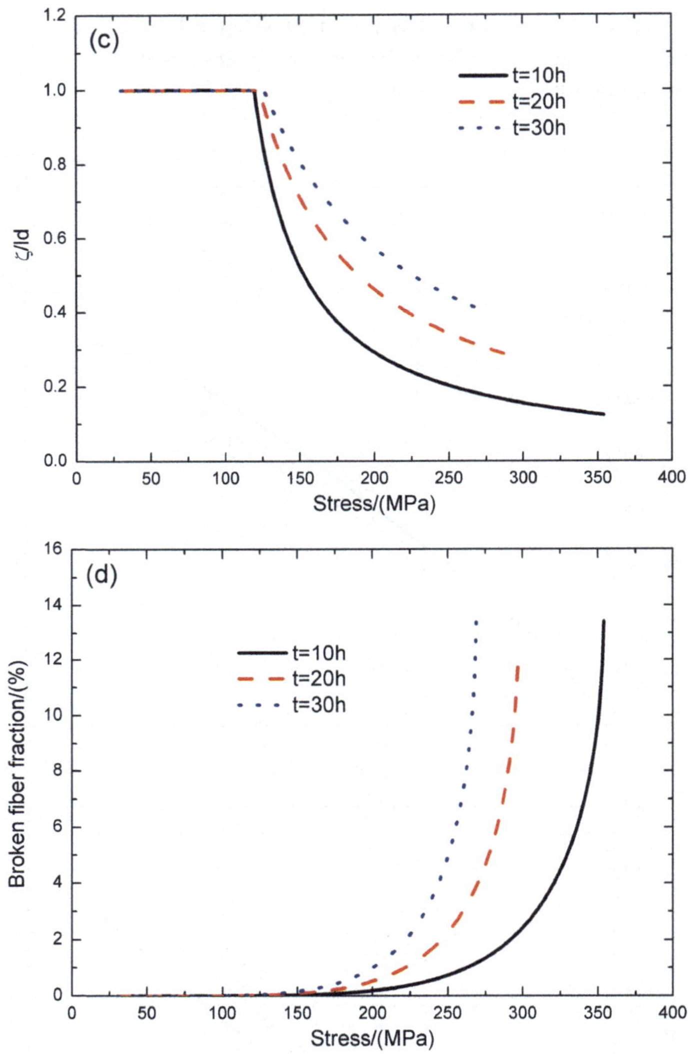

Fig. 4.2 (continued)

the pre-exposure time increases, the time-dependent composite tensile strength and failure strain both decrease, the time-dependent interface debonding length and the interface oxidation ratio both increase, and the time-dependent fiber broken fraction increases at low applied stress level.

When the pre-exposure time is $t = 10$ h, the time-dependent composite tensile strength is $\sigma_{\rm UTS} = 354$ MPa with the failure strain of $\varepsilon_{\rm f} = 0.45\%$, the time-dependent interface debonding length increases to $2l_d/l_c = 53\%$, the time-dependent interface oxidation ratio decreases to $\zeta/l_d = 12.3\%$, and the time-dependent fiber broken fraction increases to $P = 13.4\%$.

When the pre-exposure time is $t = 20$ h, the time-dependent composite tensile strength is $\sigma_{UTS} = 297$ MPa with the failure strain of $\varepsilon_f = 0.37\%$, the time-dependent interface debonding length increases to $2l_d/l_c = 48\%$, the time-dependent interface oxidation ratio decreases to $\zeta/l_d = 27.5\%$, and the time-dependent fiber broken fraction increases to $P = 12\%$.

When the pre-exposure time is $t = 30$ h, the time-dependent composite tensile strength is $\sigma_{UTS} = 269$ MPa with the failure strain of $\varepsilon_f = 0.35\%$, the time-dependent interface debonding length increases to $2l_d/l_c = 48.4\%$, the time-dependent interface oxidation ratio decreases to $\zeta/l_d = 41\%$, and the time-dependent fiber broken fraction increases to $P = 13.5\%$.

4.2.6.3 Effect of Interface Shear Stress on Time-Dependent Tensile and Damage Subjected to Pre-exposure at Elevated Temperature

The effect of the interface shear stress (i.e., $\tau_i = 5$, 10, and 15 MPa) on the tensile stress-strain curves, interface debonding and oxidation and fiber failure of C/SiC composite corresponding to pre-exposure temperature of $T = 800$ °C and pre-exposure time of $t = 20$ h is shown in Fig. 4.3. When the interface shear stress increases, the time-dependent composite failure strain decreases, the time-dependent interface debonding length decreases, and the time-dependent interface oxidation ratio increases.

When the interface shear stress is $\tau_i = 5$ MPa, the time-dependent composite failure strain is $\varepsilon_f = 0.46\%$, the time-dependent interface debonding length increases to $2l_d/l_c = 84\%$, the time-dependent interface oxidation ratio decreases to $\zeta/l_d = 15.7\%$, and the time-dependent fiber broken fraction increases to $P = 12\%$.

When the interface shear stress is $\tau_i = 10$ MPa, the time-dependent composite failure strain is $\varepsilon_f = 0.37\%$, the time-dependent interface debonding length increases to $2l_d/l_c = 48\%$, the time-dependent interface oxidation ratio decreases to $\zeta/l_d = 27\%$, and the time-dependent fiber broken fraction increases to $P = 12\%$.

When the interface shear stress is $\tau_i = 15$ MPa, the time-dependent composite failure strain is $\varepsilon_f = 0.35\%$, the time-dependent interface debonding length increases to $2l_d/l_c = 36\%$, the time-dependent interface oxidation ratio decreases to $\zeta/l_d = 36.7\%$, and the time-dependent fiber broken fraction increases to $P = 12\%$.

4.2.6.4 Effect of Fiber Strength on Tensile and Damage Subjected to Pre-exposure at Elevated Temperature

The effect of the fiber strength (i.e., $\sigma_0 = 1$ and 2 GPa) on the time-dependent tensile stress-strain curves and fiber failure of C/SiC composite corresponding to

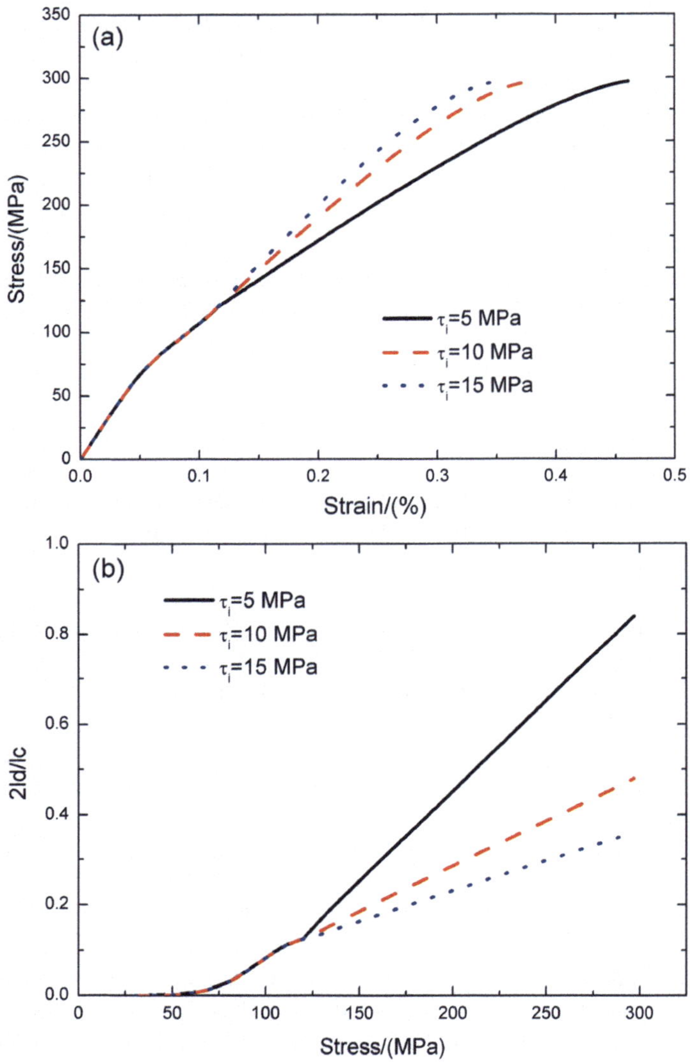

Fig. 4.3 Effect of the interface shear stress on **a** the time-dependent tensile stress-strain curves; **b** the time-dependent interface debonding length versus the applied stress curves; **c** the time-dependent interface oxidation ratio versus the applied stress curves; and **d** the time-dependent broken fiber fraction versus applied stress curves of C/SiC composite

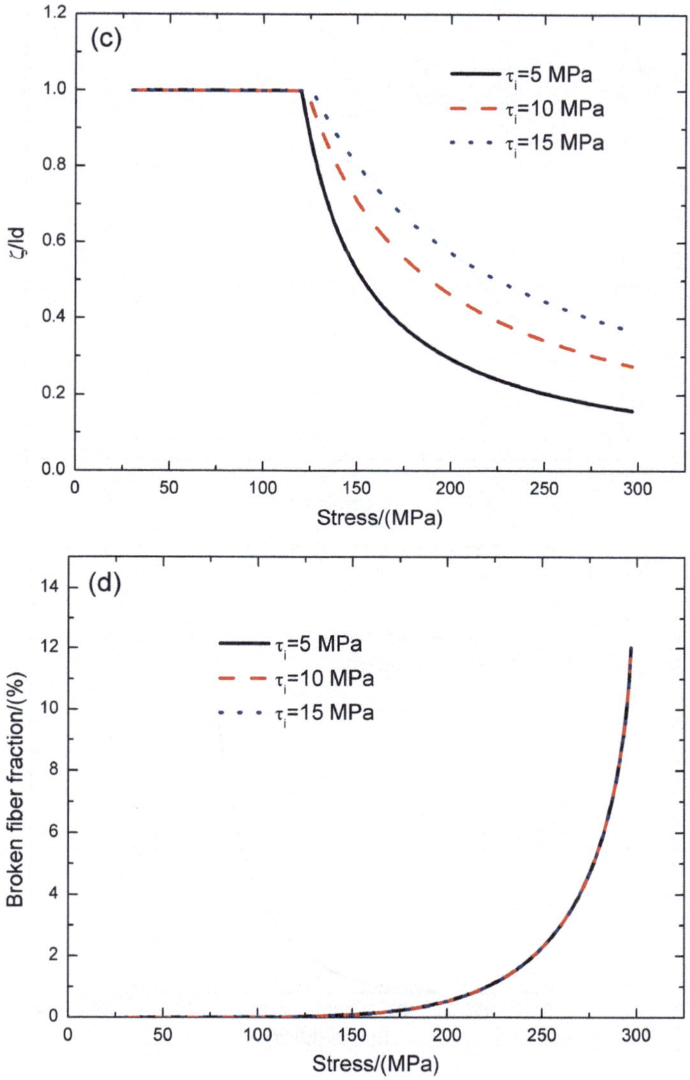

Fig. 4.3 (continued)

pre-exposure temperature of $T = 800$ °C and pre-exposure time of $t = 20$ h is shown in Fig. 4.4. When the fiber strength increases, the time-dependent composite tensile strength and failure strain both increase and the time-dependent fiber broken fraction decreases at low applied stress level.

When the fiber strength is $\sigma_0 = 1$ GPa, the time-dependent composite tensile strength is $\sigma_{\text{UTS}} = 278$ MPa with the failure strain of $\varepsilon_{\text{f}} = 0.31\%$ and the time-dependent fiber broken fraction increases to $P = 12.9\%$.

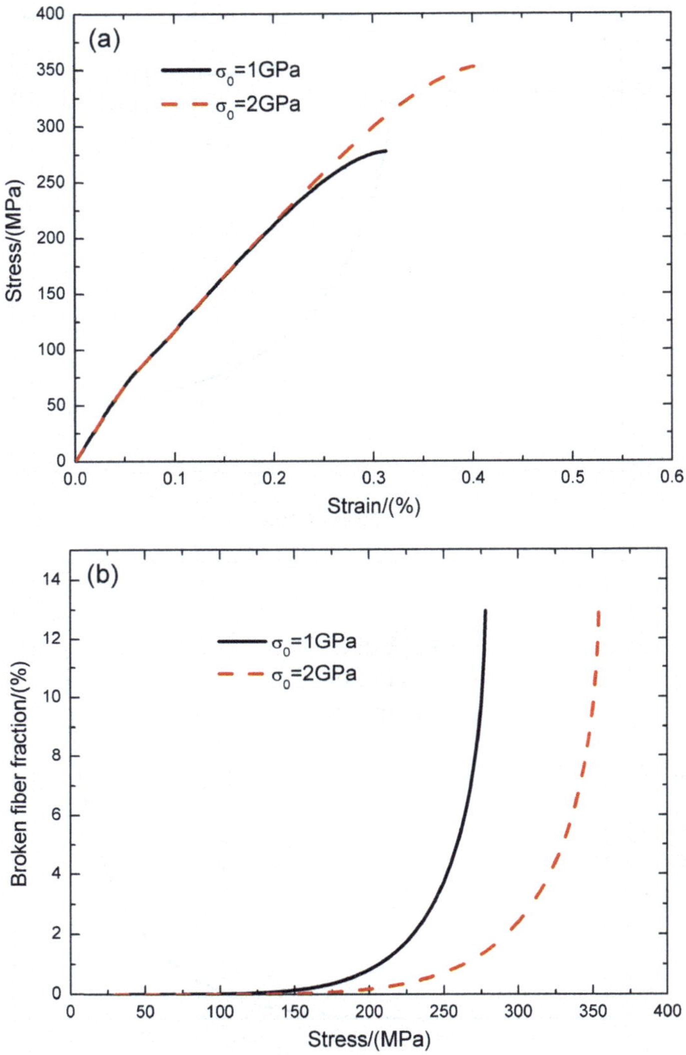

Fig. 4.4 Effect of the fiber strength on **a** the time-dependent tensile stress-strain curves and **b** the time-dependent broken fiber fraction versus applied stress curves of C/SiC composite

When the fiber strength is $\sigma_0 = 2$ GPa, the time-dependent composite tensile strength is $\sigma_{\text{UTS}} = 354$ MPa with the failure strain of $\varepsilon_{\text{f}} = 0.41\%$ and the time-dependent fiber broken fraction increases to $P = 13.4\%$.

4.2.6.5 Effect of Fiber Weibull Modulus on Time-Dependent Tensile and Damage Subjected to Pre-exposure at Elevated Temperature

The effect of fiber Weibull modulus (i.e., $m = 3$ and 5) on the time-dependent tensile stress-strain curves and fiber failure of C/SiC composite corresponding to pre-exposure temperature of $T = 800$ °C and pre-exposure time of $t = 20$ h is shown in Fig. 4.5. When the fiber Weibull modulus increases, the time-dependent composite tensile strength and failure strain both increase and the time-dependent fiber broken fraction decreases at low applied stress level.

When the fiber Weibull modulus is $m = 3$, the time-dependent composite tensile strength is $\sigma_{\text{UTS}} = 272$ MPa with the failure strain of $\varepsilon_{\text{f}} = 0.34\%$ and the time-dependent fiber broken fraction increases to $P = 18.2\%$.

When the fiber Weibull modulus is $m = 5$, the time-dependent composite tensile strength is $\sigma_{\text{UTS}} = 297$ MPa with the failure strain of $\varepsilon_{\text{f}} = 0.35\%$ and the time-dependent fiber broken fraction increases to $P = 12\%$.

4.2.7 Experimental Comparisons

Wang et al. (2013) investigated the tensile behavior of 1D, 2D, and 3D C/SiC composite at room temperature. Zhang et al. (2016) investigated the tensile behavior of 2.5D C/SiC composite after exposure at elevated temperature. The material properties of 1D, 2D, 2.5D, and 3D C/SiC composites are listed in Table 4.1.

The experimental and predicted tensile stress-strain curves, interface debonded length and oxidation length, and fiber broken fraction of 1D C/SiC composite without and with pre-exposure at $T = 800$ °C and $t = 10$, 20, and 30 h are shown in Fig. 4.6. With increase in pre-exposure time, the composite tensile strength and failure strain both decrease, the interface debonded length and interface oxidation ratio increase, and the broken fiber fraction increases at low stress level. Without pre-exposure, the composite tensile strength is $\sigma_{\text{UTS}} = 333$ MPa with the failure strain of $\varepsilon_{\text{f}} = 0.68\%$; when the pre-exposure time is $t = 10$ h, the time-dependent composite tensile strength is $\sigma_{\text{UTS}} = 314$ MPa with the failure strain of $\varepsilon_{\text{f}} = 0.67\%$, the time-dependent interface debonding length increases to $2l_{\text{d}}/l_{\text{c}} = 1$, the time-dependent interface oxidation ratio decreases to $\zeta/l_{\text{d}} = 24.8\%$, and the time-dependent broken fiber fraction increases to $P = 12\%$; when the pre-exposure time is $t = 20$ h, the time-dependent composite tensile strength is $\sigma_{\text{UTS}} = 264$ MPa with the failure strain of $\varepsilon_{\text{f}} = 0.61\%$, the time-dependent interface debonded length increases to $2l_{\text{d}}/l_{\text{c}} = 1$, the time-dependent interface oxidation ratio decreases to $\zeta/l_{\text{d}} = 49\%$, and the time-dependent broken fiber fraction increases to $P = 12\%$; when the pre-exposure time is $t = 30$ h, the time-dependent composite tensile strength is $\sigma_{\text{UTS}} = 239$ MPa with the failure strain of $\varepsilon_{\text{f}} = 0.59\%$, the time-dependent interface debonding length increases to $2l_{\text{d}}/l_{\text{c}} = 1$, the time-dependent interface oxidation ratio decreases to $\zeta/l_{\text{d}} = 74.4\%$, and the time-dependent broken fiber fraction increases to $P = 12.8\%$.

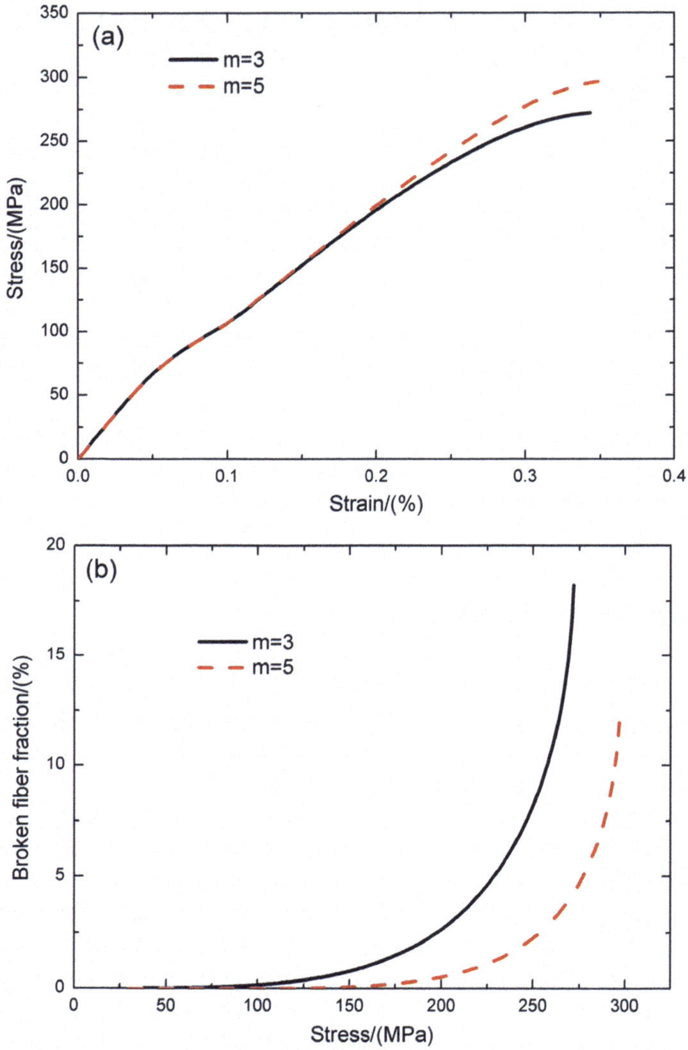

Fig. 4.5 The effect of the fiber Weibull modulus on **a** the time-dependent tensile stress-strain curves; and **b** the time-dependent broken fiber fraction versus applied stress curves of C/SiC composite

The experimental and predicted tensile stress-strain curves, interface debonding length and oxidation length, and fiber broken fraction of 2D C/SiC composite without and with pre-exposure at $T = 800$ °C and $t = 10$, 20, and 30 h are shown in Fig. 4.7. With increase in pre-exposure time, the composite tensile strength and failure strain both decrease, the interface debonded length and interface oxidation ratio increase, and the broken fiber fraction increases at low stress level. Without pre-exposure, the composite tensile strength is $\sigma_{UTS} = 148$ MPa with the failure strain of $\varepsilon_f = 0.35\%$; when the pre-exposure time is $t = 10$ h, the time-dependent composite tensile

Table 4.1 Material properties of C/SiC composites

Items	1D C/SiC	2D C/SiC	2.5D C/SiC	3D C/SiC
λ	1	0.5	0.75	0.93
r_f/(μm)	3.5	3.5	3.5	3.5
V_f/%	30	35	40	40
E_f/(GPa)	230	230	230	230
$\alpha_f/(10^{-6}/K)$	0	0	0.5	0
$\alpha_m/(10^{-6}/K)$	4.6	4.6	4.6	4.6
m	3	5	6	5
σ_R/MPa	100	40	80	80
l_{sat}/μm	120	300	80	80
τ_i/MPa	10	11	5	9
τ_f/MPa	1	1	1	1
$\zeta_d/(J/m^2)$	0.1	0.3	0.1	0.1
σ_{UTS}/(MPa)	333	149	226	206
ε_f/%	0.59	0.34	0.56	0.37
m_f	5	5	5	5

strength is $\sigma_{UTS} = 148$ MPa with the failure strain of $\varepsilon_f = 0.37\%$, the time-dependent interface debonding length increases to $2l_d/l_c = 38.4\%$, the time-dependent interface oxidation ratio decreases to $\zeta/l_d = 8.6\%$, and the time-dependent broken fiber fraction increases to $P = 11.7\%$; when the pre-exposure time is $t = 20$ h, the time-dependent composite tensile strength is $\sigma_{UTS} = 130$ MPa with the failure strain of $\varepsilon_f = 0.33\%$, the time-dependent interface debonding length increases to $2l_d/l_c = 35.4\%$, the time-dependent interface oxidation ratio decreases to $\zeta/l_d = 18.6\%$, and the time-dependent broken fiber fraction increases to $P = 12.2\%$; when the pre-exposure time is $t = 30$ h, the time-dependent composite tensile strength is $\sigma_{UTS} = 117$ MPa with the failure strain of $\varepsilon_f = 0.3\%$, the time-dependent interface debonding length increases to $2l_d/l_c = 34\%$, the time-dependent interface oxidation ratio decreases to $\zeta/l_d = 29\%$, and the time-dependent broken fiber fraction increases to $P = 10.9\%$.

The experimental and predicted tensile stress-strain curves, interface debonded length and oxidation length, and fiber broken fraction of 2.5D C/SiC composite without and with pre-exposure at $T = 900$ °C and $t = 10$ h are shown in Fig. 4.8. Without pre-exposure, the composite tensile strength is $\sigma_{UTS} = 225$ MPa with the failure strain of $\varepsilon_f = 0.54\%$; when the pre-exposure time is $t = 10$ h, the time-dependent composite tensile strength is $\sigma_{UTS} = 191$ MPa with the failure strain of $\varepsilon_f = 0.48\%$, the time-dependent interface debonding length increases to $2l_d/l_c = 1$, the time-dependent interface oxidation ratio decreases to $\zeta/l_d = 25\%$, and the time-dependent broken fiber fraction increases to $P = 23.5\%$.

The experimental and predicted tensile stress-strain curves, interface debonding length and oxidation length, and fiber broken fraction of 3D C/SiC composite without and with pre-exposure at $T = 800$ °C and $t = 10$, 20, and 30 h are shown in Fig. 4.9.

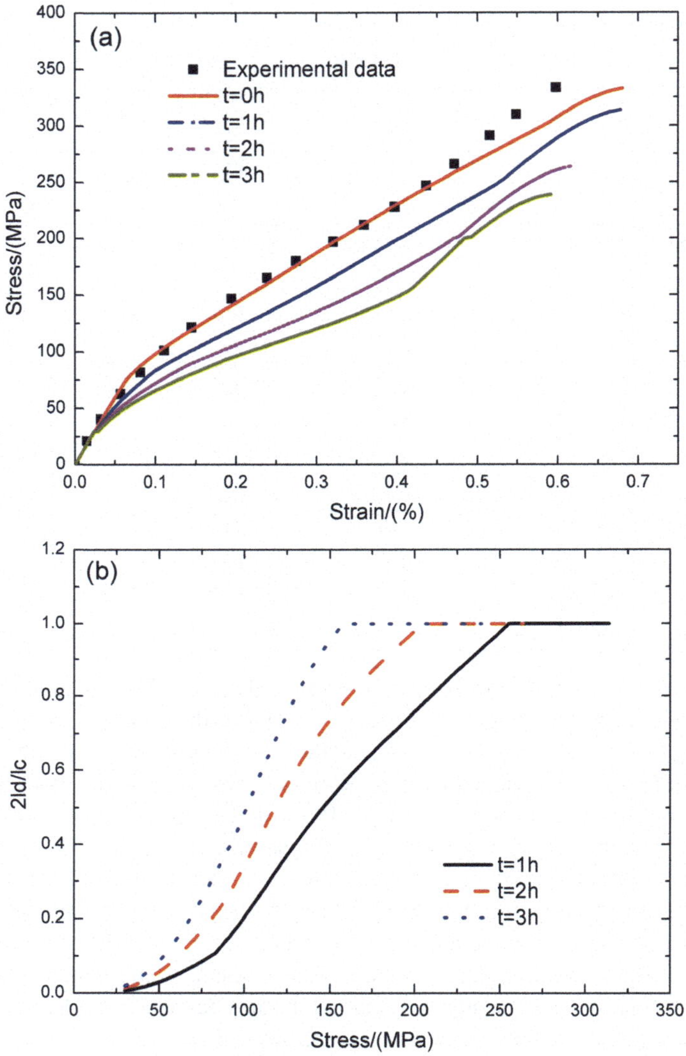

Fig. 4.6 **a** Experimental and predicted tensile stress-strain curves; **b** the interface debonded length versus the applied stress curves; **c** the interface oxidation ratio versus the applied stress curves; and **d** the broken fiber fraction versus applied stress curves of unidirectional C/SiC composite

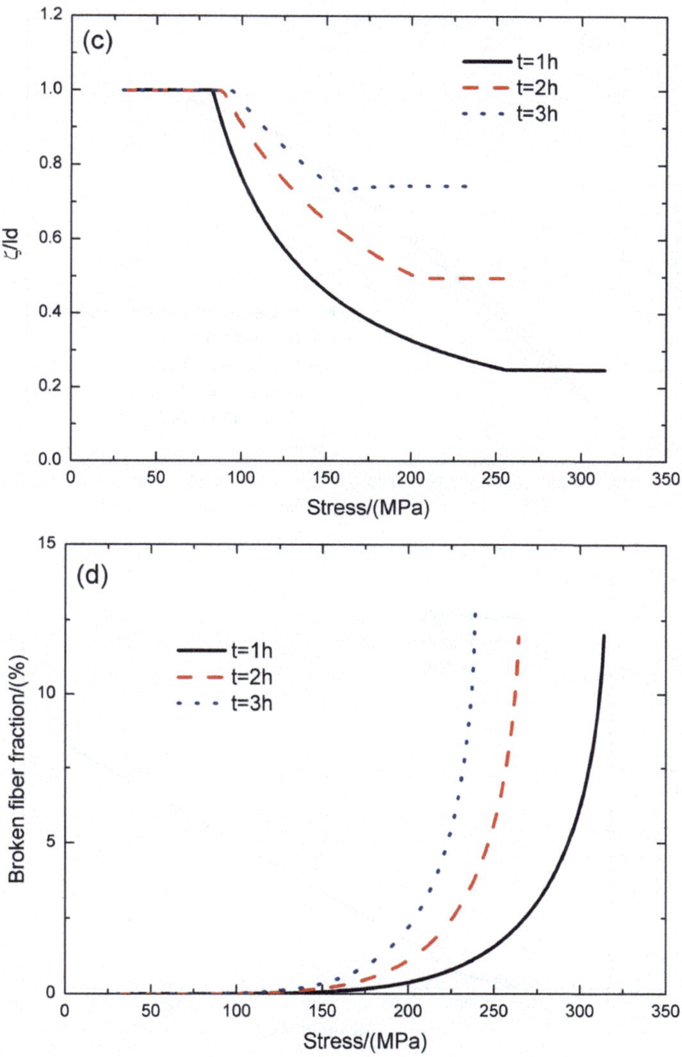

Fig. 4.6 (continued)

With increase in pre-exposure time, the composite tensile strength and failure strain both decrease, the interface debonded length and interface oxidation ratio increase, and the broken fiber fraction increases at low stress level. Without pre-exposure, the composite tensile strength is $\sigma_{UTS} = 203$ MPa with the failure strain of $\varepsilon_f = 0.38\%$; when the pre-exposure time is $t = 10$ h, the time-dependent composite tensile strength is $\sigma_{UTS} = 192$ MPa with the failure strain of $\varepsilon_f = 0.38\%$, the time-dependent interface debonding length increases to $2l_d/l_c = 83\%$, the time-dependent interface oxidation ratio decreases to $\zeta/l_d = 15\%$, and the time-dependent broken fiber fraction increases

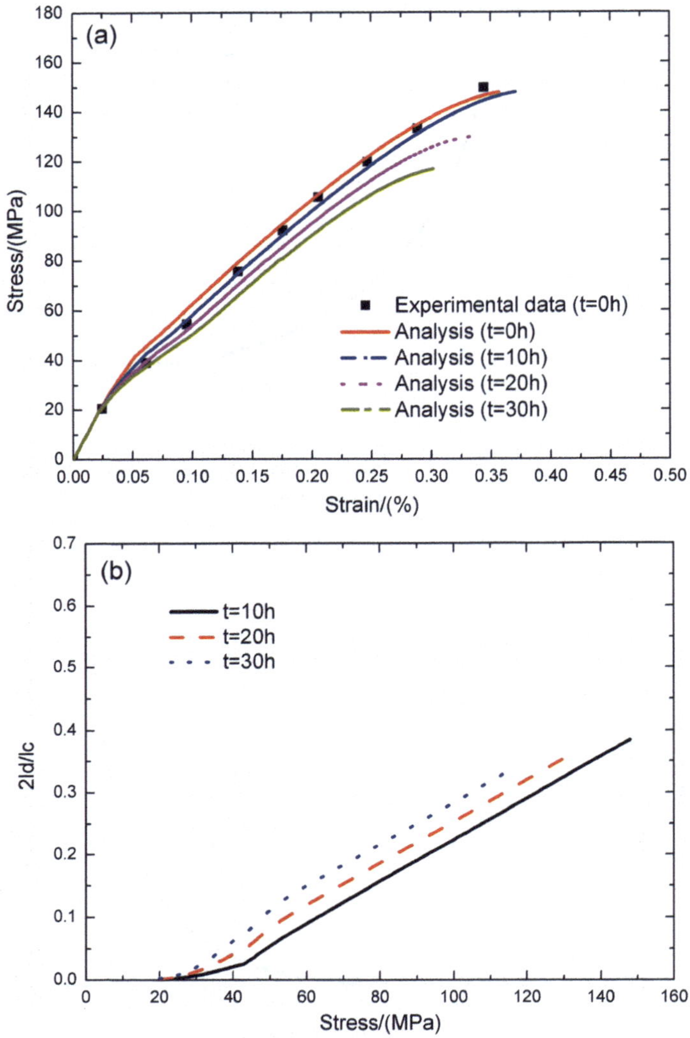

Fig. 4.7 **a** Experimental and predicted tensile stress-strain curves; **b** the interface debonded length versus the applied stress curves; **c** the interface oxidation ratio versus the applied stress curves; and **d** the broken fiber fraction versus applied stress curves of 2D C/SiC composite

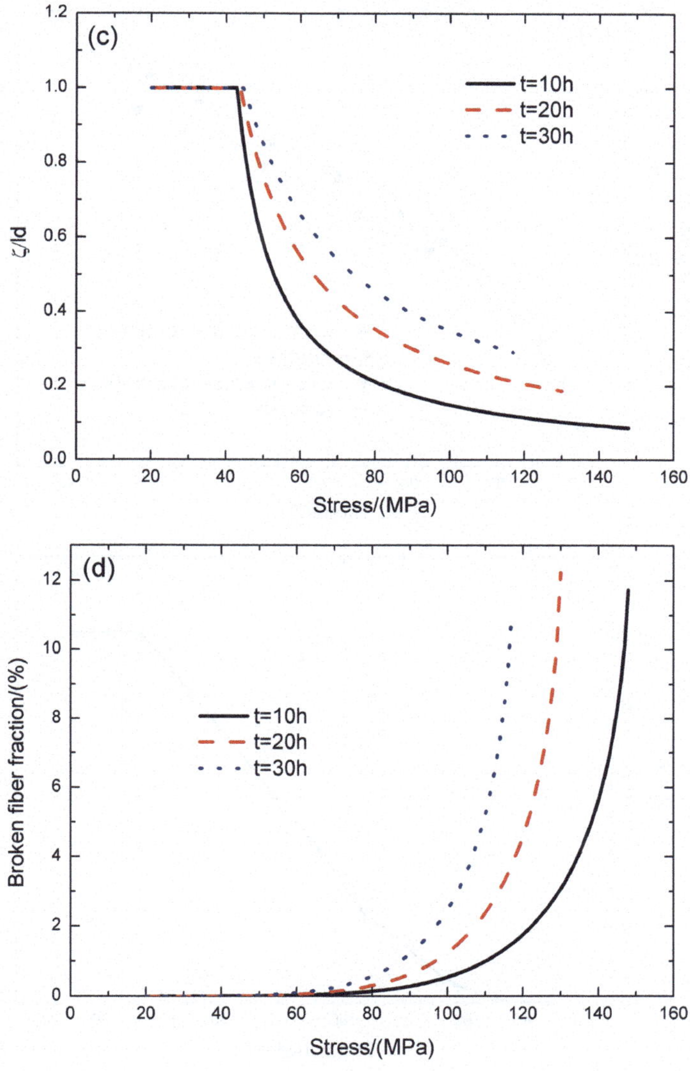

Fig. 4.7 (continued)

to $P = 13.1\%$; when the pre-exposure time is $t = 20$ h, the time-dependent composite tensile strength is $\sigma_{\text{UTS}} = 161$ MPa with the failure strain of $\varepsilon_{\text{f}} = 0.33\%$, the time-dependent interface debonding length increases to $2l_{\text{d}}/l_{\text{c}} = 79\%$, the time-dependent interface oxidation ratio decreases to $\zeta/l_{\text{d}} = 31\%$, and the time-dependent broken fiber fraction increases to $P = 11.7\%$; when the pre-exposure time is $t = 30$ h, the time-dependent composite tensile strength is $\sigma_{\text{UTS}} = 145$ MPa with the failure strain of $\varepsilon_{\text{f}} = 0.32\%$, the time-dependent interface debonding length increases to $2l_{\text{d}}/l_{\text{c}} =$

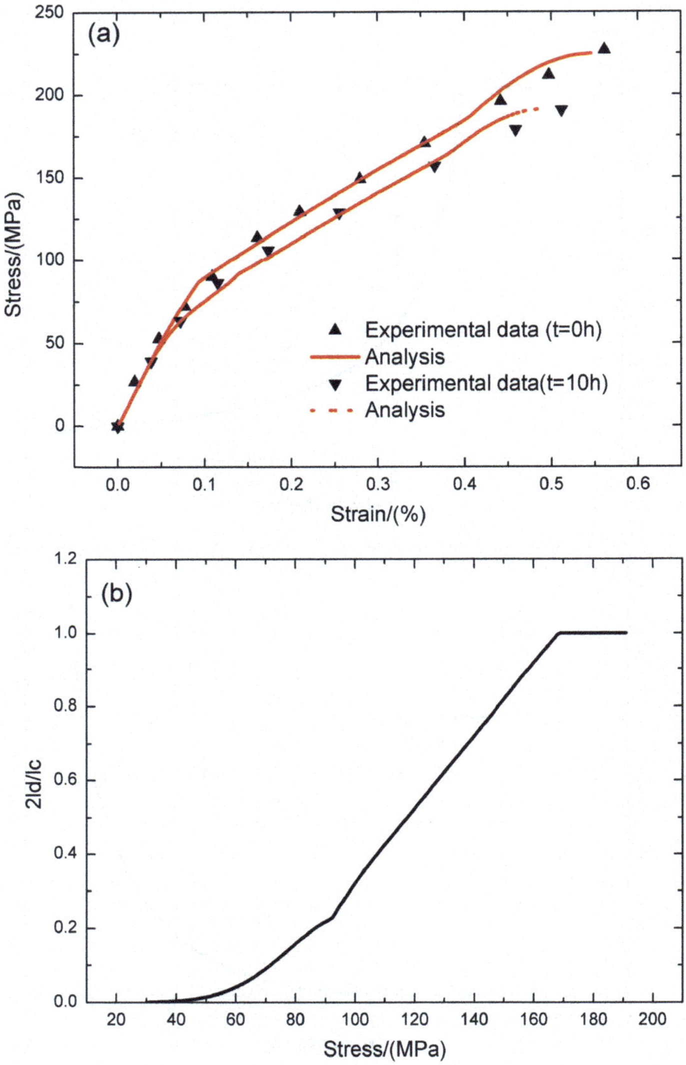

Fig. 4.8 **a** Experimental and predicted tensile stress-strain curves; **b** the interface debonded length versus the applied stress curves; **c** the interface oxidation ratio versus the applied stress curves; and **d** the broken fiber fraction versus applied stress curves of 2.5D C/SiC composite

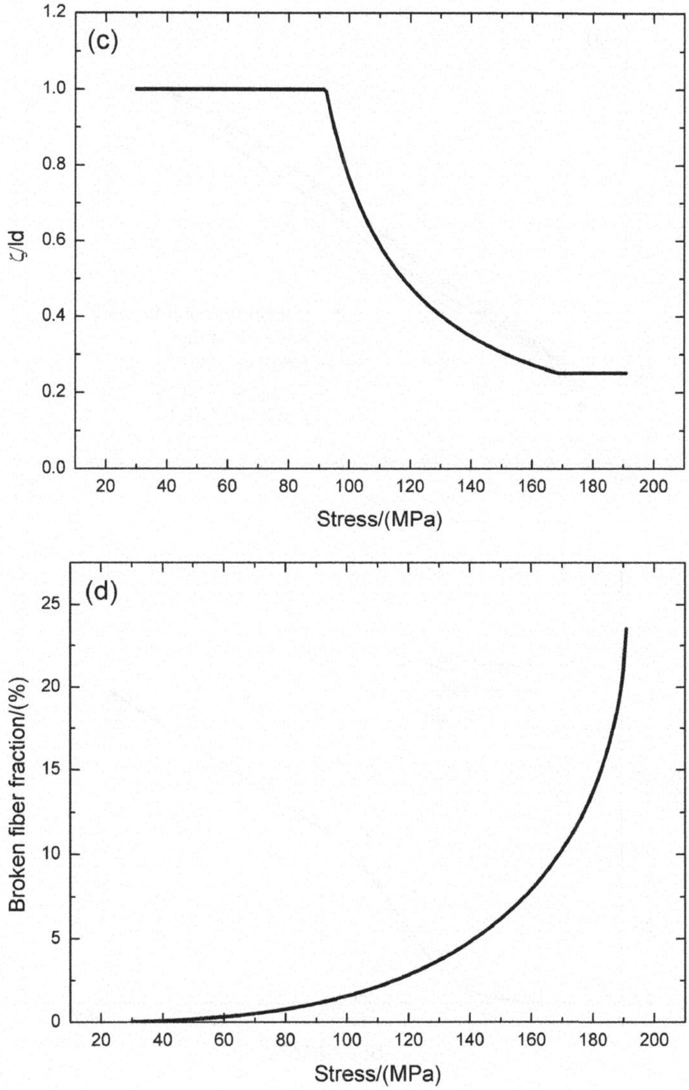

Fig. 4.8 (continued)

83%, the time-dependent interface oxidation ratio decreases to $\zeta/l_d = 44\%$, and the time-dependent broken fiber fraction increases to $P = 10.8\%$.

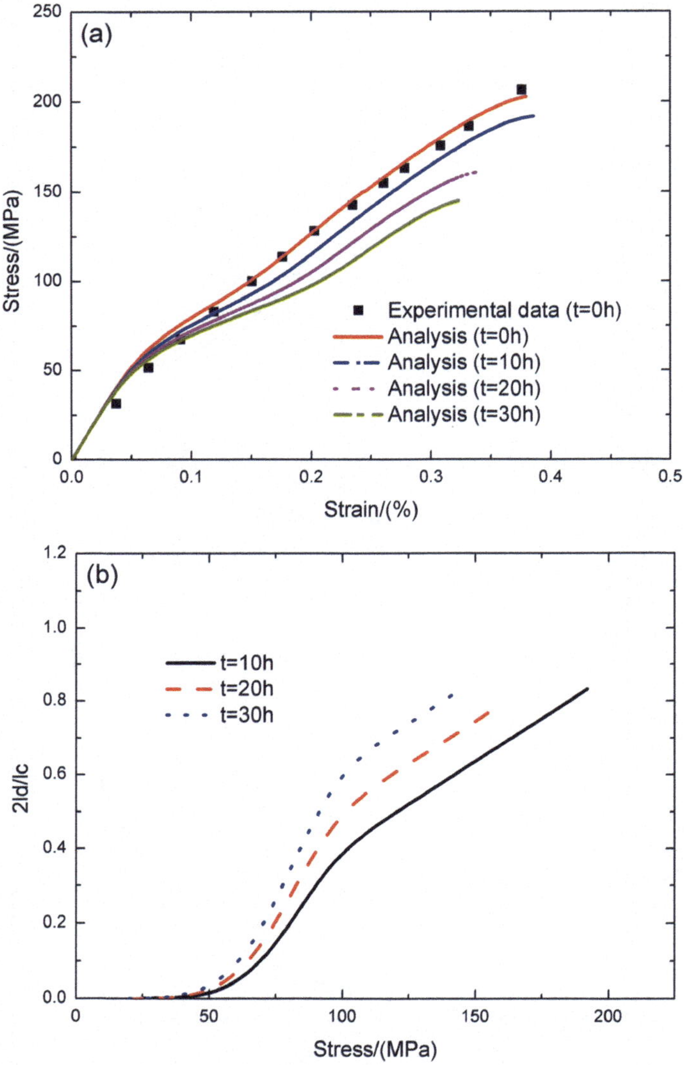

Fig. 4.9 **a** Experimental and predicted tensile stress-strain curves; **b** the interface debonded length versus the applied stress curves; **c** the interface oxidation ratio versus the applied stress curves; and **d** the broken fiber fraction versus applied stress curves of 3D C/SiC composite

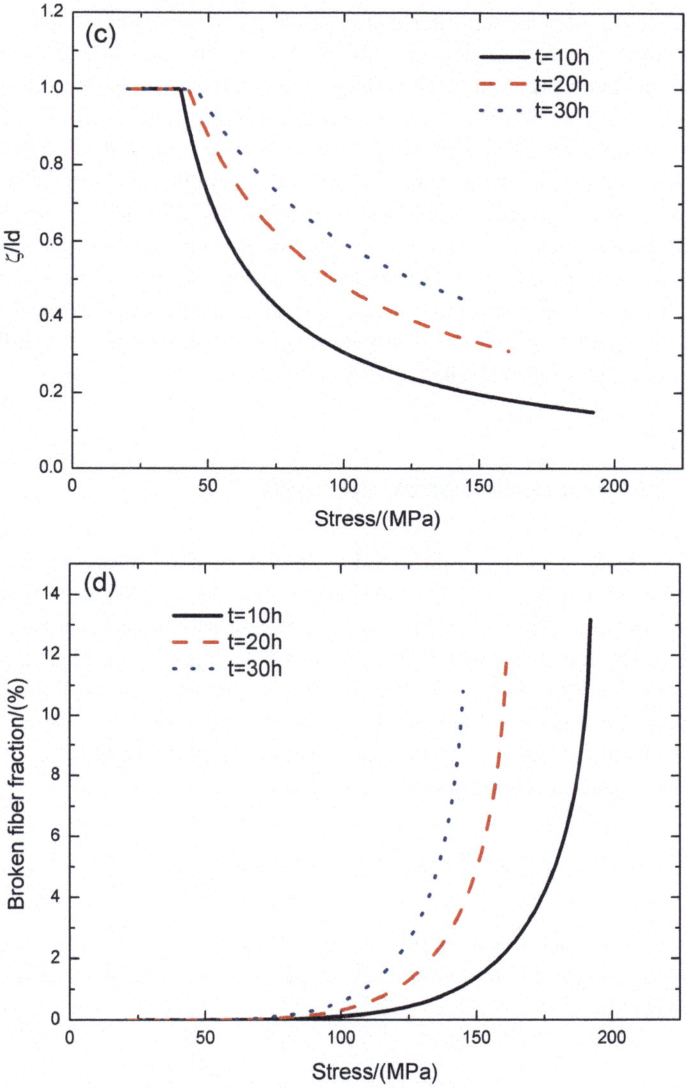

Fig. 4.9 (continued)

4.3 Cyclic-Dependent Tensile Damage and Fracture of Fiber-Reinforced Ceramic-Matrix Composites Subjected to Thermal Fatigue Loading

In this section, the cyclic-dependent tensile damage and fracture of fiber-reinforced CMCs subjected to thermal fatigue loading are investigated. The damage mechanisms of matrix cracking, fiber/matrix interface debonding and wear, interface and fiber oxidation, and fiber failure after thermal fatigue loading are considered in the tensile stress-strain behavior prediction. The relationships between the thermal fatigue temperature, applied cycle number, composite internal damage, and tensile damage and fracture are established. The effects of thermal fatigue temperature, thermal cycle number, fiber volume, interface shear stress, fiber strength, and fiber Weibull modulus on the tensile damage and fracture of fiber-reinforced CMCs subjected to the thermal fatigue loading are analyzed. The experimental tensile stress-strain curves of different C/SiC composites with and without thermal fatigue loading are predicted. When the thermal fatigue temperature and applied cycle number increase, the composite tensile strength and failure strain decrease and the interface debonding ratio, interface oxidation ratio, and broken fiber fraction increase at low applied stress level.

4.3.1 Cyclic-Dependent Stress Analysis

Under tensile loading of fiber-reinforced CMCs, the tensile stress-strain curve can be divided into four stages, i.e., (1) the linear-elastic stage, and the strain increases in direct proportion to the stress; (2) the damage stage with matrix cracking and fiber debonding at the interface, which makes the CMCs appear the characteristics of pseudo-plastic fracture and high toughness; (3) the damage stage with the saturation of matrix cracking and complete interface debonding; and (4) the fiber failure stage. Under thermal fatigue loading, the interface oxidation length can be determined using the following equation (Casas and Martinez-Esnaola 2003).

$$\zeta(t) = \varphi_1\left[1 - \exp\left(-\frac{\varphi_2 t}{b}\right)\right] \tag{4.30}$$

where b is a delay factor considering the deceleration of reduced oxygen activity and φ_1 and φ_2 are parameters dependent on temperature and described using the Arrhenius-type laws.

$$\varphi_1 = 7.021 \times 10^{-3} \times \exp\left(\frac{8231}{T}\right) \tag{4.31}$$

$$\varphi_2 = 227.1 \times \exp\left(-\frac{17{,}090}{T}\right) \tag{4.32}$$

where φ_1 is in mm and φ_2 in s^{-1}; φ_1 represents the asymptotic behavior for long times, which decreases with temperature; and the product $\varphi_1\varphi_2$ represents the initial oxidation rate, which is an increasing function of temperature.

Under the thermal fatigue loading, the fiber/matrix interface shear stress decreases due to the interface wear. The fiber/matrix interface debonding region can be divided into two regions, including:

(1) The interface oxidation region, i.e., $z \in [0, \zeta]$, the stress transfer between the fiber and the matrix is controlled by a sliding stress of $\tau_i(z) = \tau_f$.
(2) The interface wear region, i.e., $z \in [\zeta, l_d]$, the stress transfer between the fiber and the matrix is controlled by a sliding stress $\tau_i(z) = \tau_i(N)$.

Barabash et al. (2011) investigated the interface strength in NiAl-Mo composite from 3D X-ray microdiffraction and developed an approach to calculate the fiber/matrix interfacial strength. Basaran and Nie (2007) developed a thermodynamics-based interface damage model for particulate composite subjected to tensile loading. However, under cyclic loading, the interface wear is the main damage mechanisms for the interface shear stress degradation of fiber-reinforced CMCs. Evans (1997) obtained the fiber/matrix interface shear stress through the hysteresis loop analysis and developed an empirical model for the degradation of the fiber/matrix interface shear stress subjected to cyclic loading, as shown in Eq. (4.33). Li et al. (2017) investigated the tension–tension fatigue behavior of unidirectional SiC/Si_3N_4 composite and found that Evans's model can better describe the degradation of the fiber/matrix interface shear stress for fiber-reinforced CMCs.

$$(\tau_i(N) - \tau_s)/(\tau_0 - \tau_s) = (1 + b_0)\left(1 + b_0 N^j\right)^{-1} \tag{4.33}$$

where τ_0 denotes the initial fiber/matrix interface shear stress; τ_s denotes the steady-state fiber/matrix interface shear stress; b_0 is a coefficient; and j is an exponent which determines the rate at which the interface shear stress drops with the number of cycle N.

When the damages of matrix cracking and fiber/matrix interface debonding occur in fiber-reinforced CMCs, the shear-lag model can be used to analyze the microstress distributions in the interface oxidation region, interface slip region, and interface bonded region. The distributions of the fiber and matrix axial stress distribution, and the fiber/matrix interface shear stress can be determined using the following equations.

$$\sigma_f(z) = \begin{cases} \dfrac{\sigma}{V_f} - \dfrac{2\tau_f}{r_f} z, & z \in [0, \zeta(t)] \\ \dfrac{\sigma}{V_f} - \dfrac{2\tau_f}{r_f}\zeta(t) - \dfrac{2\tau_i(N)}{r_f}(z - \zeta(t)), & z \in [\zeta(t), l_d] \\ \sigma_{fo} + \left[\dfrac{V_m}{V_f}\sigma_{mo} - 2\dfrac{\tau_f}{r_f}\zeta(t) - 2\dfrac{\tau_i(N)}{r_f}(l_d - \zeta(t))\right]\exp\left(-\rho\dfrac{z - l_d}{r_f}\right), & z \in \left[l_d, \dfrac{l_c}{2}\right] \end{cases} \tag{4.34}$$

$$\sigma_m(z) = \begin{cases} 2\dfrac{V_f}{V_m}\dfrac{\tau_f}{r_f} z, & z \in [0, \zeta(t)] \\ 2\dfrac{V_f}{V_m}\dfrac{\tau_f}{r_f}\zeta(t) + 2\dfrac{V_f}{V_m}\dfrac{\tau_i(N)}{r_f}(z - \zeta(t)), & z \in [\zeta(t), l_d] \\ \sigma_{mo} - \left[\sigma_{mo} - 2\dfrac{V_f}{V_m}\dfrac{\tau_f}{r_f}\zeta(t) - 2\dfrac{V_f}{V_m}\dfrac{\tau_i(N)}{r_f}(l_d - \zeta(t))\right]\exp\left(-\rho\dfrac{z - l_d}{r_f}\right), & z \in \left[l_d, \dfrac{l_c}{2}\right] \end{cases} \tag{4.35}$$

$$\tau_{\rm i}(z) = \begin{cases} \tau_{\rm f}, & z \in [0, \zeta(t)] \\ \tau_{\rm i}(N), & z \in [\zeta(t), l_{\rm d}] \\ \frac{\rho}{2}\left[\frac{V_{\rm m}}{V_{\rm f}}\sigma_{\rm mo} - 2\frac{\tau_{\rm f}}{r_{\rm f}}\zeta(t) - \frac{2\tau_{\rm i}(N)}{r_{\rm f}}(l_{\rm d} - \zeta(t))\right]\exp\left(-\rho\frac{z - l_{\rm d}}{r_{\rm f}}\right), & z \in \left[l_{\rm d}, \frac{l_{\rm c}}{2}\right] \end{cases} \tag{4.36}$$

where $V_{\rm f}$ and $V_{\rm m}$ denote the fiber and the matrix volume fraction, respectively; ρ denotes the shear-lag model parameter; and $\sigma_{\rm fo}$ and $\sigma_{\rm mo}$ denote the fiber and matrix axial stress in the interface bonded region, respectively.

$$\sigma_{\rm fo} = \frac{E_{\rm f}}{E_{\rm c}}\sigma + E_{\rm f}(\alpha_{\rm c} - \alpha_{\rm f})\Delta T \tag{4.37}$$

$$\sigma_{\rm mo} = \frac{E_{\rm m}}{E_{\rm c}}\sigma + E_{\rm m}(\alpha_{\rm c} - \alpha_{\rm m})\Delta T \tag{4.38}$$

where $E_{\rm f}$, $E_{\rm m}$, and $E_{\rm c}$ denote the fiber, matrix, and composite elastic modulus, respectively; $\alpha_{\rm f}$, $\alpha_{\rm m}$, and $\alpha_{\rm c}$ denote the fiber, matrix, and composite thermal expansion coefficient, respectively; and ΔT denotes the temperature difference between the fabricated temperature T_0 and testing temperature T_1 ($\Delta T = T_1 - T_0$).

When the fiber failure occurs, the fiber axial stress in the interface debonding and bonded region can be determined using the following equation.

$$\sigma_{\rm f}(z) = \begin{cases} \Phi - \frac{2\tau_{\rm f}}{r_{\rm f}}z, & z \in [0, \zeta(t)] \\ \Phi - \frac{2\tau_{\rm f}}{r_{\rm f}}\zeta(t) - \frac{2\tau_{\rm i}(N)}{r_{\rm f}}[z - \zeta(t)], & z \in [\zeta(t), l_{\rm d}] \\ \sigma_{\rm fo} + \left\{\Phi - \frac{2\tau_{\rm f}}{r_{\rm f}}\zeta(t) - \frac{2\tau_{\rm i}(N)}{r_{\rm f}}[l_{\rm d} - \zeta(t)] - \sigma_{\rm fo}\right\}\exp\left(-\rho\frac{z - l_{\rm d}}{r_{\rm f}}\right), & z \in \left[l_{\rm d}, \frac{l_{\rm c}}{2}\right] \end{cases} \tag{4.39}$$

where Φ denotes the intact fiber stress.

4.3.2 *Cyclic-Dependent Interface Debonding*

When the matrix cracking propagates to the fiber/matrix interface, the fracture mechanics approach is used to determine the interface debonding length (Gao et al. 1988).

$$\xi_{\rm d} = -\frac{F}{4\pi r_{\rm f}}\frac{\partial w_{\rm f}(z=0)}{\partial l_{\rm d}} - \frac{1}{2}\int_0^{l_{\rm d}} \tau_{\rm i}\frac{\partial v(z)}{\partial l_{\rm d}}{\rm d}z \tag{4.40}$$

where ξ_d denotes the fiber/matrix interface debond energy; $F(=\pi r_f^2\sigma/V_f)$ denotes the fiber stress on the matrix cracking plane; $w_f(z=0)$ denotes the fiber axial displacement at the matrix cracking plane; and $v(z)$ denotes the relative displacement between the fiber and the matrix.

$$\begin{aligned} w_f(z) &= \int_x^{l_c/2} \frac{\sigma_f(z)}{E_f} dz \\ &= \frac{\sigma}{V_f E_f}(l_d - z) - \frac{\tau_f}{r_f E_f}\left[2\zeta(t)l_d - \zeta^2(t) - z^2\right] \\ &- \frac{\tau_i(N)}{r_f E_f}[l_d - \zeta(t)]^2 + \frac{\sigma_{fo}}{E_f}\left(\frac{l_c}{2} - l_d\right) \\ &+ \frac{r_f}{\rho E_f}\left[\frac{V_m}{V_f}\sigma_{mo} - \frac{2\tau_f}{r_f}\zeta(t) - \frac{2\tau_i(N)}{r_f}(l_d - \zeta(t))\right] \\ &\left[1 - \exp\left(-\rho\frac{l_c/2 - l_d}{r_f}\right)\right] \end{aligned} \tag{4.41}$$

$$\begin{aligned} w_m(z) &= \int_x^{l_c/2} \frac{\sigma_m(z)}{E_m} dz \\ &= \frac{V_f \tau_f}{r_f V_m E_m}\left(2\zeta(t)l_d - \zeta^2(t) - z^2\right) + \frac{V_f \tau_i(N)}{r_f V_m E_m}(l_d - \zeta(t))^2 \\ &+ \frac{\sigma_{mo}}{E_m}\left(\frac{l_c}{2} - l_d\right) \\ &- \frac{r_f}{\rho E_m}\left[\sigma_{mo} - 2\frac{V_f \tau_f}{r_f V_m}\zeta(t) - 2\frac{V_f \tau_i(N)}{r_f V_m}(l_d - \zeta(t))\right] \\ &\left[1 - \exp\left(-\rho\frac{l_c/2 - l_d}{r_f}\right)\right] \end{aligned} \tag{4.42}$$

The relative displacement $v(z)$ between the fiber and the matrix is described using the following equation.

$$\begin{aligned} v(z) &= |w_f(z) - w_m(z)| \\ &= \frac{\sigma}{V_f E_f}(l_d - x) - \frac{E_c \tau_f}{r_f V_m E_m E_f}\left(2\zeta(t)l_d - \zeta^2(t) - x^2\right) \\ &- \frac{E_c \tau_i(N)}{r_f V_m E_m E_f}(l_d - \zeta(t))^2 \\ &+ \frac{r_f E_c}{\rho V_m E_m E_f}\left[\sigma_{mo} - 2\frac{\tau_f}{r_f}\zeta(t) - 2\frac{\tau_i(N)}{r_f}(l_d - \zeta(t))\right] \\ &\left[1 - \exp\left(-\rho\frac{l_c/2 - l_d}{r_f}\right)\right] \end{aligned} \tag{4.43}$$

Substituting $w_f(x=0)$ and $v(z)$ into Eq. (4.16), it leads to the following equation.

$$\begin{aligned}&\frac{E_c\tau_i^2}{r_f V_m E_m E_f}(l_d-\zeta(t))^2+\frac{E_c\tau_i^2}{\rho V_m E_m E_f}(l_d-\zeta(t))\\&-\frac{\tau_i\sigma}{V_f E_f}(l_d-\zeta(t))+\frac{2E_c\tau_f\tau_i}{r_f V_m E_m E_f}\zeta(t)(l_d-\zeta(t))\\&-\frac{r_f\tau_i\sigma}{2\rho V_f E_f}+\frac{E_c\tau_f^2}{r_f V_m E_m E_f}\zeta^2(t)+\frac{E_c\tau_f\tau_i}{\rho V_m E_m E_f}\zeta(t)\\&-\frac{\tau_f\sigma}{V_f E_f}\zeta(t)+\frac{r_f V_m E_m\sigma^2}{4V_f^2 E_f E_c}-\xi_d=0\end{aligned}\tag{4.44}$$

Solving Eq. (4.44), the fiber/matrix interface debonding length considering interface damage after thermal fatigue loading can be determined using the following equation.

$$l_d=\left(1-\frac{\tau_f}{\tau_i(N)}\right)\zeta(t)+\frac{r_f}{2}\left(\frac{V_m E_m\sigma}{V_f E_c\tau_i(N)}-\frac{1}{\rho}\right)-\sqrt{\left(\frac{r_f}{2\rho}\right)^2+\frac{r_f V_m E_m E_f}{E_c\tau_i^2}\xi_d}\tag{4.45}$$

4.3.3 Cyclic-Dependent Fiber Failure

Basaran and Nie (2004) developed the thermodynamic framework for damage mechanics of solid materials, and the entropy production is used for the damage evolution in systems. In the present analysis, the two-parameter Weibull model is adopted to describe the fiber strength distribution, and the global load sharing criterion is used to determine the stress distributions between the intact and fracture fibers (Curtin 1991).

$$\frac{\sigma}{V_f}=\Phi(1-P)+\frac{2\tau_i(N)}{r_f}\langle L\rangle P\tag{4.46}$$

where $\langle L\rangle$ denotes the average fiber pullout length and P denotes the fiber failure probability.

$$P=1-\exp\left[-\left(\frac{\Phi}{\sigma_c}\right)^{m+1}\right]\tag{4.47}$$

where m denotes the fiber Weibull modulus and σ_c denotes the fiber characteristic strength of a length δ_c of fiber.

$$\sigma_c = \left(\frac{l_o \sigma_0^{m_f} \tau_i(N)}{r_f}\right)^{\frac{1}{m+1}}, \quad \delta_c = \left(\frac{\sigma_0 r_f l_o^{\frac{1}{m}}}{\tau_i(N)}\right)^{\frac{m}{m+1}} \tag{4.48}$$

The time-dependent fiber strength of $\sigma_0(t)$ can be determined using the following equation (Lara-Curzio 1999).

$$\sigma_0(t) = \begin{cases} \sigma_0, & t \leq \frac{1}{k}\left(\frac{K_{IC}}{Y\sigma_0}\right)^4 \\ \frac{K_{IC}}{Y\sqrt[4]{kt}}, & t > \frac{1}{k}\left(\frac{K_{IC}}{Y\sigma_0}\right)^4 \end{cases} \tag{4.49}$$

When the fiber failure probability approaches the critical value of q, the composite tensile fractures (Curtin 1991).

$$q = \frac{2}{m+2} \tag{4.50}$$

The composite tensile strength is given by the following equation (Curtin 1991).

$$\sigma_{UTS} = V_f \sigma_c \left(\frac{2}{m+2}\right)^{\frac{1}{m+1}} \left(\frac{m+1}{m+2}\right) \tag{4.51}$$

4.3.4 Cyclic-Dependent Tensile Constitutive Relationship

For the fiber-reinforced CMCs without damage, the composite strain can be determined using the following equation.

$$\varepsilon_c = \sigma / E_c \tag{4.52}$$

When damage forms inside of CMCs, the composite strain can be determined using the following equation.

$$\varepsilon_c = \frac{2}{E_f l_c} \int_{l_c/2} \sigma_f(z) \mathrm{d}z - (\alpha_c - \alpha_f)\Delta T \tag{4.53}$$

When the matrix cracking and fiber/matrix interface debonding occur, the composite strain considering interface damage after thermal fatigue loading can be determined using the following equation.

$$
\begin{aligned}
\varepsilon_{\mathrm{c}} = & \frac{2\sigma}{V_{\mathrm{f}}E_{\mathrm{f}}}\frac{l_{\mathrm{d}}}{l_{\mathrm{c}}} + \frac{2\tau_{\mathrm{f}}}{r_{\mathrm{f}}E_{\mathrm{f}}l_{\mathrm{c}}}\zeta^{2}(t) - \frac{4\tau_{\mathrm{f}}}{r_{\mathrm{f}}E_{\mathrm{f}}}\frac{l_{\mathrm{d}}}{l_{\mathrm{c}}}\zeta(t) \\
& - \frac{2\tau_{\mathrm{i}}(N)}{r_{\mathrm{f}}E_{\mathrm{f}}l_{\mathrm{c}}}[l_{\mathrm{d}} - \zeta(t)]^{2} + \frac{2\sigma_{\mathrm{fo}}}{E_{\mathrm{f}}l_{\mathrm{c}}}\left(\frac{l_{\mathrm{c}}}{2} - l_{\mathrm{d}}\right) \\
& + \frac{2r_{\mathrm{f}}}{\rho E_{\mathrm{f}}l_{\mathrm{c}}}\left\{\frac{\sigma}{V_{\mathrm{f}}} - \frac{2\tau_{\mathrm{f}}}{r_{\mathrm{f}}}\zeta(t) - \frac{2\tau_{\mathrm{i}}(N)}{r_{\mathrm{f}}}[l_{\mathrm{d}} - \zeta(t)] - \sigma_{\mathrm{fo}}\right\} \\
& \times\left[1 - \exp\left(-\rho\frac{l_{\mathrm{c}}/2 - l_{\mathrm{d}}}{r_{\mathrm{f}}}\right)\right] - (\alpha_{\mathrm{c}} - \alpha_{\mathrm{f}})\Delta T
\end{aligned}
\tag{4.54}
$$

When the fiber failure occurs, the composite strain considering interface damage after thermal fatigue loading can be determined using the following equation.

$$
\begin{aligned}
\varepsilon_{\mathrm{c}} = & \frac{\Phi}{E_{\mathrm{f}}}\frac{2l_{\mathrm{d}}}{l_{\mathrm{c}}} + \frac{2\tau_{\mathrm{f}}}{r_{\mathrm{f}}E_{\mathrm{f}}l_{\mathrm{c}}}\zeta^{2}(t) - \frac{4\tau_{\mathrm{f}}l_{\mathrm{d}}}{r_{\mathrm{f}}E_{\mathrm{f}}l_{\mathrm{c}}}\zeta(t) \\
& - \frac{2\tau_{\mathrm{i}}(N)}{r_{\mathrm{f}}E_{\mathrm{f}}l_{\mathrm{c}}}(l_{\mathrm{d}} - \zeta(t))^{2} \\
& + \frac{2\sigma_{\mathrm{fo}}}{E_{\mathrm{f}}l_{\mathrm{c}}}\left(\frac{l_{\mathrm{c}}}{2} - l_{\mathrm{d}}\right) \\
& + \frac{2r_{\mathrm{f}}}{\rho E_{\mathrm{f}}l_{\mathrm{c}}}\left\{\Phi - \frac{2\tau_{\mathrm{f}}}{r_{\mathrm{f}}}\zeta(t) - \frac{2\tau_{\mathrm{i}}(N)}{r_{\mathrm{f}}}[l_{\mathrm{d}} - \zeta(t)] - \sigma_{\mathrm{fo}}\right\} \\
& \times\left[1 - \exp\left(-\rho\frac{l_{\mathrm{c}}/2 - l_{\mathrm{d}}}{r_{\mathrm{f}}}\right)\right] - (\alpha_{\mathrm{c}} - \alpha_{\mathrm{f}})\Delta T
\end{aligned}
\tag{4.55}
$$

4.3.5 Results and Discussion

The effects of the thermal fatigue temperature, thermal cycle number, fiber volume, fiber/matrix interface shear stress, fiber strength, and fiber Weibull modulus on tensile damage and fracture after thermal fatigue loading are analyzed. The C/SiC composite is used for the case analysis, and the material properties are given by: $V_{\mathrm{f}} = 40\%$, $E_{\mathrm{f}} = 230$ GPa, $E_{\mathrm{m}} = 350$ GPa, $r_{\mathrm{f}} = 3.5$ μm, $m = 6$, $\sigma_{\mathrm{R}} = 100$ MPa, $l_{\mathrm{sat}} = 150$ μm, $\alpha_{\mathrm{f}} = 0\times10^{-6}$/K, $\alpha_{\mathrm{m}} = 4.6 \times 10^{-6}$/K, $\Delta T = -1000$ °C, $\xi_{\mathrm{d}} = 0.1$ J/m^2, $\tau_{\mathrm{i}} = 20$ MPa, $\tau_{\mathrm{f}} = 1$ MPa, $\sigma_{\mathrm{c}} = 1.6$ GPa, and $m = 5$.

4.3.5.1 Effect of the Interface Damage on Cyclic-Dependent Tensile Damage and Fracture for Different Thermal Fatigue Temperatures

The effect of the thermal fatigue temperature (i.e., $T_{\max} = 600$, 800, 1000, and 1200 °C) on the cyclic-dependent tensile stress-strain curve, fiber/matrix interface debonding and interface oxidation, and fiber failure curves of C/SiC composite after $N = 100$ thermal fatigue cycles is shown in Fig. 4.10. When the thermal fatigue temperature increases, the composite tensile strength and failure strain decrease, the interface debonding ratio, interface oxidation ratio, and broken fiber fraction all increase at low applied stress level.

When $T_{max} = 600$ °C and $N = 100$, the composite tensile strength is $\sigma_{UTS} =$ 445 MPa with the failure strain of $\varepsilon_f = 0.57\%$, the fiber/matrix interface debonding ratio increases to $2l_d/l_c = 0.67$, the fiber/matrix interface oxidation ratio decreases to $\zeta/l_d = 0.31\%$, and the critical fiber broken fraction is $P = 13.3\%$. When $T_{max} =$ 800 °C and $N = 100$, the composite tensile strength is $\sigma_{UTS} = 342$ MPa with the failure strain of $\varepsilon_f = 0.41\%$, the fiber/matrix interface debonding ratio increases to $2l_d/l_c = 0.47$, the fiber/matrix interface oxidation ratio decreases to $\zeta/l_d = 2.9\%$, and the critical fiber broken fraction is $P = 12.1\%$. When $T_{max} = 1000$ °C and $N = 100$, the composite tensile strength is $\sigma_{UTS} = 249$ MPa with the failure strain of $\varepsilon_f = 0.29\%$, the fiber/matrix interface debonding ratio increases to $2l_d/l_c = 0.31$, the fiber/matrix interface oxidation ratio decreases to $\zeta/l_d = 16\%$, and the critical fiber broken fraction is $P = 13.1\%$. When $T_{max} = 1200$ °C and $N = 100$, the composite tensile strength is $\sigma_{UTS} = 197$ MPa with the failure strain of $\varepsilon_f = 0.22\%$, the fiber/matrix interface debonding ratio increases to $2l_d/l_c = 0.25$, the fiber/matrix interface oxidation ratio decreases to $\zeta/l_d = 46\%$, and the critical fiber broken fraction is $P = 12.1\%$.

4.3.5.2 Effect of the Interface Damage on Cyclic-Dependent Tensile Damage and Fracture for Different Thermal Fatigue Cycle Numbers

The effect of thermal fatigue cycle number (i.e., $N = 10$, 50, 100, and 500) on the tensile stress-strain curve, fiber/matrix interface debonding and interface oxidation, and fiber failure curves of C/SiC composite under $T_{max} = 800$ °C thermal fatigue cycles is shown in Fig. 4.11. When the thermal fatigue cycle number increases, the composite tensile strength and failure strain decrease, and the interface debonding ratio, interface oxidation ratio, and broken fiber fraction all increase at low applied stress level.

When $N = 10$ and $T_{max} = 800$ °C, the composite tensile strength is $\sigma_{UTS} =$ 445 MPa with the failure strain of $\varepsilon_f = 0.5\%$, the fiber/matrix interface debonding ratio increases to $2l_d/l_c = 0.4$, the fiber/matrix interface oxidation ratio decreases to $\zeta/l_d = 0.3\%$, and the critical fiber broken fraction is $P = 13.3\%$. When $N = 50$ and $T_{max} = 800$ °C, the composite tensile strength is $\sigma_{UTS} = 407$ MPa with the failure strain of $\varepsilon_f = 0.48\%$, the fiber/matrix interface debonding ratio increases to $2l_d/l_c = 0.5$, the fiber/matrix interface oxidation ratio decreases to $\zeta/l_d = 1.3\%$, and the critical fiber broken fraction is $P = 12.5\%$. When $N = 100$ and $T_{max} = 800$ °C, the composite tensile strength is $\sigma_{UTS} = 342$ MPa with the failure strain of $\varepsilon_f = 0.41\%$, the fiber/matrix interface debonding ratio increases to $2l_d/l_c = 0.47$, the fiber/matrix interface oxidation ratio decreases to $\zeta/l_d = 2.9\%$, and the critical fiber broken fraction is $P = 12.1\%$. When $N = 500$ and $T_{max} = 800$ °C, the composite tensile strength

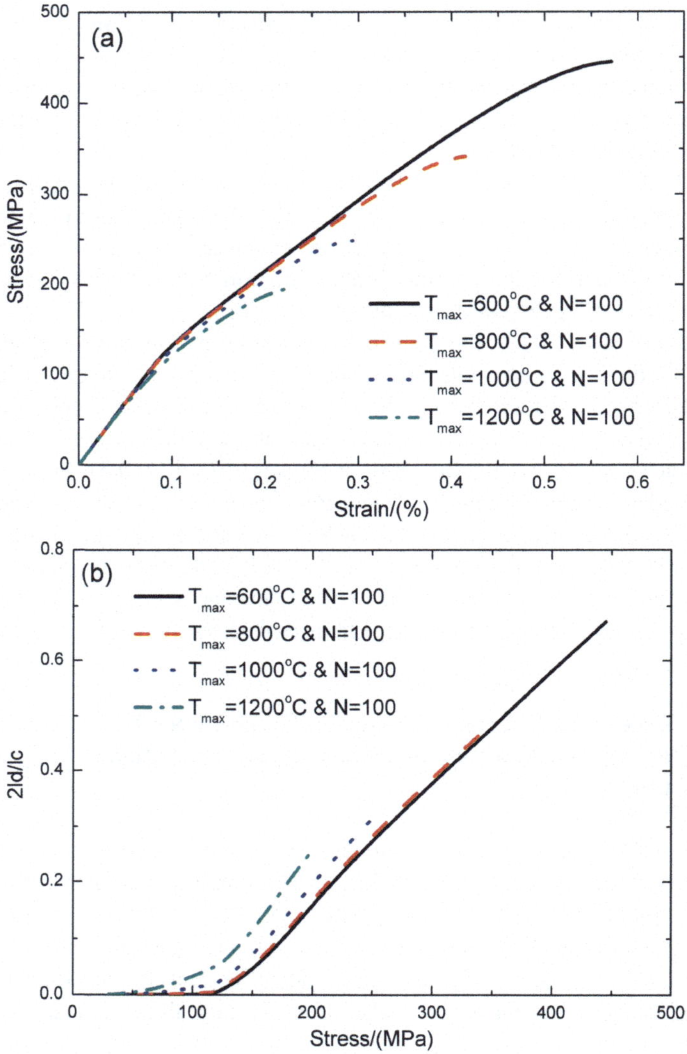

Fig. 4.10 Effect of thermal fatigue temperature on **a** the tensile stress-strain curves; **b** the fiber/matrix interface debonding length versus the applied stress curves; **c** the fiber/matrix interface oxidation ratio versus the applied stress curves; and **d** the broken fiber fraction versus the applied stress curves of C/SiC composite

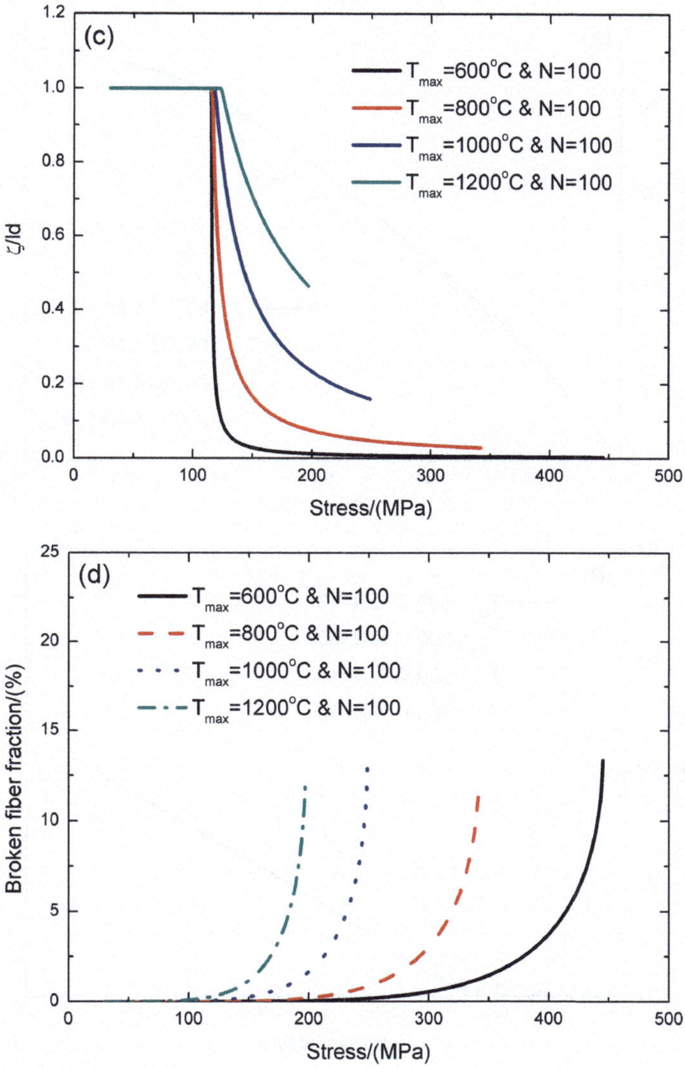

Fig. 4.10 (continued)

is $\sigma_{\text{UTS}} = 229$ MPa with the failure strain of $\varepsilon_{\text{f}} = 0.28\%$, the fiber/matrix interface debonding ratio increases to $2l_{\text{d}}/l_{\text{c}} = 0.37$, the fiber/matrix interface oxidation ratio decreases to $\zeta/l_{\text{d}} = 17.8\%$, and the critical fiber broken fraction is $P = 12.8\%$.

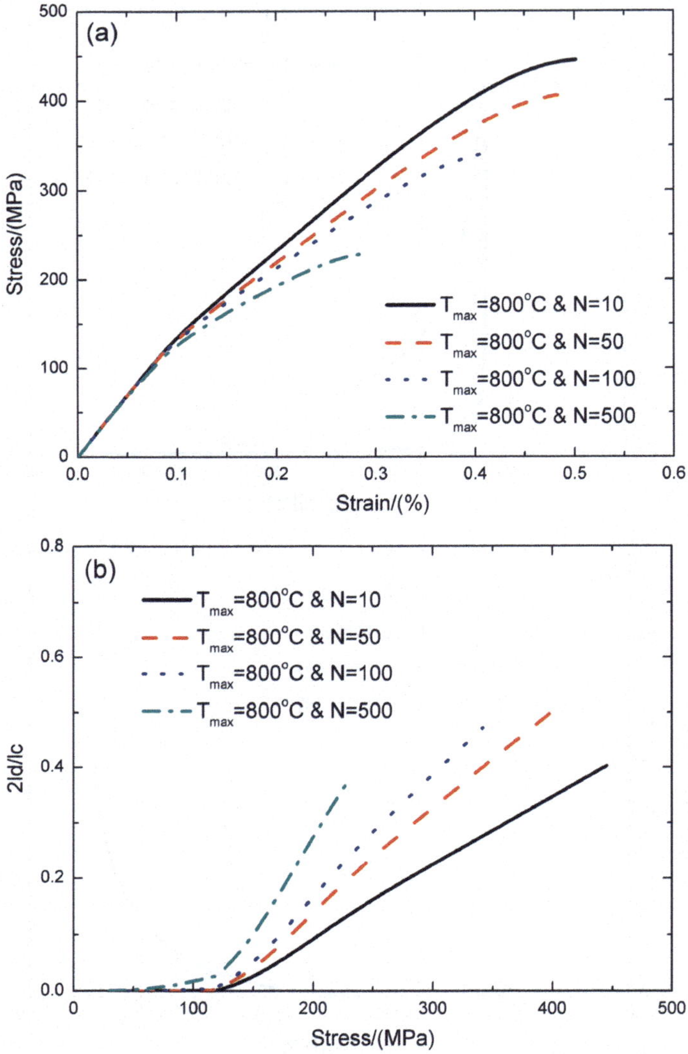

Fig. 4.11 Effect of thermal fatigue cyclic number on **a** the tensile stress-strain curves; **b** the fiber/matrix interface debonding length versus the applied stress curves; **c** the fiber/matrix interface oxidation ratio versus the applied stress curves; and **d** the broken fiber fraction versus the applied stress curves of C/SiC composite

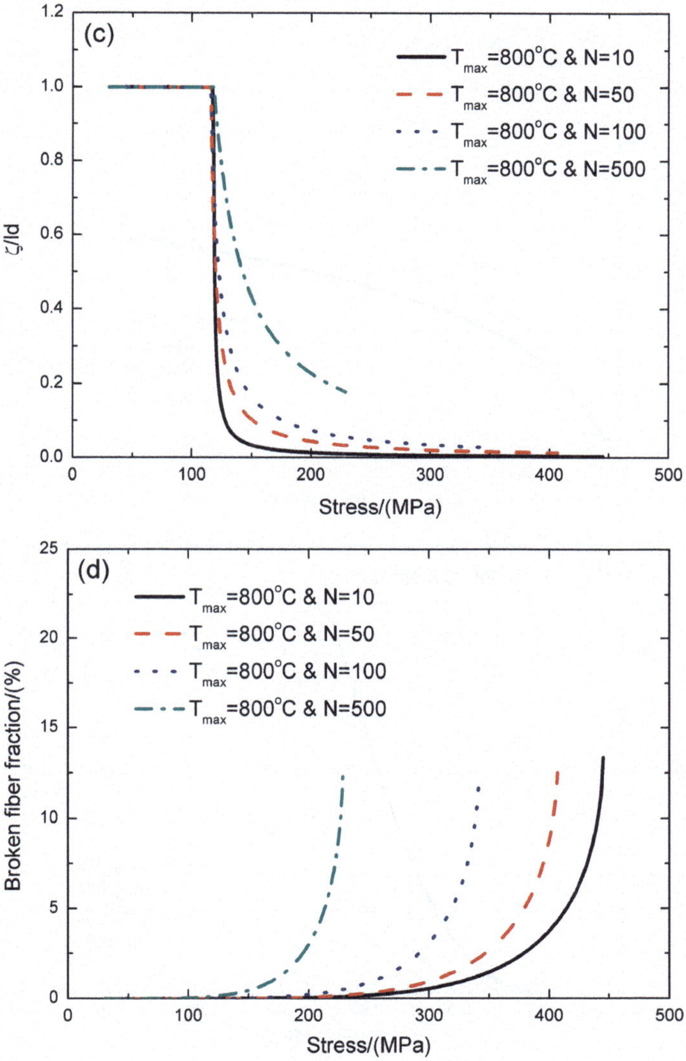

Fig. 4.11 (continued)

4.3.5.3 Effect of the Fiber Volume on Cyclic-Dependent Tensile Damage and Fracture

The effect of fiber volume (i.e., $V_f = 20, 25, 30$, and 35%) on the tensile stress-strain curve, fiber/matrix interface debonding and interface oxidation, and fiber failure curves of C/SiC composite under T_{max} = 800 °C and N = 100 thermal fatigue cycles is shown in Fig. 4.12. When the fiber volume increases, the composite tensile strength increases, the failure strain decreases, and the interface debonding ratio,

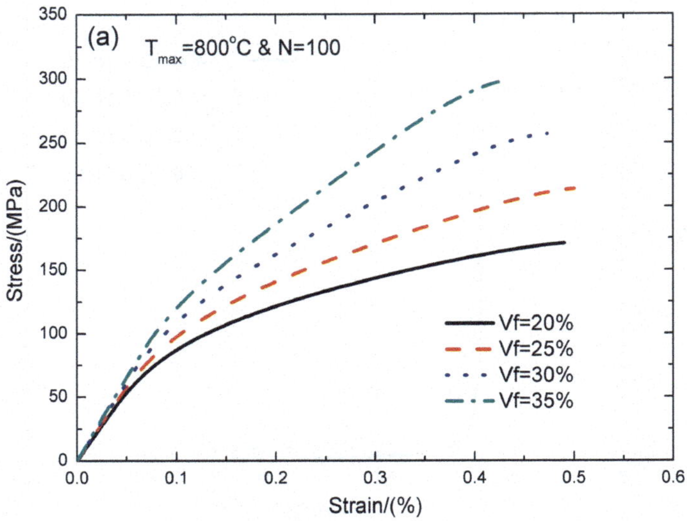

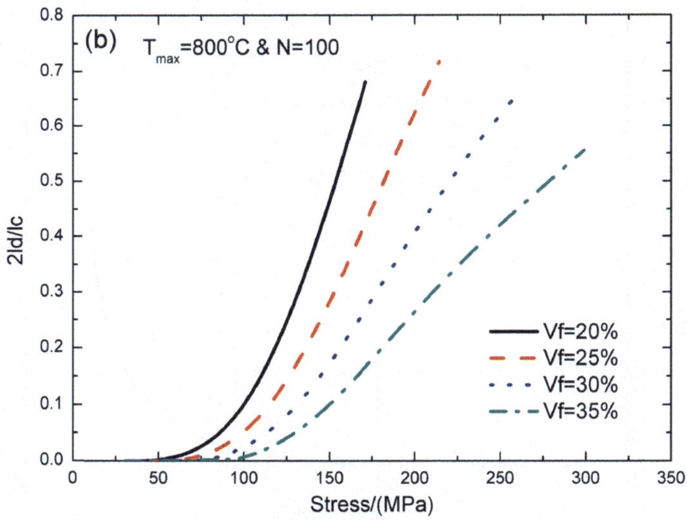

Fig. 4.12 Effect of fiber volume on **a** the tensile stress-strain curves; **b** the fiber/matrix interface debonding length versus the applied stress curves; **c** the fiber/matrix interface oxidation ratio versus the applied stress curves; and **d** the broken fiber fraction versus the applied stress curves of C/SiC composite

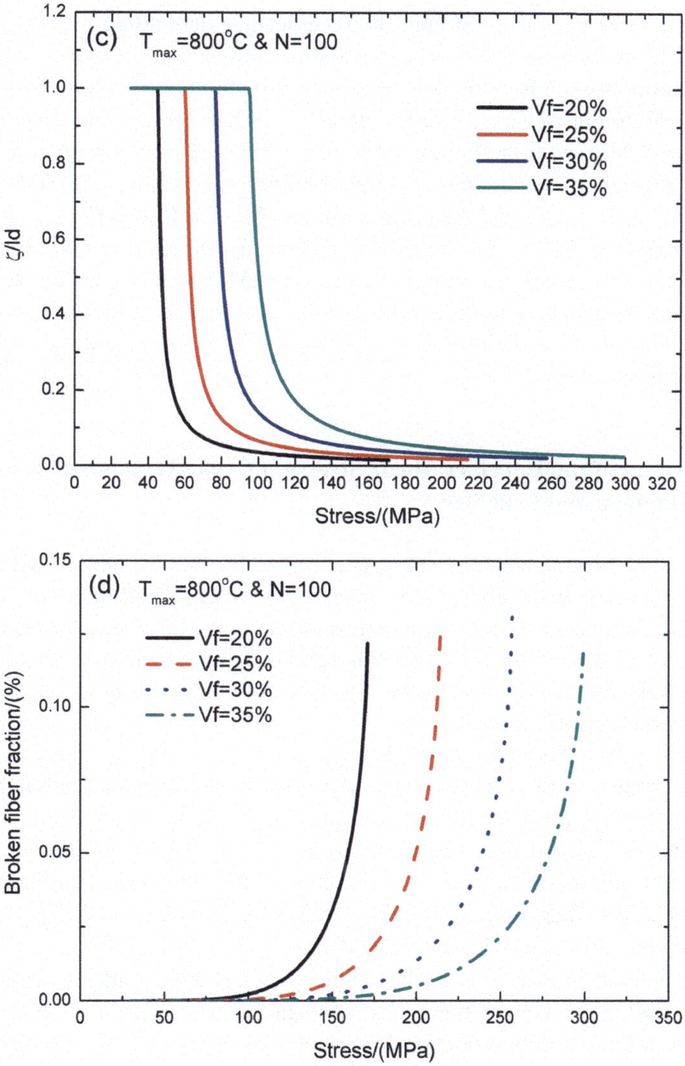

Fig. 4.12 (continued)

interface oxidation ratio, and broken fiber fraction all decrease at low applied stress level.

When the fiber volume is $V_f = 20\%$ under thermal fatigue of $T_{max} = 800\ ^\circ C$ and $N = 100$, the composite tensile strength is $\sigma_{UTS} = 171$ MPa with the failure strain of $\varepsilon_f = 0.49\%$, the fiber/matrix interface debonding ratio increases to $2l_d/l_c = 0.68$, the fiber/matrix interface oxidation ratio decreases to $\zeta/l_d = 1.5\%$, and the critical fiber broken fraction is $P = 12.1\%$. When $V_f = 25\%$ under thermal fatigue of $T_{max} = 800\ ^\circ C$ and $N = 100$, the composite tensile strength is $\sigma_{UTS} = 214$ MPa with the failure strain of $\varepsilon_f = 0.5\%$, the fiber/matrix interface debonding ratio increases to $2l_d/l_c = 0.71$, the fiber/matrix interface oxidation ratio decreases to $\zeta/l_d = 1.8\%$, and the critical fiber broken fraction is $P = 12.7\%$. When the fiber volume is $V_f = 30\%$ under thermal fatigue of $T_{max} = 800$ °C and $N = 100$, the composite tensile strength is $\sigma_{UTS} = 257$ MPa with the failure strain of $\varepsilon_f = 0.47\%$, the fiber/matrix interface debonding ratio increases to $2l_d/l_c = 0.64$, the fiber/matrix interface oxidation ratio decreases to $\zeta/l_d = 2.1\%$, and the critical fiber broken fraction is $P = 13.4\%$. When the fiber volume is $V_f = 35\%$ under thermal fatigue of $T_{max} = 800$ °C and $N = 100$, the composite tensile strength is $\sigma_{UTS} = 299$ MPa with the failure strain of $\varepsilon_f = 0.43\%$, the fiber/matrix interface debonding ratio increases to $2l_d/l_c = 0.55$, the fiber/matrix interface oxidation ratio decreases to $\zeta/l_d = 2.4\%$, and the critical fiber broken fraction is $P = 11.8\%$.

4.3.5.4 Effect of the Interface Shear Stress on Cyclic-Dependent Tensile Damage and Fracture

The effect of the interface shear stress (i.e., $\tau_0 = 10$, 20, 30, and 40 MPa) on the tensile stress-strain curve, fiber/matrix interface debonding and interface oxidation, and fiber failure curves of C/SiC composite under $T_{max} = 800$ °C and $N = 100$ thermal fatigue cycles is shown in Fig. 4.13. When the interface shear stress increases, the composite failure strain decreases, the interface debonding ratio decreases, and the interface oxidation ratio increases.

When $\tau_0 = 10$ MPa under thermal fatigue of $T_{max} = 800$ °C and $N = 100$, the composite failure strain is $\varepsilon_f = 0.51\%$, the fiber/matrix interface debonding ratio increases to $2l_d/l_c = 0.86$, the fiber/matrix interface oxidation ratio decreases to $\zeta/l_d = 1.5\%$, and the critical fiber broken fraction is $P = 12.1\%$. When $\tau_0 = 20$ MPa under thermal fatigue of $T_{max} = 800$ °C and $N = 100$, the composite failure strain is $\varepsilon_f = 0.41\%$, the fiber/matrix interface debonding ratio increases to $2l_d/l_c = 0.47$, the fiber/matrix interface oxidation ratio decreases to $\zeta/l_d = 2.9\%$, and the critical fiber broken fraction is $P = 12.1\%$. When $\tau_0 = 30$ MPa under thermal fatigue of $T_{max} = 800$ °C and $N = 100$, the composite failure strain is $\varepsilon_f = 0.37\%$, the fiber/matrix interface debonding ratio increases to $2l_d/l_c = 0.32$, the fiber/matrix interface oxidation ratio decreases to $\zeta/l_d = 4.2\%$, and the critical fiber broken fraction is $P = 12.1\%$. When $\tau_0 = 40$ MPa under thermal fatigue of $T_{max} = 800$ °C and $N = 100$, the composite failure strain is $\varepsilon_f = 0.36\%$, the fiber/matrix interface debonding ratio increases to $2l_d/l_c = 0.25$, the fiber/matrix interface oxidation ratio decreases to $\zeta/l_d = 5.5\%$, and the critical fiber broken fraction is $P = 12.1\%$.

4.3.5.5 Effect of the Fiber Strength on Cyclic-Dependent Tensile Damage and Fracture

The effect of fiber strength (i.e., $\sigma_0 = 1$ and 1.5 GPa) on the tensile stress-strain curve, fiber/matrix interface debonding and interface oxidation, and fiber failure curves of

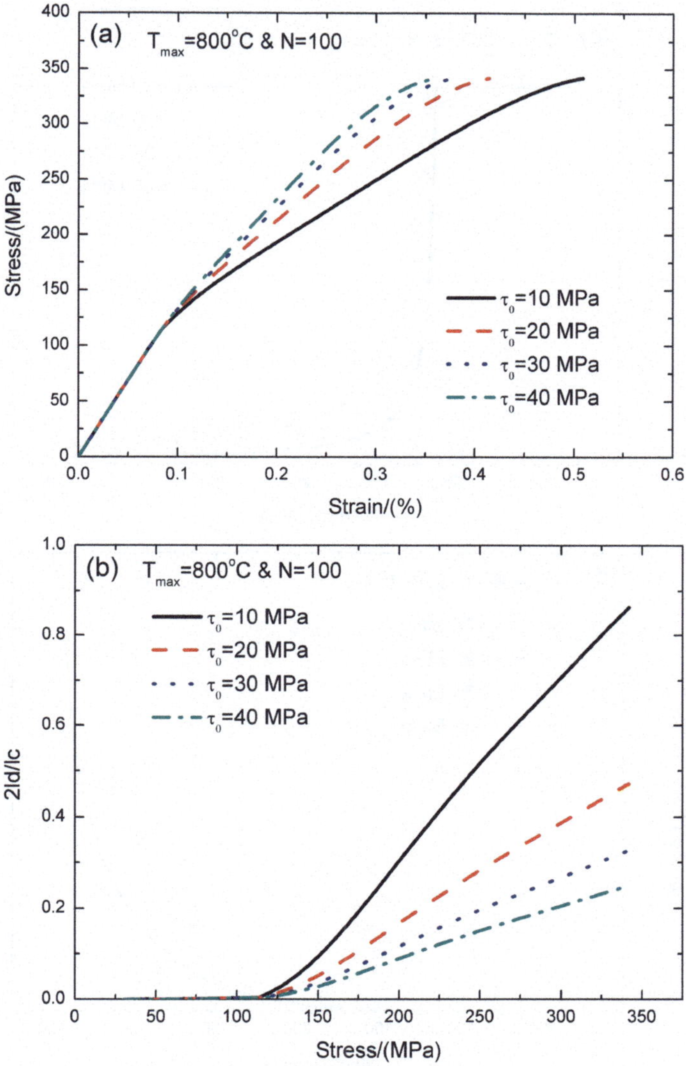

Fig. 4.13 Effect of the interface shear stress on **a** the tensile stress-strain curves; **b** the fiber/matrix interface debonding length versus the applied stress curves; **c** the fiber/matrix interface oxidation ratio versus the applied stress curves; and **d** the broken fiber fraction versus the applied stress curves of C/SiC composite

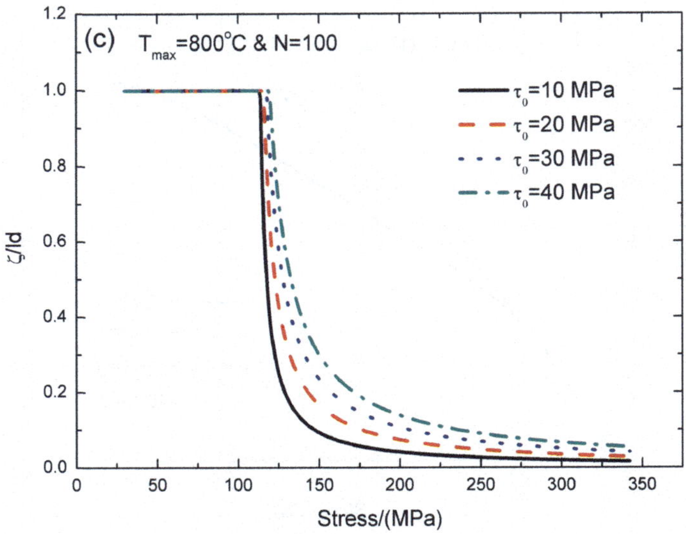

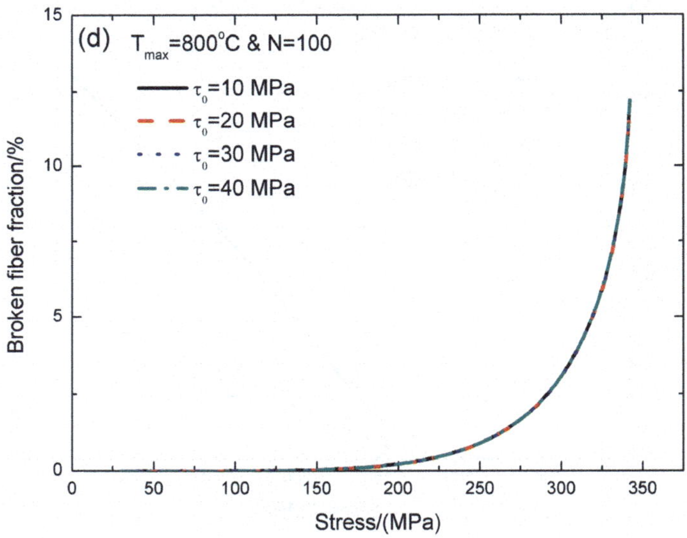

Fig. 4.13 (continued)

C/SiC composite under $T_{\max} = 800$ °C and $N = 100$ thermal fatigue cycles is shown in Fig. 4.14. When the fiber strength increases, the composite tensile strength and failure strain increase, and the fiber broken fraction at the low applied stress decreases.

When $\sigma_0 = 1$ GPa under thermal fatigue of $T_{\max} = 800$ °C and $N = 100$, the composite tensile strength is $\sigma_{\text{UTS}} = 278$ MPa with the failure strain of $\varepsilon_{\text{f}} = 0.32\%$ and the critical fiber broken fraction is $P = 12.9\%$. When $\sigma_0 = 1.5$ GPa under thermal fatigue of $T_{\max} = 800$ °C and $N = 100$, the composite tensile strength is $\sigma_{\text{UTS}} =$

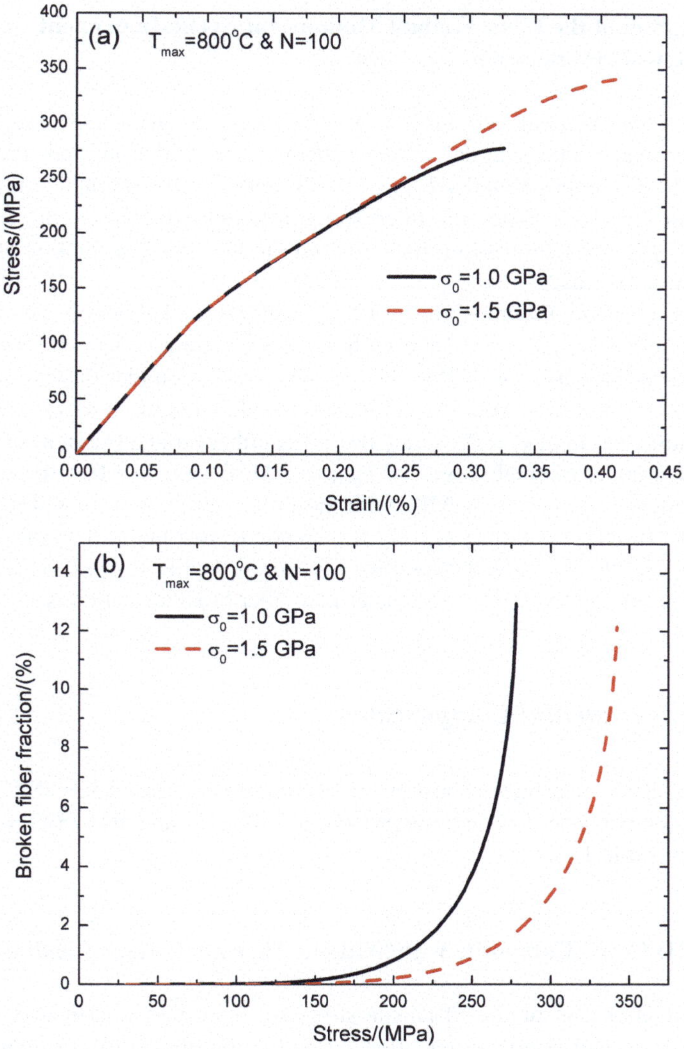

Fig. 4.14 Effect of the fiber strength on **a** the tensile stress-strain curves and **b** the broken fiber fraction versus the applied stress curves of C/SiC composite

342 MPa with the failure strain of $\varepsilon_f = 0.41\%$ and the critical fiber broken fraction is $P = 12.1\%$.

4.3.5.6 Effect of the Fiber Weibull Modulus on Cyclic-Dependent Tensile Damage and Fracture

The effect of fiber Weibull modulus (i.e., $m = 2, 3, 4$, and 5) on the tensile stress-strain curve, fiber/matrix interface debonding and interface oxidation, and fiber failure curves of C/SiC composite under T_{max} = 800 °C and N = 100 thermal fatigue cycles is shown in Fig. 4.15. When the fiber Weibull modulus increases, the composite tensile strength and failure strain increase and the fiber broken fraction at the low applied stress decreases.

When $m = 2$ under thermal fatigue of T_{max} = 800 °C and N = 100, the composite tensile strength is σ_{UTS} = 293 MPa with the failure strain of ε_f = 0.39% and the critical fiber broken fraction is P = 23.2%. When m_f = 3 under thermal fatigue of T_{max} = 800 °C and N = 100, the composite tensile strength is σ_{UTS} = 313 MPa with the failure strain of ε_f = 0.4% and the critical fiber broken fraction is P = 18%. When m = 4 under thermal fatigue of T_{max} = 800 °C and N = 100, the composite tensile strength is σ_{UTS} = 329 MPa with the failure strain of ε_f = 0.41% and the critical fiber broken fraction is P = 14.6%. When m = 5 under thermal fatigue of T_{max} = 800 °C and N = 100, the composite tensile strength is σ_{UTS} = 342 MPa with the failure strain of ε_f = 0.41% and the critical fiber broken fraction is P = 12.1%.

4.3.6 Experimental Comparisons

Wang et al. (2013) investigated the tensile behavior of 1D, 2D, and 3D C/SiC composite at room temperature. The material properties of 1D, 2D, and 3D C/SiC composites are listed in Table 1.

4.3.6.1 1D C/SiC Composite with/Without Thermal Fatigue Loading

The experimental and predicted tensile stress-strain curves, fiber/matrix interface debonding ratio and interface oxidation ratio, and the fiber broken fraction of 1D C/SiC composite without and with thermal fatigue loading at T_{max} = 800, 1000, and 1200 °C and N = 50 are shown in Fig. 4.16. With increasing thermal fatigue temperature, the interface debonding ratio, interface oxidation ratio, and broken fiber fraction all increase at low applied stress level.

Without thermal fatigue loading, the composite tensile strength is σ_{UTS} = 333 MPa with the failure strain of ε_f = 0.68%; when T_{max} = 800 °C and N = 50, the cyclic-dependent composite tensile strength is σ_{UTS} = 305 MPa with the failure strain of ε_f = 0.67%, the cyclic-dependent interface debonding length increases to $2l_d/l_c = 1$ at the applied stress of σ = 180 MPa, the cyclic-dependent interface oxidation ratio decreases to ζ/l_d = 1.1%, and the cyclic-dependent broken fiber fraction increases to P = 12.1%; when T_{max} = 1000 °C and N = 50, the cyclic-dependent composite

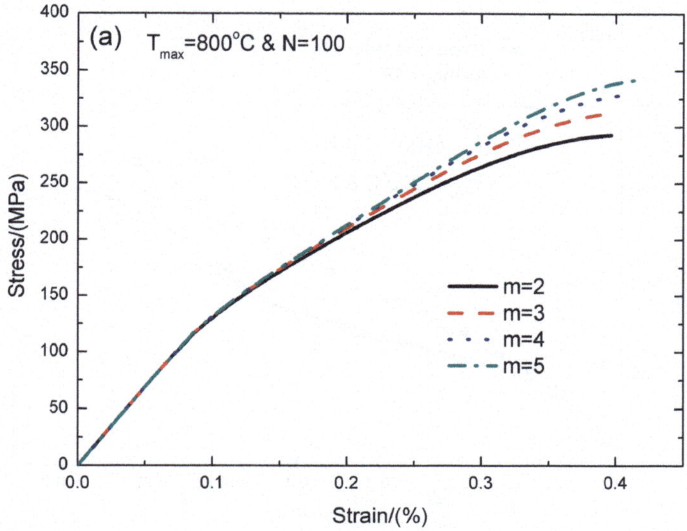

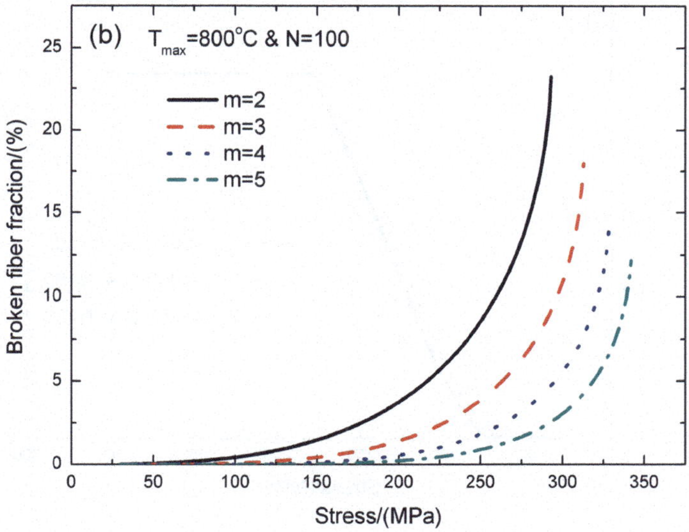

Fig. 4.15 Effect of the fiber Weibull modulus on **a** the tensile stress-strain curves and **b** the broken fiber fraction versus the applied stress curves of C/SiC composite

tensile strength is $\sigma_{UTS} = 222$ MPa with the failure strain of $\varepsilon_f = 0.54\%$, the cyclic-dependent interface debonding length increases to $2l_d/l_c = 1$ at the applied stress of $\sigma = 177$ MPa, the cyclic-dependent interface oxidation ratio decreases to $\zeta/l_d = 4.2\%$, and the cyclic-dependent broken fiber fraction increases to $P = 12.8\%$; when $T_{max} = 1200$ °C and $N = 50$, the cyclic-dependent composite tensile strength is $\sigma_{UTS} = 176$ MPa with the failure strain of $\varepsilon_f = 0.47\%$, the cyclic-dependent interface debonding length increases to $2l_d/l_c = 1$ at the applied stress of $\sigma = 172$ MPa,

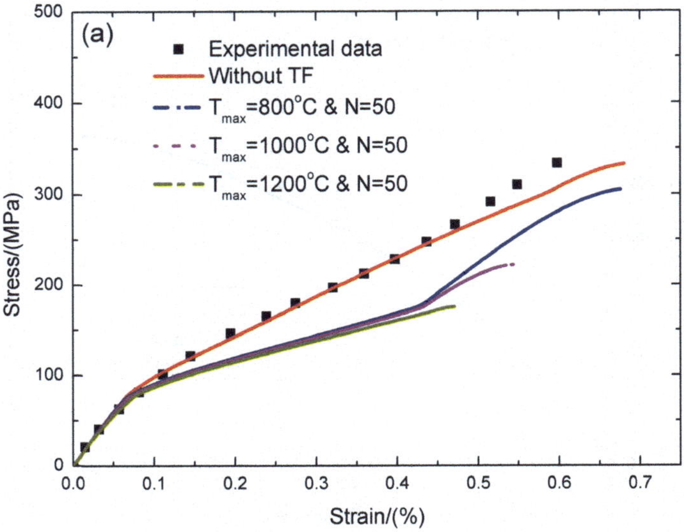

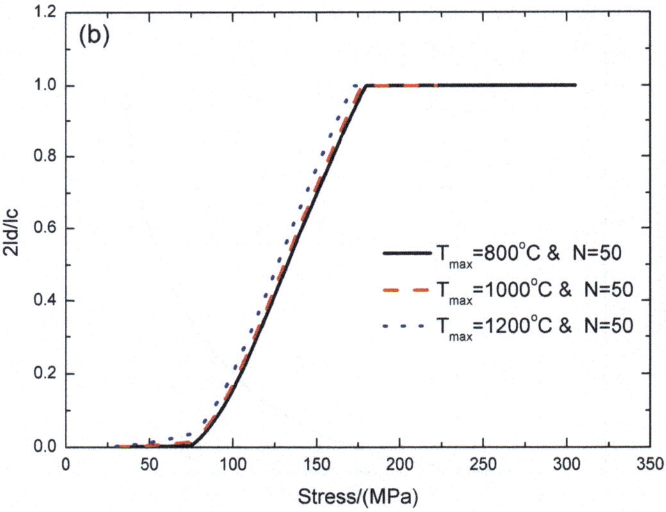

Fig. 4.16 **a** Experimental and predicted tensile stress-strain curves; **b** the fiber/matrix interface debonding length versus the applied stress curves; **c** the fiber/matrix interface oxidation ratio versus the applied stress curves; and **d** the broken fiber fraction versus the applied stress curves of 1D C/SiC composite subjected to a different thermal fatigue loading

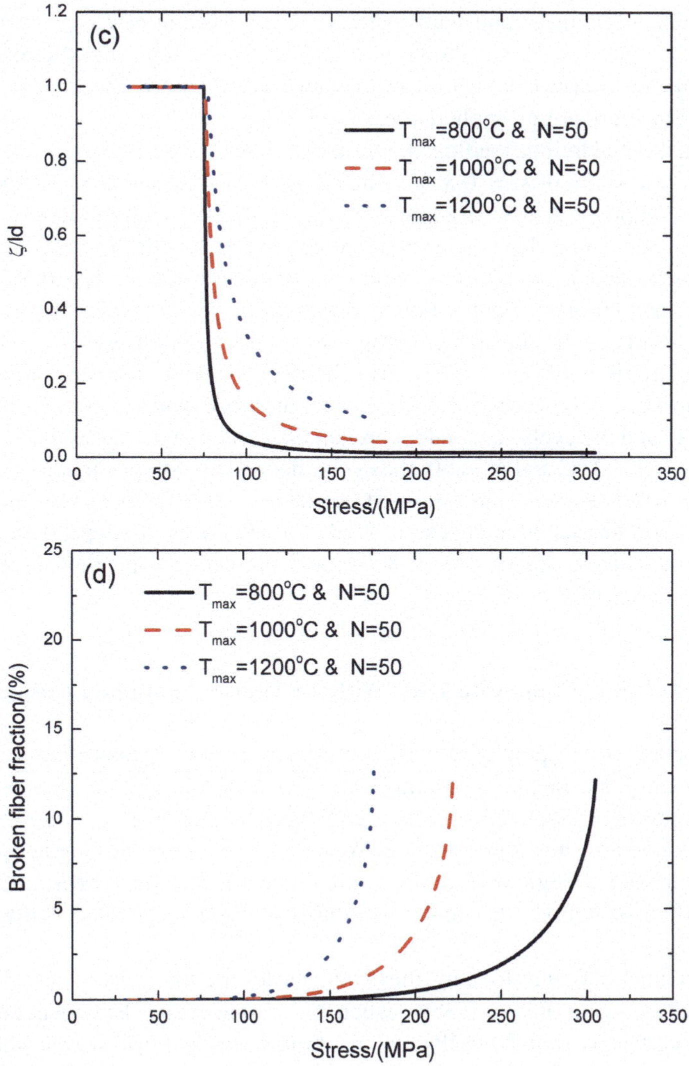

Fig. 4.16 (continued)

the cyclic-dependent interface oxidation ratio decreases to $\zeta/l_d = 10.7\%$, and the cyclic-dependent broken fiber fraction increases to $P = 13\%$.

4.3.6.2 2D C/SiC Composite with/Without Thermal Fatigue Loading

The experimental and predicted tensile stress-strain curves, fiber/matrix interface debonding ratio and interface oxidation ratio, and the fiber broken fraction of 2D

C/SiC composite without and with thermal fatigue loading at T_{max} = 800, 1000, and 1200 °C and N = 50 are shown in Fig. 4.17. With increasing thermal fatigue temperature, the interface debonding ratio, interface oxidation ratio, and broken fiber fraction all increase at low applied stress level.

Without thermal fatigue loading, the composite tensile strength is σ_{UTS} = 148 MPa with the failure strain of ε_f = 0.35%; when T_{max} = 800 °C and N = 50, the cyclic-dependent composite tensile strength is σ_{UTS} = 148 MPa with the failure strain of ε_f = 0.52%, the cyclic-dependent interface debonding length increases to $2l_d/l_c$ = 0.79, the cyclic-dependent interface oxidation ratio decreases to ζ/l_d = 0.5%, and the cyclic-dependent broken fiber fraction increases to P = 11.7%; when T_{max} = 1000 °C and N = 50, the cyclic-dependent composite tensile strength is σ_{UTS} = 129 MPa with the failure strain of ε_f = 0.44%, the cyclic-dependent interface debonding length increases to $2l_d/l_c$ = 0.66, the cyclic-dependent interface oxidation ratio decreases to ζ/l_d = 2.5%, and the cyclic-dependent broken fiber fraction increases to P = 11.3%; when T_{max} = 1200 °C and N = 50, the cyclic-dependent composite tensile strength is σ_{UTS} = 102 MPa with the failure strain of ε_f = 0.32%, the cyclic-dependent interface debonding length increases to $2l_d/l_c$ = 0.49, the cyclic-dependent interface oxidation ratio decreases to ζ/l_d = 8.7%, and the cyclic-dependent broken fiber fraction increases to P = 10.6%.

4.3.6.3 3D C/SiC Composite With/Without Thermal Fatigue Loading

The experimental and predicted tensile stress-strain curves, fiber/matrix interface debonding ratio and interface oxidation ratio, and the fiber broken fraction of 3D C/SiC composite without and with thermal fatigue loading at T_{max} = 800 °C and N = 5, 10, and 50 are shown in Fig. 4.18. With increasing thermal fatigue cycles, the composite tensile strength and failure strain decrease, and the interface debonding ratio, interface oxidation ratio, and broken fiber fraction all increase at low applied stress level.

Without thermal fatigue loading, the composite tensile strength is σ_{UTS} = 203 MPa with the failure strain of ε_f = 0.38%; when N = 5, the cyclic-dependent composite tensile strength is σ_{UTS} = 203 MPa with the failure strain of ε_f = 0.41%, the cyclic-dependent interface debonding length increases to $2l_d/l_c$ = 1 at the applied stress of σ = 192 MPa, the cyclic-dependent interface oxidation ratio decreases to ζ/l_d = 0.6%, and the cyclic-dependent broken fiber fraction increases to P = 11.6%; when N = 10, the cyclic-dependent composite tensile strength is σ_{UTS} = 180 MPa with the failure strain of ε_f = 0.38%, the cyclic-dependent interface debonding length increases to $2l_d/l_c$ = 1 at the applied stress of σ = 171 MPa, the cyclic-dependent interface oxidation ratio decreases to ζ/l_d = 1.2%, and the cyclic-dependent broken fiber fraction increases to P = 12.5%; when N = 50, the cyclic-dependent composite tensile strength is σ_{UTS} = 120 MPa with the failure strain of ε_f = 0.29%, the cyclic-dependent interface debonding length increases to $2l_d/l_c$ = 0.91, the cyclic-dependent interface oxidation ratio decreases to ζ/l_d = 6.9%, and the cyclic-dependent broken fiber fraction increases to P = 11.3%.

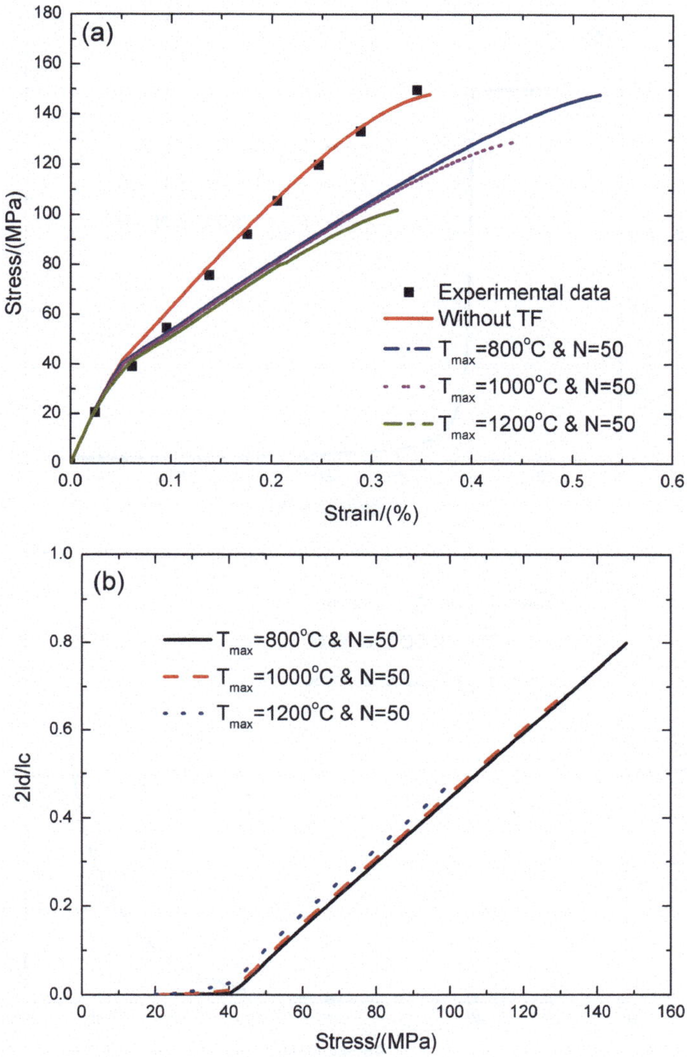

Fig. 4.17 **a** Experimental and predicted tensile stress-strain curves; **b** the fiber/matrix interface debonding length versus the applied stress curves; **c** the fiber/matrix interface oxidation ratio versus the applied stress curves; and **d** the broken fiber fraction versus the applied stress curves of 2D C/SiC composite subjected to a different thermal fatigue loading

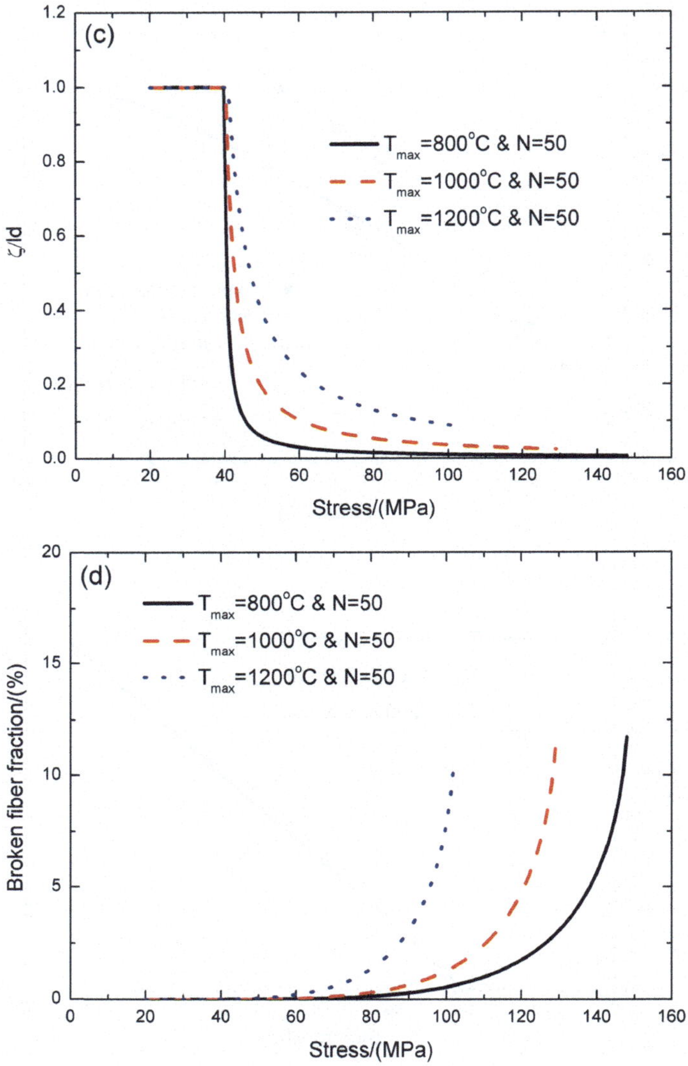

Fig. 4.17 (continued)

4.4 Conclusion

In this chapter, the tensile damage and fracture of fiber-reinforced CMCs subjected to the pre-exposure at elevated temperatures and thermal fatigue are investigated. The damage mechanisms of interface oxidation and fiber failure are considered in the stress analysis, matrix multicracking, interface debonding, and fiber failure. Combining the stress analysis and damage models, the tensile stress-strain curves of fiber-reinforced CMCs for different damage stages can be obtained. The effects of

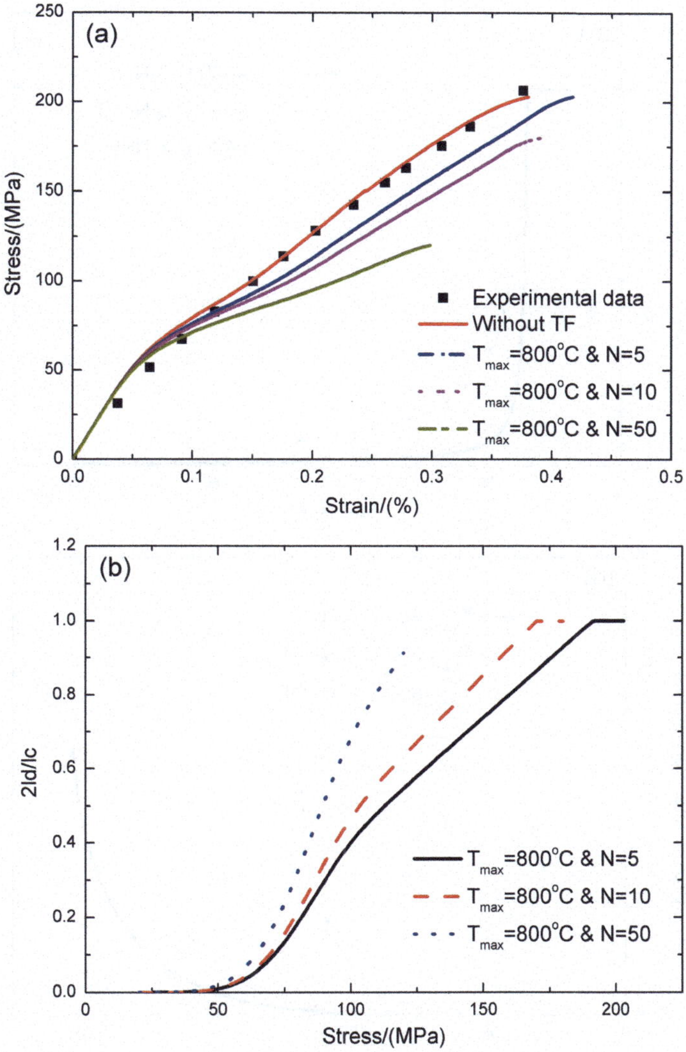

Fig. 4.18 **a** Experimental and theoretical tensile stress-strain curves; **b** the fiber/matrix interface debonding length versus the applied stress curves; **c** the fiber/matrix interface oxidation ratio versus the applied stress curves; and **d** the broken fiber fraction versus the applied stress curves of 3D C/SiC composite subjected to a different thermal fatigue loading

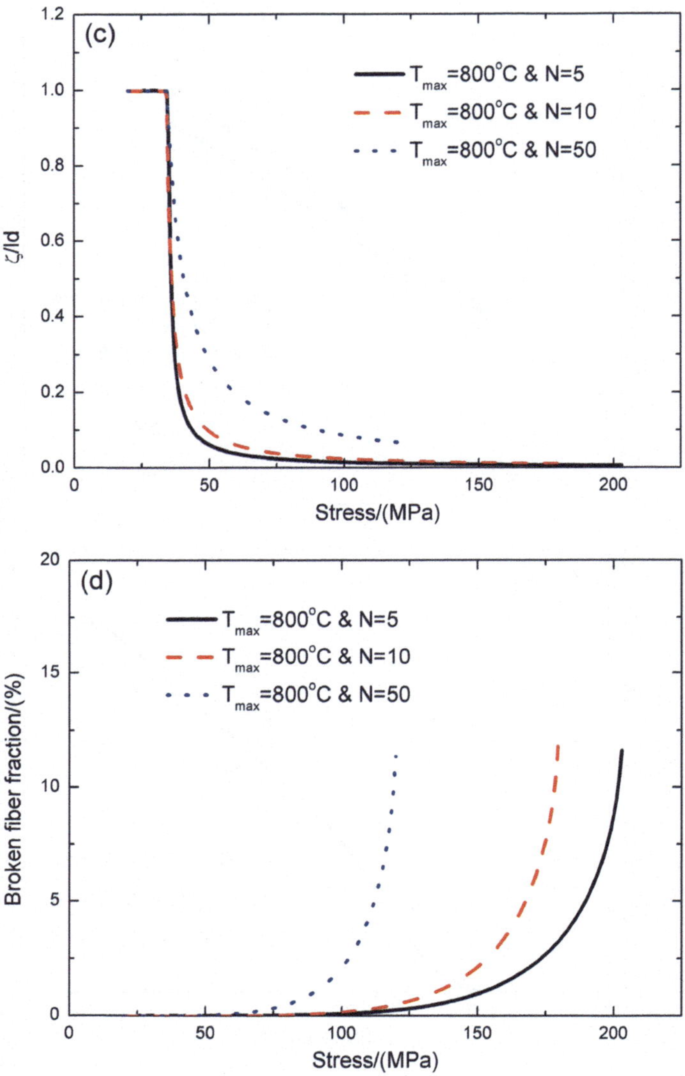

Fig. 4.18 (continued)

the pre-exposure temperature and time, thermal fatigue temperature and thermal cyclic number, the interface shear stress, fiber strength, and fiber Weibull modulus on tensile damage and fracture processes are analyzed. The experimental tensile damage and fracture process of fiber-reinforced CMCs with different fiber preforms are predicted for a different pre-exposure temperature and time.

(1) When the pre-exposure temperature and time increase, the composite tensile strength and failure strain both decrease, the interface debonded length and the

interface oxidation ratio both increase, and the fiber broken fraction increases at low applied stress level.

(2) When the thermal fatigue temperature and cycle number increase, the composite tensile strength and failure strain decrease, and the interface debonding ratio, interface oxidation ratio, and broken fiber fraction all increase at low applied stress level.

References

Barabash RI, Bei H, Cao YF, Ice GE (2011) Interface strength in NiAl-Mo composites from 3-D X-ray mocrodiffraction. Scripta Mater 64:900–903. https://doi.org/10.1016/j.scriptamat.2011.01.028

Basaran C, Nie S (2004) An irreversible thermodynamic theory for damage mechanics of solids. Int J Damage Mech 13:205–224. https://doi.org/10.1177/1056789504041058

Basaran C, Nie S (2007) A thermodynamics based damage mechanics model for particulate composites. Int J Solids Struct 44:1099–1114. https://doi.org/10.1016/j.ijsolstr.2006.06.001

Bouillon E, Ojard G, Ouyang Z, Zawada L, Habarou G, Louchet C, Feindel D, Spriet P, Logan C, Arnold T, Rogers K, Stetson D (2005) Post engine characterization and flight test experience of self sealing ceramic matrix composites for nozzle seals in gas turbine engines. In: ASME Turbo Expo 2005: power for land sea and air, 6–9 June 2005, Reno Nevada, ASME, Paper GT2005-68428. https://doi.org/10.1115/GT2005-68428

Casas L, Martinez-Esnaola JM (2003) Modeling the effect of oxidation on the creep behavior of fiber-reinforced ceramic matrix composites. Acta Mater 51:3745–3757. https://doi.org/10.1016/j.jeurceramsoc.2015.11.005

Curtin WA (1991) Theory of mechanical properties of ceramic-matrix composites. J Am Ceram Soc 74:2837–2845. https://doi.org/10.1111/j.1151-2916.1991.tb06852.x

Curtin WA (1993) Multiple matrix cracking in brittle matrix composites. Acta Metal Mater 41:1369–1377. https://doi.org/10.1016/0956-7151(93)90246-O

Evans AG (1997) Design and life prediction issues for high-temperature engineering ceramics and their composites. Acta Mater 45:23–40. https://doi.org/10.1016/S1359-6454(96)00143-7

Gao Y, Mai Y, Cotterell B (1988) Fracture of fiber-reinforced materials. J Appl Math Phys 39:550–572. https://doi.org/10.1007/BF00948962

Gowayed Y, Abouzeida E, Smyth I, Ojard G, Ahmad J, Santhosh U, Jefferson G (2015) The role of oxidation in time-dependent response of ceramic-matrix composites. Compos B 76:20–30. https://doi.org/10.1016/j.compositesb.2015.02.005

Halbig MC, Cawley JD (1999) Modeling the oxidation kinetics of continuous carbon fibers in a ceramic matrix. NASA/TM-2000-209651

Halbig MC, McGuffin-Cawley JD, Eckel AJ, Brewer DN (2008) Oxidation kinetics and stress effects for the oxidation of continuous carbon fibers within a microcracked C/SiC ceramic matrix composite. J Am Ceram Soc 91:519–526. https://doi.org/10.1111/j.1551-2916.2007.02170.x

Halbig MC, Jaskowiak MH, Kiser JD, Zhu D (2013) Evaluation of ceramic matrix composite technology for aircraft turbine engine applications. In: 51st AIAA aerospace science meeting including the new horizons forum and aerospace exposition, 07–10 Jan 2013, Grapevine, Texas. https://doi.org/10.2514/6.2013-539

Hou J, Qiao S, Zhang C, Zhang Y (2009) The influence of high temperature exposure to air on the damage to 3D-C/SiC composites. New Carbon Mater 24:173–177

Lara-Curzio E (1999) Analysis of oxidation-assisted stress-rupture of continuous fiber-reinforced ceramic matrix composites at intermediate temperatures. Compos Part A 30:549–554. https://doi.org/10.1016/S1359-835X(98)00148-1

Li L (2018a) Damage, fracture and fatigue of ceramic-matrix composites. Springer Nature Singapore Pte Ltd., 2018. ISBN: 978–981-13-1782-8. https://www.springer.com/in/book/9789811317828

Li L (2018b) Damage, fracture and fatigue of ceramic-matrix composites. Springer Nature Singapore Pte Ltd. ISBN: 978-981-13-1782-8. https://doi.org/10.1007/978-981-13-1783-5

Li L (2018c) Modeling the monotonic and cyclic tensile stress-strain behavior of 2D and 2.5D woven C/SiC ceramic-matrix composites. Mech Compos Mater 54:165–178. https://doi.org/10.1007/s11029-018-9729-5

Li L (2019a) Modeling tensile damage and fracture processes of fiber-reinforced ceramic-matrix composites under the effect of pre-exposure at elevated temperatures. Ceramics-Silikaty. https://doi.org/10.13168/cs.2019.0048

Li L (2019b) Effect of interface damage on tensile behavior of fiber-reinforced ceramic-matrix composites after thermal fatigue loading. Compos Interfaces. https://doi.org/10.1080/09276440.2019.1685281

Li L (2019) Thermomechanical fatigue of ceramic-matrix composites. Wiley-VCH. ISBN: 978-3-527-34637-0. https://onlinelibrary.wiley.com/doi/book/10.1002/9783527822614

Li L, Song Y, Sun Y (2013) Modeling the tensile behavior of unidirectional C/SiC ceramic-matrix composites. Mech Compos Mater 49:659–672. https://doi.org/10.1007/s11029-013-9382-y

Li L, Song Y, Sun Y (2015) Modeling the tensile behavior of cross-ply C/SiC ceramic-matrix composites. Mech Compos Mater 51:359–376. https://doi.org/10.1007/s11029-015-9507-6

Li L, Reynaud P, Fantozzi G (2017) Tension-tension fatigue behavior of unidirectional SiC/Si_3N_4 composite with strong and weak interface bonding at room temperature. Ceram Int 43:8769–8777. https://doi.org/10.1016/j.ceramint.2017.03.211

Mei H, Cheng L, Zhang L (2006) Thermal cycling damage mechanisms of C/SiC composites in displacement constraint and oxidizing atmosphere. J Am Ceram Soc 89:2330–2334. https://doi.org/10.1111/j.1551-2916.2006.01012.x

Ohnabe H, Masaki S, Onozuka M, Miyahara K, Sasa T (1999) Potential application of ceramic matrix composites to aero-engine components. Compos Part A 30:489–496. https://doi.org/10.1016/S1359-835X(98)00139-0

Wallentine SM (2015) Effect of prior exposure at elevated temperatures on tensile properties and stress-strain behavior of four non-oxide ceramic matrix composites. Master thesis, Air Force Institute of Technology

Wang Y, Zhang L, Cheng L (2013) Comparison of tensile behaviors of carbon/ceramic composites with various fiber architectures. Int J Appl Ceram Technol 10:266–275. https://doi.org/10.1111/j.1744-7402.2011.02727.x

Zhang C, Zhao M, Liu Y, Wang B, Wang X, Qiao S (2016) Tensile strength degradation of a 2.5D-C/SiC composite under thermal cycles in air. J Euro Ceram Soc 36:3011–3019. https://doi.org/10.1016/j.jeurceramsoc.2015.12.007

Chapter 5
Time-Dependent Fatigue Behavior of Fiber-Reinforced Ceramic-Matrix Composites at Elevated Temperatures

Abstract In this chapter, the time-dependent static fatigue and cyclic fatigue behavior of fiber-reinforced ceramic-matrix composites (CMCs) are investigated. The stress-strain relationships considering interface oxidation and interface wear in the interface debonding region under static and cyclic fatigue loading are developed to establish the relationships between the peak strain, the interface debonding length, the interface oxidation length, and the interface slip lengths. The effects of the stress-rupture time, stress levels, matrix crack spacing, fiber volume, and oxidation temperature on the peak strain and the interface slip lengths are investigated. The experimental fatigue hysteresis loops, interface slip lengths, peak strain, and interface oxidation length of C/[Si–B–C] and SiC/MAS composite at 566, 1093, and 1200 °C in air atmosphere are predicted.

Keywords Ceramic-matrix composites (CMCs) · Static fatigue · Stress-rupture · Matrix multicracking · Interface debonding · Interface oxidation

5.1 Introduction

Ceramic materials possess high strength and modulus at elevated temperature. But their use as structural components is severely limited because of their brittleness. Continuous fiber-reinforced ceramic-matrix composites, by incorporating fibers in ceramic matrices, however, not only exploit their attractive high-temperature strength but also reduce the propensity for catastrophic failure (Naslain 2004; Santhosh et al. 2013; Murthy Pappu et al. 2008). However, one of the barriers to their uses in certain long-term or reusable applications is that degradation of the carbon fibers in oxidizing environments can lead to strength reduction and component failure (Naslain et al. 2004; Li 2016, 2017).

Many researchers performed the experimental and theoretical investigations on the effects of oxidation damage on mechanical behavior of fiber-reinforced CMCs. In the experimental research area, Zhu (2006) investigated the effect of oxidation on the fatigue behavior of 2D SiC/SiC composite at elevated temperatures. It was found that the fatigue life decreased 13% after oxidation at 600 °C for 100 h due to disappearance of carbon interphase. Mall and Engesser (2006) investigated the

L. Li, *Time-Dependent Mechanical Behavior of Ceramic-Matrix Composites at Elevated Temperatures*, Advanced Ceramics and Composites 1,
https://doi.org/10.1007/978-981-15-3274-0_5

damage evolution in 2D C/SiC composite under different fatigue loading frequencies at an elevated temperature of 550 °C in air atmosphere. The oxidation of carbon fibers caused a reduction in fatigue life of C/SiC composite under lower loading frequency. However, the oxidation of carbon fibers was almost absent or negligible at higher frequency at elevated temperature. Fantozzi and Reynaud (2009) investigated the static fatigue behavior of C/[Si–B–C] composite at 1200 °C in air atmosphere. The areas of stress-strain hysteresis loops after a static fatigue of 144 h have significantly decreased, attributed to time-dependent of fiber/matrix PyC interface recession by oxidation or by a beginning of carbon fiber recession by oxidation.

In the theoretical research area, much work has been conducted to analyze and model the oxidation of fibers, matrices, and interfaces without loading by assuming steady-state diffusion of oxidation (Lamouroux et al. 1994; Sullivan 2005). Halbig et al. (2000) investigated the stressed oxidation of different fiber-reinforced CMCs, i.e., C/SiC, SiC/SiC, and SiC/SiNC, and developed a model to predict the oxidation pattern and kinetics of carbon fiber tows in a non-reactive matrix. Pailler and Lamon (2005) developed a fatigue–oxidation model to investigate the strain response of a SiC/SiC minicomposite under matrix multicracking and interface oxidation. Casas and Martinez-Esnaola (2003) developed a creep–oxidation model for fiber-reinforced CMCs at elevated temperature, including the effects of interface and matrix oxidation, creep of fibers, and degradation of fiber strength with time. The broken fiber fraction increases with time in an accelerated manner due to fiber strength degradation. Under static fatigue loading at elevated temperature, the shape, location, and area of the stress-strain hysteresis loops would evolve with increase of the oxidation time, which can be used to monitor the damage evolution inside of the damaged composite. However, there is no research work on the hysteresis loop models considering oxidation at elevated temperature.

In this chapter, the time-dependent static fatigue and cyclic fatigue behavior of fiber-reinforced CMCs are investigated. The stress-strain relationships considering interface oxidation and interface wear in the interface debonding region under static and cyclic fatigue loading are developed to establish the relationships between the peak strain, the interface debonding length, the interface oxidation length, and the interface slip lengths. The effects of stress-rupture time, stress levels, matrix crack spacing, fiber volume, and oxidation temperature on the peak strain and the interface slip lengths are investigated. The experimental fatigue hysteresis loops, interface slip lengths, peak strain, and interface oxidation length of C/[Si–B–C] and SiC/MAS composite at 566, 1093, and 1200 °C in air atmosphere are predicted.

5.2 Time-Dependent Static Fatigue Damage Evolution at Elevated Temperature

In this section, an analytical method is developed to investigate the effect of oxidation on the hysteresis loops of fiber-reinforced CMCs under static fatigue at elevated temperature. The oxidation region propagating model is adopted to analyze the oxidation effect on the hysteresis loops, which is controlled by the interface frictional slip and diffusion of oxygen gas through matrix multicrackings. Based on the damage mechanism of fiber sliding relative to the matrix, the hysteresis loop models corresponding to different interface slip cases considering the interface oxidation are established. The relationships between the hysteresis loops, the hysteresis dissipated energy, the interface slip, and oxidation time are established. The effects of the stress level, matrix crack spacing, fiber volume, and oxidation temperature on the hysteresis dissipated energy, interface debonding, oxidation, and slip lengths versus the oxidation time are analyzed. The experimental hysteresis loops of C/[Si–B–C] composite under static fatigue in air atmosphere at 1200 °C are predicted.

5.2.1 Time-Dependent Static Fatigue Hysteresis Theories

If matrix multicracking and fiber/matrix interface debonding are present upon first loading to the peak stress, the stress-strain hysteresis loops develop as a result of energy dissipation through the frictional slip between the fiber and the matrix upon unloading and subsequent reloading. At elevated temperature, matrix cracks serve as avenues for the ingress of environment atmosphere into the composite. The oxygen reacts with carbon layer along the fiber length at a certain rate of $d\zeta/dt$, where ζ is the length of carbon lost in each side of the crack. The interface shear stress in the oxidized region decreases from the initial value τ_i to τ_f due to interface oxidation. With the increase of oxidation time under static fatigue, the oxidized region would propagate along fiber/matrix interface, leading to the evolution of the shape, area, and location of fatigue hysteresis loops.

Based on the interface frictional slip cases between fibers and matrix, the stress-strain hysteresis loops under static fatigue at elevated temperature can be divided into four different cases, i.e.,

1. Case 1, i.e., the interface partial debonding and the fiber complete sliding relative to the matrix in the interface debonding region.
2. Case 2, i.e., the interface partial debonding and the fiber partial sliding relative to the matrix in the interface debonding region.
3. Case 3, i.e., the interface complete debonding and the fiber partial sliding relative to the matrix in the interface debonding region.
4. Case 4, i.e., the interface complete debonding and the fiber complete sliding relative to the matrix in the interface debonding region.

5.2.1.1 Case 1

Upon unloading to the applied stress σ, i.e., $\sigma > \sigma_{\min}$, the unit cell can be divided into the interface debonding region and the interface bonding region. The interface debonding region can be divided into three regions, i.e., interface counter-slip region with the interface shear stress τ_f ($z \in [0, \zeta]$), the interface slip region with the interface shear stress τ_i ($z \in [\zeta, l_y]$), and the interface slip region with interface shear stress τ_i ($z \in [l_y, l_d]$), in which l_y denotes the interface counter-slip length. Upon unloading to σ_{tr_pu}, the interface counter-slip length l_y approaches the interface debonding length l_d, i.e., $l_y(\sigma = \sigma_{tr\_pu}) = l_d$.

$$l_y = \frac{1}{2}\left\{l_d + \left(1 - \frac{\tau_f}{\tau_i}\right)\zeta - \left[\frac{r_f}{2}\left(\frac{V_m E_m}{V_f E_c}\frac{\sigma}{\tau_i} - \frac{1}{\rho}\right) - \sqrt{\left(\frac{r_f}{2\rho}\right)^2 + \frac{r_f V_m E_m E_f}{E_c \tau_i^2}\xi_d}\right]\right\} \tag{5.1}$$

When $\sigma < \sigma_{tr\_pu}$, the counter-slip occurs over the entire interface debonded region, i.e., $y(\sigma < \sigma_{tr\_pu}) = l_d$. The fiber axial stress distribution upon unloading for interface slip Case 1 considering interface oxidation is shown in Fig. 5.1a.

Upon reloading to the applied stress σ, slip again occurs near the matrix crack plane over a distance l_z, which is denoted to be the interface new-slip region. The interface debonding region can be divided into four regions, i.e., the interface new-slip region with the interface shear stress τ_f ($z \in [0, l_z]$), the interface counter-slip region with the interface shear stress τ_f ($z \in [l_z, \zeta]$), the interface counter-slip region with the interface shear stress τ_i ($z \in [\zeta, l_y]$), and interface slip region with interface shear stress τ_i ($z \in [l_y, l_d]$). Upon reloading to σ_{tr_pr}, the interface new-slip length z approaches the interface debonded length l_d, i.e., $l_z(\sigma_{tr\_pr}) = l_d$.

$$l_z = \frac{\tau_i}{\tau_f}\left\{l_y - \frac{1}{2}\left[l_d + \left(1 - \frac{\tau_f}{\tau_i}\right)\zeta - \left[\frac{r_f}{2}\left(\frac{V_m E_m}{V_f E_c}\frac{\sigma}{\tau_i} - \frac{1}{\rho}\right) - \sqrt{\left(\frac{r_f}{2\rho}\right)^2 + \frac{r_f V_m E_m E_f}{E_c \tau_i^2}\xi_d}\right]\right]\right\} \tag{5.2}$$

When $\sigma > \sigma_{tr\_pr}$, the new slip occurs over the entire interface debonded length, i.e., $l_z(\sigma > \sigma_{tr\_pr}) = l_d$. The fiber axial stress distribution upon reloading for the interface slip Case 1 considering the interface oxidation is shown in Fig. 5.1b.

5.2.1.2 Case 2

For the interface slip Case 2, the interface counter-slip length l_y upon complete unloading is less than the interface debonding length l_d, i.e., $l_y(\sigma = \sigma_{min}) < l_d$. The fiber axial stress distribution upon unloading for the interface slip Case 2 considering the interface oxidation is shown in Fig. 5.2a. Upon reloading to σ_{max}, the interface new-slip length l_z is less than the interface debonding length l_d, $l_z(\sigma = \sigma_{max}) < l_d$. The fiber axial stress distribution upon reloading for the interface slip Case 2 considering the interface oxidation is shown in Fig. 5.2b.

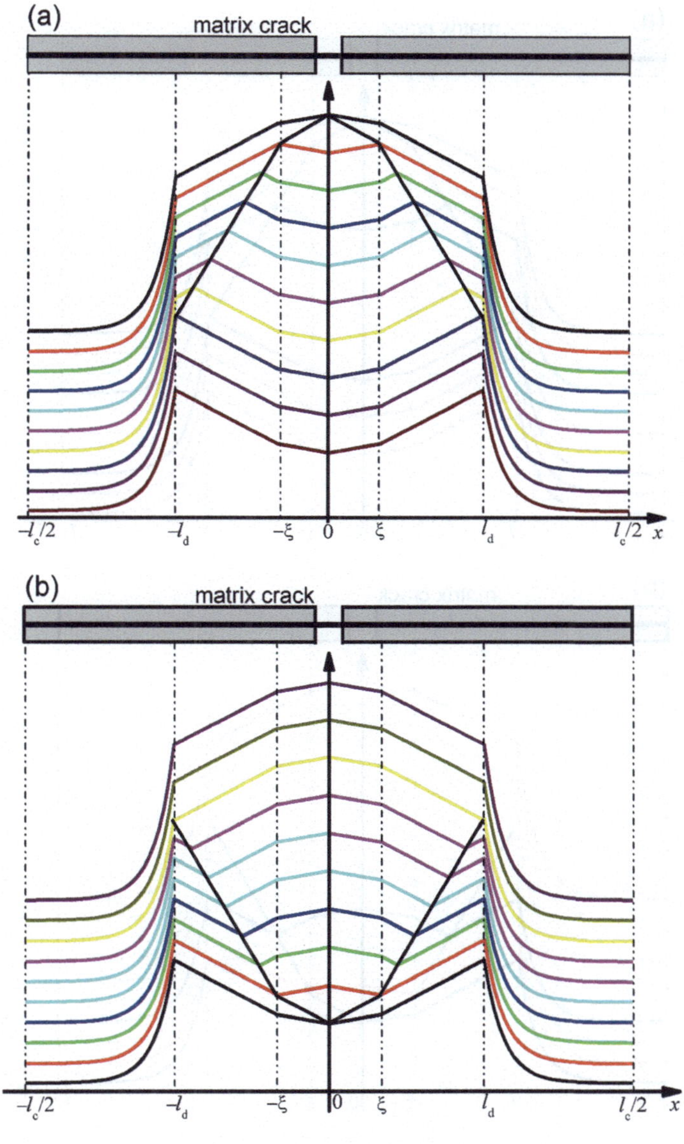

Fig. 5.1 Fiber axial stress distribution upon **a** unloading and **b** reloading for interface slip Case 1 considering interface oxidation. Reprinted with permission from Li (2016). Copyright 2015, Elsevier Ltd.

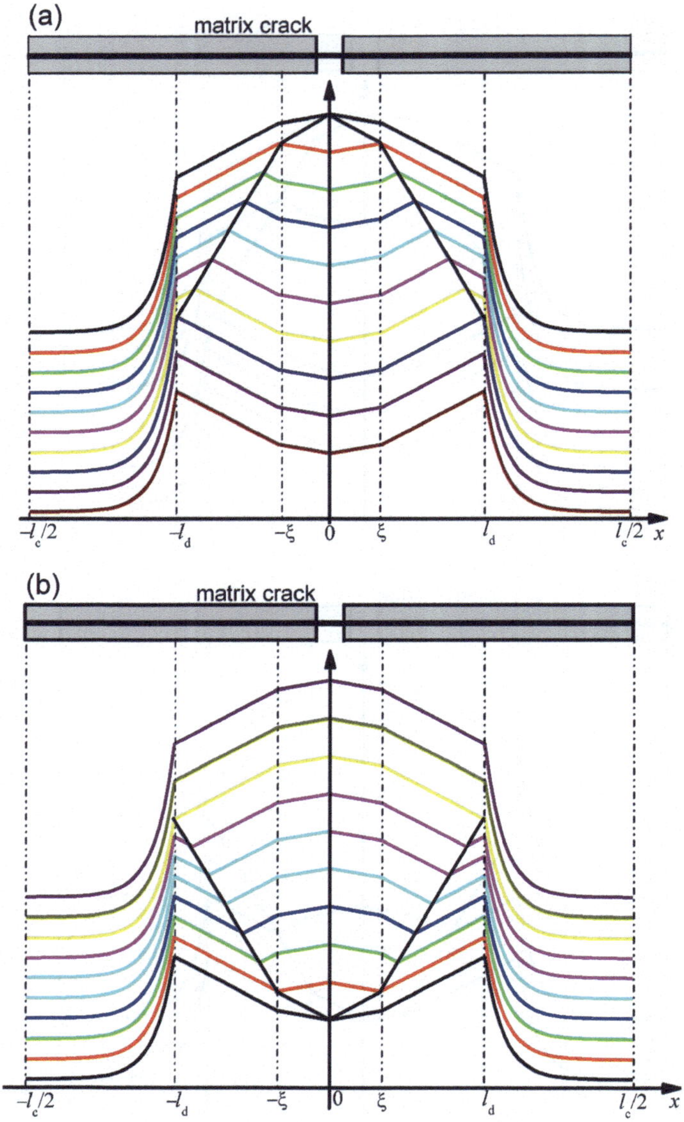

Fig. 5.2 Fiber axial stress distribution upon **a** unloading and **b** reloading for interface slip Case 1 considering interface oxidation. Reprinted with permission from Li (2016). Copyright 2015, Elsevier Ltd.

5.2.1.3 Case 3

For the interface slip Case 3, the interface complete debonds. Upon complete unloading, the interface counter-slip length l_y is less than the half matrix crack spacing $l_c/2$, i.e., $l_y(\sigma = \sigma_{\min}) < l_c/2$.

$$l_y = \left(1 - \frac{\tau_f}{\tau_i}\right)\zeta + \frac{r_f V_m E_m}{4 V_f E_c \tau_i}(\sigma_{max} - \sigma) \tag{5.3}$$

The fiber axial stress distribution upon unloading for interface slip Case 3 considering interface oxidation is shown in Fig. 5.3a.

Upon reloading to σ_{max}, the interface new-slip length l_z is less than the half matrix crack spacing $l_c/2$, i.e., $z(\sigma = \sigma_{max}) < l_c/2$.

$$l_z = l_y - \frac{r_f V_m E_m}{4 V_f E_c \tau_i}(\sigma_{max} - \sigma) \tag{5.4}$$

The fiber axial stress distribution upon reloading for the interface slip Case 3 considering the interface oxidation is shown in Fig. 5.3b.

5.2.1.4 Case 4

For the interface slip Case 4, the interface complete debonds. Upon unloading to the applied stress σ_{tr_fu}, the interface counter-slip length l_y approaches the half matrix crack spacing $l_c/2$, i.e., $l_y(\sigma = \sigma_{tr\_fu}) = l_c/2$. When $\sigma > \sigma_{tr\_fu}$, the unloading interface counter-slip length l_y is less than the half matrix crack spacing $l_c/2$, i.e., $l_y(\sigma > \sigma_{tr\_fu}) < l_c/2$. When $\sigma < \sigma_{tr\_fu}$, the unloading interface counter-slip occurs over the entire matrix crack spacing $l_c/2$, i.e., $l_y(\sigma < \sigma_{tr\_fu}) = l_c/2$. The fiber axial stress distribution upon unloading for the interface slip Case 4 considering the interface oxidation is shown in Fig. 5.4a.

Upon reloading to the applied stress σ_{tr_fr}, the interface new-slip length l_z approaches the half matrix crack spacing $l_c/2$, i.e., $l_z(\sigma = \sigma_{tr\_fr}) = l_c/2$. When $\sigma < \sigma_{tr\_fr}$, the interface new-slip length l_z is less than half matrix crack spacing $l_c/2$, i.e., $l_z(\sigma < \sigma_{tr\_fr}) < l_c/2$. When $\sigma_{tr\_fr} < \sigma < \sigma_{max}$, the interface new-slip length occurs over the entire matrix crack spacing $l_c/2$, i.e., $l_z(\sigma_{tr\_fr} < \sigma < \sigma_{max}) = l_c/2$. The fiber axial stress distribution upon reloading for the interface slip Case 4 considering the interface oxidation is shown in Fig. 5.4b.

5.2.1.5 Hysteresis Loops and Hysteresis Dissipated Energy

When damage forms within the composite, the composite strain is determined by Eq. (5.5), which assumes that the composite strain is equivalent to the average strain in an undamaged fiber.

$$\varepsilon_c = \frac{2}{E_f l_c}\int_{l_c/2} \sigma_f(z)\mathrm{d}z - (\alpha_c - \alpha_f)\Delta T \tag{5.5}$$

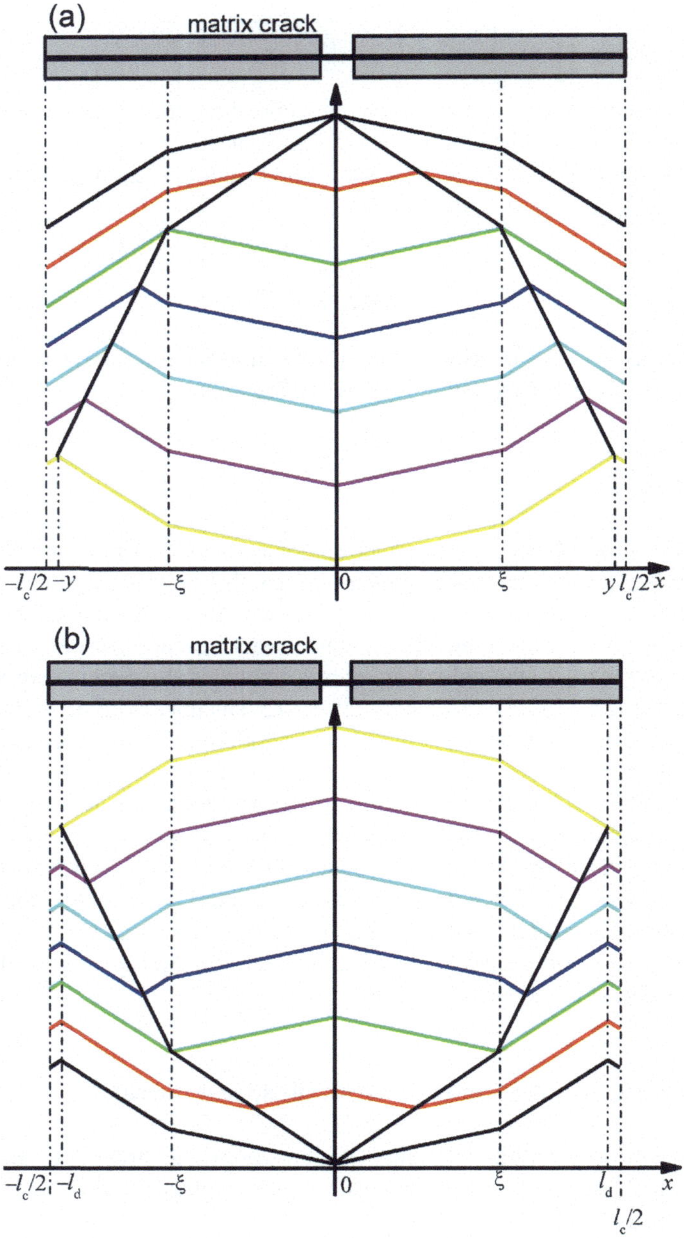

Fig. 5.3 Fiber axial stress distribution upon **a** unloading and **b** reloading for interface slip Case 3 considering interface oxidation. Reprinted with permission from Li (2016). Copyright 2015, Elsevier Ltd.

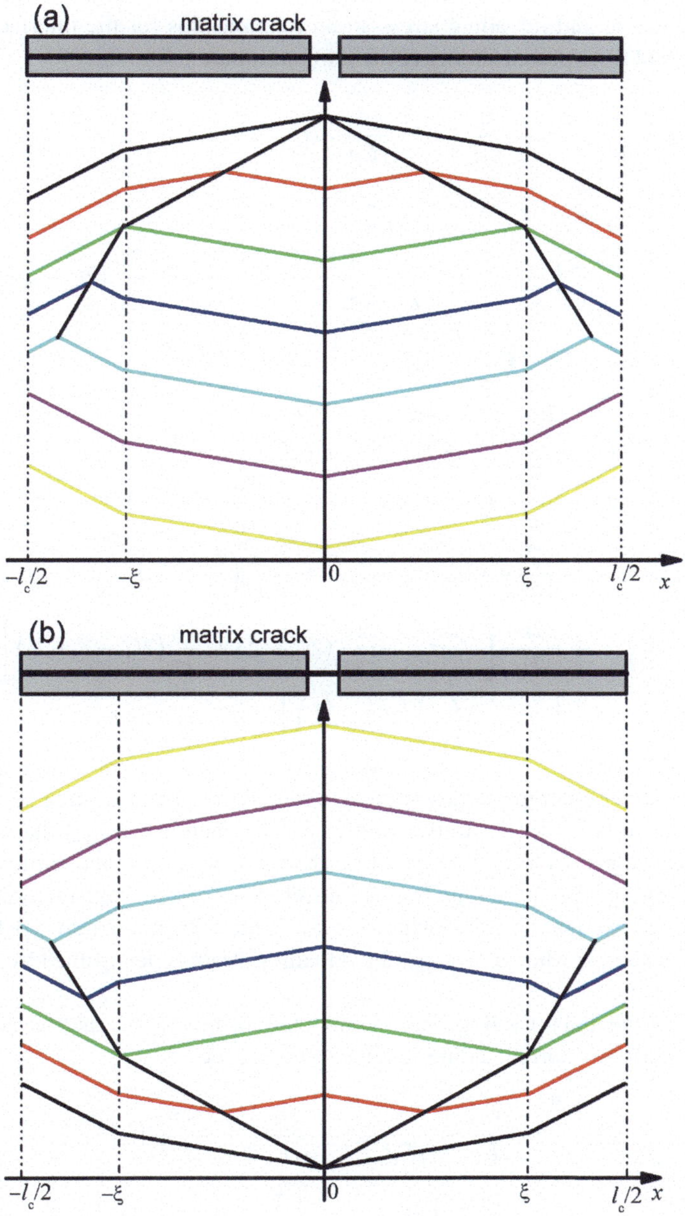

Fig. 5.4 Fiber axial stress distribution upon **a** unloading and **b** reloading for interface slip Case 4 considering interface oxidation. Reprinted with permission from Li (2016). Copyright 2015, Elsevier Ltd.

The unloading and reloading stress-strain relationships for the interface partial debonding and fiber partial sliding relative to matrix are

$$\begin{aligned}\varepsilon_{\mathrm{unloading}} = & \frac{2\sigma l_{\mathrm{d}}}{V_{\mathrm{f}}E_{\mathrm{f}}l_{\mathrm{c}}} + \frac{2\tau_{\mathrm{f}}}{r_{\mathrm{f}}E_{\mathrm{f}}l_{\mathrm{c}}}\zeta^2 + \frac{4\tau_{\mathrm{f}}}{r_{\mathrm{f}}E_{\mathrm{f}}l_{\mathrm{c}}}\zeta(l_{\mathrm{d}} - \zeta) \\ & + \frac{4\tau_{\mathrm{i}}}{r_{\mathrm{f}}E_{\mathrm{f}}l_{\mathrm{c}}}\left(l_y - \zeta\right)^2 - \frac{2\tau_{\mathrm{i}}}{r_{\mathrm{f}}E_{\mathrm{f}}l_{\mathrm{c}}}\left(2l_y - \zeta - l_{\mathrm{d}}\right)^2 \\ & + \frac{2\sigma_{\mathrm{fo}}}{E_{\mathrm{f}}l_{\mathrm{c}}}\left(\frac{l_{\mathrm{c}}}{2} - l_{\mathrm{d}}\right) + \frac{2r_{\mathrm{f}}}{\rho E_{\mathrm{f}}l_{\mathrm{c}}}\left[\frac{V_{\mathrm{m}}}{V_{\mathrm{f}}}\sigma_{\mathrm{mo}} + \frac{2\tau_{\mathrm{f}}}{r_{\mathrm{f}}}\zeta + \frac{2\tau_{\mathrm{i}}}{r_{\mathrm{f}}}\left(2l_y - \zeta - l_{\mathrm{d}}\right)\right] \\ & \times \left[1 - \exp\left(-\rho\frac{l_{\mathrm{c}}/2 - l_{\mathrm{d}}}{r_{\mathrm{f}}}\right)\right] - (\alpha_{\mathrm{c}} - \alpha_{\mathrm{f}})\Delta T \end{aligned} \tag{5.6}$$

$$\begin{aligned}\varepsilon_{\mathrm{reloading}} = & \frac{2\sigma}{V_{\mathrm{f}}E_{\mathrm{f}}l_{\mathrm{c}}}l_{\mathrm{d}} - \frac{4\tau_{\mathrm{f}}}{r_{\mathrm{f}}E_{\mathrm{f}}l_{\mathrm{c}}}l_z^2 + \frac{2\tau_{\mathrm{f}}}{r_{\mathrm{f}}E_{\mathrm{f}}l_{\mathrm{c}}}(2l_z - \zeta)^2 \\ & - \frac{4\tau_{\mathrm{f}}}{r_{\mathrm{f}}E_{\mathrm{f}}l_{\mathrm{c}}}(2l_z - \zeta)(l_{\mathrm{d}} - \zeta) + \frac{4\tau_{\mathrm{i}}}{r_{\mathrm{f}}E_{\mathrm{f}}l_{\mathrm{c}}}\left(l_y - \zeta\right)^2 \\ & - \frac{2\tau_{\mathrm{i}}}{r_{\mathrm{f}}E_{\mathrm{f}}l_{\mathrm{c}}}\left(2l_y - \zeta - l_{\mathrm{d}}\right)^2 + \frac{2\sigma_{\mathrm{fo}}}{E_{\mathrm{f}}l_c}\left(\frac{l_{\mathrm{c}}}{2} - l_{\mathrm{d}}\right) \\ & + \frac{2r_{\mathrm{f}}}{\rho E_{\mathrm{f}}l_{\mathrm{c}}}\left[\frac{V_{\mathrm{m}}}{V_{\mathrm{f}}}\sigma_{\mathrm{mo}} - \frac{2\tau_{\mathrm{f}}}{r_{\mathrm{f}}}(2l_z - \zeta) + \frac{2\tau_{\mathrm{i}}}{r_{\mathrm{f}}}\left(2l_y - \zeta - l_{\mathrm{d}}\right)\right] \\ & \times \left[1 - \exp\left(-\rho\frac{l_{\mathrm{c}}/2 - l_{\mathrm{d}}}{r_{\mathrm{f}}}\right)\right] - (\alpha_{\mathrm{c}} - \alpha_{\mathrm{f}})\Delta T \end{aligned} \tag{5.7}$$

When the fiber complete slides relative to the matrix, the unloading stress-strain relationship can be divided into two parts, i.e., (1) when $\sigma > \sigma_{\mathrm{tr\_pu}}$, the unloading strain is determined by Eq. (5.6), and (2) when $\sigma < \sigma_{\mathrm{tr\_pu}}$, the unloading strain is determined by Eq. (5.6) by setting $l_y = l_{\mathrm{d}}$. The reloading stress-strain relationship can be divided into two parts, i.e., (1) when $\sigma < \sigma_{\mathrm{tr\_pr}}$, the reloading strain is determined by Eq. (5.7), and (2) when $\sigma > \sigma_{\mathrm{tr\_pr}}$, the reloading strain is determined by Eq. (5.7) by setting $l_z = l_{\mathrm{d}}$.

The unloading and reloading stress-strain relationships for the interface complete debonding and fiber partial sliding relative to matrix are

$$\begin{aligned}\varepsilon_{\mathrm{unloading}} = & \frac{\sigma}{V_{\mathrm{f}}E_{\mathrm{f}}} - \frac{2\tau_{\mathrm{f}}}{r_{\mathrm{f}}E_{\mathrm{f}}l_{\mathrm{c}}}\zeta^2 + \frac{2\tau_{\mathrm{f}}}{r_{\mathrm{f}}E_{\mathrm{f}}}\zeta + \frac{4\tau_{\mathrm{i}}}{r_{\mathrm{f}}E_{\mathrm{f}}l_{\mathrm{c}}}\left(l_y - \zeta\right)^2 \\ & - \frac{2\tau_{\mathrm{i}}}{r_{\mathrm{f}}E_{\mathrm{f}}l_{\mathrm{c}}}\left(2l_y - \zeta - \frac{l_{\mathrm{c}}}{2}\right)^2 - (\alpha_{\mathrm{c}} - \alpha_{\mathrm{f}})\Delta T \end{aligned} \tag{5.8}$$

$$\begin{aligned}\varepsilon_{\mathrm{reloading}} = & \frac{\sigma}{V_{\mathrm{f}}E_{\mathrm{f}}} - \frac{4\tau_{\mathrm{f}}}{r_{\mathrm{f}}E_{\mathrm{f}}l_{\mathrm{c}}}l_z^2 + \frac{2\tau_{\mathrm{f}}}{r_{\mathrm{f}}E_{\mathrm{f}}l_{\mathrm{c}}}(2l_z - \zeta)^2 \\ & - \frac{4\tau_{\mathrm{f}}}{r_{\mathrm{f}}E_{\mathrm{f}}l_{\mathrm{c}}}(2l_z - \zeta)\left(l_y - \zeta\right) + \frac{4\tau_{\mathrm{i}}}{r_{\mathrm{f}}E_{\mathrm{f}}l_c}\left(l_y - \zeta\right)^2 \end{aligned}$$

$$-\frac{4\tau_{\mathrm{f}}}{r_{\mathrm{f}}E_{\mathrm{f}}l_{\mathrm{c}}}(2l_z-\zeta)\left(\frac{l_{\mathrm{c}}}{2}-l_y\right)-\frac{2\tau_{\mathrm{i}}}{r_{\mathrm{f}}E_{\mathrm{f}}l_{\mathrm{c}}}\left(2l_y-\zeta-\frac{l_{\mathrm{c}}}{2}\right)^2-(\alpha_{\mathrm{c}}-\alpha_{\mathrm{f}})\Delta T \tag{5.9}$$

When the fiber complete slides relative to the matrix, the unloading stress-strain relationship can be divided into two parts, i.e., (1) when $\sigma > \sigma_{\mathrm{tr\_fu}}$, the unloading strain is determined by Eq. (5.8), and (2) when $\sigma < \sigma_{\mathrm{tr\_fu}}$, the unloading strain is determined by Eq. (5.8) by setting $l_y = l_c/2$. The reloading stress-strain relationship can be divided into two parts, i.e., (1) when $\sigma < \sigma_{\mathrm{tr\_fr}}$, the reloading strain is determined by Eq. (5.9), and (2) when $\sigma > \sigma_{\mathrm{tr\_fr}}$, the reloading strain is determined by Eq. (5.9) by setting $l_z = l_c/2$.

Under cyclic loading, the area associated with the stress-strain hysteresis loops is the energy lost during corresponding cycle, which is defined as

$$W = \int_{\sigma_{\min}}^{\sigma_{\max}} \left[\varepsilon_{\mathrm{unloading}}(\sigma) - \varepsilon_{\mathrm{reloading}}(\sigma)\right] \mathrm{d}\sigma \tag{5.10}$$

The hysteresis dissipated energy of four different interface slip cases can be derived by inserting the corresponding unloading and reloading strains into Eq. (5.10).

5.2.2 *Results and Discussion*

The stress-strain hysteresis loops, hysteresis dissipated energy, and interface slip of C/SiC composite under static fatigue at $T = 800$ °C in air atmosphere and a stress applied of $\sigma_{\max} = 180$ MPa are shown in Fig. 5.5. The stress-strain hysteresis loops corresponding to a different oxidation time of $t = 1, 3, 5, 10, 15, 20, 25$, and 30 h are shown in Fig. 5.5a, in which the shape, location, and area of the stress-strain hysteresis loops change with the oxidation time, corresponding to different interface slip cases mentioned above due to interface oxidation. With the increase of oxidation time, the hysteresis dissipated energy increases from $W = 10.4$–22.6 kPa, i.e., A–B part in Fig. 5.5b, corresponding to the interface partial debonding, i.e., A–B part in Fig. 5.5c, interface partial oxidation, i.e., A–B part in Fig. 5.5d, and interface partial slipping between the fiber and the matrix, i.e., A–B part in Fig. 5.5e, and increases to the peak value, then decreases with the increase of oxidation time, i.e., B–C–D part in Fig. 5.5b, corresponding to the interface complete debonding, i.e., B–C part in Fig. 5.5c, interface partial oxidation, i.e., B–C part in Fig. 5.5d, and the interface partial slipping between the fiber and the matrix, i.e., B–C part in Fig. 5.5e, and then remains to be the constant value with the increase of oxidation time, i.e., D–E part in Fig. 5.5b, corresponding to the interface complete debonding, i.e., B–C part in Fig. 5.5c, interface complete oxidation, i.e., C–D part in Fig. 5.5d, and interface partial slipping between the fiber and the matrix, i.e., C–D part in Fig. 5.5e.

The effects of stress level, matrix crack spacing, fiber volume, and oxidation temperature on time-dependent static fatigue damage evolution are analyzed.

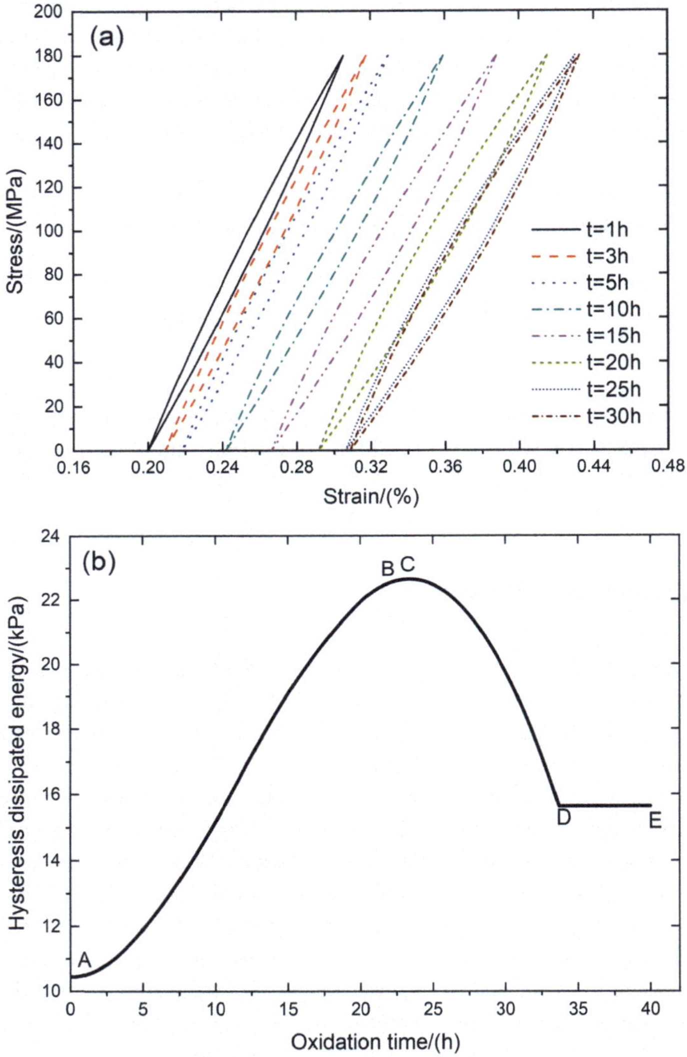

Fig. 5.5 **a** Stress-strain hysteresis loops corresponding to a different oxidation time; **b** the hysteresis dissipated energy versus oxidation time; **c** the interface debonding length versus oxidation time; **d** the oxidation region length versus oxidation time; and **e** the interface slip length versus oxidation time of C/SiC composite at an elevated temperature of $T = 800$ °C in air condition

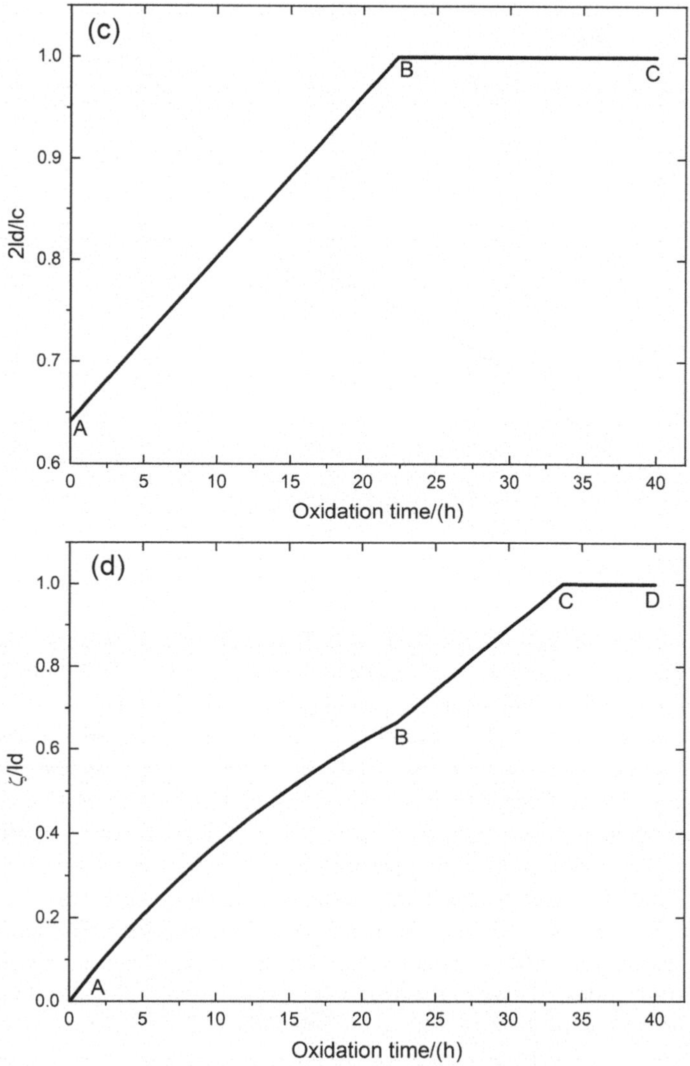

Fig. 5.5 (continued)

5.2.2.1 Effect of Stress Level on Time-Dependent Static Fatigue Damage Evolution

The hysteresis dissipated energy, interface debonding, oxidation and slip lengths versus the oxidation time corresponding to different static stress levels of σ_{max} = 180 and 200 MPa are shown in Fig. 5.6.

When the fatigue peak stress is $\sigma_{max} = 180$ MPa and with increasing oxidation time, the hysteresis dissipated energy versus the oxidation time curve can be divided

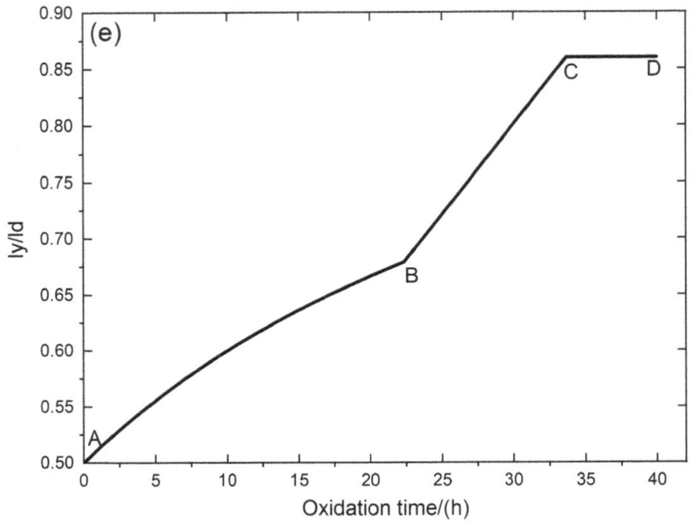

Fig. 5.5 (continued)

into three regions. In region I, the hysteresis dissipated energy increases from $W = 9.5$ kPa to the peak value of $W = 20.5$ kPa and decreases to $W = 18.9$ kPa (i.e., A_1–B_1–C_1 part in Fig. 5.6a), the interface partial debonds (i.e., A_1–B_1 part in Fig. 5.6b) and partial oxidizes (i.e., A–B_1 part in Fig. 5.6c), and the fiber partial slides relative to the matrix (i.e., A–B_1 part in Fig. 5.6d). In region II, the hysteresis dissipated energy continually decreases to $W = 9.3$ kPa (i.e., C_1–D_1 part in Fig. 5.6a), the interface complete debonds (i.e., B_1–C part in Fig. 5.6b) and partial oxidizes (i.e., B_1–C part in Fig. 5.6c), and the fiber partial slides relative to the matrix (i.e., B_1–C_1 part in Fig. 5.6d). In region III, the hysteresis dissipated energy remains to be constant value of $W = 9.3$ kPa (i.e., D_1–E_1 part in Fig. 5.6a), the interface complete debonds (i.e., B_1–C part in Fig. 5.6b) and complete oxidizes (i.e., C–D part in Fig. 5.6c), and the fiber partial slides relative to the matrix (i.e., C_1–D_1 part in Fig. 5.6d).

When the fatigue peak stress is $\sigma_{max} = 200$ MPa and with increasing oxidation time, the hysteresis dissipated energy versus the oxidation time curve is divided into three regions. In region I, the hysteresis dissipated energy increases from $W = 13$ kPa to the peak value of $W = 28.2$ kPa and decreases to $W = 28$ kPa (i.e., A_2–B_2–C_2 part in Fig. 5.6a), the interface partial debonds (i.e., A_2–B_2 part in Fig. 5.6b) and partial oxidizes (i.e., A–B_2 part in Fig. 5.6c), and the fiber partial slides relative to the matrix (i.e., A–B_2 part in Fig. 5.6d). In region II, the hysteresis dissipated energy continually decreases to $W = 20$ kPa (i.e., C_2–D_2 part in Fig. 5.6a), the interface complete debonds (i.e., B_2–C part in Fig. 5.6b) and partial oxidizes (i.e., B_2–C part in Fig. 5.6c), and the fiber partial slides relative to the matrix (i.e., B_2–C_2 part in Fig. 5.6d). In region III, the hysteresis dissipated energy remains to be constant value of $W = 20$ kPa (i.e., D_2–E_2 part in Fig. 5.6a). The interface complete debonds (i.e., B_2–C part in Fig. 5.6b) and complete oxidizes (i.e., C–D part in Fig. 5.6c), and the fiber partial slides relative to the matrix (i.e., C_2–D_2 part in Fig. 5.6d).

When the static peak stress increases, the hysteresis dissipated energy increases for a different oxidation time, and the oxidation time corresponding to the interface complete debonding decreases.

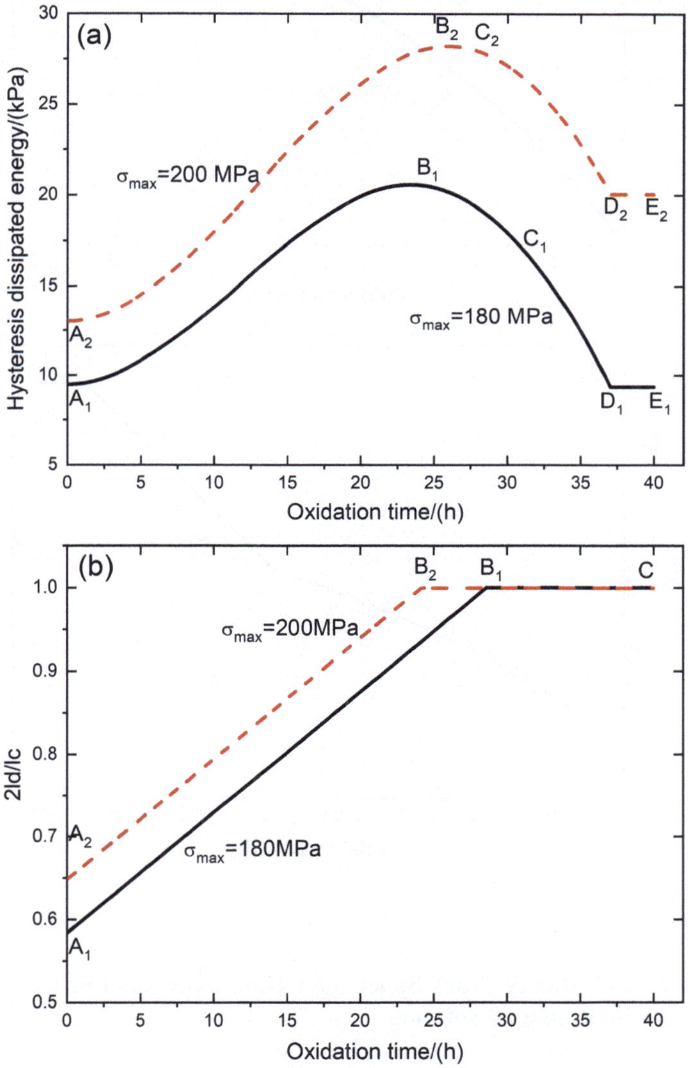

Fig. 5.6 **a** Hysteresis dissipated energy versus the oxidation time; **b** the interface debonding length versus the oxidation time; **c** the interface oxidation length versus the oxidation time; and **d** the interface counter-slip length versus the oxidation time corresponding to different static peak stresses of $\sigma_{\max} = 180$ and 200 MPa

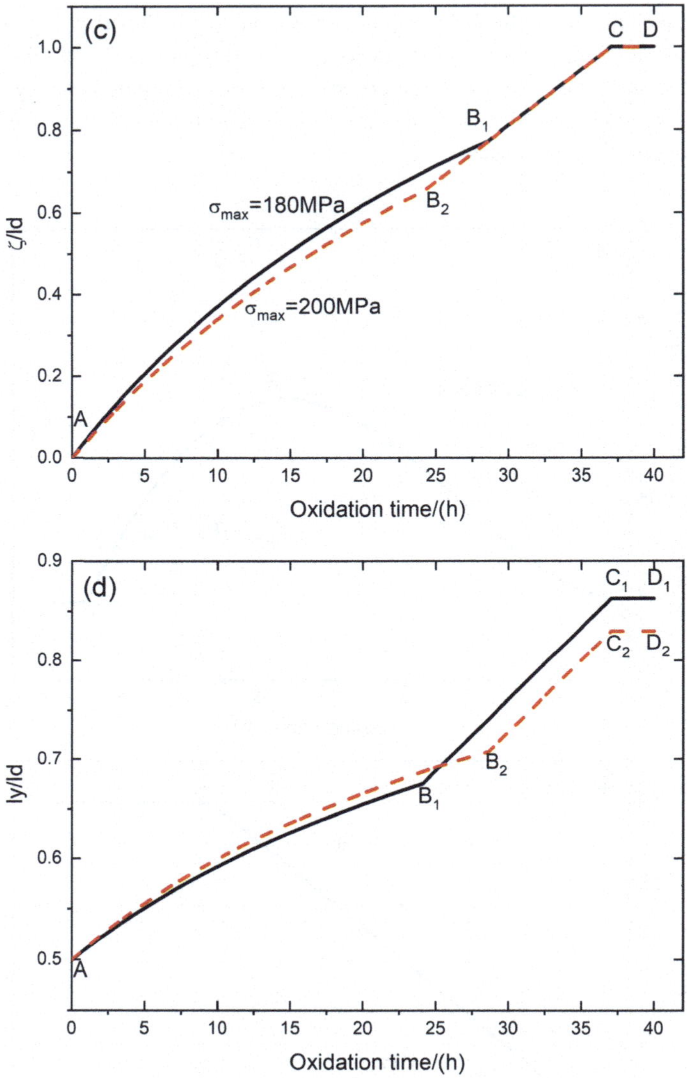

Fig. 5.6 (continued)

5.2.2.2 Effect of Matrix Crack Spacing on Time-Dependent Static Fatigue Damage Evolution

The hysteresis dissipated energy, the interface debonding, the oxidation and slip lengths versus the oxidation time corresponding to a different matrix crack spacing of $l_c = 200$ and 240 μm are shown in Fig. 5.7.

When the matrix crack spacing is $l_c = 200$ μm and with increasing oxidation time, the hysteresis dissipated energy versus the oxidation time curve is divided into three regions. In region I, the hysteresis dissipated energy increases from $W = 10.4$ kPa to the peak value of $W = 22.6$ kPa and decreases to $W = 22.5$ kPa (i.e., A_1–B_1–C_1 part in Fig. 5.7a), the interface partial debonds (i.e., A_1–B_1 part in Fig. 5.7b) and partial oxidizes (i.e., A–B_1 part in Fig 5.7c), and the fiber partial slides relative to the matrix (i.e., A–B_1 part in Fig. 5.7d). In region II, the hysteresis dissipated energy

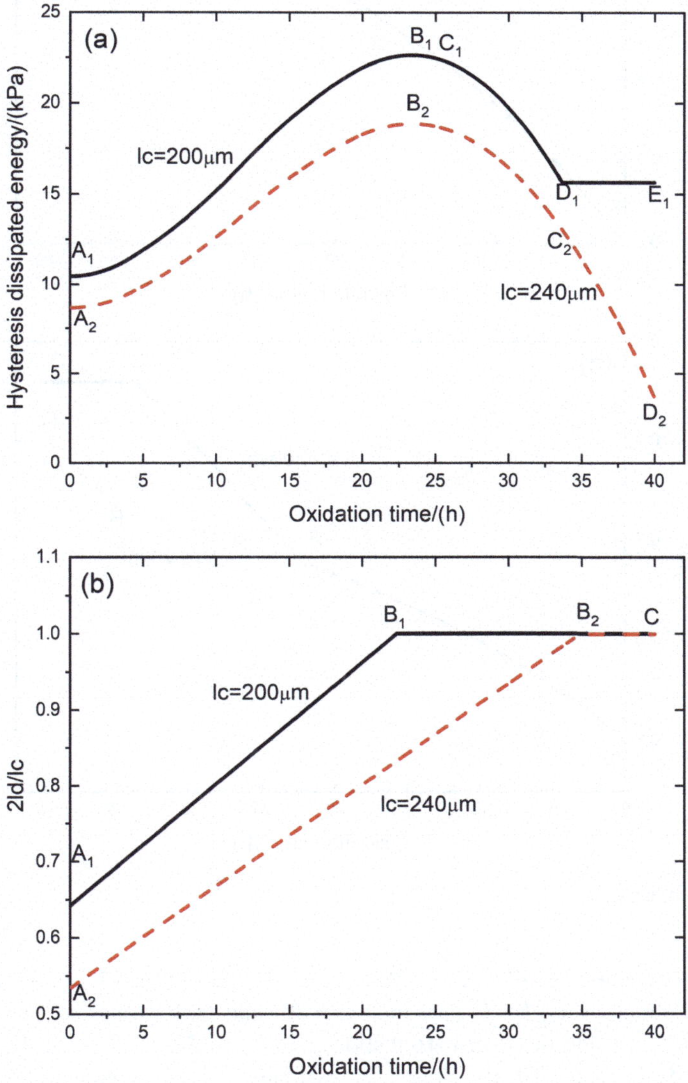

Fig. 5.7 **a** Hysteresis dissipated energy versus the oxidation time; **b** the interface debonding length versus the oxidation time; **c** the interface oxidation length versus the oxidation time; and **d** the interface counter-slip length versus the oxidation time corresponding to a different matrix crack spacing of $l_c = 200$ and 240 μm

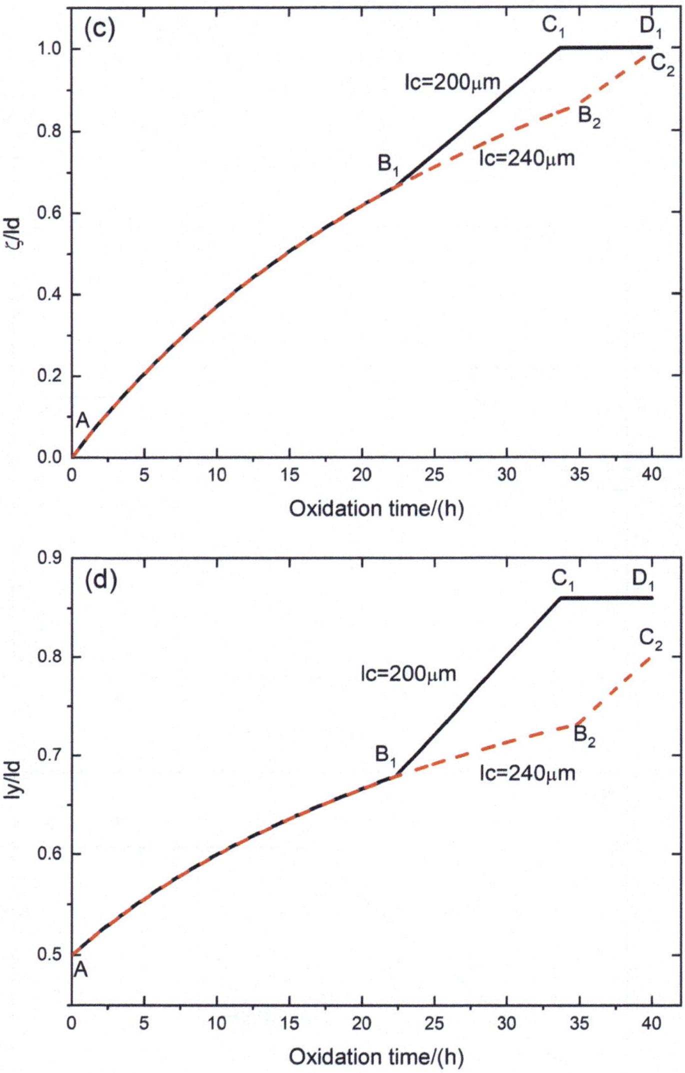

Fig. 5.7 (continued)

continually decreases to $W = 15.6$ kPa (i.e., C_1–D_1 part in Fig. 5.7a), the interface complete debonds (i.e., B_1–C part in Fig. 5.7b) and partial oxidizes (i.e., B_1–C_1 part in Fig. 5.7c), and the fiber partial slides relative to the matrix (i.e., B_1–C_1 part in Fig. 5.7d). In region III, the hysteresis dissipated energy remains to be constant value of $W = 15.6$ kPa (i.e., D_1–E_1 part in Fig. 5.7a), the interface complete debonds (i.e., B_1–C part in Fig. 5.7b) and complete oxidizes (i.e., C_1–D_1 part in Fig. 5.7c), and the fiber partial slides relative to the matrix (i.e., C_1–D_1 part in Fig. 5.7d).

When the matrix crack spacing is $l_c = 240$ μm and with increasing oxidation time, the hysteresis dissipated energy versus the oxidation time curve is divided into two regions. In region I, the hysteresis dissipated energy increases from $W = 8.7$ kPa to the peak value of $W = 18.8$ kPa and decreases to $W = 11.5$ kPa (i.e., A_2–B_2–C_2 part in Fig. 5.7a), the interface partial debonds (i.e., A_2–B_2 part in Fig. 5.7b) and partial oxidizes (i.e., A–B_2 part in Fig. 5.7c), and fiber partial slides relative to the matrix (i.e., A–B_2 part in Fig. 5.7d). In region II, the hysteresis dissipated energy continually decreases to $W = 3.6$ kPa (i.e., C_2–D_2 part in Fig. 5.7a), the interface complete debonds (i.e., B_2–C part in Fig. 5.7b) and partial oxidizes (i.e., B_2–C_2 part in Fig. 5.7c), and the fiber partial slides relative to the matrix (i.e., B_2–C_2 part in Fig. 5.7d).

When the matrix crack spacing increases, the hysteresis dissipated energy decreases for a different oxidation time, and the oxidation time corresponding to the interface complete debonding increases.

5.2.2.3 Effect of Fiber Volume Content on Time-Dependent Static Fatigue Damage Evolution

The hysteresis dissipated energy, the interface debonding, the oxidation and slip lengths versus the oxidation time corresponding to different fiber volumes of $V_f =$ 35 and 40% are shown in Fig. 5.8.

When the fiber volume is $V_f = 35\%$ and with increasing oxidation time, the hysteresis dissipated energy versus the oxidation time curve is divided into three regions. In region I, the hysteresis dissipated energy increases from $W = 15.9$ kPa to $W = 29.4$ kPa (i.e., A_1–B_1 part in Fig. 5.8a), the interface partial debonds (i.e., A_1–B_1 part in Fig. 5.8b) and partial oxidizes (i.e., A–B_1 part in Fig. 5.8c), and the fiber partial slides relative to the matrix (i.e., A–B_1 part in Fig. 5.8d). In region II, the hysteresis dissipated energy continually increases to the peak value of $W = 34.5$ kPa (i.e., B_1–C_1 part in Fig. 5.8a), the interface complete debonds (i.e., B_1–C part in Fig. 5.8b) and partial oxidizes (i.e., B_1–C part in Fig. 5.8c), and the fiber partial slides relative to the matrix (i.e., B_1–C_1 part in Fig. 5.8d). In region III, the hysteresis dissipated energy decreases to $W = 30.7$ kPa (i.e., C_1–D_1 part in Fig. 5.8a), the interface complete debonds (i.e., B_1–C part in Fig. 5.8b) and partial oxidizes (i.e., B_1–C part in Fig. 5.8c), and the fiber partial slides relative to the matrix (i.e., B_1–C_1 part in Fig. 5.8d).

When the fiber volume is $V_f = 40\%$ and with increasing oxidation time, the hysteresis dissipated energy versus the oxidation time curve is divided into two regions. In region I, the hysteresis dissipated energy increases from $W = 10.3$ kPa to the peak value of $W = 22.3$ kPa and decreases to $W = 20.4$ kPa (i.e., A_2–B_2–C_2 part in Fig. 5.8a), the interface partial debonds (i.e., A_2–B_2 part in Fig. 5.8b) and partial oxidizes (i.e., A–B_2 part in Fig. 5.8c), and the fiber partial slides relative to the matrix (i.e., A–B_2 part in Fig. 5.8d). In region II, the hysteresis dissipated energy continually decreases to $W = 10.6$ kPa (i.e., C_2–D_2 part in Fig. 5.8a), the interface complete debonds (i.e., B_2–C part in Fig. 5.8b) and partial oxidizes (i.e., B_2–C part

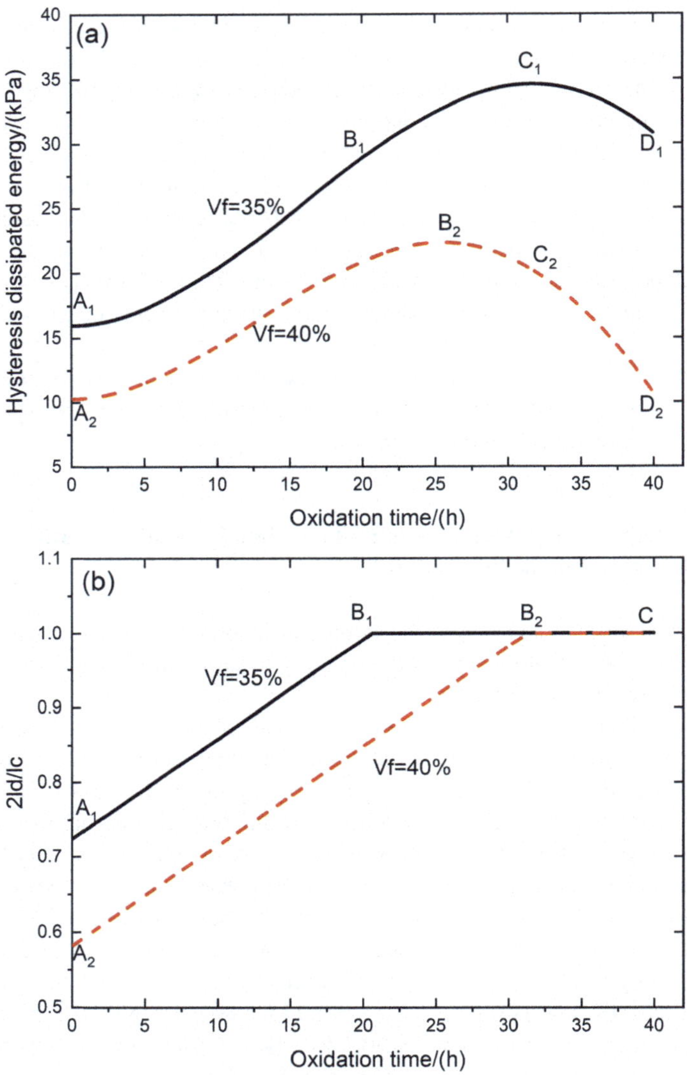

Fig. 5.8 **a** Hysteresis dissipated energy versus the oxidation time; **b** the interface debonding length versus the oxidation time; **c** the interface oxidation length versus the oxidation time; and **d** the interface counter-slip length versus the oxidation time corresponding to different fiber volumes of $V_f = 35$ and 40%

in Fig. 5.8c), and the fiber partial slides relative to the matrix (i.e., B_2–C_2 part in Fig. 5.8d).

When the fiber volume increases, the hysteresis dissipated energy decreases for a different oxidation time, and the oxidation time corresponding to the interface complete debonding increases.

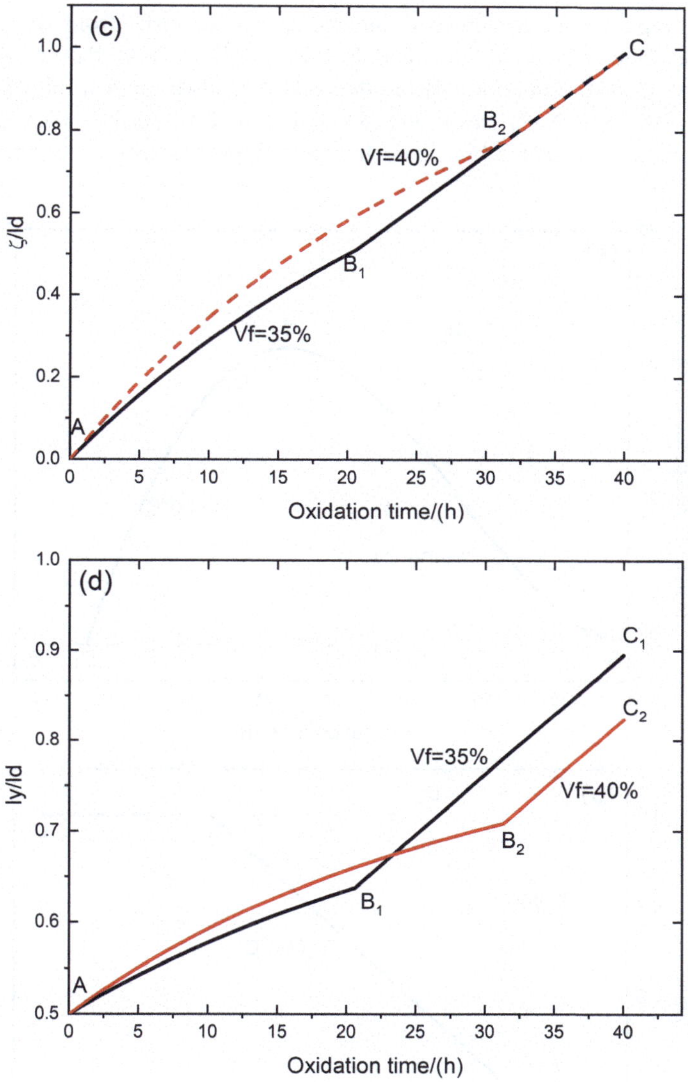

Fig. 5.8 (continued)

5.2.2.4 Effect of Oxidation Temperature on Time-Dependent Static Fatigue Damage Evolution

The hysteresis dissipated energy, the interface debonding, the oxidation and slip lengths versus the oxidation time corresponding to different oxidation temperatures of $T = 800$ and 900 °C are shown in Fig. 5.9.

When the oxidation temperature is $T = 800$ °C and with increasing oxidation time, the hysteresis dissipated energy versus the oxidation time curve is divided into three regions. In region I, the hysteresis dissipated energy increases from $W = 9.5$ kPa to the peak value of $W = 20.5$ kPa and decreases to $W = 18.9$ kPa (i.e., A–B_1–C_1 part in Fig. 5.9a), the interface partial debonds (i.e., A–B_1 part in Fig. 5.9b) and partial oxidizes (i.e., A–B_1 part in Fig. 5.9c), and the fiber partial slides relative to the matrix (i.e., A–B_1 part in Fig. 5.9d). In region II, the hysteresis dissipated energy

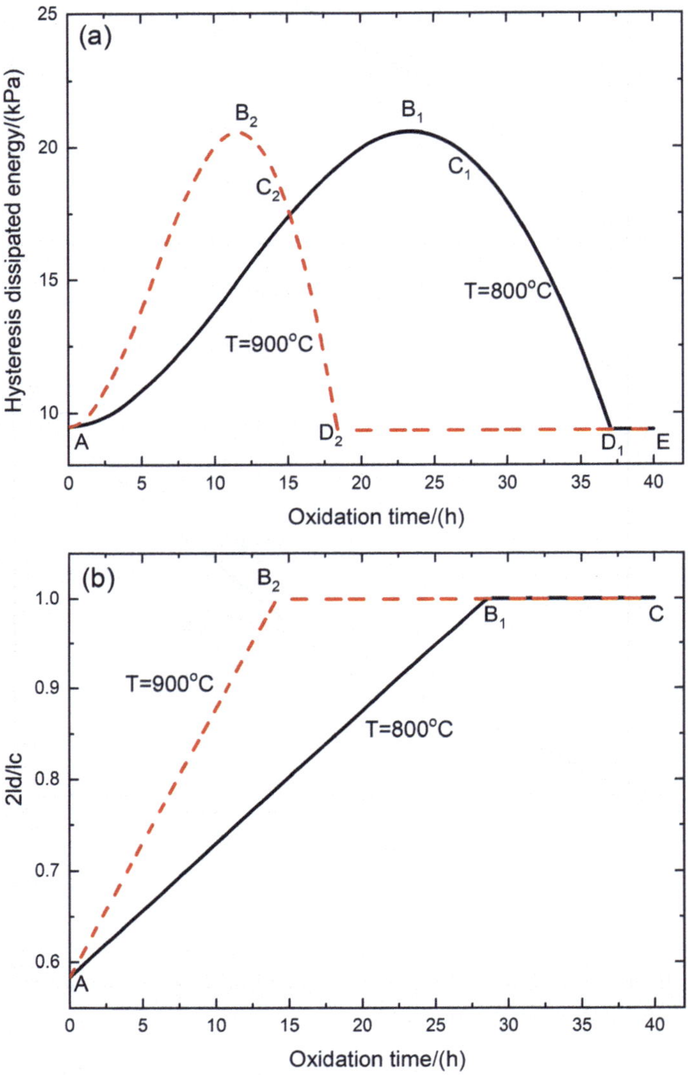

Fig. 5.9 **a** Hysteresis dissipated energy versus the oxidation time; **b** the interface debonded length versus the oxidation time; **c** the interface oxidation length versus the oxidation time; and **d** the interface counter-slip length versus the oxidation time corresponding to different oxidation temperatures of $T = 800$ and 900 °C

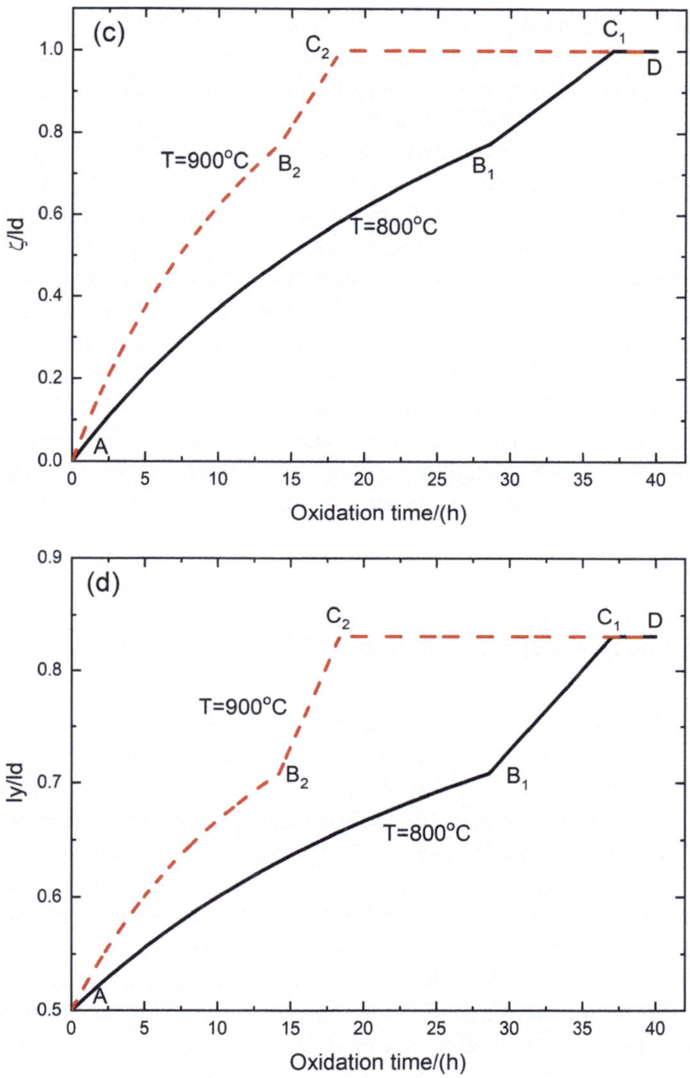

Fig. 5.9 (continued)

continually decreases to $W = 9.3$ kPa (i.e., C_1–D_1 part in Fig. 5.9a), the interface complete debonds (i.e., B_1–C part in Fig. 5.9b) and partial oxidizes (i.e., B_1–C_1 part in Fig. 5.9c), and the fiber partial slides relative to the matrix (i.e., B_1–C_1 part in Fig. 5.9d). In region III, the hysteresis dissipated energy remains to be constant value of $W = 9.3$ kPa (i.e., D_1–E part in Fig. 5.9a), the interface complete debonds (i.e., B_1–C part in Fig. 5.9b) and complete oxidizes (i.e., C_1–D part in Fig. 5.9c), and the fiber partial slides relative to the matrix (i.e., C_1–D part in Fig. 5.9d).

When the oxidation temperature is $T = 900$ °C and with increasing oxidation time, the hysteresis dissipated energy versus the oxidation time curve is divided into three regions. In region I, the hysteresis dissipated energy increases from $W = 9.5$ kPa to the peak value of $W = 20.5$ kPa and decreases to $W = 18.9$ kPa (i.e., A–B_2–C_2 part in Fig. 5.9a), the interface partial debonds (i.e., A–B_2 part in Fig. 5.9b) and partial oxidizes (i.e., A–B_2 part in Fig. 5.9c), and the fiber partial slips relative to the matrix (i.e., A–B_2 part in Fig. 5.9d). In region II, the hysteresis dissipated energy continually decreases to $W = 9.3$ kPa (i.e., C_2–D_2 part in Fig. 5.9a), the interface complete debonds (i.e., B_2–C part in Fig. 5.9b) and partial oxidizes (i.e., B_2–C_2 part in Fig. 5.9c), and the fiber partial slides relative to the matrix (i.e., B_2–C_2 part in Fig. 5.9d). In region III, the hysteresis dissipated energy remains to be constant value of $W = 9.3$ kPa (i.e., D_2–E part in Fig. 5.9a), the interface complete debonds (i.e., B_2–C part in Fig. 5.9b) and complete oxidizes (i.e., C_2–D part in Fig. 5.9c), and the fiber partial slides relative to the matrix (i.e., C_2–D part in Fig. 5.9d).

When the oxidation temperature increases, the oxidation time corresponding to interface complete debonding decreases, and the peak hysteresis dissipated energy remains constant.

5.2.3 *Experimental Comparisons*

Fantozzi and Reynaud (Fantozzi and Reynaud 2009) investigated the stress-strain hysteresis loops of C/[Si–B–C] composite during static fatigue at $T = 1200$ °C in air atmosphere and a stress applied of $\sigma_{max} = 170$ MPa. The load applied on the composite is steady and periodically performed an unloading/reloading sequence to obtain the stress-strain hysteresis loops. The experimental and predicted stress-strain hysteresis loops corresponding to a different oxidation time, i.e., from first loading to 144 h static fatigue in air, are shown in Fig. 5.10a. The hysteresis dissipated energy versus the oxidation time curve is shown in Fig. 5.10b, in which the hysteresis dissipated energy increases from $W = 25.3$ kPa to the peak value $W = 26.3$ first and then decreases with increasing oxidation time, to $W = 9.4$ kPa at 144 h, attributed to the propagation of oxidation region with decreasing interface shear stress in the oxidized region. The stress-strain hysteresis loops correspond to the interface slip Case 1 upon first loading, i.e., the interface partial debonds and the fiber partial slips relative to the matrix in the interface debonding region; then with increasing oxidation time, the hysteresis loops correspond to the interface slip Case 4, i.e., the interface complete debonds and the fiber complete slips relative to the matrix in the interface debonding region. The predicted stress-strain hysteresis loops and the evolution of hysteresis dissipated energy versus oxidation time are agreed with experimental data.

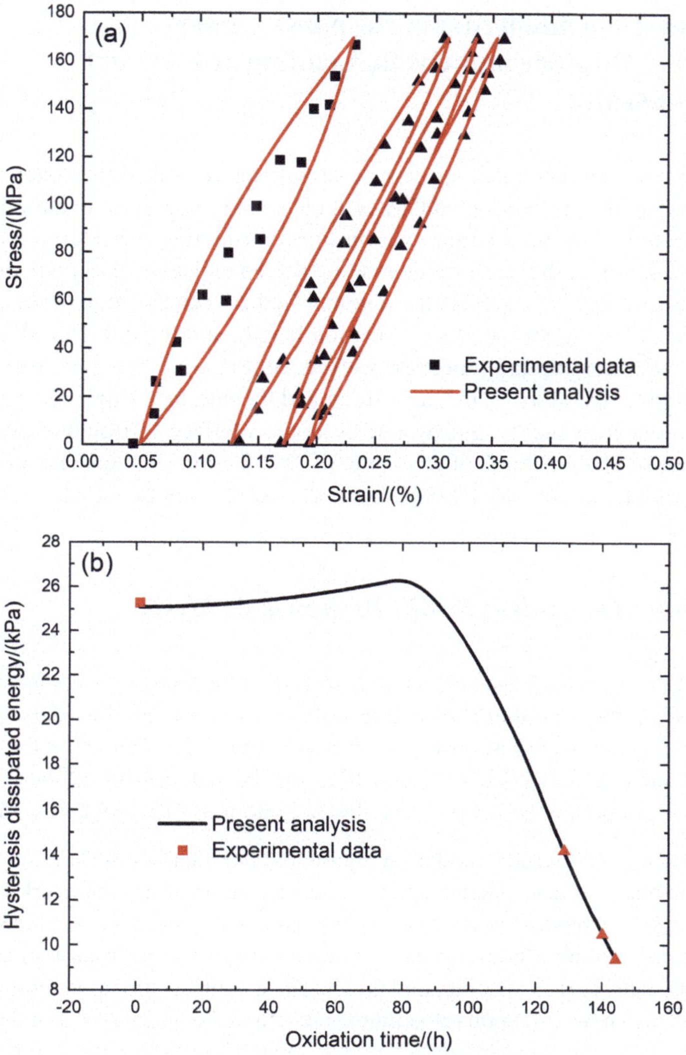

Fig. 5.10 **a** Experimental and predicted stress-strain hysteresis loops corresponding to a different oxidation time and **b** the experimental and predicted hysteresis dissipated energy versus oxidation time of C/[Si–B–C] composite during static fatigue under air at 1200 °C and a stress applied of 170 MPa

5.3 Time-Dependent Strain Response Under Stress-Rupture and Cyclic Loading at Elevated Temperature

In this section, the synergistic effects of stress-rupture and cyclic loading on the strain response of fiber-reinforced CMCs at elevated temperature in air atmosphere are investigated. The stress-strain relationships considering the interface wear and interface oxidation in the interface debonding region under stress-rupture and cyclic loading are developed to establish the relationship between the peak strain, the interface debonded length, the interface oxidation length, and the interface slip lengths. The effects of stress-rupture time, stress levels, matrix crack spacing, fiber volume, and oxidation temperature on the peak strain and the interface slip lengths are investigated. The experimental fatigue hysteresis loops, interface slip lengths, peak strain, and interface oxidation length of cross-ply SiC/MAS composite under cyclic fatigue and stress-rupture at 566 and 1093 °C in air atmosphere are predicted.

5.3.1 Time-Dependent Strain Response Analysis

Under cyclic fatigue loading at elevated temperature, the interface wear and interface oxidation affect the degradation of interface shear stress, the interface debonding and slip length, and the strain response of fiber-reinforced CMCs. Based on the interface debonding and interface slip between the fiber and the matrix inside of the composite, the interface debonding and slip can be divided into four different cases, including:

1. Case 1, i.e., the interface oxidation region and the interface wear region are less than the matrix crack spacing, and the interface counter-slip upon unloading and the interface new-slip upon reloading are equal to the interface debonded length.
2. Case 2, i.e., the interface oxidation region and the interface wear region are less than the matrix crack spacing, and the interface counter-slip upon unloading and the interface new-slip upon reloading are less than the interface debonded length.
3. Case 3, i.e., the interface oxidation region and the interface wear region are equal to the matrix crack spacing, and the interface counter-slip upon unloading and the interface new-slip upon reloading are less than the matrix crack spacing.
4. Case 4, i.e., the interface oxidation region and the interface wear region are equal to the matrix crack spacing, and the interface counter-slip upon unloading and the interface new-slip upon reloading are equal to the matrix crack spacing.

5.3.1.1 Case 1

When the interface oxidation region and the interface wear region are less than matrix crack spacing, upon unloading, the interface debonded region can be divided into three regions, i.e., the interface counter-slip region with the interface shear stress of τ_f ($z \in [0, \zeta]$), the interface counter-slip region with the interface shear stress of $\tau_i(N)$ ($z \in [\zeta, l_y]$), and the interface slip region with the interface shear stress of $\tau_i(N)$ ($z \in [l_y, l_d]$), in which l_y denotes the interface counter-slip length. Upon unloading to the unloading transition stress of σ_{tr_pu}, the interface counter-slip length approaches the interface debonding length, i.e., $l_y(\sigma = \sigma_{tr\_pu}) = l_d$. When $\sigma < \sigma_{tr\_pu}$, the counter-slip occurs over the entire interface debonding region, i.e., $l_y(\sigma < \sigma_{tr\_pu}) = l_d$. The unloading strain is divided into two regions, as follows:

$$\begin{aligned}\varepsilon_{\text{unloading}} = & \frac{2\sigma l_d}{V_f E_f l_c} + \frac{2\tau_f}{r_f E_f l_c}\zeta^2 + \frac{4\tau_f}{r_f E_f l_c}\zeta(l_d - \zeta) \\ & + \frac{4\tau_i(N)}{r_f E_f l_c}(l_y - \zeta)^2 - \frac{2\tau_i(N)}{r_f E_f l_c}(2l_y - \zeta - l_d)^2 \\ & + \frac{2\sigma_{fo}}{E_f l_c}\left(\frac{l_c}{2} - l_d\right) + \frac{2r_f}{\rho E_f l_c}\left[\frac{V_m}{V_f}\sigma_{mo} + \frac{2\tau_f}{r_f}\zeta + \frac{2\tau_i(N)}{r_f}(2l_y - \zeta - l_d)\right] \\ & \times \left[1 - \exp\left(-\rho\frac{l_c/2 - l_d}{r_f}\right)\right] - (\alpha_c - \alpha_f)\Delta T, \sigma > \sigma_{tr\_pu}\end{aligned} \tag{5.11}$$

$$\begin{aligned}\varepsilon_{\text{unloading}} = & \frac{2\sigma l_d}{V_f E_f l_c} + \frac{2\tau_f}{r_f E_f l_c}\zeta^2 + \frac{4\tau_f}{r_f E_f l_c}\zeta(l_d - \zeta) \\ & + \frac{4\tau_i(N)}{r_f E_f l_c}(l_d - \zeta)^2 - \frac{2\tau_i(N)}{r_f E_f l_c}(l_d - \zeta)^2 \\ & + \frac{2\sigma_{fo}}{E_f l_c}\left(\frac{l_c}{2} - l_d\right) + \frac{2r_f}{\rho E_f l_c}\left[\frac{V_m}{V_f}\sigma_{mo} + \frac{2\tau_f}{r_f}\zeta + \frac{2\tau_i(N)}{r_f}(l_d - \zeta)\right] \\ & \times \left[1 - \exp\left(-\rho\frac{l_c/2 - l_d}{r_f}\right)\right] - (\alpha_c - \alpha_f)\Delta T, \sigma < \sigma_{tr\_pu}\end{aligned} \tag{5.12}$$

where

$$\begin{aligned}l_y = & \frac{1}{2}\left\{l_d + \left(1 - \frac{\tau_f}{\tau_i(N)}\right)\zeta \right. \\ & \left. - \left[\frac{r_f}{2}\left(\frac{V_m E_m}{V_f E_c}\frac{\sigma}{\tau_i(N)} - \frac{1}{\rho}\right) - \sqrt{\left(\frac{r_f}{2\rho}\right)^2 + \frac{r_f V_m E_m E_f}{E_c[\tau_i(N)]^2}\xi_d}\right]\right\}\end{aligned} \tag{5.13}$$

Upon reloading, the interface debonded region can be divided into four regions, i.e., the interface new-slip region with the interface shear stress of τ_f ($z \in [0, l_z]$), the interface counter-slip region with the interface shear stress of τ_f ($z \in [l_z, \zeta]$), the interface counter-slip region with the interface shear stress of $\tau_i(N)$ ($z \in [\zeta, l_y]$), and the interface slip region with the interface shear stress of $\tau_i(N)$ ($z \in [l_y, l_d]$). Upon reloading to the reloading transition stress of σ_{tr_pr}, the interface new-slip length approaches the interface debonded length, i.e., $l_z(\sigma_{tr\_pr}) = l_d$. When $\sigma > \sigma_{tr\_pr}$, new slip occurs over the entire interface debonded length, i.e., $l_z(\sigma > \sigma_{tr\_pr}) = l_d$. The reloading strain is divided into two regions, as follows:

$$
\begin{aligned}
\varepsilon_{\text{reloading}} = & \frac{2\sigma}{V_{\mathrm{f}}E_{\mathrm{f}}l_{\mathrm{c}}}l_{\mathrm{d}} - \frac{4\tau_{\mathrm{f}}}{r_{\mathrm{f}}E_{\mathrm{f}}l_{\mathrm{c}}}l_z^2 + \frac{2\tau_{\mathrm{f}}}{r_{\mathrm{f}}E_{\mathrm{f}}l_{\mathrm{c}}}(2l_z - \zeta)^2 \\
& - \frac{4\tau_{\mathrm{f}}}{r_{\mathrm{f}}E_{\mathrm{f}}l_{\mathrm{c}}}(2l_z - \zeta)(l_{\mathrm{d}} - \zeta) + \frac{4\tau_{\mathrm{i}}(N)}{r_{\mathrm{f}}E_{\mathrm{f}}l_{\mathrm{c}}}(l_y - \zeta)^2 \\
& - \frac{2\tau_{\mathrm{i}}(N)}{r_{\mathrm{f}}E_{\mathrm{f}}l_{\mathrm{c}}}(2l_y - \zeta - l_{\mathrm{d}})^2 + \frac{2\sigma_{\mathrm{fo}}}{E_{\mathrm{f}}l_{\mathrm{c}}}\left(\frac{l_{\mathrm{c}}}{2} - l_{\mathrm{d}}\right) \\
& + \frac{2r_{\mathrm{f}}}{\rho E_{\mathrm{f}}l_{\mathrm{c}}}\left[\frac{V_{\mathrm{m}}}{V_{\mathrm{f}}}\sigma_{\mathrm{mo}} - \frac{2\tau_{\mathrm{f}}}{r_{\mathrm{f}}}(2l_z - \zeta) + \frac{2\tau_{\mathrm{i}}(N)}{r_{\mathrm{f}}}(2l_y - \zeta - l_{\mathrm{d}})\right] \\
& \times \left[1 - \exp\left(-\rho\frac{l_{\mathrm{c}}/2 - l_{\mathrm{d}}}{r_{\mathrm{f}}}\right)\right] - (\alpha_{\mathrm{c}} - \alpha_{\mathrm{f}})\Delta T, \sigma < \sigma_{\text{tr\_pr}}
\end{aligned} \tag{5.14}
$$

$$
\begin{aligned}
\varepsilon_{\text{reloading}} = & \frac{2\sigma}{V_{\mathrm{f}}E_{\mathrm{f}}l_{\mathrm{c}}}l_{\mathrm{d}} - \frac{4\tau_{\mathrm{f}}}{r_{\mathrm{f}}E_{\mathrm{f}}l_{\mathrm{c}}}l_{\mathrm{d}}^2 + \frac{2\tau_{\mathrm{f}}}{r_{\mathrm{f}}E_{\mathrm{f}}l_{\mathrm{c}}}(2l_{\mathrm{d}} - \zeta)^2 \\
& - \frac{4\tau_{\mathrm{f}}}{r_{\mathrm{f}}E_{\mathrm{f}}l_{\mathrm{c}}}(2l_{\mathrm{d}} - \zeta)(l_{\mathrm{d}} - \zeta) + \frac{4\tau_{\mathrm{i}}(N)}{r_{\mathrm{f}}E_{\mathrm{f}}l_{\mathrm{c}}}(l_{\mathrm{d}} - \zeta)^2 \\
& - \frac{2\tau_{\mathrm{i}}(N)}{r_{\mathrm{f}}E_{\mathrm{f}}l_{\mathrm{c}}}(l_{\mathrm{d}} - \zeta)^2 + \frac{2\sigma_{\mathrm{fo}}}{E_{\mathrm{f}}l_{\mathrm{c}}}\left(\frac{l_{\mathrm{c}}}{2} - l_{\mathrm{d}}\right) \\
& + \frac{2r_{\mathrm{f}}}{\rho E_{\mathrm{f}}l_{\mathrm{c}}}\left[\frac{V_{\mathrm{m}}}{V_{\mathrm{f}}}\sigma_{\mathrm{mo}} - \frac{2\tau_{\mathrm{f}}}{r_{\mathrm{f}}}(2l_{\mathrm{d}} - \zeta) + \frac{2\tau_{\mathrm{i}}(N)}{r_{\mathrm{f}}}(l_{\mathrm{d}} - \zeta)\right] \\
& \times \left[1 - \exp\left(-\rho\frac{l_{\mathrm{c}}/2 - l_{\mathrm{d}}}{r_{\mathrm{f}}}\right)\right] - (\alpha_{\mathrm{c}} - \alpha_{\mathrm{f}})\Delta T, \sigma > \sigma_{\text{tr\_pr}}
\end{aligned} \tag{5.15}
$$

where

$$
\begin{aligned}
l_z = & \frac{\tau_{\mathrm{i}}(N)}{\tau_{\mathrm{f}}}\left\{l_y - \frac{1}{2}\left[l_{\mathrm{d}} + \left(1 - \frac{\tau_{\mathrm{f}}}{\tau_{\mathrm{i}}(N)}\right)\zeta\right.\right. \\
& \left.\left. - \left[\frac{r_{\mathrm{f}}}{2}\left(\frac{V_{\mathrm{m}}E_{\mathrm{m}}}{V_{\mathrm{f}}E_{\mathrm{c}}}\frac{\sigma}{\tau_{\mathrm{i}}(N)} - \frac{1}{\rho}\right) - \sqrt{\left(\frac{r_{\mathrm{f}}}{2\rho}\right)^2 + \frac{r_{\mathrm{f}}V_{\mathrm{m}}E_{\mathrm{m}}E_{\mathrm{f}}}{E_{\mathrm{c}}[\tau_i(N)]^2}\xi_{\mathrm{d}}}\right]\right]\right\}
\end{aligned} \tag{5.16}
$$

5.3.1.2 Case 2

When the interface oxidation region and the interface wear region are less than the matrix crack spacing, upon unloading to the fatigue valley stress, the interface counter-slip length is less than the interface debonding length, i.e., $l_y(\sigma = \sigma_{\min}) < l_{\mathrm{d}}$, and the unloading strain is determined by Eq. (5.12). Upon reloading to the fatigue peak stress, the interface new-slip length is less than the interface debonding length, $l_z(\sigma = \sigma_{\max}) < l_{\mathrm{d}}$, and the reloading strain is determined by Eq. (5.14).

5.3.1.3 Case 3

When the interface oxidation region and the interface wear region are equal to the matrix crack spacing, upon unloading to the fatigue valley stress, the interface counter-slip length is less than half matrix crack spacing, i.e., $l_y(\sigma = \sigma_{\min}) < l_c/2$. The unloading strain is determined by the following equation.

$$\begin{aligned} \varepsilon_{\text{unloading}} = & \frac{\sigma}{V_f E_f} - \frac{2\tau_f}{r_f E_f l_c}\zeta^2 + \frac{2\tau_f}{r_f E_f}\zeta + \frac{4\tau_i(N)}{r_f E_f l_c}(l_y - \zeta)^2 \\ & - \frac{2\tau_i(N)}{r_f E_f l_c}\left(2l_y - \zeta - \frac{l_c}{2}\right)^2 - (\alpha_c - \alpha_f)\Delta T \end{aligned} \tag{5.17}$$

where

$$l_y = \left[1 - \frac{\tau_f}{\tau_i(N)}\right]\zeta + \frac{r_f V_m E_m}{4 V_f E_c \tau_i(N)}(\sigma_{\max} - \sigma) \tag{5.18}$$

Upon reloading to the fatigue peak stress, the interface new-slip length is less than the half matrix crack spacing, i.e., $l_z(\sigma = \sigma_{\max}) < l_c/2$. The reloading strain is determined by the following equation.

$$\begin{aligned} \varepsilon_{\text{reloading}} = & \frac{\sigma}{V_f E_f} - \frac{4\tau_f}{r_f E_f l_c} l_z^2 + \frac{2\tau_f}{r_f E_f l_c}(2l_z - \zeta)^2 \\ & - \frac{4\tau_f}{r_f E_f l_c}(2l_z - \zeta)(l_y - \zeta) + \frac{4\tau_i(N)}{r_f E_f l_c}(l_y - \zeta)^2 \\ & - \frac{4\tau_f}{r_f E_f l_c}(2l_z - \zeta)\left(\frac{l_c}{2} - l_y\right) \\ & - \frac{2\tau_i(N)}{r_f E_f l_c}\left(2l_y - \zeta - \frac{l_c}{2}\right)^2 - (\alpha_c - \alpha_f)\Delta T \end{aligned} \tag{5.19}$$

where

$$l_z = l_y(\sigma_{\min}) - \frac{r_f V_m E_m}{4 V_f E_c \tau_i(N)}(\sigma_{\max} - \sigma) \tag{5.20}$$

5.3.1.4 Case 4

When the interface oxidation region and the interface wear region are equal to the matrix crack spacing, upon unloading to the transition stress of $\sigma_{\text{tr\_fu}}$, the interface counter-slip length approaches half matrix crack spacing, i.e., $l_y(\sigma = \sigma_{\text{tr\_fu}}) = l_c/2$. When $\sigma > \sigma_{\text{tr\_fu}}$, the interface counter-slip length is less than half matrix crack spacing, i.e., $l_y(\sigma > \sigma_{\text{tr\_fu}}) < l_c/2$, and the unloading strain is determined by Eq. (5.17).

When $\sigma < \sigma_{\text{tr\_fu}}$, the unloading interface counter-slip occurs over the entire matrix crack spacing, i.e., $y(\sigma < \sigma_{\text{tr\_fu}}) = l_c/2$, and the unloading strain is determined by the following equation.

$$\varepsilon_{\text{unloading}} = \frac{\sigma}{V_f E_f} - \frac{2\tau_f}{r_f E_f l_c}\zeta^2 + \frac{2\tau_f}{r_f E_f}\zeta + \frac{4\tau_i(N)}{r_f E_f l_c}\left(\frac{l_c}{2} - \zeta\right)^2 - \frac{2\tau_i(N)}{r_f E_f l_c}\left(\frac{l_c}{2} - \zeta\right)^2 - (\alpha_c - \alpha_f)\Delta T \tag{5.21}$$

Upon reloading to the transition stress of $\sigma_{\text{tr\_fr}}$, the interface new-slip length approaches the half matrix crack spacing, i.e., $l_z(\sigma = \sigma_{\text{tr\_fr}}) = l_c/2$. When $\sigma < \sigma_{\text{tr\_fr}}$, the interface new-slip length is less than the half matrix crack spacing, i.e., $l_z(\sigma < \sigma_{\text{tr\_fr}}) < l_c/2$, and the reloading strain is determined by Eq. (5.19). When $\sigma > \sigma_{\text{tr\_fr}}$, the interface new-slip length occurs over the entire matrix crack spacing, i.e., $l_z(\sigma_{\text{tr\_fr}} < \sigma) = l_c/2$, and the reloading strain is determined by the following equation.

$$\begin{aligned}\varepsilon_{\text{reloading}} = &\frac{\sigma}{V_f E_f} - \frac{4\tau_f}{r_f E_f l_c} l_z^2 + \frac{2\tau_f}{r_f E_f l_c}(l_c - \zeta)^2 \\ &- \frac{4\tau_f}{r_f E_f l_c}(l_c - \zeta)\left(\frac{l_c}{2} - \zeta\right) + \frac{4\tau_i(N)}{r_f E_f l_c}\left(\frac{l_c}{2} - \zeta\right)^2 \\ &- \frac{4\tau_f}{r_f E_f l_c}(l_c - \zeta)\left(\frac{l_c}{2} - l_y\right) - \frac{2\tau_i(N)}{r_f E_f l_c}\left(\frac{l_c}{2} - \zeta\right)^2 - (\alpha_c - \alpha_f)\Delta T\end{aligned} \tag{5.22}$$

5.3.2 Results and Discussion

Under cyclic loading at elevated temperature, there are two types of loading sequences considered as shown in Fig. 5.11, including:

1. Case 1: Cyclic fatigue loading without stress-rupture, and the interface debonding and frictional slipping between the fiber and the matrix are mainly affected by the interface wear.
2. Case 2: Cyclic fatigue loading with stress-rupture, and the interface debonding and frictional slip between the fiber and the matrix are mainly affected by the interface oxidation.

The synergistic effects of stress-rupture and cyclic loading on the strain response of fiber-reinforced CMCs are investigated, considering different fatigue peak stresses, matrix crack spacing, fiber volume, oxidation temperature, and stress-rupture time. The ceramic composite system of unidirectional SiC/MAS (Grant 1994) is used for the case study, and its basic material properties are given by: $V_f = 40\%$, $r_f = 7.5\ \mu\text{m}$,

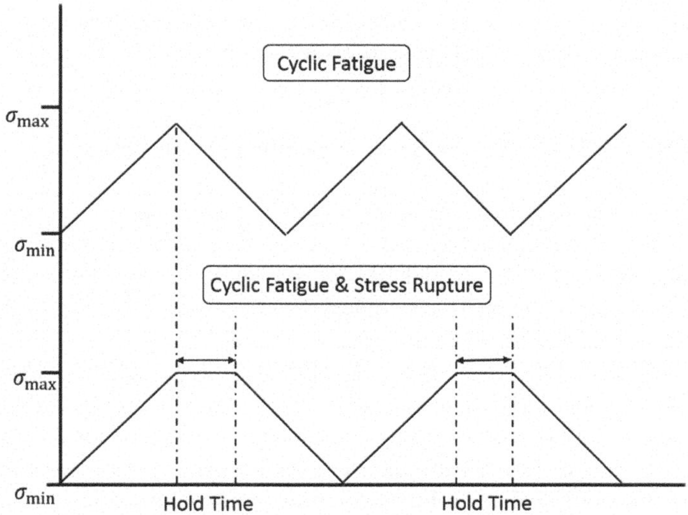

Fig. 5.11 Schematic of cyclic fatigue loading and cyclic fatigue loading with hold time

$E_f = 200$ GPa, $E_m = 138$ GPa, $\alpha_f = 4\times10^{-6}$/°C, $\alpha_m = 2.4 \times 10^{-6}$/°C, $\Delta T = -1000$ °C, $\tau_i = 20$ MPa, $\tau_f = 5$ MPa, and $\xi_d = 0.1$ J/m^2.

The interface shear stress versus cycle number curve is illustrated in Fig. 5.12, in which the parameters of the interface shear stress degradation model are given by:

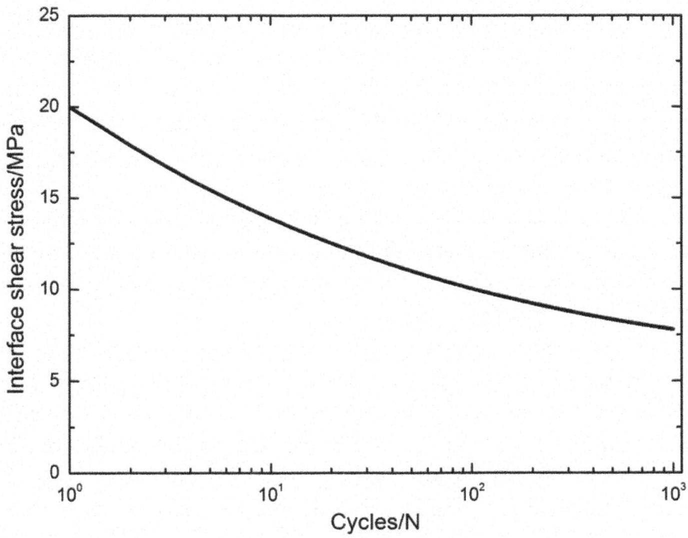

Fig. 5.12 Interface shear stress degradation model

$\tau_0 = 20$ MPa, $\tau_s = 5$ MPa, $b_0 = 4$, and $j = 0.27$. The interface shear stress degrades from 20 MPa at the first applied cycle to 7.8 MPa at the 1000th applied cycle.

5.3.2.1 Effect of Hold Time on Time-Dependent Strain Response

The peak strain, the interface debonding length, and the interface oxidation length versus applied cycle number curves under $\sigma_{max} = 200$ MPa corresponding to different stress-rupture times of $t = 1$, 5, and 10 s are shown in Fig. 5.13. When the stress-rupture time increases, the interface slip lengths increase with applied cycles due to interface oxidation, leading to the increase of the peak strain.

When the stress-rupture time is $t = 1$ s, the peak strain increases from $\varepsilon_{max} = 0.13\%$ at the first applied cycle to $\varepsilon_{max} = 0.22\%$ at the 33816th applied cycle (i.e., A–B_1 part in Fig. 5.13a), corresponding to the interface partial debonding (i.e., $2l_d/l_c < 1$ and A–B_1 part in Fig. 5.13b) and the interface partial oxidation (i.e., $\zeta/l_d < 1$ and A–B_1 part in Fig. 5.13c); increases to the peak value of $\varepsilon_{max} = 0.223\%$ at the 41301th applied cycle (i.e., the B_1–C_1 part in Fig. 5.13a), corresponding to the interface complete debonding (i.e., $2l_d/l_c = 1$ and B_1–C_1 part in Fig. 5.13b) and the interface partial oxidation (i.e., $\zeta/l_d < 1$ and B_1–C_1 part in Fig. 5.13c); and then remains constant of $\varepsilon_{max} = 0.223\%$ with increasing applied cycle number (i.e., the C_1–D_1 part in Fig. 5.13a), corresponding to the interface complete debonding (i.e., $2l_d/l_c = 1$ and B_1–C_1 part in Fig. 5.13b) and the interface complete oxidation (i.e., $\zeta/l_d = 1$ and C_1–D_1 part in Fig. 5.13c).

When the stress-rupture time is $t = 5$ s, the peak strain increases from $\varepsilon_{max} = 0.13\%$ at the first applied cycle to $\varepsilon_{max} = 0.22\%$ at the 6923th applied cycle, i.e., A–B_2 part in Fig. 5.13a, corresponding to the interface partial debonding (i.e., $2l_d/l_c < 1$ and A–B_2 part in Fig. 5.13b) and the interface partial oxidation (i.e., $\zeta/l_d < 1$ and A–B_2 part in Fig. 5.13c); increases to the peak value of $\varepsilon_{max} = 0.223\%$ at the 8261th applied cycle (i.e., the B_2–C_2 part in Fig. 5.13a), corresponding to the interface complete debonding (i.e., $2l_d/l_c = 1$ and B_2–C_2 part in Fig. 5.13b) and the interface partial oxidation (i.e., $\zeta/l_d < 1$ and B_2–C_2 part in Fig. 5.13c); and then remains constant of $\varepsilon_{max} = 0.223\%$ with increasing applied cycle number (i.e., the C_2–D_2 part in Fig. 5.13a), corresponding to the interface complete debonding (i.e., $2l_d/l_c = 1$ and B_2–C_2 part in Fig. 5.13b) and the interface complete oxidation (i.e., $\zeta/l_d = 1$ and C_2–D_2 part in Fig. 5.13c).

When the stress-rupture time is $t = 10$ s, the peak strain increases from $\varepsilon_{max} = 0.13\%$ at the first applied cycle to $\varepsilon_{max} = 0.22\%$ at the 3501th applied cycle (i.e., A–B_3 part in Fig. 5.13a, corresponding to the interface partial debonding (i.e., $2l_d/l_c < 1$ and A–B_3 part in Fig. 5.13b) and the interface partial oxidation (i.e., $\zeta/l_d < 1$ and A–B_3 part in Fig. 5.13c); increases to the peak value of $\varepsilon_{max} = 0.223\%$ at the 4131th applied cycle (i.e., the B_3–C_3 part in Fig. 5.13a), corresponding to the interface complete debonding (i.e., $2l_d/l_c = 1$ and B_3–C_3 part in Fig. 5.13b) and the interface partial oxidation (i.e., $\zeta/l_d < 1$ and B_3–C_3 part in Fig. 5.13c); and then remains constant of $\varepsilon_{max} = 0.223\%$ with increasing applied cycle number (i.e., the C_3–D_3 part in Fig. 5.13a), corresponding to the interface complete debonding (i.e.,

$2l_d/l_c = 1$ and B_3–C_3 part in Fig. 5.13b) and the interface complete oxidation (i.e., $\zeta/l_d = 1$ and C_3–D_3 part in Fig. 5.13c).

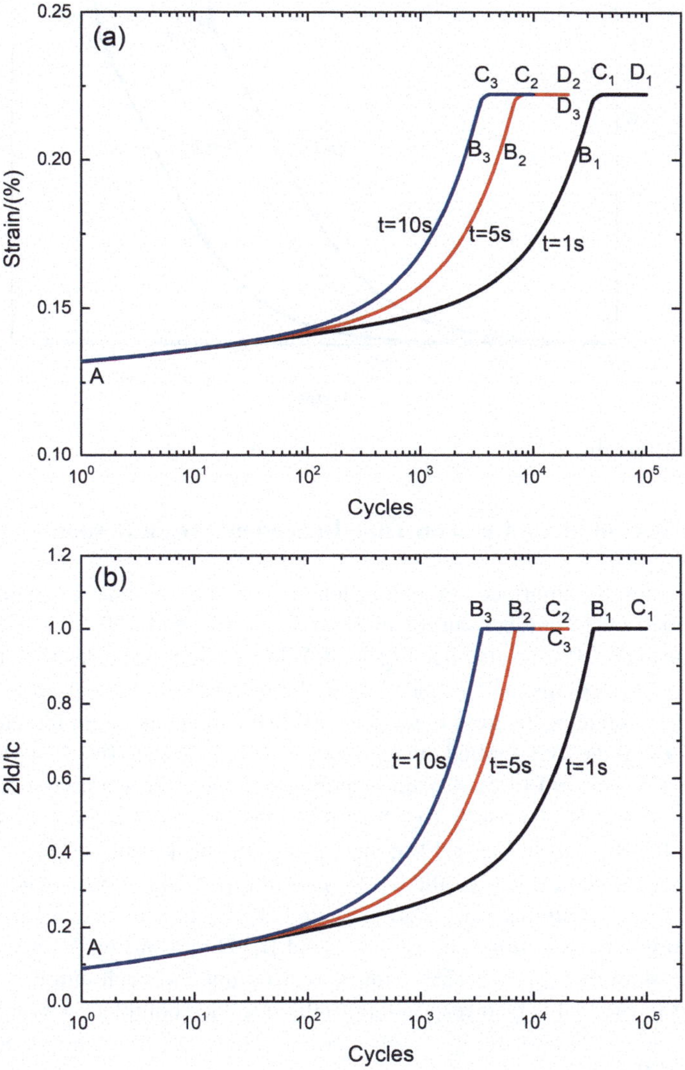

Fig. 5.13 **a** Peak strain versus applied cycle number curves; **b** the interface debonding length versus applied cycle number curves; and **c** the interface oxidation length versus applied cycle number curves under $\sigma_{max} = 200$ MPa corresponding to different stress-rupture times $t = 1$, 5, and 10 s at the oxidation temperature of $T = 800$ °C

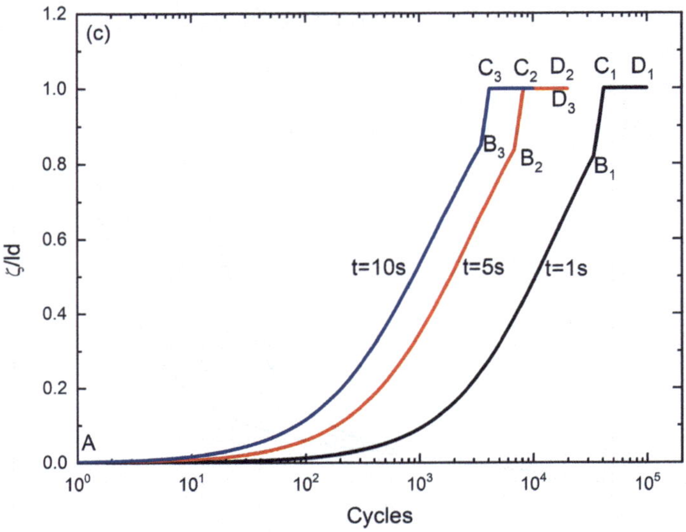

Fig. 5.13 (continued)

5.3.2.2 Effect of Stress Level on Time-Dependent Strain Response

The peak strain, the interface debonding length, and the interface oxidation length versus applied cycle number curves under $\sigma_{max} = 150$ and 250 MPa with stress-rupture time of $t = 10$ s are shown in Fig. 5.14. When the fatigue peak stress increases, the interface slip lengths increase, leading to the increase of the peak strain.

When the fatigue peak stress is $\sigma_{max} = 150$ MPa, the peak strain increases from $\varepsilon_{max} = 0.09\%$ at the first applied cycle to $\varepsilon_{max} = 0.13\%$ at the 2091th applied cycle (i.e., the A_1–B_1 part in Fig. 5.14a), corresponding to the interface partial debonding (i.e., $2l_d/l_c < 1$ and A_1–B_1 part in Fig. 5.14b) and the interface partial oxidation (i.e., $\zeta/l_d < 1$ and A–B_1 part in Fig. 5.14c); increases to the peak value of $\varepsilon_{max} = 0.15\%$, at the 4131th applied cycle (i.e., the B_1–C_1 part in Fig. 5.14a), corresponding to the interface partial debonding (i.e., $2l_d/l_c < 1$ and B_1–C_1 part in Fig. 5.14b) and the interface complete oxidation (i.e., $\zeta/l_d = 1$ and B_1–D part in Fig. 5.14c); and then remains constant of $\varepsilon_{max} = 0.15\%$ with increasing applied cycle number (i.e., the C_1–D_1 part in Fig. 5.14a), corresponding to the interface complete debonding (i.e., $2l_d/l_c = 1$ and C_1–D part in Fig. 5.14b) and the interface complete oxidation (i.e., $\zeta/l_d = 1$ and B_1–D part in Fig. 5.14c).

When the fatigue peak stress is $\sigma_{max} = 250$ MPa, the peak strain increases from $\varepsilon_{max} = 0.17\%$ at the first applied cycle to $\varepsilon_{max} = 0.27\%$ at the 2580th applied cycle (i.e., the A_2–B_2 part in Fig. 5.14a), corresponding to the interface partial debonding (i.e., $2l_d/l_c < 1$ and A_2–B_2 part in Fig. 5.14b) and the interface partial oxidation (i.e., $\zeta/l_d < 1$ and A–B_2 part in Fig. 5.14c); increases to the peak value of $\varepsilon_{max} = 0.28\%$ at the 4131th applied cycle (i.e., the B_2–C_2 part in Fig. 5.14a), corresponding to the interface complete debonding (i.e., $2l_d/l_c = 1$ and B_2–D part in Fig. 5.14b) and the interface partial oxidation (i.e., $\zeta/l_d < 1$ and B_2–C_2 part in Fig. 5.14c); and then remains constant of $\varepsilon_{max} = 0.28\%$ with increasing applied cycle number (i.e., the C_2–D_2 part in Fig. 5.14a), corresponding to the interface complete debonding (i.e., $2l_d/l_c = 1$ and B_2–D part in Fig. 5.14b) and the interface complete oxidation (i.e., $\zeta/l_d = 1$ and C_2–D part in Fig. 5.14c).

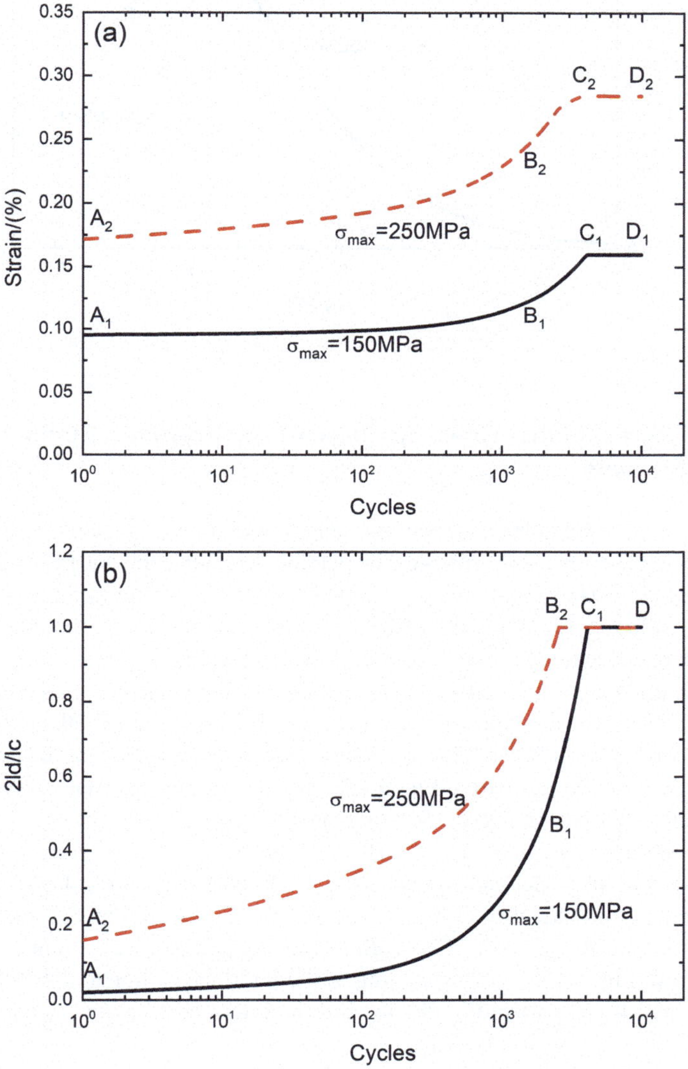

Fig. 5.14 **a** Peak strain versus cycle number curves; **b** the interface debonded length versus applied cycle number curves; and **c** the interface oxidation length versus applied cycle number curves under fatigue peak stresses of $\sigma_{max} = 150$ and 250 MPa with stress-rupture time $t = 10$ s at the oxidation temperature of $T = 800$ °C

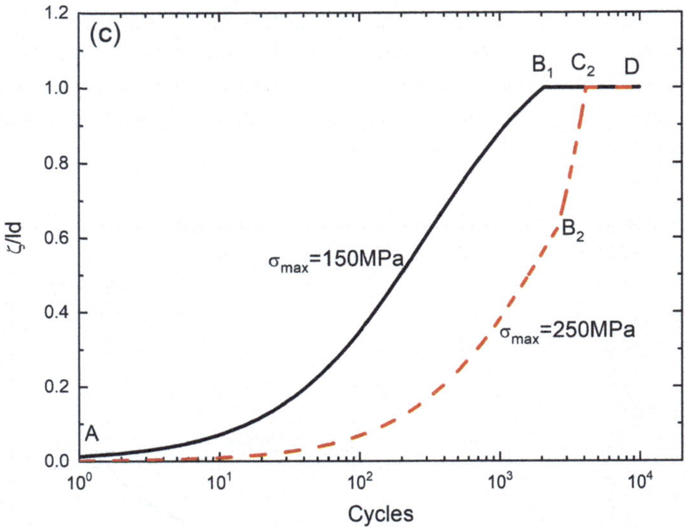

Fig. 5.14 (continued)

5.3.2.3 Effect of Matrix Crack Spacing on Time-Dependent Strain Response

The peak strain, the interface debonding length, and the interface oxidation length versus applied cycle number curves corresponding to a different matrix crack spacing of $l_c = 200$ and 300 μm under $\sigma_{max} = 200$ MPa with stress-rupture time of $t = 10$ s are shown in Fig. 5.15. When the matrix crack spacing increases, the interface slip lengths decrease, leading to the decrease of the peak strain.

When the matrix crack spacing is $l_c = 200$ μm, the peak strain increases from $\varepsilon_{max} = 0.138\%$ at the first applied cycle to $\varepsilon_{max} = 0.22\%$ at the 1583th applied cycle (i.e., the A_1–B_1 part in Fig. 5.15a), corresponding to the interface partial debonding (i.e., $2l_d/l_c < 1$ and A_1–B_1 part in Fig. 5.15b) and the interface partial oxidation (i.e., $\zeta/l_d < 1$ and A–B_1 part in Fig. 5.15c); increases to the peak value of $\varepsilon_{max} = 0.227\%$ at the 2424th applied cycle (i.e., the B_1–C_1 part in Fig. 5.15a), corresponding to the interface complete debonding (i.e., $2l_d/l_c = 1$ and B_1–C part in Fig. 5.15b) and the interface partial oxidation (i.e., $\zeta/l_d < 1$ and B_1–C_1 part in Fig. 5.15c); and then remains constant of $\varepsilon_{max} = 0.227\%$ with increasing applied cycle number (i.e., the C_1–D_1 part in Fig. 5.15a), corresponding to the interface complete debonding (i.e., $2l_d/l_c = 1$ and B_1–C part in Fig. 5.15b) and the interface complete oxidation (i.e., $\zeta/l_d = 1$ and C_1–D part in Fig. 5.15c).

When the matrix crack spacing is $l_c = 300$ μm, the peak strain increases from $\varepsilon_{max} = 0.133\%$ at the first applied cycle to $\varepsilon_{max} = 0.22\%$ at the 2944th applied cycle (i.e., the A_2–B_2 part in Fig. 5.15a), corresponding to the interface partial debonding (i.e., $2l_d/l_c < 1$ and A_2–B_2 part in Fig. 5.15b) and the interface partial oxidation (i.e., $\zeta/l_d < 1$ and A–B_2 part in Fig. 5.15c); increases to the peak value of $\varepsilon_{max} = 0.223\%$

at the 3642th applied cycle (i.e., the B_2–C_2 part in Fig. 5.15a), corresponding to the interface complete debonding (i.e., $2l_d/l_c = 1$ and B_2–C part in Fig. 5.15b) and the interface partial oxidation (i.e., $\zeta/l_d < 1$ and B_2–C_2 part in Fig. 5.15c); and then remains constant of $\varepsilon_{max} = 0.223\%$ with increasing applied cycle number (i.e., the C_2–D_2 part in Fig. 5.15a), corresponding to the interface complete debonding (i.e.,

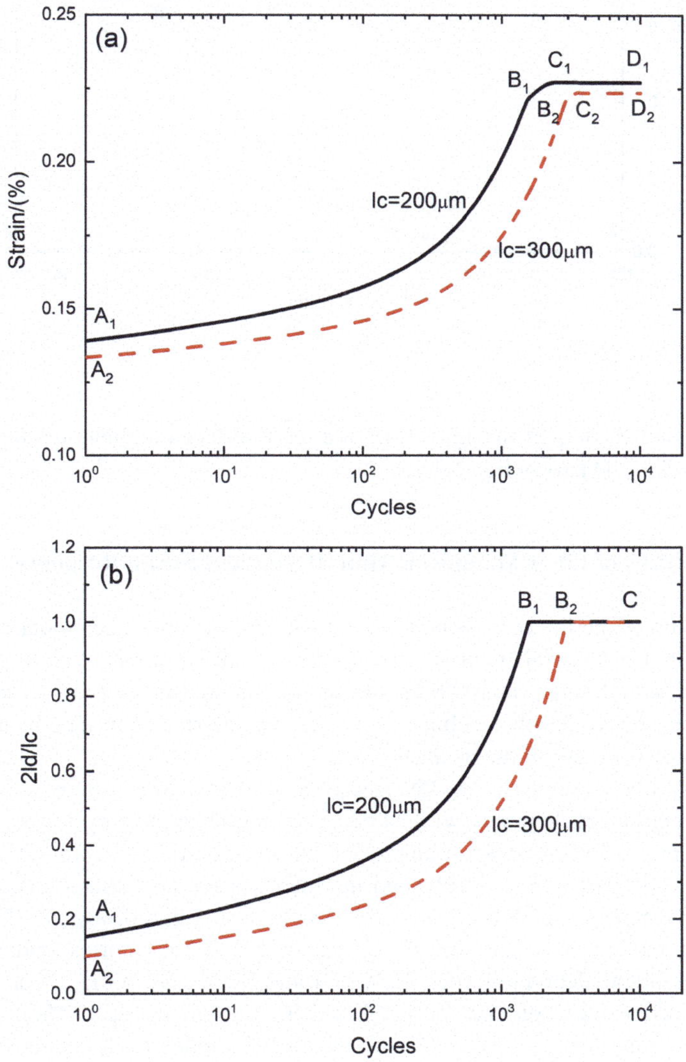

Fig. 5.15 **a** Peak strain versus cycle number curves; **b** the interface debonded length versus cycle number curves; and **c** the interface oxidation length versus cycle number curves corresponding to a different matrix crack spacing of $l_c = 200$ and 300 μm under $\sigma_{max} = 200$ MPa with stress-rupture time of t = 10 s at the oxidation temperature of $T = 800$ °C

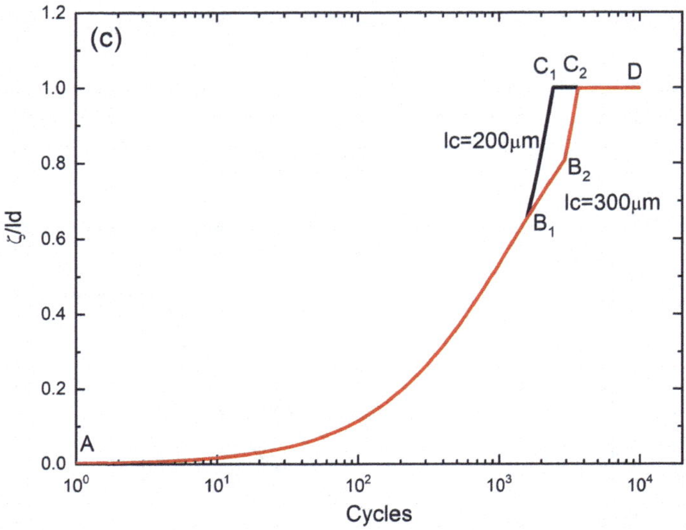

Fig. 5.15 (continued)

$2l_d/l_c = 1$ and B_2–C part in Fig. 5.15b) and the interface complete oxidation (i.e., $\zeta/l_d = 1$ and C_2–D part in Fig. 5.15c).

5.3.2.4 Effect of Fiber Volume on Time-Dependent Strain Response

The peak strain, the interface debonding length, and the interface oxidation length versus applied cycle number curves corresponding to different fiber volumes of $V_f =$ 35% and 45% under $\sigma_{max} = 200$ MPa with stress-rupture time of $t = 10$ s are shown in Fig. 5.16. When the fiber volume increases, the interface slip lengths decrease, leading to the decrease of the peak strain.

When the fiber volume is $V_f = 35\%$, the peak strain increases from $\varepsilon_{max} = 0.143\%$ at the first applied cycle to $\varepsilon_{max} = 0.247\%$ at the 2660th applied cycle (i.e., the A_1–B_1 part in Fig. 5.16a), corresponding to the interface partial debonding (i.e., $2l_d/l_c < 1$ and A_1–B_1 part in Fig. 5.16b) and the interface partial oxidation (i.e., $\zeta/l_d <$ 1 and A–B_1 part in Fig. 5.16c); increases to the peak value of $\varepsilon_{max} = 0.256\%$ at the 4131th applied cycle (i.e., the B_1–C_1 part in Fig. 5.16a), corresponding to the interface complete debonding (i.e., $2l_d/l_c = 1$ and B_1–D part in Fig. 5.16b) and the interface partial oxidation (i.e., $\zeta/l_d < 1$ and B_1–C_1 part in Fig. 5.16c); and then remains constant of $\varepsilon_{max} = 0.256\%$ with increasing applied cycle number (i.e., the C_1–D_1 part in Fig. 5.16a), corresponding to the interface complete debonding (i.e., $2l_d/l_c = 1$ and B_1–D part in Fig. 5.16b) and the interface complete oxidation (i.e., $\zeta/l_d = 1$ and C_1–D part in Fig. 5.16c).

When the fiber volume is $V_f = 45\%$, the peak strain increases from $\varepsilon_{max} = 0.125\%$ at the first applied cycle to $\varepsilon_{max} = 0.195\%$ at the 4066th applied cycle (i.e., the A_2–B_2 part in Fig. 5.16a), corresponding to the interface partial debonding (i.e., $2l_d/l_c < 1$ and A_2–B_2 part in Fig. 5.16b) and the interface partial oxidation (i.e., $\zeta/l_d < 1$ and A–B_2 part in Fig. 5.16c); increases to the peak value of $\varepsilon_{max} = 0.196\%$ at the 4131th applied cycle (i.e., the B_2–C_2 part in Fig. 5.16a), corresponding to the

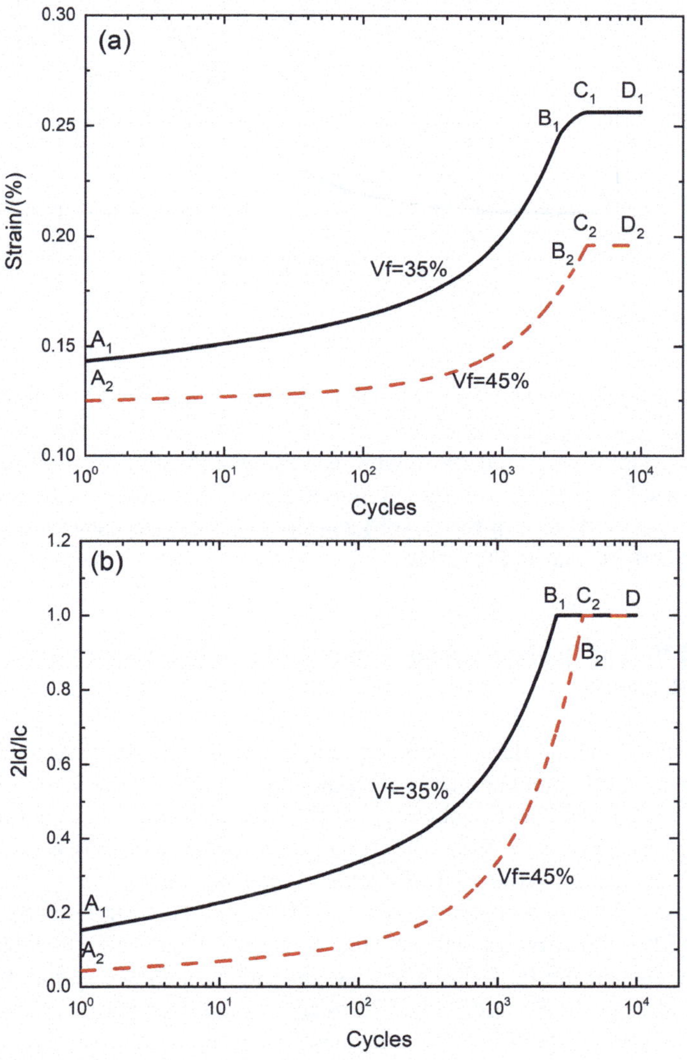

Fig. 5.16 **a** Peak strain versus applied cycle number curves; **b** the interface debonding length versus applied cycle number curves; and **c** the interface oxidation length versus applied cycle number curves, corresponding to different fiber volumes of $V_f = 35$ and 45% under $\sigma_{max} = 200$ MPa with stress-rupture time of 10 s at the oxidation temperature of $T = 800$ °C

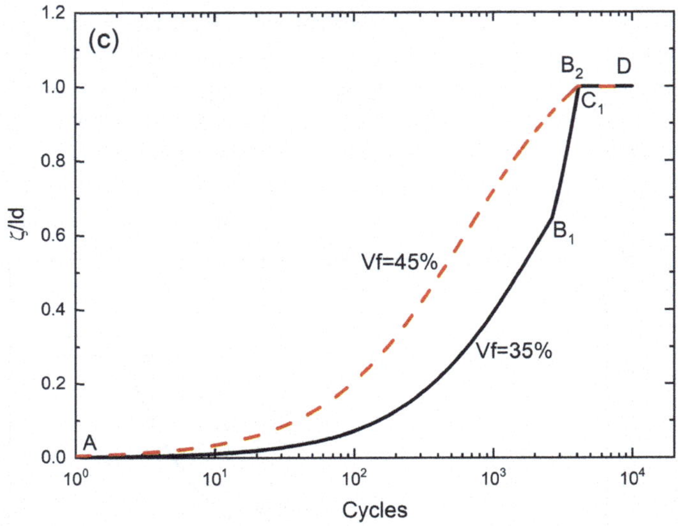

Fig. 5.16 (continued)

interface partial debonding (i.e., $2l_d/l_c < 1$ and B_c–C_2 part in Fig. 5.16b) and the interface complete oxidation (i.e., $\zeta/l_d = 1$ and B_2–D part in Fig. 5.16c); and then remains constant of $\varepsilon_{max} = 0.196\%$ with increasing applied cycle number (i.e., the C_2–D_2 part in Fig. 5.16a), corresponding to the interface complete debonding (i.e., $2l_d/l_c = 1$ and C_2–D part in Fig. 5.16b) and the interface complete oxidation (i.e., $\zeta/l_d = 1$ and B_2–D part in Fig. 5.16c).

5.3.2.5 Effect of Oxidation Temperature on Time-Dependent Strain Response

The peak strain, the interface debonding length, and the interface oxidation length versus applied cycle number curves corresponding to different oxidation temperatures of $T = 700$ and 900 °C under $\sigma_{max} = 200$ MPa with stress-rupture time of $t =$ 10 s are shown in Fig. 5.17. When the oxidation temperature increases, the interface slip lengths increase, leading to the increase of the peak strain.

When the oxidation temperature is $T = 700$ °C, the peak strain increases from $\varepsilon_{max} = 0.132\%$ at the first applied cycle to $\varepsilon_{max} = 0.22\%$ at the 8047th applied cycle (i.e., the A–B_1 part in Fig. 5.17a), corresponding to the interface partial debonding (i.e., $2l_d/l_c < 1$ and A–B_1 part in Fig. 5.17b) and the interface partial oxidation (i.e., $\zeta/l_d < 1$ and A–B_1 part in Fig. 5.17c); increases to the peak value of $\varepsilon_{max} = 0.222\%$ at the 9619th applied cycle (i.e., the B_1–C_1 part in Fig. 5.17a), corresponding to the interface complete debonding (i.e., $2l_d/l_c = 1$ and B_1–C part in Fig. 5.17b) and the interface partial oxidation (i.e., $\zeta/l_d < 1$ and B_1–C_1 part in Fig. 5.17c); and then remains constant of $\varepsilon_{max} = 0.222\%$ with increasing applied cycle number (i.e., the

C_1–D_1 part in Fig. 5.17a), corresponding to the interface complete debonding (i.e., $2l_d/l_c = 1$ and B_1–C part in Fig. 5.17b) and the interface complete oxidation (i.e., $\zeta/l_d = 1$ and C_1–D part in Fig. 5.17c).

When the oxidation temperature is T = 900 °C, the peak strain increases from $\varepsilon_{max} = 0.132\%$ at the first applied cycle to $\varepsilon_{max} = 0.22\%$ at the 1760th applied cycle (i.e., the A–B_2 part in Fig. 5.17a), corresponding to the interface partial debonding

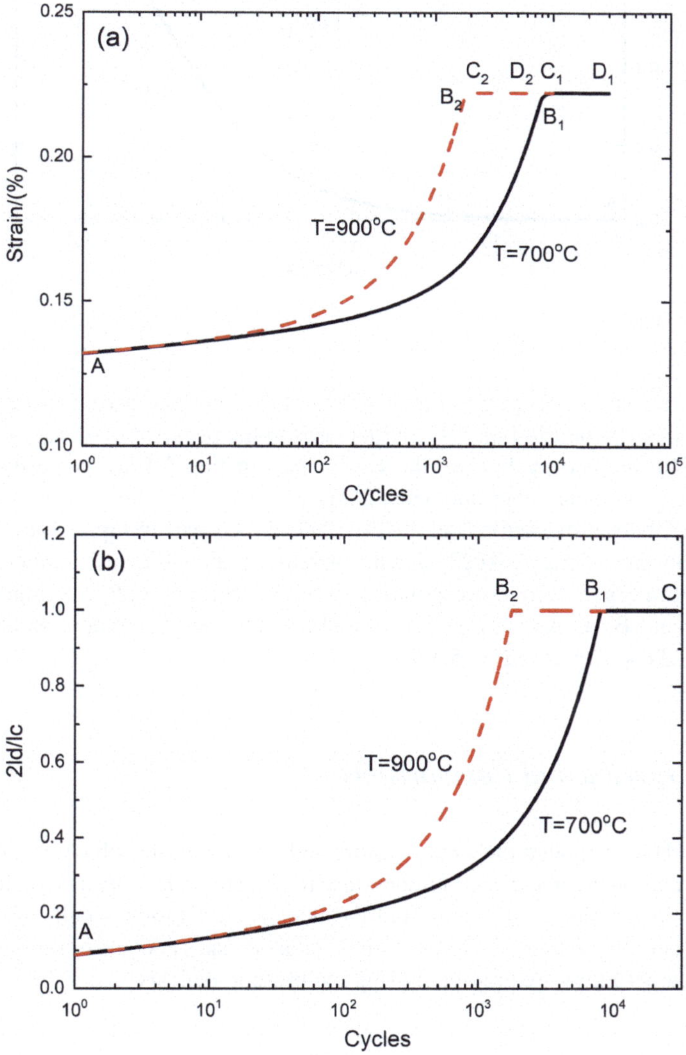

Fig. 5.17 **a** Peak strain versus applied cycle number curves; **b** the interface debonded length versus applied cycle number curves; and **c** the interface oxidation length versus applied cycle number curves corresponding to different oxidation temperatures of T = 700 and 900 °C under σ_{max} = 200 MPa with stress-rupture time of 10 s

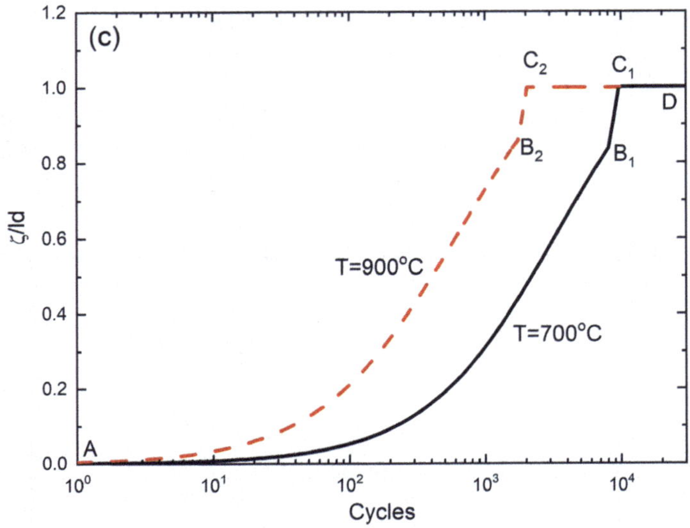

Fig. 5.17 (continued)

(i.e., $2l_d/l_c < 1$ and A–B$_2$ part in Fig. 5.17b) and the interface partial oxidation (i.e., $\zeta/l_d < 1$ and A–B$_2$ part in Fig. 5.17c); increases to the peak value of $\varepsilon_{max} = 0.222\%$ at the 2054th applied cycle (i.e., the B$_2$–C$_2$ part in Fig. 5.17a), corresponding to the interface complete debonding (i.e., $2l_d/l_c = 1$ and B$_2$–C part in Fig. 5.17b) and the interface partial oxidation (i.e., $\zeta/l_d < 1$ and B$_2$–C$_2$ part in Fig. 5.17c); and then remains constant of $\varepsilon_{max} = 0.222\%$ with increasing applied cycle number (i.e., the C$_2$–D$_2$ part in Fig. 5.17a), corresponding to the interface complete debonding (i.e., $2l_d/l_c = 1$ and B$_2$–C part in Fig. 5.17b) and the interface complete oxidation (i.e., $\zeta/l_d = 1$ and C$_2$–D part in Fig. 5.17c).

5.3.3 Experimental Comparisons

Grant (1994) investigated the stress-rupture and cyclic fatigue behavior of cross-ply SiC/MAS composite at elevated temperature in air. The fatigue hysteresis loops, the interface slip lengths, peak strain, and the interface oxidation length of cross-ply SiC/MAS composite under different fatigue peak stresses and stress-rupture time at elevated temperatures are predicted using the present analysis.

5.3.3.1 Strain Response Under Cyclic Fatigue and Stress-Rupture at 566 °C in Air Atmosphere

The experimental and predicted fatigue hysteresis loops under the fatigue peak stress of $\sigma_{max} = 138$ MPa with the stress-rupture time $t = 10$ s are given in Fig. 5.18a. The fatigue hysteresis loops at the cycle number of $N = 1$, 133, and 265 correspond to

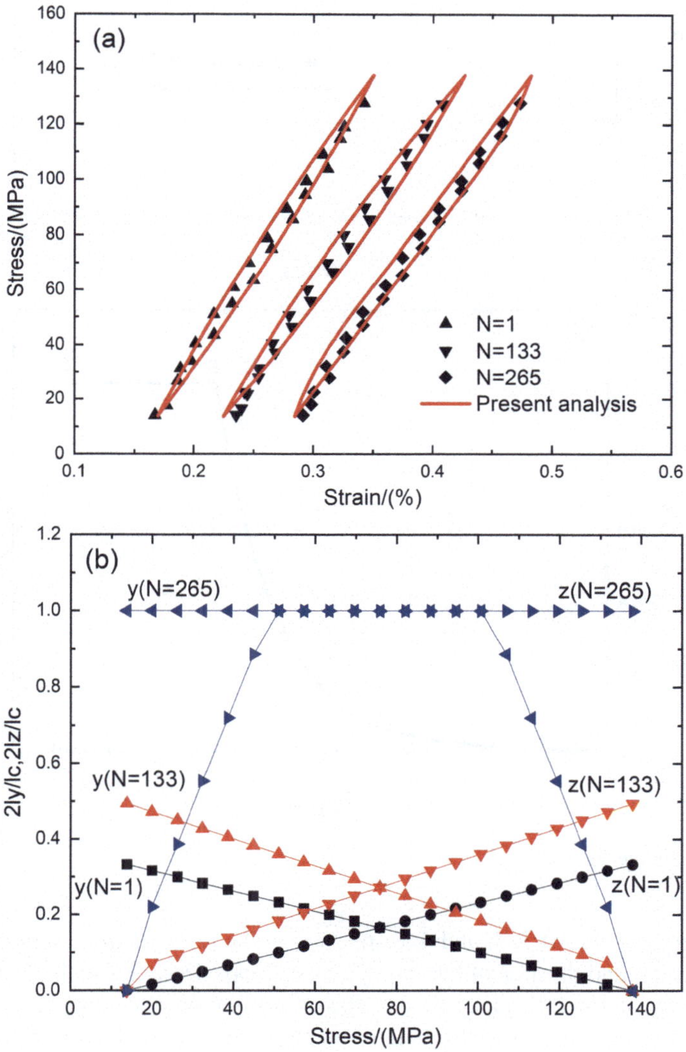

Fig. 5.18 **a** Fatigue hysteresis loops corresponding to different applied cycles; **b** the interface slip lengths versus applied stress corresponding to different cycle numbers; **c** the peak strain versus cycle number curves; and **d** the interface oxidation length versus cycle number curve of cross-ply SiC/MAS composite under $\sigma_{max} = 138$ MPa at $T = 566$ °C with stress-rupture time of $t = 10$ s

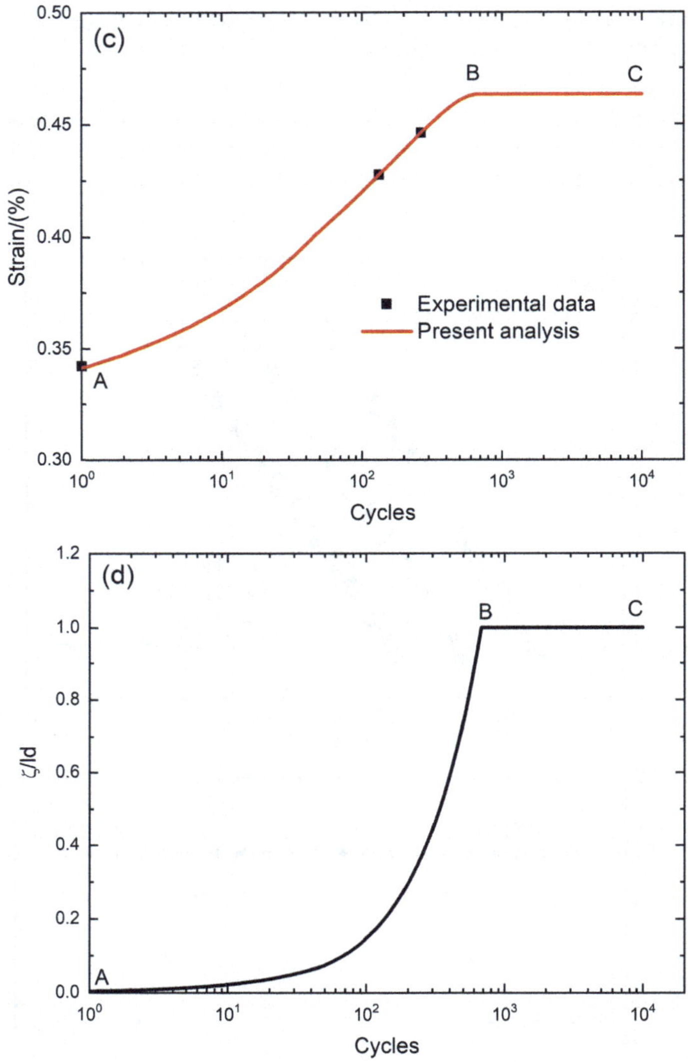

Fig. 5.18 (continued)

the interface slip Cases 3, 3, and 4, respectively. The residual strain increases with applied cycles, and the area of fatigue hysteresis loops decreases with applied cycle number. The interface slip lengths, i.e., the unloading interface counter-slip length and reloading interface new-slip length, increase with applied cycle number, i.e., from $2l_y/l_c = 2l_z/l_c = 0.33$ at the first applied cycle to $2l_y/l_c = 2z/l_c = 1.0$ at the 265th applied cycle, as shown in Fig. 5.18b. The experimental and predicted peak strain versus applied cycle number curves is shown in Fig. 5.18c. The experimental peak strain increases from $\varepsilon_{max} = 0.34\%$ at $N = 1$ to $\varepsilon_{max} = 0.446\%$ at $N = 265$;

the predicted peak strain increases from $\varepsilon_{max} = 0.34\%$ at $N = 1$ to the peak value of $\varepsilon_{max} = 0.463\%$ at $N = 677$ (i.e., the A–B part in Fig. 5.18c), corresponding to the interface complete debonding and the interface partial oxidation (i.e., the A–B part in Fig. 5.18d), and remains constant of $\varepsilon_{max} = 0.463\%$ with increasing applied cycles (i.e., the B–C part in Fig. 5.18c), corresponding to the interface complete debonding and the interface complete oxidation (i.e., the B–C part in Fig. 5.18d).

5.3.3.2 Strain Response Under Cyclic Fatigue and Stress-Rupture at 1093 °C in Air Atmosphere

The experimental and predicted fatigue hysteresis loops under the fatigue peak stress of $\sigma_{max} = 103$ MPa with the stress-rupture time $t = 10$ s are given in Fig. 5.19a. The fatigue hysteresis loops at the cycle number of $N = 1$, 108, and 216 correspond to the interface slip Cases 3, 3, and 4, respectively. The residual strain increases with applied cycles, and the area of fatigue hysteresis loops decreases with applied cycle number. The interface slip lengths, i.e., the unloading interface counter-slip length and reloading interface new-slip length, increase with cycle number, i.e., from $2l_y/l_c = 2z/l_c = 0.51$ at the first applied cycle to $2l_y/l_c = 2z/l_c = 1$ at the 216th applied cycle, as shown in Fig. 5.19b. The experimental and theoretical peak strain versus cycle number curves is shown in Fig. 5.19c. The experimental peak strain increases from $\varepsilon_{max} = 0.46\%$ at the first applied cycle to $\varepsilon_{max} = 0.52\%$ at the 216th applied cycle; the theoretical peak strain increases from $\varepsilon_{max} = 0.46\%$ at $N = 1$ to the peak value of $\varepsilon_{max} = 0.546\%$ at $N = 590$ (i.e., the A–B part in Fig. 5.19c), corresponding to the interface complete debonding and the interface partial oxidation (i.e., the A–B part in Fig. 5.19d), and remains constant of $\varepsilon_{max} = 0.546\%$ with increasing applied cycles (i.e., the B–C part in Fig. 5.19c), corresponding to the interface complete debonding and the interface complete oxidation (i.e., the B–C part in Fig. 5.19d).

5.4 Conclusion

In this chapter, the time-dependent static fatigue and cyclic fatigue behavior of fiber-reinforced CMCs are investigated. The stress-strain relationships considering the interface oxidation and interface wear in the interface debonding region under static and cyclic fatigue loading are developed to establish the relationships between the peak strain, the interface debonding length, the interface oxidation length, and the interface slip lengths. The effects of the stress-rupture time, stress level, matrix crack spacing, fiber volume, and oxidation temperature on the peak strain and the interface slip lengths are investigated. The experimental fatigue hysteresis loops, interface slip lengths, peak strain, and interface oxidation length of C/[Si–B–C] and SiC/MAS composite at elevated temperatures of 566, 1093, and 1200 °C in air atmosphere are predicted.

1. With increasing stress-rupture time, fatigue peak stress, and oxidation temperature, the interface slip lengths increase with applied cycles due to interface oxidation, leading to the increase of peak strain.
2. With increasing matrix crack spacing and fiber volume fraction, the interface slip lengths decrease, leading to the decrease of peak strain.

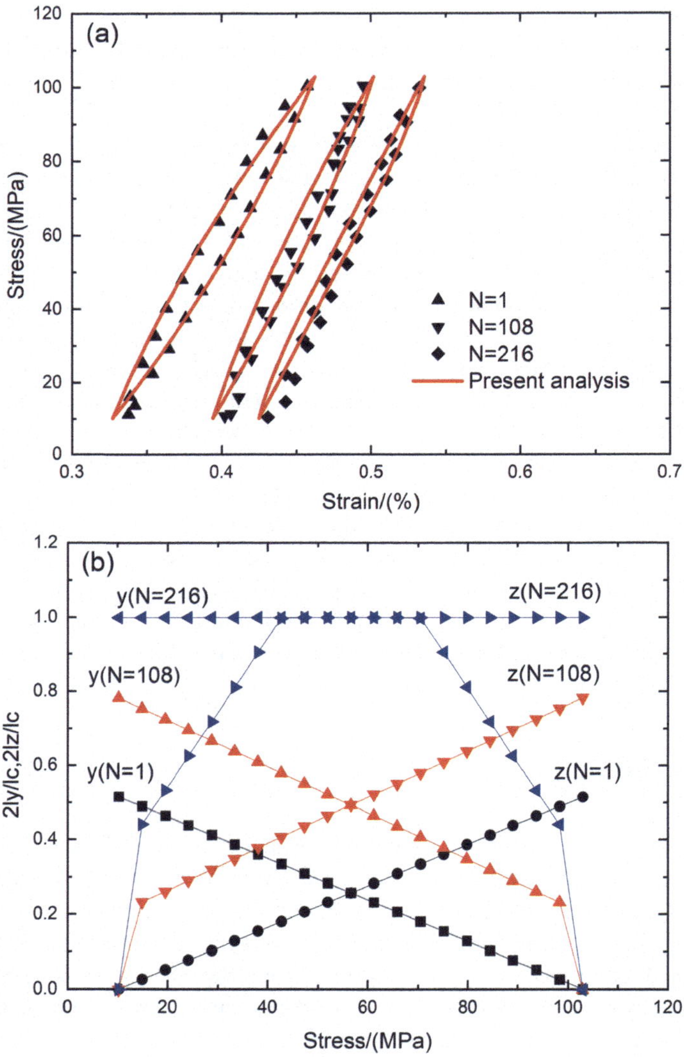

Fig. 5.19 **a** Fatigue hysteresis loops corresponding to different applied cycles; **b** the interface slip lengths versus applied stress corresponding to different cycle numbers; **c** the peak strain versus cycle number curves; and **d** the interface oxidation length versus cycle number curve of cross-ply SiC/MAS composite under $\sigma_{max} = 103$ MPa at 1093 °C with stress-rupture time of $t = 10$ s

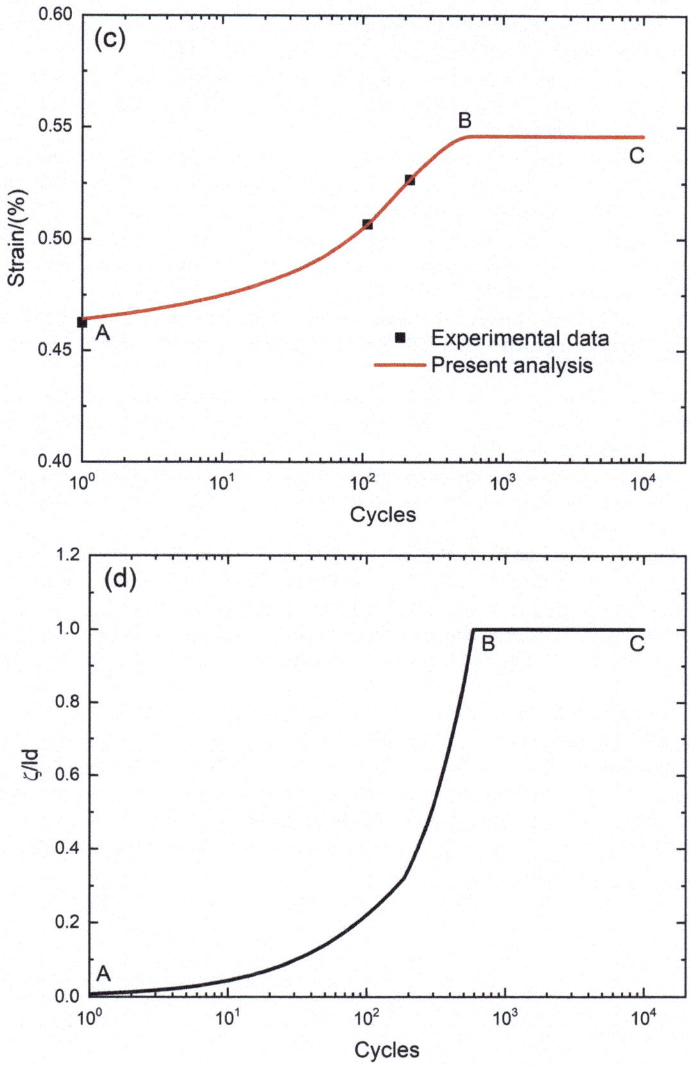

Fig. 5.19 (continued)

References

Casas L, Martinez-Esnaola JM (2003) Modelling the effect of oxidation on the creep behavior of fiber-reinforced ceramic matrix composites. Acta Mater 51(13):3745–3757. https://doi.org/10.1016/S1359-6454(03)00189-7

Fantozzi G, Reynaud P (2009) Mechanical hysteresis in ceramic matrix composites. Mater Sci Eng A 521–522:18–23. https://doi.org/10.1016/j.msea.2008.09.128

Grant SA (1994) Fatigue behavior of a cross-ply ceramic matrix composite subjected to tension-tension cycling with hold time. AFIT/GAE/ENY/94D-16

Halbig MC, Brewer DN, Eckel AJ (2000) Degradation of continuous fiber ceramic matrix composites under constant-load conditions. NASA/TM-2000-209681

Lamouroux F, Naslain R, Jouin JM (1994) Kinetics and mechanisms of oxidation of 2D woven C/SiC composites: II, theoretical approach. J Am Ceram Soc 77(8):2058–2068. https://doi.org/10.1111/j.1151-2916.1994.tb07097.x

Li L (2016) Modeling the effect of oxidation on hysteresis loops of carbon fiber-reinforced ceramic-matrix composites under static fatigue at elevated temperature. J Eur Ceram Soc 36:465–480. https://doi.org/10.1016/j.jeurceramsoc.2015.11.005

Li L (2017) Synergistic effects of stress-rupture and cyclic loading on strain response of fiber-reinforced ceramic-matrix composites at elevated temperature in oxidizing atmosphere. Materials 10:182. https://doi.org/10.3390/ma10020182

Mall S, Engesser JM (2006) Effects of frequency on fatigue behavior of CVI C/SiC at elevated temperature. Compos Sci Technol 66(7–8):863–874. https://doi.org/10.1016/j.compscitech.2005.06.020

Murthy Pappu LN, Nemeth NN, Brewer DN, Mital S (2008) Probabilistic analysis of a SiC/SiC ceramic matrix composite turbine vane. Compos Part B: Eng 39:694–703. https://doi.org/10.1016/j.compositesb.2007.05.006

Naslain R (2004) Design, preparation and properties of non-oxide CMCs for application in engines and nuclear reactors: an overview. Compos Sci Technol 64(2):155–170. https://doi.org/10.1016/S0266-3538(03)00230-6

Naslain R, Guette A, Rebillat F, Gallet S, Lamouroux F, Filipuzzi L, Louchet C (2004) Oxidation mechanisms and kinetics of SiC-matrix composites and their constituents. J Mater Sci 39(24):7303–7316. https://doi.org/10.1023/B:JMSC.0000048745.18938.d5

Pailler F, Lamon J (2005) Micromechanics based model of fatigue/oxidation for ceramic matrix composites. Compos Sci Technol 65(3–4):369–374. https://doi.org/10.1016/j.compscitech.2004.09.029

Santhosh U, Ahmad J, John R, Ojard G, Miller R, Gowayed Y (2013) Modeling of stress concentration in ceramic matrix composites. Compos Part B: Eng 45(1):1156–1163. https://doi.org/10.1016/j.compositesb.2012.07.034

Sullivan RM (2005) A model for the oxidation of carbon silicon carbide composite structures. Carbon 43(2):275–285. https://doi.org/10.1016/j.carbon.2004.09.010

Zhu SJ (2006) Fatigue behavior of ceramic matrix composite oxidized at intermediate temperature. Mater Trans 47(8):1965–1967

The manufacturer's authorised representative in the EU is Springer Nature Customer Service Centre GmbH, Europaplatz 3, 69115 Heidelberg, Germany. If you have any concerns regarding our products, please contact ProductSafety@springernature.com

Printed and bound by CPI Group (UK) Ltd, Croydon, CR0 4YY
30/09/2026
02225556-0008